7-24-89

D0893316

ELECTRONIC COMMUNICATIONS HANDBOOK

Other McGraw-Hill Reference Books of Interest

Handbooks

Avallone and Baumeister · STANDARD HANDBOOK FOR MECHANICAL ENGINEERS

Beeman · INDUSTRIAL POWER SYSTEMS HANDBOOK

Benson · AUDIO ENGINEERING HANDBOOK

Benson · TELEVISION ENGINEERING HANDBOOK

Coombs · BASIC ELECTRONIC INSTRUMENT HANDBOOK

Coombs · PRINTED CIRCUITS HANDBOOK

Croft and Summers · AMERICAN ELECTRICIANS' HANDBOOK

Di Giacomo · VLSI HANDBOOK

Fink and Beaty · STANDARD HANDBOOK FOR ELECTRICAL ENGINEERS

Fink and Christiansen · ELECTRONICS ENGINEERS' HANDBOOK

Harper · HANDBOOK OF ELECTRONIC SYSTEMS DESIGN

Harper · HANDBOOK OF COMPONENTS FOR ELECTRONICS

Harper · HANDBOOK OF THICK FILM HYBRID MICROELECTRONICS

Harper · HANDBOOK OF WIRING, CABLING, AND INTERCONNECTION FOR ELECTRONICS

Hicks · STANDARD HANDBOOK OF ENGINEERING CALCULATIONS

Johnson and Jasik · ANTENNA ENGINEERING HANDBOOK

Juran · QUALITY CONTROL HANDBOOK

Kaufman and Seidman · HANDBOOK OF ELECTRONICS CALCULATIONS

Kaufman and Seidman · HANDBOOK FOR ELECTRONICS ENGINEERING TECHNICIANS

Kurtz · HANDBOOK OF ENGINEERING ECONOMICS

Perry · ENGINEERING MANUAL

Stout · HANDBOOK OF MICROPROCESSOR DESIGN AND APPLICATIONS

Stout and Kaufman · HANDBOOK OF MICROCIRCUIT DESIGN AND APPLICATION

Stout and Kaufman · HANDBOOK OF OPERATIONAL AMPLIFIER DESIGN

Tuma · ENGINEERING MATHEMATICS HANDBOOK

Williams · DESIGNER'S HANDBOOK OF INTEGRATED CIRCUITS

Williams and Taylor · ELECTRONIC FILTER DESIGN HANDBOOK

Encyclopedias

CONCISE ENCYCLOPEDIA OF SCIENCE AND TECHNOLOGY

ENCYCLOPEDIA OF ELECTRONICS AND COMPUTERS

ENCYCLOPEDIA OF ENGINEERING

Dictionaries

DICTIONARY OF SCIENTIFIC AND TECHNICAL TERMS

DICTIONARY OF COMPUTERS

DICTIONARY OF ELECTRICAL AND ELECTRONIC ENGINEERING

DICTIONARY OF ENGINEERING

Markus · ELECTRONICS DICTIONARY

ELECTRONIC COMMUNICATIONS HANDBOOK

Andrew F. Inglis *Editor-in-Chief*

*President and Chief Executive Officer,
RCA American Communications, Inc.;
Vice President, RCA Corporation;
President of RCA Network Services.
(retired)*

MCGRAW-HILL BOOK COMPANY

New York   St. Louis   San Francisco   Auckland
Bogotá   Hamburg   London   Madrid   Mexico
Milan   Montreal   New Delhi   Panama
Paris   Sâo Paulo   Singapore
Sydney   Tokyo   Toronto

Library of Congress Cataloging-in-Publication Data
Electronic communications handbook.
    Includes bibliographies and index.
    1. Telecommunication—Handbooks, manuals, etc.
I. Inglis, Andrew F.
TK5102.5.E39   1988   621.38′0413   88-669
ISBN 0-07-031711-9

1234567890   DOCDOC   8921098

ISBN 0-07-031711-9

*The editors for this book were Daniel A. Gonneau and Caroline Levine,
the designer was Naomi Auerbach, and the production supervisor was
Suzanne Babeuf. It was set in Times Roman by the McGraw-Hill Book
Company Professional and Reference Book Division composition unit.*

Printed and bound by R. R. Donnelley & Sons.

CONTENTS

CONTRIBUTORS

Thomas J. Aprille *Supervisor, AT&T Bell Laboratories* (CHAP. 7)

H. Charles Baker *President, Telecommunications Engineering, Inc.; Professor of Electrical Engineering, Southern Methodist University* (CHAP. 16)

A. David Boomstein *Manager, Product Marketing, Compression Labs, Inc.* (CHAP. 18)

John W. Bowker *Director of Frequency Management, RCA Corporation (retired)* (CHAP. 3)

Jack H. Branch *Chief Executive Officer, CW Incotel Ltd.*(CHAP. 13)

Carl J. Cangelosi *Vice President and General Counsel, GE American Communications, Inc.* (CHAP. 6)

James W. Cofer *Antenna Product Line Manager, Scientific-Atlanta, Inc.* (CHAP. 2)

Virgil Conanan *Senior Systems Engineer, Home Box Office, Inc.* (CHAP. 17)

James H. Cook, Jr. *Principal Engineer, Scientific-Atlanta, Inc.* (CHAP. 2)

H. Allen Ecker *Vice President, Corporate Development, Scientific-Atlanta, Inc.* (CHAP. 2)

Forrest F. Fulton, Jr. *Vice President, Advanced Engineering, Avantek, Inc.* (CHAP. 4)

Michael C. Goldstein *General Electric Information Services Company* (CHAP. 15)

Edward D. Horowitz *Senior Vice President, Technology and Operations, Home Box Office, Inc.* (CHAP. 17)

Ira Jacobs *Department of Electrical Engineering, Virginia Polytechnic Institute and State University; formerly, AT&T Bell Laboratories* (CHAP. 8)

Alauddin Javed *Director, Radio Systems, Bell Northern Research* (CHAP. 11)

Amos E. Joel, Jr. *Consultant; formerly AT&T Bell Laboratories* (CHAP. 12)

V. H. MacDonald *Staff Scientist, AT&T Bell Laboratories* (CHAP. 22)

Stuart F. Meyer *Director, Government and Industry Relations, E. F. Johnson Company (retired); Consultant* (CHAP. 20)

Peter H. Plush *Director, Facilities Planning, RCA Communications and Electronic Services* (CHAP. 19)

Jim Potts *Consultant; formerly, Chief Engineer, COMSAT World Systems Division* (CHAP. 5)

Kerns H. Powers *David Sarnoff Research Center, Stanford Research Institute* (CHAP. 10)

Sang B. Rhee *Staff Scientist, AT&T Bell Laboratories* (CHAP. 22)

Arthur R. Roberts *Vice President, Microwave Engineering and Construction, MCI Telecommunications Corporation* (CHAP. 9)

David V. Rogers *Manager, Propagation Studies Department, COMSAT Laboratories* (CHAP. 1)

Thomas H. Scholl   *Senior Vice President, Hughes Network Systems* (CHAP. 14)

Curtis J. Schultz   *President, SCS Telesystems International, Inc.* (CHAP. 21)

John I. Smith   *Staff Scientist (retired), AT&T Bell Laboratories* (CHAP. 22)

John Tyson   *President, Compression Labs, Inc.* (CHAP. 18)

Mark S. Whitty   *Advisory Engineer, MCI Telecommunications Corporation* (CHAP. 9)

Preface

The extraordinary growth in the size and complexity of the communications industry during the past decade has resulted from the confluence of three major forces: technical progress and breakthroughs, the market demand for an ever-widening range of communication services, and, in the United States, deregulation.

Technical advances have been particularly rapid in fiber optics, satellite communications, cellular radio, digital transmission, digital switching, and large-scale integrated circuits. These newer technologies, in combination with their more mature counterparts—cable and microwave, conventional land mobile systems, analog signal formats, circuit switching, and message switching—have enormously increased the capability and complexity of modern communications systems.

Concurrently, the demand for communications services has escalated both in quantity and diversity. Whereas communications once meant telegraphy and "plain old telephone service" (POTS), it now encompasses a bewildering variety of voice, video, and data services.

For nearly a century, the highly regulated communications industry in the United States stated correctly and with justifiable pride that it provided the "world's finest telephone service." Now, as the result of deregulation and related governmental policies, the industry is experiencing the leavening effect of competition. Competition has been the stimulus for technical innovation, improved efficiency, and creative new services. United States communications users have available not only the superb physical plant and services provided by the regulated segment of the industry but also a wide variety of alternative choices of technologies and suppliers.

These three forces—technology, market demand, and deregulation—have produced an order-of-magnitude enlargement of the responsibilities of corporate, institutional, and governmental communications managers. One measure of this growth is the attendance at the annual conference of the International Communications Association (ICA), which increased more than ninefold from 1980 to 1986.

The *Electronic Communications Handbook* provides a compendium of reference data for the engineers and executives who have the responsibility for planning, designing, constructing, and operating communications systems in this environment. It will be useful to industry hardware and software suppliers as well. While it is primarily a technical handbook, economic and regulatory factors are also considered—particularly where technical decisions involve cost and regulatory trade-offs.

The handbook is divided into two parts:

Part One, Transmission and Switching Technologies, provides basic technical and regulatory information concerning the most commonly used transmission media and switching systems. Chapter 9 provides a comprehensive comparison of transmission media based on cost, performance, reliability, and security.

Part Two, Electronic Communications Systems, provides reference data for the planning, design, and construction of practical communications systems.

ix

The contributors to this volume, to whom the editor is particularly grateful, are all distinguished in their specialties. Each combines a solid professional background with years of practical experience. The quality and usefulness of the handbook are largely due to their knowledge and dedication.

ANDREW F. INGLIS

P · A · R · T · 1

TRANSMISSION AND SWITCHING TECHNOLOGIES

CHAPTER 1
RADIO-WAVE PROPAGATION

David V. Rogers
Manager, Propagation Studies Department, COMSAT Laboratories

CONTENTS

INTRODUCTION

1. Significance of Radio-Wave Propagation

Many telecommunications systems use the atmosphere as the transmission medium between transmitter and receiver. Electromagnetic waves interact with the earth's surface and constituents of the atmosphere, altering the wave amplitude, phase, or both, possibly causing degradations in system performance. The effects of propagation impairments on a telecommunications system are related to wave frequency and polarization, configuration of the propagation path, atmospheric conditions, available operating margins, and specifics of the system itself.

2. Propagation Phenomena

Propagation phenomena of interest include free-space attenuation; reflection, blockage, and diffraction by the earth's surface and by objects (vegetation, buildings) near the surface; gaseous attenuation, refraction, and precipitation effects in the troposphere; and effects induced by the ion plasma in the ionosphere. As frequency increases, tropospheric impairments typically increase in severity, whereas ionospheric effects decrease. In this handbook, attention is restricted to frequencies of 30 MHz and above (VHF band and above). Above 30 MHz, the ground-wave and sky-wave (ionospheric reflection) propagation modes become secondary in importance, and consideration of ionospheric effects may usually be limited to anomalous effects (sporadic E), Faraday rotation, and ionospheric scintillations (on earth-satellite paths).

FREE-SPACE PROPAGATION

3. Flux Density from an Isotropic Radiator

In the absence of any atmosphere, the intensity of a propagating wave decreases with distance from the transmitting source as the energy "spreads" over the surface of a radiating spherical surface. For a point source radiating isotropically, the *power flux density* Ψ_D at distance D from the source is the ratio of the transmit power, p_t (W), divided by the area of a sphere of radius D (m):

$$\Psi_D = \frac{p_t}{4\pi D^2} \quad (\text{W/m}^2) \tag{1.1}$$

which is independent of frequency. The "spreading factor" $4\pi D^2$ increases rapidly with D, being 91 dB with respect to 1 m^2 for $D = 10$ km, and greater than 162 dB for paths from the earth to geostationary orbit. If the transmit antenna has a power gain of g_t in a given direction, the power flux density in that direction is $\Psi_D g_t$.

4. Calculation of Free-Space Loss

The free-space loss between a transmit and a receive antenna is of special interest. The received power p_r for a receive antenna with gain g_r and effective area A_e located a distance D from a transmit antenna with gain g_t is[1]

$$p_r = \Psi_D g_t A_e = \frac{p_t\, g_t}{4\pi D^2}\; \frac{g_r \lambda^2}{4\pi} \tag{1.2}$$

The *free-space path loss*, defined as p_t/p_r, is

$$L_f = \frac{(4\pi D/\lambda)^2}{g_t\, g_r} \tag{1.3a}$$

which is commonly referred to isotropic antennas ($g_t = g_r = 1$) and called the *basic transmission loss*:

$$L_{bf} = \left(\frac{4\pi D}{\lambda}\right)^2 = \left(\frac{4\pi}{c}\right)^2 D^2 f^2 \tag{1.3b}$$

With D in kilometers and f in gigahertz, L_{bf} expressed in decibels is

$$L_{bf} = 92.44 + 20 \log D + 20 \log f \quad (\text{dB}) \tag{1.3c}$$

and is plotted in Fig. 1.1 for several frequencies.

It is useful to note that Eq. 1.1 is independent of frequency, whereas Eqs. 1.3 are not. The decrease in power flux density for a radio wave is directly proportional to the square of the propagation distance; however, the ratio of the effective area to the gain of any antenna is frequency-dependent. It is the latter feature that introduces the frequency dependence into L_{bf}.

5. Calculation of Field Strength

In some services, particularly those (broadcasting, mobile) concerned with areal coverage, it is conventional to specify *field strength* at a point instead of path

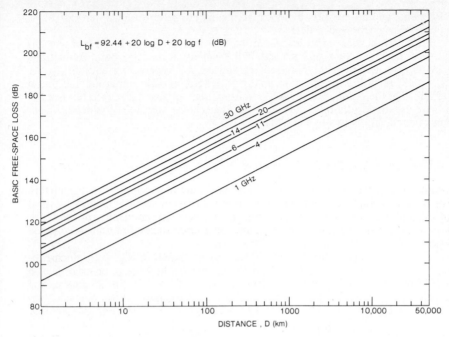

Fig. 1.1 Basic free-space transmission loss.

loss. The rms field strength E (V/m) is related to power flux density Ψ, W/m^2, for plane or nearly plane waves by

$$\Psi = \frac{E^2}{120\pi} \tag{1.4}$$

E may be related to p_r and p_t for isotropic antennas (or to $g_r p_r$ and $g_t p_t$ for antennas with gains referenced to isotropic). The free-space rms field strength a distance D (m) from a transmitter is

$$E = \frac{\sqrt{30 g_t p_t}}{D} \quad \text{(V/m)} \tag{1.5}$$

A practical engineering relation is

$$E = 105 + P_t + G_t - 20 \log D \quad \text{(dB}\mu\text{V/m)} \tag{1.6}$$

with D in kilometers, $P_t = 10 \log p_t$ in dBkW, and $G_t = 10 \log g_t$.

EARTH EFFECTS

6. Types of Earth Effects

The surface of the earth (including terrain irregularities and bodies of water) and objects (buildings, vegetation) on or near the surface can cause diffraction, signal reflections, and attenuation by blockage, leading to electric fields that differ significantly from free-space fields.[2]

Diffraction causes signal loss from the *line-of-sight* (LOS) beam and may increase or decrease the signal strength in the vicinity of the LOS path. Reflections provide propagation paths between transmitter and receiver in addition to the LOS path, resulting in reflection multipath with attendant amplitude and phase variations of the composite signal. Terrain scatter can also cause interference between terrestrial radio-relay systems. Blockage causes signal loss, which is sometimes beneficial, as in site shielding.

7. Diffraction and Fresnel Zones

An obstacle in the path of an electromagnetic wave does not cast a completely sharp shadow. Instead, some of the energy in the wave front is bent away from the LOS into the shadowed area behind the obstacle. In the vicinity of the LOS, the energy density or signal intensity may be greater or less than the free-space value. This phenomenon is called *diffraction*.

The essential characteristics of diffraction are shown in Fig. 1.2, which provides the transmission loss over a smooth sphere for a knife-edge obstruction (no reflection) and for an intermediate case of "average" terrain. In this figure, h is the distance of the LOS path from the obstruction (negative if the obstacle protrudes above the LOS) and F_1 is the radius of the first *Fresnel zone* (see below). Except for the knife edge, diffraction effects are dominant only in the "obstruction zone" ($h/F_1 \lesssim 0.6$), being supplanted by reflection multipath at larger clearances (see Sec. 8). Figure 1.2 demonstrates that in the obstruction zone, transmission loss is primarily determined by the nature of the obstacle.[2]

The concept of Fresnel zones is illustrated for a terrestrial path[3] in Fig. 1.3. These zones, concentric ellipsoids centered on the direct ray path, are defined such that the first zone bounds the primary region of constructive interference between separate rays, the second encompasses the first region of destructive interference, and succeeding zones similarly alternate.

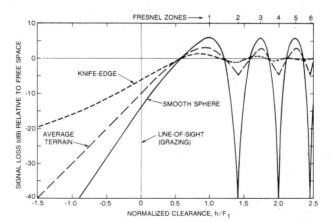

FIG. 1.2 Attenuation due to obstructions and signal reflection as a function of normalized path clearance. (Negative h/F_1 indicates obstruction of LOS.)

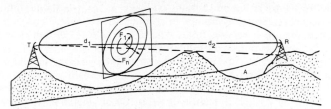

FIG. 1.3 Fresnel ellipsoid geometry. Fresnel zones (radii F_n) are illustrated at distance d_1 from transmitter T and d_2 from receiver R. Obstructed region A between terrain and the dashed line is the geometrical shadow region for T. (Adapted from Reference 3.)

For a path of length D, the radius of the nth zone at a distance d from either endpoint (transmit or receive) is

$$F_n = K\sqrt{\frac{nd(D - d)}{fD}} \tag{1.7}$$

with f in gigahertz. If D and d are in kilometers, $K = 17.3$ gives F_n in meters; with D and d in statute miles, $K = 72.1$ gives F_n in feet. The first Fresnel zone is usually the most significant in the design of communications systems. F_1 is plotted versus D for several values of d at a frequency of 4 GHz in Fig. 1.4.

Because of tropospheric refraction (see Sec. 11), the trajectory of a radio wave usually somewhat follows the earth's curvature, effectively increasing the distance to the radio horizon and the clearance h of a terrestrial path above the earth. The quantitative effect of this wave bending can be estimated by assuming that the earth has an effective radius equal to its actual radius multiplied by a factor k, which is determined by prevailing atmospheric conditions. Under "standard" conditions $k = 4/3$, but large deviations from this value can occur.

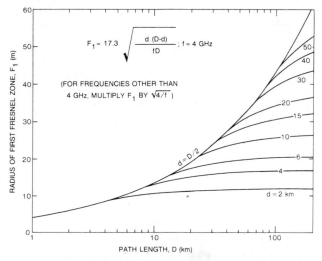

FIG. 1.4 Radius of first Fresnel zone as function of path length D and distance d from endpoint of path for a frequency of 4 GHz.

In designing terrestrial point-to-point radio circuits, it is critically important to provide adequate LOS path clearance to compensate for the effects of diffraction and refraction. Figure 1.2 shows that the LOS diffraction loss can exceed 20 dB. A typical criterion[2] specifies a path clearance of at least $0.3F_1$ for $k = 2/3$ or $1.0F_1$ for $k = 4/3$, whichever is greater. The design of line-of-sight microwave paths is covered in more detail in Chap. 4.

In contrast to point-to-point circuits, where sufficient path clearance can usually be provided by design, transmission paths for services relying on areal coverage are often in the obstruction zone. Calculations for these paths based on Fig. 1.2 are often overly optimistic. A number of methods exist for estimating diffraction attenuation.[2-6] For terrestrial paths over average terrain in areas where losses are in excess of 15 dB, the attenuation can be approximated[6] by

$$A = -20\left(\frac{h}{F_1}\right) + 10 \quad \text{(dB)} \tag{1.8}$$

which corresponds to the "average terrain" curve of Fig. 1.2 in the obstruction zone.

8. Reflection Multipath

Multipath caused by reflections from the surface of the earth or from terrestrial objects can be a major impairment for mobile services (terrestrial and satellite), broadcast services, and for some terrestrial LOS paths.[7] The reflected field usually has both *specular* (coherent) and *diffuse* (incoherent) components. The specular component combines with the direct wave to create a quasi-standing wave pattern, often with deep nulls every half-wavelength. Added to the net coherent field is any diffuse reflection, which appears as noise at the receiver. The relative magnitudes of the specular and diffuse components and their impact on system performance are determined by the local terrain, vegetation, and the occurrence and positions of reflecting objects.

Because of the difficulty in describing the terrain in mathematical terms, the effects of reflection multipath are usually described on an empirical and statistical basis. Further, reflections often occur concurrently with diffraction, tropospheric refraction, and attenuation by vegetation and terrain, and can therefore be described only quasi-quantitatively.

Mobile Services.   The standing-wave pattern that results from specular reflection can cause major changes in signal level in relatively short distances.[4] At 150 MHz, for example, field intensity minima occur approximately at 1-m intervals. In combination with shadowing, reflection multipath typically dominates performance because the moving terminal encounters points where the reflected wave interferes destructively with the direct wave, causing large signal losses. The overall impact of the various propagation factors that determine the coverage of VHF and UHF mobile systems is discussed in Chap. 21.

Broadcast Services.   Reflection multipath creates coverage problems in the FM and TV broadcast services similar to those in the mobile services, with the additional problem of spurious images ("ghosts"), which are often seen in VHF and UHF television reception.

LOS (Point-to-Point) Services.   The effect of different terrain reflectivities on a LOS service is illustrated in Fig. 1.2. In this figure, the reflectivity for the knife edge may be assumed to be zero (no reflection from the edge). For a smooth sphere at low incident angles, the reflection coefficient is -1, indicating complete reflection with a phase reversal (minus sign) at the reflecting surface. The reflection coefficient for "average" terrain is typically about -0.4 to -0.3. Severe degradations can occur in overwater paths where the reflection is quasi-specular, the reflectivity is high, and clearance is good. Under these conditions, the path trajectories may shift up and down as the refractivity of the troposphere changes, causing deep frequency-selective fading as depicted in Fig. 1.2 for $h/F_1 > 1$.

9. Attenuation by Vegetation and Buildings

Objects near the earth's surface can effectively block a radiation path or act as diffraction obstacles. *Blockage* and *shadowing* are serious sources of performance degradation in many mobile systems.[4] Their effects on land mobile systems are discussed in Chap. 21.

Earth station antennas for fixed routes are located to provide a clear LOS to the satellite, but attenuation by terrestrial objects in the LOS can be a problem for mobile-satellite terminals. Also, poor penetration of foliage and building walls by radio waves is a serious problem for the direct-to-home broadcast service and for portable telephone and paging systems.[8] Penetration losses are a function of building construction, number of windows, floor level, etc. The median receive field strength in the vicinity of buildings usually shows large variations from point to point.

TROPOSPHERIC EFFECTS

10. Properties of the Troposphere

The *troposphere* is that portion of the atmosphere from the earth's surface up to the tropopause, within which temperature normally decreases with increasing height. It contains the bulk of the atmospheric gases (molecular oxygen and water vapor) that determine the radio refractive index of the neutral (nonionized) atmosphere, and also bounds the region of precipitation formation. The height of the troposphere varies with latitude, season of the year, and local weather conditions, but averages about 8 km at the poles and 16 km at the equator. Properties of the troposphere are quite variable, both temporally and spatially, complicating the estimation of impairment statistics.

Tropospheric propagation effects are generally of prime importance for terrestrial LOS and earth-satellite paths at frequencies above 1 GHz, and are nearly always dominant for frequencies above 10 GHz.[7] Major effects include bulk and small-scale refractive phenomena (absorption, refraction, atmospheric multipath, scintillation, and ducting), precipitation and cloud impairments (attenuation, depolarization), and additional sky noise associated with path losses such as gaseous absorption and precipitation attenuation.

11. Tropospheric Refraction

The *tropospheric refractive index* is given with good accuracy by[9]

$$n = 1 + \left[\frac{77.6}{T} \left(p + 4810 \frac{e}{T} \right) \right] 10^{-6} \qquad (1.9)$$

where T is the absolute temperature, K; p the barometric pressure, mbar; and e the partial pressure, mbar, of water vapor. The bracketed expression is the *radio refractivity N*, quantified in (dimensionless) N units. It is essentially independent of frequency, the dispersion being less than 0.5 percent for frequencies up to 100 GHz.

Typically, N is on the order of 300 N units at the surface (i.e., $n \cong 1.0003$) and decreases approximately exponentially with increasing height. Radio waves are therefore refracted, with the curvature normally slightly toward the earth, somewhat extending the radio horizon beyond the geometrical horizon. Short-term variations in vertical N gradient near the surface cause angle-of-arrival fluctuations about the median ray trajectory. For slant paths, the apparent elevation angle is slightly greater than the nonrefracted angle, with typical values[10] of ray bending as shown in Table 1.1.

Ray bending varies with changes in n and particularly with changes in vertical refractivity structure (profile), and is usually classified in terms of k factors:[11]

$$k \simeq \frac{157}{157 + (dN/dh)} \qquad (1.10)$$

where dN/dh is the vertical gradient of N. The ray trajectory has a curvature of kR_E over a flat earth, where R_E is the actual earth radius (6370 km), or conversely it appears to be linear with respect to an earth with an equivalent radius $R_{eq} = kR_E$.

The median and extreme values of k for a specific path are of critical importance in establishing path clearance and reflection limitations for terrestrial LOS paths.[2] A typical median value of k is 4/3 (corresponding to $dN/dh \simeq 39$ N units/km), which defines the "standard atmosphere." Substantial departures from 4/3 can occur. If $dN/dh \geq -157$ N units/km, k becomes infinite or negative, and the transmitted energy may propagate in a quasi-waveguide mode (duct) far beyond the normal range. When $k > 4/3$, conditions are said to be *superrefractive*, and when $0 < k < 4/3$, conditions are *subrefractive*.

Since n is greater than unity, the wave propagation velocity is less than that of free space and varies with atmospheric conditions. The associated propagation delays and delay variations are important for some precise timing and geopositioning systems.

TABLE 1.1   Average Tropospheric Ray Bending and Day-to-Day Variations on Slant Paths

Elevation angle, θ	Average total ray-bending, $\Delta\theta$	
	Polar continental air	Tropical maritime air
1°	0.45°	0.65°
2°	0.32°	0.47°
4°	0.21°	0.27°
10°	0.10°	0.14°
	Day-to-day variation in $\Delta\theta$	
1°	0.1° rms	
10°	0.007° rms	

Source: Reference 10.

12. Tropospheric Multipath Fading

Tropospheric multipath occurs when refractivity conditions are such as to simultaneously support more than one propagation path between a transmitter and receiver.[11] On terrestrial paths, this phenomenon is more common on warm, still, humid summer nights when refractive layers and ducts can form, and may coexist with reflection multipath on many paths.

Multipath fading, caused by interference among the several rays, is generally the dominant propagation impairment on terrestrial LOS links at frequencies below 10 GHz. It is also important at higher frequencies. Multipath fading is frequency-selective, and the corresponding amplitude and delay distortion across a frequency band can seriously degrade the bit-error-rate performance of digital systems. Terrestrial multipath is often accompanied by signal depolarization,[12] important for systems either employing dual polarizations or sharing spectrum with other orthogonally polarized systems. On slant paths at low elevation angles, tropospheric multipath ("low-angle fading") can also be quite important, particularly in the summer months.

Representative 4-GHz fading statistics[13] for terrestrial paths in northwest Europe are provided in Fig. 1.5, illustrating worst-month fading as a function of path length D for average rolling terrain in temperate climates. For frequencies other than 4 GHz, D may be replaced by an equivalent path length $D_{eq} = D(f/4)^{0.25}$

Multipath is frequency- and distance-selective; decreases in path unavailabil-

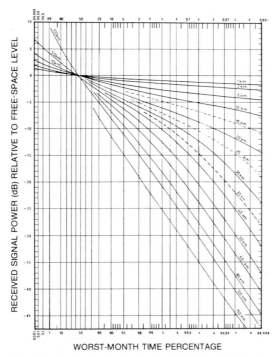

FIG. 1.5 Tropospheric multipath fading at 4 GHz for worst month (temperate climate, rolling terrain) in northwest Europe *(from Reference 13).*

ity by a factor of 10 or so can be achieved by implementing *frequency diversity* (capability of switching to a less-impaired frequency in the communications band) or *space diversity* (two vertically separated antennas). Digital systems may also employ amplitude and/or phase equalization to control distortion in the band.

Signal depolarization is also observed on terrestrial paths in clear air conditions and is often important, especially on longer links. The CCIR[13] provisionally relates path cross-polarization discrimination XPD (dB) to multipath fades A ≥ 15 dB with

$$\text{XPD} = -A + \text{XPD}_o + Q \text{ (dB)} \tag{1.11}$$

XPD_o is the unfaded system cross-polar isolation, and Q is an improvement factor related to the antenna cross-polar patterns in the vertical plane (with typical values of about 5 to 20 dB).

13. Tropospheric Scintillation

Radio-wave scatter from small-scale refractive irregularities generates *scintillations* (fluctuations) in received amplitude and phase.[14] The same mechanism supports the over-the-horizon troposcatter communications mode.[15] Scintillation activity usually peaks in the summer and increases in magnitude as frequency increases and path elevation angle (on slant paths) decreases. Amplitude scintillations associated with summertime clouds are also observed at high angles.[16] On terrestrial paths, the fluctuations in received power, which seldom exceed a few decibels peak to peak, usually have negligible effects compared to multipath (and rain) fading.

Satellite systems have small fade margins, and scintillations (and low-angle fading) can cause degradations in performance and even loss of service on some low-angle paths, particularly for low-margin business and broadcast-satellite systems.[17] The fading rate of the scintillations seldom exceeds 1 Hz, and is usually less than 0.3 Hz. (Rates for low-angle fading are often on the order of minutes.)

The rms amplitude fluctuations $\sigma_A(f, \theta)$ depend on frequency and elevation angle approximately as[18]

$$\sigma_A(f, \theta) \propto f^{7/12} (\csc \theta)^{11/12} \tag{1.12}$$

disregarding antenna aperture-averaging effects. Measured statistics indicate that at 11 GHz and an elevation angle of 10°, the summertime rms fluctuations for 0.01 percent of the time are about 2 dB in northern latitudes and 2.5 dB in midlatitudes and may be larger in humid tropical regions.[17] At a 5° elevation angle, these scintillation amplitudes will almost double.

14. Gaseous Attenuation

Molecular oxygen and water vapor in the atmosphere absorb electromagnetic energy. Attenuations are small and often ignored for $f \leq 10$ GHz, but can be large at higher frequencies, particularly those near molecular absorption lines (e.g., the water vapor line at 22.3 GHz).

Total gaseous attenuation varies as atmospheric water vapor concentration varies. Statistically, estimates of gaseous attenuation can be based on mean or median values of surface-water-vapor concentration ρ_w, g/m^3, for monthly or annual periods. Global maps of ρ_w for February and August are available.[9,19] Values for a specific location can also be estimated from monthly or annual averages of surface-water-vapor pressure e, mbar, and absolute temperature T (K), with $\rho_w = 216.7\, e/T$.

For $f \leq 50$ GHz, the specific attenuations (dB/km) of oxygen and water vapor, respectively, at the earth surface are approximated by[20]

$$\gamma_o = \left[0.0072 + \frac{6.1}{f^2 + 0.23} + \frac{4.8}{(f - 57)^2 + 1.5} \right] f^2 \times 10^{-3} \qquad (1.13a)$$

$$\gamma_w \simeq \left[0.07 + \frac{3}{(f - 22.3)^2 + 7.3} \right] \rho_w f^2 \times 10^{-4} \qquad (1.13b)$$

(with f expressed in GHz and ρ_w in g/m^3). Equation 1.13b progressively underestimates γ_w as ρ_w increases beyond 12 g/m^3, with an error of about 30 percent for $\rho_w = 20$ g/m^3.

The *gaseous attenuation* for a terrestrial path of length D, km, is

$$A_g = (\gamma_o + \gamma_w)D \qquad \text{(dB)} \qquad (1.14)$$

and for a slant path with elevation angle $\theta > 10°$

$$A_g = \frac{\gamma_w h_w + 6\gamma_o e^{-h_s/6}}{\sin \theta} \qquad (dB) \qquad (1.15a)$$

where h_s is earth-station height, km, above mean sea level (often approximated by zero) and h_w is an equivalent water-vapor height, km, given by

$$h_w \simeq 2.2 + \frac{3}{(f - 22.3)^2 + 3} \qquad (1.15b)$$

For elevation angles $\theta \leq 10°$, the gaseous attenuation is computed from

$$A_g = \frac{6\gamma_o e^{-h_s/6}}{g(h_o)} + \frac{\gamma_w h_w}{g(h_w)} \qquad \text{(dB)} \qquad (1.16a)$$

where

$$g(h) \simeq 0.66 \sin \theta + 0.34 \sqrt{\sin^2 \theta + 5.5(h/R_{eq})} \qquad (1.16b)$$

In Eq. 1.16b, h is replaced by $h_o = 6$ or by h_w of Eq. 1.15b, as required. As noted in Sec. 11, R_{eq} is the equivalent earth radius after accounting for refraction (typically, $R_{eq} \cong 8500$ km).

A surface air temperature of 15°C (288 K) is assumed in the above equations. For other temperatures, a correction to A_g of -0.6 percent per °C for water vapor and -1.0 percent per °C for oxygen may be applied if necessary. Calculations of A_g for slant paths at several elevation angles are shown in Fig. 1.6 for $\rho_w = 10$ g/m^3 and a surface air temperature of 15°C.

15. Precipitation Attenuation

At frequencies above about 10 GHz, attenuation by precipitation (especially rain) is a prime concern for many propagation paths. Other hydrometeors (snow, hail,

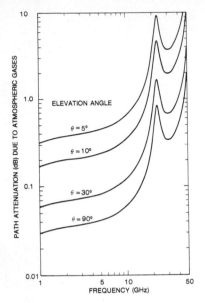

FIG. 1.6 Estimated slant-path gaseous attenuation for surface water vapor concentration of 10 g/m³ and surface temperature of 15°C.

fog, cloud) can also cause system degradations, but are usually of minor importance compared to rain for frequencies up to about 30 GHz.

For systems design, it is desirable (but uncommon) to have rain-rate statistics measured over several years on which to base impairment estimates. In the absence of measurements, rain-rate models[21] or rain climate zones[9,22] are often used. The CCIR has developed a rain attenuation model (described below) that requires only the 1-min rain rate exceeded for 0.01 percent of the year, available from rain zones or contour maps[9] such as the map for North America provided in Fig. 1.7. Both the contour maps and rain zones are approximate because of the shortage of long-term rain-rate data.

Most rain attenuation models convert rain rate R, mm/h, to specific rain attenuation γ_R, dB/km, with the expression[23]

$$\gamma_R = kR^\alpha \qquad (1.17)$$

Coefficients k and α are available[24] as a function of frequency for horizontal (H) and vertical (V) polarization, but the following polarization-independent expressions[23] for spherical rain drops are adequate for most applications in the range 8 to 50 GHz:

$$k = (4.21 \times 10^{-5})f^{2.42} \qquad (1.18a)$$

$$\alpha = \begin{cases} 1.41f^{-0.0779} & 8 \le f < 25 \text{ GHz} \\ 2.63f^{-0.272} & 25 \le f \le 50 \text{ GHz} \end{cases} \qquad (1.18b)$$

Figure 1.8 shows γ_R versus f for several rain rates, based on rigorous scattering calculations.[23,24]

To compute path rain attenuation, the specific attenuation is multiplied by an "effective" path length L_e, km. For a slant path, the length of the path through rain depends on the rain height h_R, km. This height is approximated for an annual time percentage of 0.01 percent in terms of site latitude ϕ, deg, as[20]

$$h_R = \begin{cases} 4.0 & 0 \le \phi < 36° \\ 4.0 - 0.075(\phi - 36) & \phi \ge 36° \end{cases} \qquad (1.19)$$

The slant rain path length L_s (km) is

$$L_s = \begin{cases} \dfrac{(h_R - h_s)}{\sin \theta} & \theta \ge 5° \\[4ex] \dfrac{2(h_R - h_s)}{\left[\sin^2 \theta + \dfrac{2(h_R - h_s)}{R_{eq}}\right]^{1/2} + \sin \theta} & \theta < 5° \end{cases} \qquad (1.20)$$

where θ is the path elevation angle; h_s the station height, km, above mean sea level; and R_{eq} the equivalent earth radius, km, typically about 8500 km (see Sec. 11). For terrestrial paths, the length of the rain path is simply the LOS path length D, km.

The rain path length must be multiplied by a reduction factor r to account for spatial inhomogeneity of the rain. An estimate of this factor for 0.01 percent of the time for slant paths is

$$r_s = \frac{1}{1 + 0.045L_s \cos \theta} \tag{1.21a}$$

and for terrestrial paths

$$r_t = \frac{1}{1 + 0.045D} \tag{1.21b}$$

The estimate of the *rain attenuation* A_R exceeded for 0.01 percent of the year on slant paths is

$$A_R = \gamma_R L_s r_s \quad \text{(dB)} \tag{1.22a}$$

and for terrestrial paths

$$A_R = \gamma_R D r_t \quad \text{(dB)} \tag{1.22b}$$

where γ_R is computed for the 0.01 percent rain rate. To convert to other time percentages p ($0.001 \leq p \leq 1$ percent), the following equation is used:

$$A_R(p) = A_R(0.01) \times 0.12p^{-(0.546 + 0.043 \log p)} \tag{1.23}$$

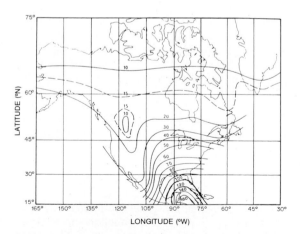

FIG. 1.7 Approximate contours of rain rate (mm/h) exceeded for 0.01 percent of an average year in North America *(from Reference 9).*

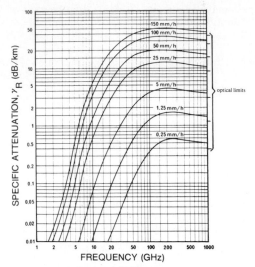

FIG. 1.8 Calculated specific attenuation for rain, also illustrating high-frequency (optical) limits of the respective curves *(from Reference 24).*

Equations 1.17 to 1.23 permit the prediction of a mean annual attenuation distribution simply by inputting a measured or estimated value of the rain rate exceeded for 0.01 percent of the year.

For some services, path availabilities are specified in terms of worst-month time percentages p_w. An approximate relation between annual time percentage p and worst-month time percentage for rain impairments is[20]

$$p \simeq 0.3 \, p_w^{1.15} \tag{1.24}$$

For example, the rain attenuation exceeded for 1 percent of the worst month corresponds approximately to the attenuation exceeded for 0.3 percent of the average year.

Information on the number and durations of rain fades exceeding a specified threshold is useful for some design applications.[16] For example, performance objectives for the *integrated services digital network* (ISDN) state that fading periods for which the system outage threshold [*bit error rate* (BER) < 10^{-3}] is exceeded for less than 10 s are classed as performance degradations ("severely errored seconds") during system available time, whereas outages exceeding 10 s contribute to system unavailable time. At K_u band, "severely errored seconds" appear to account for about 3 to 15 percent of the total time that the outage threshold is exceeded for typical fade margins.[20]

16. Site Diversity

Individual areas of intense rainfall are generally limited in extent to a few kilometers, and severe fades do not often occur simultaneously on parallel paths separated by distances on the order of 10 or more km.[25] This property can be exploited to increase system availability by installing spatially separated terminals with the capability of switching to the less-impaired path as required, usually called *path diversity* in terrestrial systems and *site diversity* in satellite systems.

Spatial diversity may be of more utility in earth-satellite systems, where attenuation margins are typically small. An empirical analysis of two-site diversity[26] for slant paths has been performed that relates diversity gain (defined as the difference between the single-path and minimum joint-path rain attenuations for a fixed time percentage p) to the single-path rain attenuation and parameters of the diversity configuration. As anticipated, site separation is the controlling factor in performance for most configurations.

17. Radio (Sky Noise) Emission

Attenuation processes involving energy absorption (e.g., gaseous absorption, rain attenuation) are accompanied by increased thermal radio noise emission, or "sky noise," because the absorbed energy must reradiate to maintain thermal equilibrium. (Conversely, multipath and scintillation are not accompanied by sky noise increases.) Sky noise appears to a communications receiver as system noise that degrades detection sensitivity (i.e., degrades the G/T figure of merit). Other sources of noise are also important for some systems.[27]

An atmospheric noise source can be treated as an equivalent blackbody medium of temperature T_m, K. *The sky noise temperature T_n, K,* is approximately related to the corresponding path attenuation A, dB, through the medium by

$$T_n \simeq T_m(1 - 10^{-A/10}) \tag{1.25}$$

Typical values of T_m for atmospheric noise are 260 to 280 K, approaching 300 K in heavy rain and in clear air at very low elevation angles.

Figure 1.9 shows computed clear-sky noise emission[27] for path elevation angles from 0° (terrestrial case) to 90° for frequencies up to 60 GHz, assuming a moderate atmosphere ($\rho_w = 7.5$ g/m³, $T = 288$ K). For real antennas, part of the

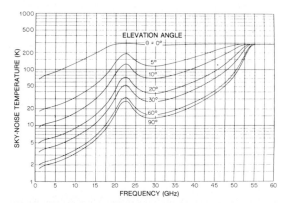

FIG. 1.9 Clear-sky thermal noise temperature at several elevation angles for typical atmospheric conditions: surface $\rho_w = 7.5$ g/m³, surface temperature 15°C, pressure 1,023 mbar *(from Reference 27).*

antenna pattern intersects the earth and thus affects the total noise temperature perceived by a receive system.

18. Depolarization by Precipitation

Rain and cloudborne ice crystals can generate significant cross-polarization interference between orthogonally polarized channels.[28–30] *Rain depolarization* can be the most significant path impairment in dual-polarized C-band (6/4-GHz) satellite systems. At frequencies from about 10 to 15 GHz, some slant path availabilities are dominated by attenuation and others by depolarization, depending on climate, path geometry, and system sensitivity to cross-polar interference. At frequencies above K_u-band, rain attenuation is usually dominant.

The isolation between orthogonal polarizations can be expressed in terms of *cross-polarization discrimination* (XPD), defined as the ratio of the received copolar and cross-polar electric field components of a transmitted wave. XPD is infinite for perfect isolation and decreases as isolation degrades because of wave depolarization.

For slant paths at elevation angles $\theta \le 60°$ and frequencies $8 \le f \le 35$ GHz, the rain-induced degradation in cross-polarization discrimination XPD_R, dB, can be related to the rain attenuation A_R (dB) with the expression[20]

$$XPD_R = 30 \log f - V \log A_R \qquad (1.26a)$$
$$- 40 \log (\cos \theta) + 0.0052\sigma^2 + C_\tau$$

with $V = 20$ for $f \le 15$ GHz and $V = 23$ for $f > 15$ GHz; σ is the standard deviation, deg, of the raindrop canting angle distribution, empirically established as $\sigma = -5 \log p$ for time percentages $0.001 \le p \le 1$; and

$$C_\tau = -10 \log [1 - 0.484(1 + \cos 4\tau)] \qquad (1.26b)$$

where τ is the tilt angle of the plane of linear polarization with respect to local horizontal at the earth station ($\tau = 45°$ for circular polarization). An empirical adjustment for ice crystal depolarization, dependent on time percentage p, can be calculated from

$$C_I = \left(\frac{0.3 + 0.1 \log p}{2}\right) XPD_R \qquad (1.26c)$$

The total path XPD is then

$$XPD = XPD_R - C_I \qquad \text{(dB)} \qquad (1.26d)$$

On terrestrial paths, a different expression is used for the same (8 to 35 GHz) frequency range[13]

$$XPD_R = U_o + 30 \log f - 20 \log A_R \qquad (1.27)$$

where nearly horizontal or vertical polarization is assumed. U_o is on average about 15 dB, with an empirical lower bound of about 9 dB for rain attenuations exceeding 15 dB. XPD values at frequency f_1 can be scaled to another frequency f_2 (for f_1 and f_2 between 4 and 30 GHz) with

$$XPD_2 = XPD_1 - 20 \log \left(\frac{f_2}{f_1}\right) \tag{1.28}$$

for both terrestrial and slant paths.

IONOSPHERIC EFFECTS

19. Properties of the Ionosphere

The *ionosphere* is the ionized region of the earth's atmosphere, beginning at a height of about 50 km and extending outward to about 2000 km. It has solar-dependent characteristics that vary diurnally, seasonally, and with the 11-year solar cycle. There are three distinct regions or layers, classed in order of increasing altitude as D (about 50 to 90 km in height), E (90 to 130 km), and F (above 130 km), which may subdivide in some circumstances (e.g., the nighttime F layer divides into layers F1 and F2 in daytime). Ionospheric structure is strongly influenced by the earth's magnetic field, and the geomagnetic latitude,[31] illustrated in Fig. 1.10, differs from the geographic latitude.

At frequencies above 30 MHz, wave reflection by the ionosphere is usually weak and ionospheric effects on terrestrial propagation are of secondary importance. However, near the geomagnetic equator and, secondarily, in polar auroral regions (and elsewhere under some conditions), ionospheric effects at VHF and UHF can be intense.

On slant paths, ionospheric scintillations occur on occasion at frequencies as high as a few gigahertz.[32] Near the peak of a solar cycle, reflections from the F2 layer occur for significant time percentages at frequencies up to 60 MHz. Faraday rotation of the plane of linear polarization can also be important.

20. Sporadic E

Sporadic E (E_s) refers to the intermittent formation of thin ($\sim$1 km) layers of intense ionization, with horizontal extents of several kilometers up to 1000 km,

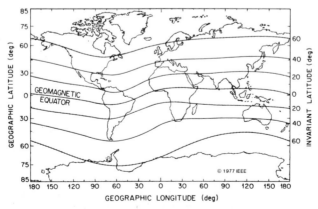

FIG. 1.10 Geomagnetic (invariant) latitude at height of 300 km *(from Reference 31).*

within the E region of the ionosphere.[33] E_s is more frequent and intense near the geomagnetic equator and in the auroral zones and occurs primarily in daylight hours except in the auroral zones, where it is a nighttime phenomenon. In temperate zones, E_s occurs predominantly in the summer and shows little correlation with solar activity, being thought related to wind shear in the upper atmosphere.

Intense E_s layers are highly reflective at frequencies up to about 100 MHz (and sometimes higher) and can cause strong interference fields in the lower VHF band (not uncommon, for example, in VHF television reception in fringe areas). At VHF the reflected field is typically on the order of 30 dB below the incident field but is sufficient to be of concern. Shortwave amateurs take advantage of E_s for long-distance communications in the 50-MHz band and, occasionally, in the 144-MHz band.[34] Methods are available for prediction of long-term sporadic-E effects[35] but not for short-term (hourly, daily) effects.

21. Ionospheric Scintillation

Electron density irregularities in the ionosphere, mainly in the F region, cause amplitude, phase, and angle-of-arrival scintillations on transionospheric radio paths, particularly those traversing the ionosphere within about 20° of the geomagnetic equator or at geomagnetic latitudes greater than about 55° (Fig. 1.10). Scintillation intensity increases with solar activity,[36] as indicated in Fig. 1.11. Near the solar maximum, amplitude scintillations exceeding 30 dB peak-to-peak have been observed at L band (1.5 GHz) in midlatitude regions.[37] Near the geomagnetic equator, values exceeding 10 dB have been seen at 4 GHz in iono-

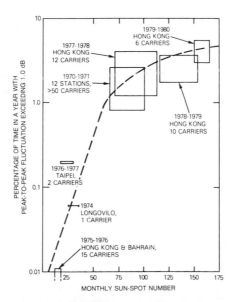

FIG. 1.11 Dependence of 4-GHz equatorial ionospheric scintillation on average monthly sunspot number *(from Reference 36)*. Boxes bound observations for the locations and time periods noted on figure.

spheric anomaly regions. Such scintillations are especially adverse for digital communications systems and for antenna autotracking systems.

The frequency dependence of scintillation amplitude is about $f^{-3/2}$. Fading rates are normally 0.1 to 10 Hz. In the equatorial zone, scintillation activity peaks diurnally about 1 h after local sunset and peaks seasonally near the spring and fall equinoxes. Near the solar minimum, scintillation activity dramatically diminishes (Fig. 1.11). In midlatitudes, ionospheric scintillations are generally of minor concern except for some VHF/UHF earth-satellite systems with small impairment margins.[38]

22. Faraday Rotation and Propagation Delay

Several ionospheric propagation mechanisms, including polarization rotation of linearly polarized waves and propagation delay in excess of the free-space delay, depend directly on the columnar electron density N_e, electrons/m², integrated along the path. The excess propagation delay is given by

$$\tau \quad 1.33 \times 10^{-19} \frac{N_e}{f^2} \quad \text{(s)} \tag{1.29}$$

where f is the frequency, MHz. Minimum and maximum values of N_e are about 10^{16} to 10^{19} electrons/m², respectively.

Faraday rotation of the plane of linear polarization occurs because the ionospheric plasma is birefringent in the presence of the earth's magnetic field, i.e., has different effective indices of refraction for incident *right-hand* (RH) and *left-hand circularly polarized* (LHCP) waves. The separate RHCP and LHCP components of an incident linearly polarized wave undergo different phase shifts in the ionosphere, resulting in rotation of the plane of polarization. For a total electron content of 10^{18} electrons/m² and path elevation angle of 30°, the maximum rotation is about 108° at 1 GHz.[32] Scaling by f^2 yields a rotation of 10,800° (30 rotations of 360°) at 100 MHz and 48° at 1.5 GHz. Rotations of these magnitudes cause serious impairments to linearly polarized receiving systems. Usually, the problem is avoided by employing circular polarization, for which effects are negligible.

REFERENCES

1. Kraus, J. D., *Electromagnetics,* Chap. 13, pp. 484–518, McGraw-Hill, New York, 1953.

2. *Engineering Considerations for Microwave Communications Systems,* GTE Lenkurt, Inc., San Carlos, Calif., 1981.

3. CCIR Report 715-1 (MOD I), "Propagation by Diffraction," CCIR Doc. 5/209, pp. 14–18, International Telecommunications Union, Geneva, 6 February 1984.

4. Lee, W. C. Y., *Mobile Communications Engineering,* Chap. 4, pp. 115–143, McGraw-Hill, New York, 1982.

5. Boithias, L., *Propagation des Ondes Radioélectroniques dans l'Environment Terrestre,* Chap. 7, pp. 161–185, Dunod, Paris, 1983.

6. Vigants, A., "Microwave Radio Obstruction Fading," *Bell System Tech. J.,* vol. 60, pp. 785–801, 1981.

7. Hall, M. P. M., *Effects of the Troposphere on Radio Communication,* Peter Peregrinus, Ltd., Stevenage, U.K., 1979.

8. Cox, D. C., R. R. Murray, and A. W. Norris, "800 MHz Attenuation Measured in and Around Suburban Houses," *Bell System Tech. J.,* vol. 63, pp. 921–954, 1984.

9. CCIR Report 563-3, "Radiometeorological Data," *Recommendations and Reports of the CCIR,* vol. 5, pp. 108–142, International Telecommunications Union, Geneva, 1986.

10. CCIR Report 718-1, "Effects of Large-Scale Tropospheric Refraction on Radiowave Propagation," *Recommendations and Reports of the CCIR,* vol. 5, pp. 123–131, International Telecommunications Union, Geneva, 1982.

11. Stephansen, E. T., "Clear-Air Propagation on Line-of-Sight Radio Paths: A Review," *Radio Science,* vol. 16, pp. 609–629, 1981.

12. Olsen, R. L., "Cross Polarization During Clear-Air Conditions on Terrestrial Links: A Review," *Radio Science,* vol. 16, pp. 631–647, 1981.

13. CCIR Report 338-5, *"Propagation Data and Prediction Methods Required for Line-of-Sight Radio-Relay Systems,"* *Recommendations and Reports of the CCIR,* vol. 5, pp. 325–366, International Telecommunications Union, Geneva, 1986.

14. Moulsley, T. J., and E. Vilar, "Experimental and Theoretical Statistics of Microwave Amplitude Scintillations on Satellite Down-Links," *IEEE Trans. on Antennas and Propagation,* vol. AP-30, pp. 1099–1106, 1982.

15. Crane, R. K., "A Review of Transhorizon Propagation Phenomena," *Radio Science,* vol. 16, pp. 649–669, 1981,

16. Cox, D. C., and H. W. Arnold, "Results From the 19- and 28-GHz COMSTAR Satellite Propagation Experiments at Crawford Hill," *Proc. IEEE,* vol. 70, pp. 458–488, 1982.

17. Rogers, D. V., "Propagation Considerations for Satellite Broadcasting at Frequencies Above 10 GHz," *IEEE J. on Selected Areas in Communications,* vol. SAC-3, pp. 100–110, 1985,

18. Tatarski, V. I., *Wave Propagation in a Turbulent Medium,* McGraw-Hill, New York, 1961.

19. Bean, B. R., and E. J. Dutton, *Radio Meteorology,* Chap. 7, pp. 269–309, Dover Publications, New York, 1968.

20. CCIR Report 564-3 , "Propagation Data and Prediction Methods Required for Earth-Space Telecommunication Systems," *Recommendations and Reports of the CCIR,* vol. 5, pp. 389–431, International Telecommunications Union, Geneva, 1986.

21. Tattelman, P., and D. D. Grantham, "A Review of Models for Estimating 1 min Rainfall Rates for Microwave Attenuation Calculations," *IEEE Trans. on Communications,* vol. COM-33, pp. 361–372, 1985.

22. Crane, R. K., "Evaluation of Global and CCIR Models for Estimation of Rain Rate Statistics," *Radio Science,* vol. 20, pp. 865–879, 1985.

23. Olsen, R. L., D. V. Rogers, and D. B. Hodge, "The aR^b Relation in the Calculation of Rain Attenuation," *IEEE Trans. on Antennas and Propagation,* vol. AP-26, pp. 318–329, 1978.

24. CCIR Report 721-1, "Attenuation by Hydrometeors, in Particular Precipitation, and Other Atmospheric Particles," *Recommendations and Reports of the CCIR,* vol. 5, pp. 167–181. International Telecommunications Union, Geneva, 1982.

25. Allnutt, J. E., "Nature of Space Diversity in Microwave Communications via Geostationary Satellites: A Review," *IEE Proc.,* vol. 125, pp. 369–376, 1978.

26. Hodge, D. B., "An Improved Model for Diversity Gain on Earth-Space Paths," *Radio Science,* vol. 17, pp. 1393–1399, 1982.

27. CCIR Report 720-1, "Radio-Emission From Natural Sources Above 50 MHz," *Rec-*

ommendations and Reports of the CCIR, vol. 5, pp. 151–166, International Telecommunications Union (ITU), Geneva, 1982.

28. Olsen, R. L., "Cross-Polarization During Precipitation on Terrestrial Links: A Review," *Radio Science,* vol. 16, pp. 761–779, 1981.

29. Cox, D. C., "Depolarization of Radio Waves by Atmospheric Hydrometeors: A Review," *Radio Science,* vol. 16, pp. 781–812, 1981.

30. Oguchi, T., and M. Yamada, "Frequency Characteristics of Attenuation, Phase Shift, and Cross-Polarization Due to Rain Within Communication Bands: Calculations at 4, 6, 11, and 14 GHz Bands for INTELSAT Satellite Communication System," *J. Radio Research Laboratories* (Japan), vol. 28, pp. 97–131, 1981.

31. Crane, R. K., "Ionospheric Scintillation," *Proc. IEEE,* vol. 65, pp. 180–199, 1977.

32. CCIR Report 263-5, "Ionospheric Effects Upon Earth-Space Propagation," *Recommendations and Reports of the CCIR,* vol. 6, pp. 124–149, International Telecommunications Union, Geneva, 1982.

33. CCIR Report 725-1, "Ionospheric Properties," *Recommendations and Reports of the CCIR,* vol. 6, pp. 1–15, International Telecommunications Union, Geneva, 1982.

34. Jacobs, G., and T. J. Cohen, *The Short-Wave Propagation Handbook,* Chap. 6, pp. 133–155, Cowan Publishing Corporation, Port Washington, N.Y., 1979.

35. CCIR Recommendation 534, "Method for Calculating Sporadic-E Field Strength," *Recommendations and Reports of the CCIR,* vol. 6, pp. 242–257, International Telecommunications Union, Geneva, 1982.

36. Fang, D. J., and M. S. Pontes, "4/6-GHz Ionospheric Scintillation Measurements During the Peak of Sunspot Cycle 21," *COMSAT Technical Review,* vol. 11, pp. 293–320, 1981.

37. Karasawa, Y., K. Yasukawa, and M. Yamada, "Ionospheric Scintillation Measurements at 1.5 GHz in Mid-Latitude Region," *Radio Science,* pp. 643–651, 1985.

38. Aarons, J., "Global Morphology of Ionospheric Scintillations," *Proc. IEEE,* vol. 70, pp. 360–378, 1982.

CHAPTER 2
ANTENNAS

H. Allen Ecker
Vice President, Corporate Development, Scientific-Atlanta, Inc.

James H. Cook, Jr.
Principal Engineer, Scientific-Atlanta, Inc.

James W. Cofer
Antenna Product Line Manager, Scientific-Atlanta, Inc.

CONTENTS

INTRODUCTION

1. Antenna Functions and Requirements

The antenna is a basic component of any electronic communications system not limited to cable or fiber transmission. An *antenna* is a device for accomplishing a transition between a guided electromagnetic wave and a wave which propagates in free space. It provides the means to collect the energy from an incident electromagnetic field or to radiate electromagnetic energy to other points. The antenna must provide the gain necessary to allow proper transmission and reception and also must have radiation characteristics appropriate to discriminate against unwanted signals. The applications in this chapter deal chiefly with antennas for point-to-point microwave and satellite communications systems; however, the definitions and principles apply to antennas in general.

Antenna requirements can be grouped into several major categories, namely, electrical or RF, control system, structural, pointing and tracking accuracy, environmental, and miscellaneous requirements, such as radiation hazard and primary power requirements.

ANTENNA DEFINITIONS AND PARAMETERS

2. Basic Parameters

The basic electrical parameters of an antenna are radiation pattern, gain, directivity, impedance, polarization, frequency or bandwidth, and noise temperature. These parameters determine antenna performance in a communications system, and tradeoffs among parameter values are usually necessary to meet the requirements of practical communications systems. In the following sections, the various factors that affect the parameter values are described to provide information to conduct system tradeoff analyses.

3. Radiation Pattern

The *radiation pattern* of an antenna is its spatial distribution of radiated energy. Common types of radiation patterns are the following:

Omnidirectional

Unidirectional

Pencil beam

Fan beam

Shaped beam

The *omnidirectional* (or *isotropic*) antenna in the strictest sense is one having an exactly spherical radiation pattern. In a practical sense, for antennas located near a ground plane (usually the earth), it is an antenna which has a hemispherical radiation-intensity pattern. In two-dimensional considerations an antenna whose radiation intensity is uniform in the horizontal plane (such as a TV broadcast antenna) is called *omnidirectional*. The broadcast antenna typically has a circular pattern in the horizontal plane while the vertical-plane pattern may be shaped to increase the energy radiated in the horizontal plane.

The *bidirectional* pattern antenna has a doughnut-shaped pattern and can be considered as a special case of the omnidirectional pattern. Typical antenna types which have this characteristic are dipoles, loops, and monopoles.

The *unidirectional* pattern antenna has a radiation maximum for specific angular coordinates (θ_1, ϕ_1) as illustrated in Fig. 2.1. The radiation intensity is defined by a $\sin^n \theta \sin^m \phi$, where n and m are any real numbers. When $n = m$, the pattern is symmetrical, and conversely, when $n \neq m$, the pattern has different beamwidths in the $\theta = 90°$ and $\phi = 90°$ planes.

The *pencil beam* is a modified unidirectional pattern and is typical of reflector and array antennas. The radiation pattern concentrates energy in a narrow angular sector, and the energy radiated in that sector is maximized. The beamwidths in the two principal planes ($\theta = 0°$, $\phi = 90°$) are essentially equal. A variation of the pencil beam is the *fan beam* in which the beamwidth in one plane is much broader than the beamwidth in the orthogonal plane. The beam cross section is elliptical rather than circular.

The *shaped-beam* pattern is designed when the antenna radiation characteristics must meet specified angular requirements. An example is the so-called

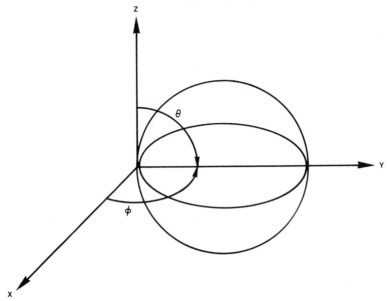

FIG. 2.1 Field intensity pattern of unidirectional source with maximum radiation in the direction $\theta = 90°$, $\phi = 90°$.

cosecant-squared beam that is used in search radars to produce an essentially constant radar return for targets approaching at fixed altitudes.

The far-field pattern of a reflector-type antenna can be calculated from a known distribution of electric fields (amplitude and phase) over the actual aperture of the reflector. This *aperture distribution* is usually characterized by the taper of the amplitude from the center of the reflector to the edge. Attempts are made to keep the phase distribution uniform to maximize antenna efficiency.

For purposes of analysis, it is useful to approximate the actual aperture distribution by a simple mathematical expression which can be easily subjected to integration and other mathematical operations. Many distributions for circular apertures can be approximated by one of the symmetric Spencer functions:[1]

$$F(r) = [\, 1 - (r/a)^2 \,]^{\,p} \quad \text{for} \quad p = 0, 1, 2, 3, 4, \ldots \tag{2.1}$$

where $F(r)$ is the aperture distribution, a is the aperture radius, and r is the aperture radial variable.

Two of the more important parameters for defining the radiation pattern of antennas are the half-power beamwidth and the first sidelobe level. The *half-power beamwidth* is defined as the angular separation between the points on the beam where the power density is 3 dB below that at the beam center. The *first sidelobe level* is the ratio of pattern power density at the first local pattern maximum separated in angle from main beam to that at the center of the main beam. Both the beamwidth and the first sidelobe level are direct functions of the antenna size and the manner of feed illumination. For example, the first sidelobe levels corresponding to the Spencer aperture distributions above are −17.6 dB ($p = 0$), −24.6 dB ($p = 1$), −30.6 dB ($p = 2$), −36.0 dB ($p = 3$), and −40.7 dB ($p = 4$).

Estimates of the *half-power beamwidths* (HPBW) of aperture and array antennas can be made using the equation

$$\text{HPBW} = k(\lambda/D) \quad \text{deg} \tag{2.2}$$

where k is a beamwidth constant, λ is the wavelength, and D is the aperture dimension in the plane of the pattern. The value of k is determined by the type of antenna and the feed illumination. For typical reflector-type antennas, the value of k is approximately 70. Although this value is sufficiently accurate for many applications, a more precise estimate can be made by considering the aperture illumination function.

Komen[2] determined empirically that the value of the beamwidth constant k is determined by the ratio of the aperture illumination at the edge of the dish to that at the center and that the relationship between beamwidth and edge illumination holds regardless of frequency, reflector size, reflector type, or feed type. Mathematically, this relationship is expressed as

$$k = 1.0524I + 55.949 \tag{2.3}$$

where I is the value in decibels of edge illumination (including space attenuation) relative to the center and k is in degrees.

The beamwidth constant k and approximate sidelobe level versus edge illumination for reflector-type antennas and a specific type of illumination function are illustrated in Fig. 2.2. The beamwidth constant and sidelobe level are greatly influenced as aperture blockage (by feed, subreflector, support structure, etc.) is increased; therefore, Fig. 2.2 should be used when the central blockage is less than 10 percent of the aperture area. A typical edge illumination is -13 dB. As shown in Fig. 2.2, this illumination results in a k of 70 and a first sidelobe level of -23 dB. From Eq. 2.2, a k of 70 yields a half-power beamwidth of $\leq 1°$ for a 1.8-m antenna at a frequency of 12.0 GHz.

4. Gain and Directivity

The power *gain* $G(\theta, \phi)$ of an antenna in a specified direction (θ, ϕ) is defined by the ratio of the power density $\Phi(\theta, \phi)$ radiated in that direction to the power density which would be radiated by a lossless isotropic antenna with the same input power P_O accepted at its terminals. Mathematically, the gain is expressed as

$$G(\theta, \phi) = \frac{4\pi\Phi(\theta, \phi)}{P_o} \tag{2.4}$$

the product of its directivity and its efficiency.

The *directivity* of the antenna is a measure of the degree to which it concentrates the radiated power in a single direction. For *directive antennas* (those with reasonably narrow main beams), directivity is inversely proportional to the product $\theta_3\phi_3$, where θ_3 and ϕ_3 are the half-power beamwidths in two orthogomal planes. Typical efficiencies of practical directive antennas range from 50 to 80 percent.

A convenient rule of thumb for predicting the gain of a low-loss directive antenna is

$$G = 10 \log (C/(\theta_3\phi_3) \quad \text{dB} \tag{2.5}$$

where C is a constant and θ_3 and ϕ_3 are as defined above. The correct value of C for an actual antenna depends upon the antenna efficiency, but typical values

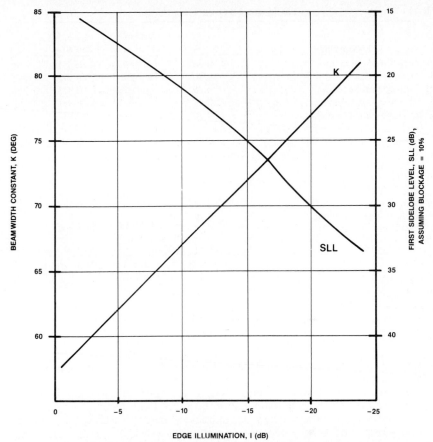

FIG. 2.2 Beamwidth constant and approximate first sidelobe level versus edge illumination for reflector-type antennas with plane-wave feeds.

vary from 27,000 to 35,000. This gain approximation is derived from the assumption that most of the radiated power of an antenna is within the main beam. The factor C would have a value of $41,253^{o2}$ for an antenna with 100 percent efficiency.

5. Gain Budgeting

Numerous techniques exist for determining the gain of an antenna. The more common ones include *direct measurement* (comparison with a reference antenna of known gain), *pattern integration* (determination of P_O for use in Eq. 2.4), and *gain budgeting*. Gain budgeting is particularly useful to the system designer because it does not require physical models and measurements. Also it allows the designer to see the gain impact of system decisions (e.g., low sidelobe requirements reduce the gain). A sample gain budget is illustrated along with typical entries in Table 2.1. As the project progresses, calculated values are replaced by

TABLE 2.1 Typical Gain Budget

Loss mechanism	Loss value dB	Numerical
Aperture illumination	0.70	0.851
Phase	0.10	0.977
Primary feed spillover	0.35	0.923
Secondary spillover	0.08	0.981
Surface roughness (0.03 in rms)	0.10	0.977
Cross-polarization (conversion)	0.02	0.995
Subreflector blockage ($D_1/D_2 = 0.13$)	0.10	0.977
Spar blockage	0.25	0.944
Feed ohmic loss	0.02	0.995
Mode transducer ohmic	0.10	0.977
VSWR (1.2:1)	0.05	0.988
Total losses	1.87	0.65

	Gain, dBi	Efficiency, %
Ideal gain ($4\pi A/\lambda^2$)	48.03	100
Expected gain (Ideal − losses)	46.2	65

measured ones until, in the final stage, the budgeted gain should approach the measured gain very closely.

The two dominant losses in most designs are *aperture illumination* and *primary spillover*. As the edge illumination becomes more tapered, sidelobes decrease; however, the illumination loss increases, thereby lowering the gain. This loss is partially offset by a decrease in spillover loss which decreases for larger tapers. In fact, these two factors tend to compensate for each other as shown in Fig. 2.2. The *optimum taper,* that is, the one which gives the maximum product of the two factors, occurs for edge illumination between 10 and 15 dB.

6. Beam Mispointing

Often the systems engineer needs to model the main beam behavior to levels 10 dB below maximum gain without having detailed knowledge of the measured pattern. Requirements to calculate beam mispointing effects or for tracking parameters produce this need. Within the 10 dB points, the main beam can usually be described adequately by the parabolic approximation

$$P(\theta) = -K(\theta/\theta_K)^2 \quad \text{dB} \tag{2.6}$$

where θ is the angle off boresight and θ_K is the angle at which the beam is K dB below the peak. As an example of the use of this equation, consider a requirement to calculate the gain loss due to a wind-induced mispointing of 0.2° for an antenna with a half-power beamwidth of 1°. From Eq. 2.5, with $K = 3$ dB, $\theta = 0.2°$, and $\theta_K = 0.5°$, the gain loss is found to be 0.48 dB.

7. Aperture Blockage

The discussion of aperture distribution thus far has been based on a clear field of view immediately in front of the antenna. Except in the case of the offset antenna, this situation rarely occurs in practice. Reflector antennas with subreflectors, primary feed horns, and spars to support the feed and/or subreflector will experience *aperture blockage* that reduces the gain and degrades the radiation pattern.

The effect of *circular central blockage* is demonstrated in Fig. 2.3 where gain losses and sidelobe degradation are plotted as a function of blockage ratio B (blockage diameter/aperture diameter).[3] The effect is shown for several members of the Spencer function family ($p = 0, 1, 2, 3, 4$) for assumed paraboloidal reflectors. The degradation is much more severe for the antennas having very low sidelobes before blockage. In dual-reflector systems, techniques have been devised for minimizing the effect of *subreflector blockage*. The subreflector can be designed to direct scattered energy away from its own region of the aperture.

8. Impedance

An antenna must be coupled to a transmitter or receiver by a transmission line. Therefore, the input impedance of the antenna is important because maximum power transfer will be possible only if the antenna and transmission line are "matched." The antenna input impedance will generally have a resistive and reactive part. The resistive part of the input impedance has contributions from all mechanisms that give rise to a loss of energy. Of course, the radiated power is the desired loss mechanism, but other losses (e.g., ohmic) are also present.

9. Polarization

At large distances from a radiating antenna, the electric, **E**, and magnetic, **H**, field vectors of the radiated field are orthogonal to each other and to the direction of

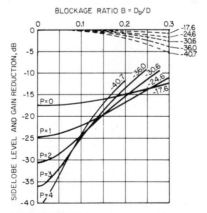

FIG. 2.3 Gain/loss (---) and resulting sidelobe level (—) for a centrally blocked circular-aperture distribution having a specified unblocked sidelobe level.

propagation. The two fields oscillate in a constant time-phase relationship, and the ratio of their magnitudes E/H, or η, is a constant 120 Ω, the intrinsic impedance of free space.

The *polarization* of an electromagnetic field is described in terms of the direction in space of the electric field vector **E**. If the vector which describes the electric field at a point in space is always parallel to a line of fixed orientation, the field is said to be *linearly polarized*. In general, however, the angular direction of the electric vector varies with time, and its terminus traces an ellipse. In this case the field is said to be *elliptically polarized. Circular polarization* is a special case of elliptical polarization in which the trace of the **E** field vector is a circle. The field is said to be *clockwise* or *counterclockwise circular,* depending on the direction of rotation of the **E** field vector when the field is viewed toward the direction of propagation.

Antennas exhibit polarization properties in relation to the fields they radiate or receive. If an antenna is in a receive mode, it will not, in general, be matched in polarization to the incident field. The polarization must be matched to extract maximum power from the field and to have a polarization efficiency of unity. If the antenna polarization is orthogonal to the field polarization, the antenna will extract zero power, and its polarization efficiency is consequently zero.

One measure of the quality of an antenna is the level of the orthogonal or *cross-polarization component* in the radiation patterns. *Low cross-polarization* is required to minimize the effects of interfering signals and to maximize efficiency.

10. Operating Frequency

Generally, the required performance of an antenna is specified over some frequency band. The frequency band is defined by stating upper and lower frequency limits, or by a center frequency and percent bandwidth. Many antenna types are inherently narrowband while others are broadband. A narrowband antenna may only operate over a few percent bandwidth. An example is a resonant waveguide slot array. On the other hand, some antenna types operate over multiple-octave frequency bands, for example, a log periodic dipole array antenna.

11. Noise Temperature

The measure of sensitivity of an antenna receiving system includes the effects of noise generated in the receiving system itself and the effects of noise generated in the entire environment (terrestrial and extraterrestrial) in which the above system is immersed. The noise power generated by these elements can be collectively represented by an "effective" *noise temperature* at some convenient reference point in the system.

Generally, the system noise temperature T_x may be regarded as the sum of three components: (1) the contribution of the antenna resulting from reception of noise from external sources, (2) the thermal noise generated by dissipative losses in the receiving transmission line itself, and (3) the noise from internal sources such as the amplifier and receiver. The contribution of the antenna, T_A, from reception of noise from external sources, is composed primarily of noise from the cosmos, the ionosphere, the troposphere, and the earth, which includes the sea as well as solid ground. Other forms of external noise, i.e., artificial noise (from ignition systems, arcs, spark discharges, etc.), are usually associated with low-frequency components and therefore are only of interest for antennas operating below the microwave frequency bands.

The effective antenna noise temperature T_A may be computed in a particular environment by well-known methods.[4] The exact procedure for computing the effective noise temperature of a particular antenna is quite complicated. Fortunately, if an antenna has a unidirectional pattern that is not extremely broad, as do the majority of antennas above 100 MHz, the beamwidth and gain have little effect on its noise temperature (averaged over all galactic directions) under ordinary circumstances. Therefore, it is possible to calculate an approximate antenna noise temperature, as a function of frequency, that is applicable to a typical directional antenna. The noise temperature is dependent on the elevation angle above the local horizon due to the thermal noise generated in the absorptive gases of the atmosphere. Figure 2.4 shows representative noise temperatures of typical directional antennas.

The quantitative effect of antenna noise temperature on transmission performance is described in Chap. 4 for microwave systems and Chap. 5 for satellite systems.

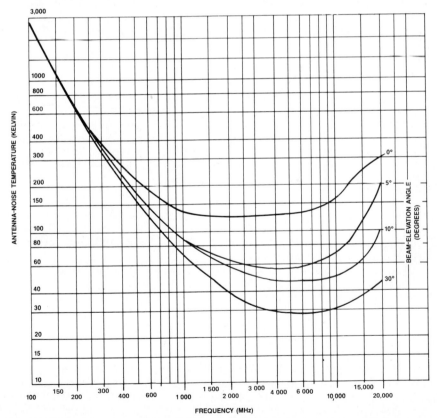

FIG. 2.4 Noise temperature of a typical directive antenna for representative environmental conditions.

ANTENNA CONFIGURATIONS

12. Basic Reflector Antennas

The *reflector antenna* is the predominant antenna in use above 100 MHz. Radar, point-to-point microwave, tropospheric communication, and satellite communications routinely employ reflector antennas. The most common of these antennas is the *paraboloidal reflector*. The paraboloidal reflector provides the basis for axisymmetric and offset-fed antennas including single-beam and multiple-beam configurations.

Significant advances in the performance of reflector antennas have occurred in the last two decades as a result of computer-aided design. These advances have resulted in improved antenna efficiency, low cross-polarization, and pattern control, along with the development of special-purpose antennas to provide beam scanning, multiple beams, and/or shaped beams. Important to these advances has been the development of new feeds and new types of reflector surfaces, accurate prediction and measurement of antenna performance, the measurement of the surface accuracy of reflectors, and the development of new manufacturing techniques. Because of the predominance of reflector antennas, the remainder of this section is devoted to this type.

Reflector antennas can be grouped into two broad categories—*single-beam antennas* and *multiple-beam antennas*. A single-beam reflector antenna generates a single beam; a multiple-beam reflector antenna generates multiple beams by employing a common reflector aperture with multiple feeds illuminating that aperture. The axes of the beams are determined by the location of the feeds. The individual beams may be steered by moving the feed without moving the reflector.

Prime-Focus-Fed Paraboloidal Antenna. The prime-focus-fed paraboloidal reflector antenna is the basic reflector antenna. For moderate-to-large aperture sizes, this type of antenna has excellent sidelobe performance in all angular regions except the spillover region around the edge of the reflector. The spillover radiation can be reduced by either placing a shroud around the periphery of the reflector or by underilluminating the edge of the reflector by the feed. Unfortunately, the latter method also results in decreased gain.

The predominant advantage of the prime-focus-fed paraboloidal is in its simplicity and lower cost than other reflector geometries. Its disadvantage is that the antenna designer is constrained by the small feed for the reflector, resulting in less flexibility in its radiation properties.

The geometry of a paraboloidal reflector antenna is shown in Fig. 2.5. The distance from the vertex of the reflector to the focal point is defined as the focal length. The focal point position defines a subtended angle between the focal point and the edge ray of the reflector. The equation of the surface is given by

$$x^2 + y^2 = 4fz \qquad (2.7)$$

where f is the focal length.

An important characteristic of this surface is that an incident plane wave parallel to its axis is collimated in-phase at the focal point. Therefore the antenna is considered *unaberrated*. If a feed is placed at the focal point, it collects the energy optimally. When the feed is located off the reflector axis, the beam is steered off the paraboloidal axis and phase aberrations occur which reduce the antenna's gain and degrade the sidelobe performance as compared to the on-axis condition.

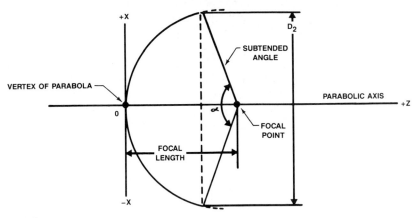

FIG. 2.5 Geometry of parabolic reflector.

13. Dual-Reflector Antenna Types

The dual reflector antenna offers additional flexibility in design as compared to the prime-focus-fed paraboloidal reflector. This flexibility includes enhancement of gain, control of near-in sidelobes, beam shaping, and control of spillover energy, and additional complexity of feed components without degrading electrical characteristics. These characteristics have made the dual-reflector antenna the predominant choice for the earth-station antenna and for terrestrial microwave systems and many other applications.

Figure 2.6 shows the geometric-optics ray tracing for three dual-reflector geometries, the conventional Cassegrain, the conventional Gregorian, and the Gregux or uncrossed Gregorian. The Gregux antenna has to date been used primarily in terrestrial microwave systems with recent applications in satellite communications. The Gregux antenna offers advantages in minimizing central blockage and VSWR effects and typically yields excellent aperture efficiencies of 65 to 80 percent, but has the disadvantage of using a nonparaboloidal main reflector because it has a ring focus rather than a point focus.

Because the basic principles of the designs of these three antennas are similar, the Cassegrain antenna will be examined in more detail as representative of dual reflectors in general.

The Cassegrain Antenna.   Cassegrain antennas can be subdivided into three primary types:

1. The classical Cassegrain geometry[5] employs a paraboloidal contour for the main reflector and a hyperboloidal contour for the subreflector as shown in Fig. 2.7. The paraboloidal reflector is a point-focus device with a diameter D_p and a focal length f_p. The hyperboloidal subreflector has two foci. For proper operation, one of the two foci is the real focal point of the system and is coincident with the phase center of the feed; the other focus, the virtual focal point, is coincident with the focal point of the main reflector.

2. A geometry consisting of a paraboloidal main reflector and a special shaped, quasi-hyperboloidal subreflector.[6,7] The main difference between this design

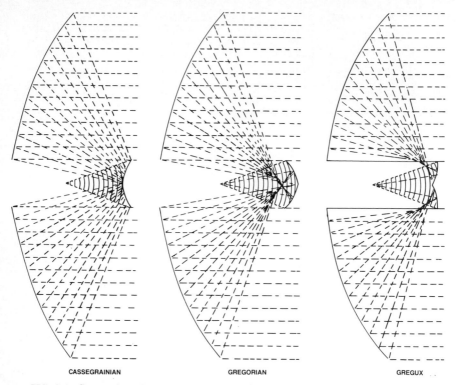

CASSEGRAINIAN   GREGORIAN   GREGUX

FIG. 2.6   Geometric optics ray traces of dual-reflector antennas.

and the classical Cassegrain above is that the subreflector is shaped so that the overall efficiency of the antenna has been enhanced. This technique is especially useful with antenna diameters of approximately 60 to 300 wavelengths. The subreflector shape may be solved for any *geometrical optics* (GO) or by

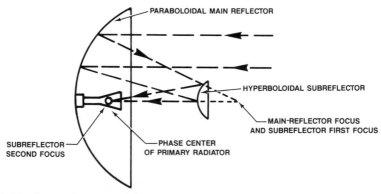

FIG. 2.7   Cassegrain antenna geometry.

diffraction optimization; then, by comparing the required main reflector surface to a paraboloidal surface, the best-fit paraboloidal surface is found. Aperture efficiencies of 65 to 75 percent can be realized by a GO design and 70 to 80 percent by diffraction optimization of the subreflector.

3. A generalization of the Cassegrain geometry consisting of a special-shaped, quasi-paraboloidal main reflector and a shaped quasi-hyperboloidal subreflector.[8-12] The dual-shaped-reflector system controls both the amplitude and phase of the aperture illumination function, thereby yielding better results in both gain and sidelobes than the previous two dual reflector configurations. Two basic approaches may be taken for this design, namely a geometric optics approach or a diffraction-optimized approach. Although the GO approach is adequate for most applications, improvements in performance can be achieved when diffraction effects[13] are included in the design synthesis. For instance, Clarricoats and Poulton[14] reported a gain increase of 0.5 dB for a diffraction-optimized design over a GO design with a 400-wavelength-diameter main reflector and a 40-wavelength-diameter subreflector.

Figure 2.8 compares the relative aperture efficiencies of the three dual-reflector-axisymmetric geometries.

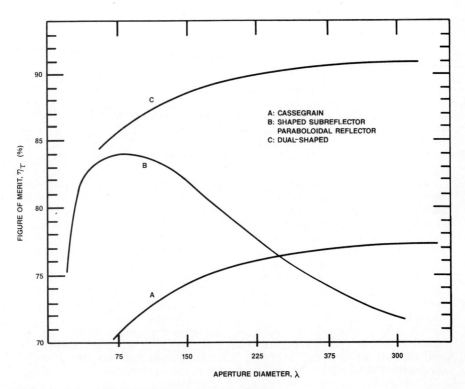

FIG. 2.8 Antenna figure of merit versus aperture diameter.

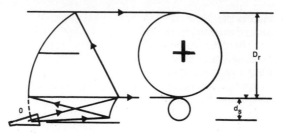

A. DOUBLE-OFFSET GEOMETRY (FEED PHASE CENTER AND
   PARABOLOIDAL VERTEX AT 0)

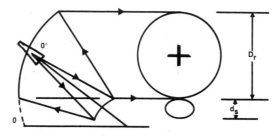

B. OPEN CASSEGRAINIAN GEOMETRY (FEED PHASE CENTER
   LOCATED AT 0′; PARABOLOIDAL VERTEX AT 0)

FIG. 2.9   Offset dual-reflector geometries.

14. Offset-Fed Reflector Antennas

The offset-fed reflector antenna geometry has been known for many years, but has been used only during the last decade or so because it has been difficult to analyze. Now that large computers allow the antenna engineer to analyze the offset-fed reflector's performance theoretically, this type of antenna will become more common. The offset-fed reflector antenna can employ a single reflector or multiple reflectors, with two reflectors the more prevalent of the multiple-reflector designs.

The *offset, front-fed reflector,* consisting of a section of a paraboloidal surface, minimizes diffraction scattering by eliminating the aperture blockage of the feed and feed-support structure. Sidelobe levels of (29-25 log θ) dBi can be expected from this type of antenna, with aperture efficiencies of 65 to 80 percent. The increase in aperture efficiency as compared to an axisymmetric, prime-focus-fed antenna results from the elimination of direct blockage. For a detailed discussion of this antenna, see Reference 15.

Offset-fed, dual-reflector antennas exhibit sidelobe performance similar to that of the front-fed, offset reflector. Two offset-fed, dual-reflector geometries are used: the *double-offset* geometry shown in Fig. 2.9a and the *open Cassegrain* geometry introduced by Cook et al.[16] of Bell Labs shown in Fig. 2.9b. In the double-offset geometry, the feed is located below the main reflector, and no blocking of the optical path occurs. In contrast, the open Cassegrain geometry is such that the primary feed protrudes through the main reflector; thus it is not completely blockage-free. Nevertheless, both of these geometries have the capability of excellent sidelobe and efficiency performance.

The disadvantage of the offset-geometry antennas is that they are asymmetric. This leads to increased manufacturing cost and also has some effect on the electrical performance. The offset-geometry antenna, when used for linear polarization, has a depolarizing effect on the primary feed radiation and produces two cross-polarized lobes within the main beam in the plane of symmetry. When used for circular polarization, a small amount of beam squint is introduced whose direction is dependent upon the sense of polarization.[17]

15. Multiple-Beam Earth-Station Antennas

During the past few years there has been an increasing interest in receiving signals simultaneously from several satellites with a single antenna. This interest has prompted the development of several multibeam antenna configurations which employ fixed reflectors and multiple feeds. The spherical reflector, the torus reflector, and the dual-reflector geometries, all using multiple feeds, have been offered as antennas with simultaneous, multibeam capability. Chu[18] in 1969 addressed the multiple-beam spherical reflector antenna for satellite communications; Hyde[19] introduced the multiple-beam torus antenna in 1970; Ohm[20] presented a novel multiple-beam Cassegrain geometry antenna in 1974. All three of these approaches are discussed below, as well as variations of scan techniques for the spherical reflector.

Spherical Reflector.   The popularity of the spherical antenna is primarily the result of the large angle through which the radiated beam can be scanned by translation and orientation of the primary feed. This wide-angle property results from the symmetry of the surface. Multiple-beam operation is realized by placing multiple feeds along the focal surface. In the conventional use of the reflector surface, the minimum angular separation between adjacent beams is determined by the feed aperture size. The maximum number of beams is determined by the percentage of the total sphere covered by the reflector. In the alternative configuration described below, the number of beams is determined by the f/D ratio and by the allowable degradation in the radiation pattern.

In the conventional use of the spherical reflector, each feed illuminates a portion of the reflector surface such that a beam is formed coincident to the axis of the feed, Fig. 2.10. All the beams have similar radiation patterns and gains, although there is degradation in performance in comparison to that of a paraboloid. The advantage of this antenna is that the reflector area illuminated by the individual feeds overlaps, reducing the surface area for a given number of beams in comparison with individual single-beam antennas.

The alternative multibeam spherical reflector geometry is shown in Fig. 2.11. For this geometry, each of the feed elements points toward the center of the reflector with the beam steering accomplished by the feed position. This method of beam generation leads to considerable increase in aberration, including coma; therefore, the radiation patterns of the off-axis beams are degraded with respect to the on-axis beam. This approach does not take advantage of the spherical reflector properties that exist in the conventional approach. In fact, somewhat similar results could be achieved with a paraboloidal reflector with a large f/D.

Torus Antenna.   The torus antenna is a dual-curvature reflector, capable of multibeam operation with a feed system similar to the conventional spherical re-

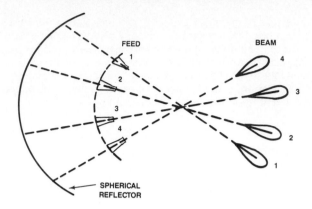

FIG. 2.10 Conventional spherical multibeam antenna using extended reflector and multiple feeds.

flector geometry. The plane of the feed can be inclined to be in the orbital-arc plane, allowing the use of a fixed reflector to view geosynchronous satellites. The reflector has a circular contour in the plane of the feeds and a parabolic contour in the orthogonal plane (see Fig. 2.12). It can be fed either in an axisymmetric or an offset-fed configuration. The offset geometry has been successfully demonstrated by Comsat Labs.[21] The radiation patterns meet a (29-25 log θ) dBi envelope.

The torus antenna has less phase aberration than the spherical antenna because of the focusing in the parabolic plane. Because of its circular symmetry, feeds placed anywhere on the feed arc form identical beams. Therefore, no performance degradation is incurred when multiple beams are placed on the focal arc. Point focus feeds may be used to feed the torus up to aperture diameters of approximately 200 wavelengths. For larger apertures, it is recommended that aberration correcting feeds be used.

The multiple-beam operation of a torus requires an oversized aperture to accommodate the scanning. For example, a reflector surface area of approximately

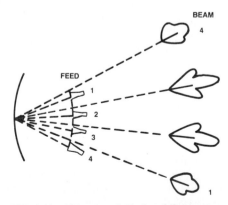

FIG. 2.11 Alternate spherical multibeam antenna using minimum reflector aperture with scanned beam feeds.

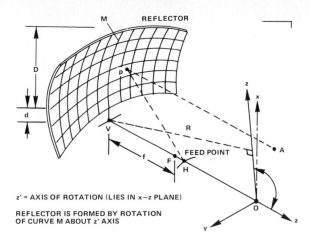

FIG. 2.12 Torus antenna geometry (COMSAT Laboratories).

214 m² will allow a field of view (i.e., orbital arc) of 30° with a gain of approximately 50.5 dB at 4 GHz (equivalent to the gain of a 9.65-m reflector antenna). This surface area is equivalent to approximately three 9.65-m antennas.

Offset-Fed, Multibeam Cassegrain Antenna. The offset-fed, multibeam Cassegrain antenna is composed of a paraboloidal main reflector, a hyperboloidal oversized subreflector, and multiple feeds located along the scan axis, as shown in Fig. 2.13. The offset geometry essentially eliminates beam blockage, thus allowing a significant reduction in sidelobes and antenna noise temperature. The Cassegrain feed system is compact and has a large focal-length-to-diameter ratio (*f/D*) which reduces aberrations to an acceptable level, even when the beam is moderately far off-axis. The low sidelobe performance is achieved by using a corrugated feed horn which produces a gaussian beam.

A typical antenna design consisting of a 10-m projected aperture would yield half-power beamwidths and gain commensurate with an axisymmetric 10-m antenna, 0.5° HPBW, and 51 dBi gain at 4 GHz. The subreflector would need to be approximately a 3- by 4.5-m elliptical aperture. The feed apertures would be approximately 0.5 m in diameter. The minimum beam separation would be less than 2°, more than sufficient to allow use with synchronous satellites with orbital spacing of 2° or greater. For a +15° scan, the gain degradation would be approximately 1 dB, and the first sidelobe would be approximately 20 to 25 dB below the main beam peak.

ANTENNA PARAMETER TRADEOFFS

16. Scope

This section demonstrates some of the broad tradeoffs (in configurations and/or performance) frequently encountered by the system designer.

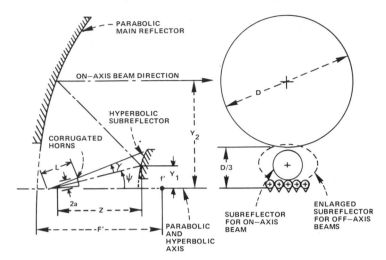

FIG. 2.13 Geometry of the offset-fed multibeam Cassegrain antenna (Reprinted with permission from the *Bell System Technical J.*, Copyright 1974 AT&T).

17. Prime Focus versus Cassegrain

Two of the most commonly used antenna configurations for microwaves and satellite earth-terminal antennas are the prime-focus paraboloid and the dual-reflector Cassegrain.

The prime-focus geometry is particularly well-suited for achieving low sidelobes in medium-sized antennas (e.g., 4 to 6 m) and lower frequencies (C band, 4 to 6 GHz). This suitability results from low aperture blockage and spillover. A Cassegrain system requires a subreflector which is several wavelengths in diameter in order to control the shape of the aperture distribution. In C band a blockage of as much as 20 to 30 in. in diameter can result. Such a large center blockage can completely govern the behavior of near-in sidelobes and in general causes them to be higher than most desired standards (e.g., 29-25 log θ) . The Cassegrain configuration also suffers in the area of primary spillover around the edge of the subreflector. This energy usually falls in the area of 20 to 30° away from the main-beam axis. Thus the Cassegrain suffers at C band from high lobes near boresight and at 30° away. Granted, the prime focus also experiences main reflector spillover; however, it occurs at angles of 100 to 120° from boresight. The prime focus designer also has the latitude to utilize a primary pattern which produces low diffraction lobes without undue sacrifice of gain.

Larger antennas such as 7 to 11 m operating at C band are able to produce low sidelobes from the Cassegrain geometry since a 30-in or larger subreflector corresponds to a smaller percentage of blockage. Similarly, K_u-band antennas in the size range of 3 to 6 m can be implemented as Cassegrains since a 10 wavelength subreflector would produce a blockage diameter of only 10 in—again a small percentage of area.

Other factors may enter into the tradeoff analysis. For a receive-only system, the LNB should always be as close to the feed as possible, thereby making the prime focus a good choice. For transmit-receive systems, the designer must consider the blockage size if the transmitter is to go at the focal point or consider the waveguide losses if the rf energy is to be piped out to the prime-focus feed.

Suppose that the designer anticipates a specific beam-mispointing angle due to initial setup, wind deflections, controller repeatability, etc., and that the on-axis gain requirement allows some flexibility in choosing aperture size. The designer may choose different implementations depending on the magnitude of the mispointing angle compared to the crossover of the beams from two different reflector sizes, as shown in Fig. 2.14. The true parameter of interest is the beam gain in a given direction after mispointing has occurred. If the beam error θ_a is less than the crossover angle θ_c, then diameter D_1 is clearly the optimum choice. Beam errors (θ_b) greater than θ_c would indicate D_2 is the better size.

18. Gain versus Sidelobe Level

Two basic performance parameters which are often in conflict are gain and sidelobe levels. Higher gain and lower sidelobes are almost always desired. High gain requires aperture illuminations which are nearly uniform (edge tapers near 0 dB), whereas low sidelobes require large edge tapers (12 to 20 dB). A choice of 10 to 15 dB may represent a suitable compromise.

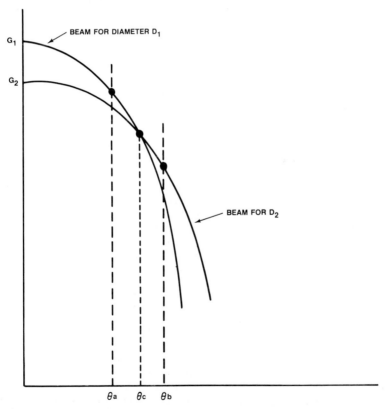

FIG. 2.14   Mispointing loss as a function of beam mispointing for two different beam errors, θ_a and θ_b, and two antenna diameters, $D_1 > D_2$.

REGULATORY CONSIDERATIONS

19. Scope of Regulations

The installation and operation of a communications antenna that is intended to transmit is subject to regulatory control in the areas of: (1) *electromagnetic interference* (EMI), (2) nonionizing radiation exposure to humans, and (3) structural safety. Receive-only antennas are also subject to structural safety regulations. Guidelines are suggested for minimizing interference on receive antennas, but the actual performance specifications are left to the discretion of the operator unless protection through licensed operation is intended.

20. Electromagnetic Interference Controls

The *envelope of antenna sidelobes* and the *isolation between orthogonal polarizations* are the two major specifications that characterize the interference potential of an antenna radiation pattern. The antenna sidelobe envelope is usually specified in the major angular regions of an antenna radiation pattern depending on the dominant contributing mechanism to sidelobes in that region (i.e., defraction, spillover, blockage, etc.). Initially, the interpretation of compliance with the sidelobe envelope specifications allowed some sidelobe peaks to exceed the envelope and some averaging of the pattern. This interpretation is still used with the international CCIR (Comité Consultatif International Rediffusion) specifications. However, the current Federal Communications Commission (FCC) specifications require that no sidelobe exceed the designated envelope within certain angular regions.

Polarization isolation becomes a major factor in applications such as frequency-reuse satellites in which both frequency and polarization separation are necessary to achieve isolation between adjacent channels. Although there currently is not a federal regulation for polarization isolation, most satellite operators with frequency-reuse satellites require that the orthogonal polarization on-axis for a transmit antenna be at least 30 dB below the main-beam radiation pattern.

The current FCC satellite earth-station sidelobe envelope requirements for 2° satellite spacing are given in Part 25, Paragraph 209, of the FCC *Rules and Regulations*. The maximum sidelobe envelope can be summarized as follows:

$$\text{(Peak) } (29\text{-}25 \log \theta) \text{ dBi } 1° < \theta \leq 7° \tag{2.8}$$
$$\text{(Average) } +8 \text{ dBi } \quad 7° < \theta \leq 9.2°$$
$$\text{(Average) } (32\text{-}25 \log \theta) \text{ dBi } \quad 9.2° < \theta \leq 48°$$
$$\text{(Average) } -10 \text{ dBi } \quad 48° < \theta \leq 180°$$

where dBi is the sidelobe level in decibels referenced to an isotropic radiator. "Peak" implies no sidelobe peak can exceed the envelope and "Average" implies some averaging can be used if an occasional peak exceeds the envelope. The same sidelobe envelopes are suggested but not required for receive antennas also.

The current international standard for satellite earth-station transmit antennas is less stringent than the FCC regulation. At present the CCIR satellite antenna transmit sidelobe envelope can be summarized as follows:

$$(29\text{-}25 \log \theta) \text{ dBi } 1° < \theta \leq 20° \ D/\lambda > 150 \text{ (Average 90\%)} \tag{2.9}$$

For D/λ between 100 and 150, higher sidelobes than indicated above are permissible only until 1990. For D/λ less than 100, less stringent requirements apply. When interpreting the radiation pattern relative to the CCIR envelope, 90 percent of the sidelobes must fall below the envelope. The reference document for current international sidelobe requirements is Recommendation 580 of CCIR Working Group 4-B.

Standards for antennas used in terrestrial microwave communications fall into the following two categories: (1) fixed point-to-point microwave under Part 21, Paragraph 108 of the FCC *Rules and Regulations* and (2) private operational-fixed microwave under Part 94, paragraph 75. In both categories the maximum sidelobe levels are given for specific frequency bands and angular regions. Table 2.2 summarizes the maximum allowable sidelobe levels for antennas operating above 2500 MHz for fixed point-to-point microwave systems. The sidelobe requirements for private operational fixed microwave systems are somewhat less stringent.

21. Nonionizing Radiation Exposure

At present there is an Occupational Safety and Health Administration (OSHA) regulation on the safe level of nonionizing radiation for human exposure that also applies to transmitting antennas. The current OSHA Standard 1910.97, "Nonionizing Radiation," states that a power density of 10 mW/cm² for a period of 0.1 h or more of an energy density of 1 mW.h/cm² during 0.1 h is the radiation protection guide. This exposure level applies for frequencies from 10 MHz to 100 GHz. The American National Standards Institute (ANSI) also publishes a standard for human exposure as ANSI C95.1, "Safety Levels with Respect to Human Exposure to Radio-Frequency Electromagnetic Fields, 300 kHz to 100 GHz." The levels and frequency range are somewhat different from those of OSHA. The OSHA standard defines the governing levels for legal purposes. A more detailed discussion of radiation exposure standards is given in Chap. 19.

22. Structural Standards

The design, fabrication, and installation of antennas, towers, equipment shelters, and other structures may come under the jurisdiction of a local municipality that enforces a building code. Although building codes can take any form, they usu-

TABLE 2.2 Maximum Allowable Sidelobe Levels for Fixed Point-to-Point Microwave Systems

Angle from center line of main lobe	Minimum radiation suppression	
	Standard A	Standard B
$5° < \theta \le 10°$	25 dB	20 dB
$10° < \theta \le 15°$	29 dB	24 dB
$15° < \theta \le 20°$	33 dB	28 dB
$20° < \theta \le 25°$	36 dB	32 dB
$30° < \theta \le 100°$	42 dB	35 dB
$100° < \theta \le 80°$	55 dB	36 dB

ally follow a pattern of specifying specific requirements for use, construction materials, maximum and minimum sizes, heights, clearances, and safety requirements, as well as local environmental conditions for which the design must be adequate. The specific requirements cannot be generalized; however, the "National Electrical Code," by the National Fire Protection Association is almost universal, and many local governments base their environmental conditions on an American National Standards Institute document called "Minimum Design Loads for Buildings and Other Structures," ANSI A58.1. The use of ANSI A58.1 in turn infers (and the local authorities usually require) the use of the following design codes:

1. "Specification for the Design, Fabrication, and Erection of Structural Steel Buildings," issued by the American Institute for Steel Construction (AISC).

2. "Specification for the Design of Cold Formed Steel Structural Members," issued by the American Iron and Steel Institute (AISI).

3. "Specifications for Aluminum Structures, Aluminum Construction Manual," published by the Aluminum Association.

4. "Building Code Requirements for Reinforced Concrete," published by the American Concrete Institute (ACI).

The use of materials not specifically listed in the specifications should be supported by a rational analysis which addresses fatigue, brittle fracture, corrosion, environmental effects, material aging, and other aspects of structural integrity, including anticipated misuse or overload.

REFERENCES

23. Cited References

1. Spencer, R. C., "Paraboloidal Diffraction Patterns From the Standpoint of Physical Optics," RL Report T-7, October 21, 1942.

2. Komen, M. J., "Use Simple Equations to Calculate Beamwidth," *Microwaves,* December 1981, pp. 61–63.

3. Bodnar, D. G., "Materials and Design Data," Chap. 46, *Antenna Engineering Handbook,* R. C. Johnson and H. Jasik, eds., McGraw-Hill, New York 1984.

4. Blake, L. V., "Antenna and Receiving-System Noise-Temperature Calculation," NRL Report No. 5668, U.S. Naval Research Laboratory, Washington, D.C., September 19, 1961.

5. Hannah, P. W., "Microwave Antennas Derived from the Cassegrain Telescope," *IEEE Trans. Ant. and Prop.,* vol. AP-9, no. 2, March 1961, pp. 140–153.

6. Bruning, J. W., "A 'Best Fit Paraboloid' Solution the Shaped Dual Reflector Antenna," U.S.A.F. Antenna Research and Development Symposium, Univ. of Illinois, November 1967.

7. Parekh, S.: "On the Solution of Best Fit Paraboloid as Applied to Shaped Dual Reflector Antennas," *IEEE Trans. Ant. and Prop.,* vol. AP-28, no. 4, July 1980, pp. 560–562.

8. Ye Kinbar, B., "On Two Reflector Antennas," *Radio Eng. Electron. Physics,* vol. 6, June 1962.

9. Green, K. A., "Modified Cassegrain Antenna for Arbitrary Aperture Illumination," *IEEE Trans. Ant. and Prop.,* vol. AP-11, no. 5, September 1963.

10. Galindo, V., "Design of Dual Reflector Antenna with Arbitrary Phase and Amplitude Distributions," PTGAP Intl. Symp., Boulder, Colo., July 1963.

11. Galindo, V., "Synthesis of Dual Reflector Antenna," Elec. Research Lab., Univ. of Calif., Report No. 64–22, July 30, 1964.

12. Williams, W. F., "High Efficiency Antenna Reflector," *Microwave Journal,* July 1965, pp. 79–82.

13. Wood, P. J., "Reflector Profiles for the Pencil-Beam Cassegrain Antenna," *Marconi Review,* vol. 135, no. 185, 1972, pp. 121–138.

14. Clarricoats, P. J. B., and G. T. Poulton, "High Efficiency Microwave Reflector Antennas—A Review," *Proc. IEEE,* vol. 65, no. 10, pp. 1470–1504, October 1977.

15. Mentzer, C. A., "Analysis and Design of High Beam Efficiency Aperture Antennas," doctoral thesis, 74-24, 370, Ohio State Univ., 1974.

16. Cook, J. S., E. M. Elam, and H. Zucker, "The Open Cassegrain Antenna: Part 1, Electromagnetic Design and Analysis," *Bell System Tech. J.,* September 1965, pp. 1255–1300.

17. Rudge, A. W., and N. A. Adatia, "Offset-Parabolic-Reflector Antennas: A Review," *Proc. IEEE,* vol. 66, no. 12, December 1978, pp. 1592–1611.

18. Chu, T. S., "A Multibeam Spherical Reflector Antenna," IEEE Antenna and Propagation International Symposium, *Program and Digest,* Dec. 9, 1969, pp. 94–101.

19. Hyde, G., "A Novel Multiple-Beam Earth Terminal Antenna for Satellite Communication," Intl. Conf. on Comm., *Record,* June 1970, Conf. Proc. Paper No. 70-CP-386-COM, pp. 38–24 to 38–33.

20. Ohm, E. A., "A Proposed Multiple-Beam Microwave Antenna for Earth Stations and Satellites," *Bell System Tech. J.,* vol. 53, pp. 1567–1665, October 1974.

21. Hyde, G., R. W. Kreutel, and L. V. Smith, "The Unattended Earth Terminal Multiple-Beam Torus Antenna," *COMSAT Tech. Rev.,* vol. 4, no. 2, Fall 1974, pp. 231–264.

22. Johnson, R. C., and H. Jasik, *Antenna Engineering Handbook,* McGraw-Hill, New York, 1984.

23. E.I.A. 411A Standard, "Electrical and Mechanical Characteristics of Earth Station Antennas for Satellite Communications," Washington, D.C., September 1986.

CHAPTER 3

RADIO-FREQUENCY ALLOCATION AND ASSIGNMENT

John D. Bowker

Director, Frequency Management, RCA Corporation
(Retired)

CONTENTS

INTRODUCTION

1. Scope of Chapter

Despite imperfections, the regulatory processes for allocating frequencies to communications radio systems perform remarkably well. The purpose of this chapter is to explain how they work, to provide reference material on communications service frequency allocations, and to outline procedures for obtaining frequency assignments.* Broadcasting allocations, except for those to broadcasting satellites, are not treated. The tables and discussion in this chapter cover alloca-

---

*In this chapter, the term *assignment* is used to designate a specific frequency for use by a particular station. An *allocation* designates frequencies to which stations in a specified service may be assigned.

tions between 30 and 275 GHz. Most modern communications systems are being designed to operate on frequencies above 30 MHz to take advantage of predictable propagation and smaller antennas and to accommodate the enormous bandwidths available only on the higher frequencies. There is a burgeoning demand for communications frequencies worldwide, requiring an ever-increasing search for available channel capacity above 30 MHz.

INTERNATIONAL REGULATORY AGENCIES

2. Radio-Frequency Band Segments

The radio-frequency spectrum has been divided into band segments as shown in Table 3.1. Frequencies above 30 MHz are shown along with their standard nomenclature.

3. International Telecommunications Union

The present system of frequency allocations is administered by the International Telecommunications Union (ITU), the telecommunications arm of the United Nations. Chief among the ITU's activities is the periodic convening of Administrative Radio Conferences on a regional (RARC) or on a worldwide (WARC) basis. Allocations of radio frequencies between 9 kHz and 275 GHz are decided at the conferences for each of three ITU regions; see the world map, Fig. 3.1. Regulations drafted at Administrative Radio Conferences are submitted to each nation for formal ratification as a treaty. Each nation's domestic radio policies are based on the *International Radio Regulations*. A copy may be obtained from the International Telecommunications Union, Sales Section, Place des Nations, CH-1211 Geneva 20, Switzerland. The two-volume set costs 131 Swiss francs (1985).

INTERNATIONAL FREQUENCY ASSIGNMENT PRACTICES

4. Regulatory Control

Within its borders, nearly every country exercises regulatory control for the assignment or licensing of radio stations. In the United States this authority is

TABLE 3.1   Frequency Band Segments

Frequency range	Band designation	Band number
30–300 MHz	VHF = very high frequency	8
300–3000 MHz	UHF = ultra high frequency	9
3–30 GHz	SHF = super high frequency	10
30–300 GHz	EHF = extremely high frequency	11

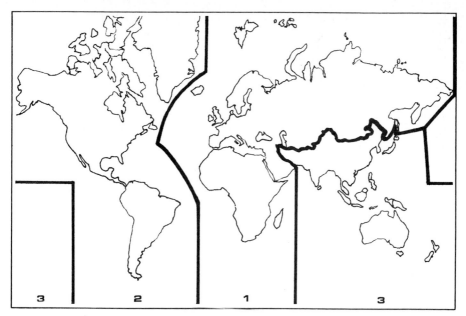

FIG. 3.1   World map showing the three ITU world regions used in allocating frequencies to the various radio services.

vested in the NTIA and the FCC; in most foreign countries in the Posts, Telegraph, and Telephone agency or in the Ministry of Communications or Transport. Assignments of frequencies by national regulatory agencies are made to facilitate the rendition of communications service on an interference-free basis in each geographic area. Ideally, each frequency assigned to a station would be for its exclusive use in the area, but the increasing need for communications demands greater densities of stations. Greater signal discrimination between systems is therefore required in order to avoid or minimize harmful interference. Higher operating frequencies, along with the creative use of highly directional antennas and electrically polarized transmissions are among the popular solutions in achieving satisfactory performance in the crowded spectrum.

5.   International Frequency Registration Board

Sophisticated international procedures have been established to assist in selecting specific carrier frequencies for most radio services. In accordance with ITU regulations, proposals for radio systems must be submitted to the International Frequency Registration Board (IFRB), an agency of the ITU, if the system is planned for international service or might cause harmful international interference. Such submissions are handled by each country's governmental regulatory agency. The IFRB then distributes engineering details of proposed new radio systems worldwide and provides a forum for all interested parties to work out any interference conflicts. This procedure is called *notification and frequency coordination*.

6. Table of Frequency Allocations

Band segment allocations for each region are shown in the ITU Table of Frequency Allocations; see Fig. 3.2. The table shows band segments for each recognized radio service and indicates, through footnotes, limits or conditions on the allocations. Frequency assignments in each country must conform to the allocation table as printed in the *International Radio Regulations* published by the ITU; the table also appears in Part 2 of the FCC *Rules and Regulations* and in Chap. 4 of the NTIA *Manual of Regulations and Procedures for Radio Frequency Management.*

INTERNATIONAL TABLE MHz		
Allocation to Services		
Region 1	Region 2	Region 3

136 - 137	AERONAUTICAL MOBILE (R)
	Fixed
	Mobile
	  except aeronautical
	  mobile (R)
	
	591, 595	

137 - 138	SPACE OPERATION
	  (space-to-earth)
	METEOROLOGICAL-SATELLITE
	  (space-to-earth)
	SPACE RESEARCH
	  (space-to-earth)
	Fixed
	Mobile
	  except aeronautical
	  mobile (R)
	
	596, 597, 598, 599	

138 - 143.6
 	
AERONAUTICAL
  MOBILE (OR)
 	
600, 601, 602, 604
	138 - 143.6
	
	FIXED
	MOBILE
	/RADIOLOCATION/
	Space Research
	  (space-to-earth)
		138 - 143.6
		
		FIXED
		MOBILE
		Space Research
		  (space-to-earth)
		
		599, 603

143.6 - 143.65
 	
AERONAUTICAL
  MOBILE (OR)
SPACE RESEARCH
  (space-to-earth)
 	
601, 602, 604
	143.6 - 143.65
	
	FIXED
	MOBILE
	SPACE RESEARCH
	  (space-to-earth)
	/RADIOLOCATION/
		143.6 - 143.65
		
		FIXED
		MOBILE
		SPACE RESEARCH
		  (space-to-earth)
		
		599, 603

FIG. 3.2 Portion of Table of Allocations (Reprinted courtesy RCA Corporation).

7. Fixed Service and Mobile Service

All two-way radio communications stations are categorized as operating in either fixed service or mobile service. *Fixed service* is defined as a radio communications service between specified fixed points. Note that this excludes other services such as broadcasting where the location of only one end of the system is specified. Frequency allocations and application-processing procedures for subcategories of these commercial services are described later in this chapter. *Mobile service* provides for radio communications between mobile stations, or between mobile and nonmobile stations. Mobile service needs for additional spectrum have grown tremendously in the past decade and there is every indication the trend will continue. Regulatory agencies are constantly reviewing the needs for mobile communications and occasionally making changes in domestic allocations to satisfy the increasing demand. Many of the allocations to specific mobile radio services are discussed in later paragraphs.

UNITED STATES REGULATORY AGENCIES

8. Communications Act of 1934

The Communications Act of 1934, as amended, provides for the regulation of interstate and foreign commerce in communication by wire or radio. As printed in Title 47 of the United States Code, the act establishes the Federal Communications Commission (FCC) and holds it responsible for the regulation of interstate and foreign telecommunications operated by non-federal-government entities. For systems operated by the federal government, the act vests regulation in the National Telecommunications and Information Administration (NTIA), an operating unit of the U.S. Department of Commerce. The FCC and the NTIA each regulate about 30 percent of the allocated frequencies for radio stations operating in the United States, but the responsibility for about 40 percent of the spectrum is shared by the two organizations as explained in later paragraphs.

9. Federal Communications Commission

The *Rules and Regulations* of the FCC are contained in Title 47 of the Code of Federal Regulations. Once a year, usually in March, editions of the *Rules* are published, updated to the preceding October 1. Copies can be purchased from the Government Printing Office, Washington, D.C., for $54 U.S. (1985 cost). A listing of the commission's technical rules is shown in Table 3.2.

On a daily basis, the FCC issues public notices and news items of its actions, including announcements of changes it is proposing or making to its rules. Arrangements for obtaining the daily releases can be made through the FCC Press Office; call 202 254-7674 for details.

Most rule-change proposals, particularly those relating to frequency allocations, are subject to public comment before action is taken. Further, in order to solicit the widest possible comment on its proposed rule changes or additions, the FCC establishes advisory committees from time to time. Experts from the private sector are invited to consider specific technical matters and report back to the commission in a specified time. Advisory committees and their meeting dates are

TABLE 3.2 Federal Communications Commission Technical *Rules* Parts

Part	Title
1	Practice and Procedure
2	Frequency Allocations and Radio Treaty Matters; General Rules and Regulations
5	Experimental Radio Services (Other than Broadcast)
13	Commercial Radio Operators
15	Radio Frequency Devices
17	Construction, Marking, and Lighting of Antenna Structures
18	Industrial, Scientific, and Medical Equipment
21	Domestic Public Fixed Radio Service
22	Domestic Public Mobile Radio Service
23	International Fixed Public Radiocommunications Services
25	Satellite Communications
68	Connection of Terminal Equipment to the Telephone Network
73	Radio Broadcast Services
74	Experimental, Auxiliary, and Special Broadcast, and Other Program Distributional Services
76	Cable Television Service
78	Cable Television Relay Service
81	Stations on Land in the Maritime Services and Alaska Public-Fixed Stations
83	Stations on Shipboard in the Maritime Services
94	Private Operational-Fixed Microwave Service
95	Personal Radio Service (General Mobile, Radio Control, Citizens Band)
97	Amateur Radio Service
99	Disaster Communications Service
100	Satellite Broadcasting Service

announced in the FCC's daily releases and in the *Federal Register* (available in many libraries). The public is invited to attend the meetings and to take an active role in the proceedings at the instruction of the chairperson.

10. National Telecommunications and Information Agency (NTIA)

Section 305 of the Communications Act of 1934, as amended, assigns responsibility for the frequency management of all radio stations belonging to and operated by the United States to the President. This authority has been delegated to the Assistant Secretary of Commerce for Communications and Information, who serves as head of the NTIA. The NTIA publishes its rules in the *Manual of Regulations and Procedures for Federal Radio Frequency Management*. Details of government-only frequency assignment procedures and limitations are contained in the manual. Copies are available from the Superintendent of Documents, Washington, D.C. 20402 for $112 U.S. (1985 cost).

The assignment of frequencies in support of many industrial government contracts is handled directly by the contracting agency through the NTIA, not through the FCC.

11. Interdepartment Radio Advisory Committee (IRAC)

The Table of Frequency Allocations in Sec. 2.106 of the commission's *Rules,* also in Chap. 4 of the NTIA *Manual,* identifies the bands of frequencies for which the

NTIA and FCC share regulatory responsibility. Allocations for government or nongovernment stations may be authorized in the bands of shared frequencies only after coordination has been effected with all branches of the federal government through IRAC, an advisory committee to the NTIA. IRAC participants are representatives of federal departments and agencies. A representative of the FCC, while not a formal member of IRAC, attends IRAC meetings in a liaison capacity for the private sector. Thus the needs of each branch of government and the public are considered before any frequency assignments are made in the shared frequency bands. IRAC review of such requests usually takes 2 to 4 months depending upon the complexity of the request. For the nongovernment operator, IRAC review is initiated by the FCC upon receiving an application to use shared (or government-only) frequencies.

UNITED STATES FREQUENCY ASSIGNMENT PRACTICES

12. Selecting an Operating Frequency

For the nongovernment system designer, the sensible first step in selecting an operating frequency is to study the part of the FCC *Rules* dealing with the planned radio service. The system designer can thus become familiar with specific channeling plans and other service details such as permitted power and modulation levels, minimum service, and message-loading requirements. Section 2.106 of the commission's *Rules* identifies all service allocations in the spectrum so an evaluation can be made of the compatibility with existing or possible future radio stations in the area.

The procedure for selecting an appropriate operating frequency band is relatively straightforward, provided the proposed radio system falls within one of the service definitions listed in the ITU Table of Frequency Allocations, and a satisfactory site can be found that meets local zoning requirements, FAA rules, and environmental impact regulations. The ITU table divides the spectrum into discrete frequency bands for each communications service so the choice of an operating band for a new communications system can be selected to match the desired propagational characteristics for the proposed system. The specific operating frequency is then chosen in accordance with established domestic channeling plans as shown in the NTIA manual, or in the part of the FCC *Rules* covering the planned service.

A more difficult set of problems confronts the system designer planning a new communications system for which there is no frequency allocation in the international or domestic tables. Virtually every frequency in the radio spectrum has been allocated for use so the designer, in such cases, must "borrow" some frequencies from an allocated radio service. Usually it is possible to define any new system in terms compatible with an existing allocated service. At the very least, the system designer should describe a proposed operation by highlighting its similarities to existing cochannel services. An application, to be acceptable, must show that the new station will not cause harmful interference to its cochannel and adjacent-channel neighbors. This is accomplished either by strict adherence to an established channeling plan, or by submitting a technical document clearly demonstrating that no harmful interferences would be expected.

13. Frequency Coordination

In many radio services, the selection of a specific operating frequency is often handled by an equipment manufacturer's representative. In others a frequency coordination committee or firm will analyze a station proposal and work with the applicant and other station operators to avoid potential interferences. Such organizations maintain a comprehensive data bank showing the operating characteristics of all stations in a particular service area and can thereby anticipate possible interferences. This process can consume from a day or two to several months depending upon the radio service to be implemented, the complexity of the system, and the frequency congestion in the locale under consideration. Since there are many coordinating organizations, and since the FCC specifies which organization is to be used in some cases, it is usually best to check with the FCC (telephone 202 634-2443) or with a potential equipment supplier for guidance in selecting an appropriate coordinator for the service of interest.

14. Filing Fees

Most applications to the FCC require simultaneous submission of a filing fee. In order to determine the proper amount to remit, call the FCC on a special filing-fee telephone number, 202 632-3337.

PRIVATE RADIO SERVICES

15. Definitions

Regulatory distinctions are made between communications systems built for use by the public or for use primarily by the company operating the system. If one's primary purpose is to provide communications service to the public, the operation will be licensed under common-carrier regulations; the next section of this chapter is devoted to such communications systems. If at least 50 percent of the system capacity will be for the operator's own communications, the system is classified as *private*.

16. Land Mobile Services

The FCC has allocated frequency bands for specific land mobile service categories (which include fixed base stations). These categories are as follows:

1. Public safety radio services
   a. Local government
   b. Police radio services
   c. Fire radio services
   d. Highway maintenance radio services
   e. Forestry conservation radio services
2. Industrial radio services
   a. Power radio service
   b. Petroleum radio service
   c. Forest products radio service
   d. Motion picture radio service

    e. Special industrial radio service
    f. Business radio service
    g. Manufacturers radio service
    h. Telephone maintenance
3. Land Transportation Radio Services
    a. Motor-carrier radio service
    b. Railroad radio service
    c. Taxicab radio service
    d. Automobile emergency radio service
4. Paging operations
5. Special emergency radio services

A tabulation of the frequency bands allocated to each of these services and a description of licensee eligibility requirements and application procedures are given in Chap. 20.

17. Aviation Radio Services

The aviation services are allocated primarily to enhance the safe, expeditious, and economical operation of aircraft. Eligibility is generally limited to citizens or companies not controlled by aliens or foreign interests.

Applications should be submitted on FCC Form 404 for aircraft stations, and on FCC Form 406 for most fixed ground stations. Civil Air Patrol or ground mobile station applications should be submitted on FCC Form 480. Frequency bands are assigned to stations according to their specific function or business at the time of transmission and, as appropriate, their position with respect to airports as shown in Table 3.3.

18. Maritime Radio Services

Most communications with ships at sea occur either by means of geostationary satellites, discussed in Chap. 6, or on frequencies below 30 MHz not covered in this handbook. While in port, or while traversing coastal, lake, or river waters, VHF and UHF are used with shore stations, and for bridge-to-bridge as necessity dictates. While at sea, the higher frequencies are used for intraship and, on occasion, intership radio communications. Frequencies are reserved for specific purposes in this service as shown in Table 3.4.

PUBLIC MOBILE SERVICES

19. Public Land Mobile Radio Service (PLMRS)

PLMRS provides communications for hire between land mobile stations and their associated base stations. *Communications common carriers* are eligible to furnish this service. Also, *airborne mobile stations* (air-to-ground stations) may be licensed to individual users. Other mobile stations are licensed as part of base station authorizations. FCC *Rules* Sec. 22.501 contains channeling, power, and administrative details for each of the listed services.

The application procedure is to submit a completed FCC Form 401 with any attachments to the Secretary, Federal Communications Commission, Washington, D.C. 20554. When construction is completed, submit FCC Form 489.

TABLE 3.3 Aviation Radio Frequencies

Frequency, MHz	Primary purpose
118–121.4	Air traffic control
121.5	Distress or condition of emergency
121.6–123.075	Air traffic control
122.75	Air-to-air
122.85	Utility communications on airport apron
122.9	Search and rescue
123.025	Helicopters, air-to-air
123.1	Search and rescue
123.6–128.8	Air traffic control
132.025–135.975	Air traffic control
156.3	Safety purposes only
156.375–156.625	Air-to-ship communications
156.8	Distress, safety, and calling only
156.9	Air-to-ship communications
243	Emergency and distress
454.675–454.975	Air-to-ground public telephone
960–1215	Airborne aids to navigation
1300–1350	Aeronautical radionavigation
1535–1557.5	Satellite-to-aircraft
1557.5–1567.5	Glide-path stations
1567.5–1592.5	Airborne aids to navigation
1592.5–1622.5	Collision avoidance
1622.5–1637.5	Airborne aids to navigation
1637.5–1660	Aircraft-to-satellite
2700–2900	Aeronautical radionavigation
4200–4400	Radio altimeters
5000–5200	Airborne aids to navigation
5350–5470	Airborne radars and associated beacons
8750–8850	Airborne doppler radar
9000–9200	Aeronautical radionavigation
9300–9500	Airborne radars and associated beacons
13,250–13,400	Airborne doppler radar
14,000–14,400	Aeronautical radionavigation
15,400–15,700	Airborne aids to navigation

Common carriers may be assigned frequencies from the following table for two-way PLMRS; one-way service is permitted provided two-way PLMRS service is offered.

Base station frequency, MHz	Mobile, dispatch, and auxiliary test station frequency, MHz
Wireline common carriers	
152.51–152.81	157.77–158.07
454.375–454.60	459.375–459.650
Radio common carriers	
152.03–152.21	158.49–158.67
454.025–454.350	459.025–459.350

Control and repeater frequencies may also be assigned to channels in the UHF bands.

In areas where no UHF broadcast television interference may be encountered, certain frequencies allocated to television broadcasting may be assigned to PLMRS stations. The regulations, while under constant review, presently list up to 12 PLMRS channels for base stations on frequencies in the 470 to 512 MHz band, and up to 12 more frequencies in that band for mobile, dispatch, or auxil-

TABLE 3.4 Maritime Radio Frequencies (MHz)

Channel number	Coast frequency	Ship frequency
Distress, safety, and calling		
16	156.800	156.800
Port operations		
01	156.050	156.050
05	156.250	156.250
12	156.600	156.600
14	156.700	156.700
20	161.600	157.000
63	156.175	156.175
65	156.275	156.275
66	156.325	156.325
73	156.675	156.675
74	156.725	156.725
Navigational		
13	156.650	156.650
Environmental		
15	156.750	—
State control		
17	156.850	156.850
Commercial		
01	156.050	156.050
07	156.350	156.350
09	156.450	156.450
10	156.500	156.500
11	156.550	156.550
18	156.900	156.900
19	156.950	156.950
63	156.175	156.175
79	156.975	156.975
80	157.025	157.025
Noncommercial		
09	156.450	156.450
68	156.425	156.425
69	156.475	156.475
71	156.575	156.575
78	156.925	156.925

iary test stations in each of the following cities: Boston, Chicago, Cleveland, Dallas, Detroit, Houston, Los Angeles, Miami, New York, Philadelphia, Pittsburgh, San Francisco, and Washington. Channeling, power limitations, and other details are in FCC *Rules* Sec. 22.501.

20. Paging Radio Service

Common carriers may be authorized to offer a one-way signalling or paging system to the public on selected frequencies within the bands shown here; see FCC *Rules* Sec. 22.501 for channeling, power, and administrative details.

Base station frequency, MHz	Paging service
35.20–35.66	One way
43.20–43.66	One way
152.03–152.21	Two way for radio common carriers; mobile channels are spaced 6.46 MHz higher.
152.51–152.81	Two way for wireline common carriers; mobile channels are spaced 5.26 MHz higher.
454.025–454.350	Two way for radio common carriers; mobile channels are spaced 5.0 MHz higher.
454.375–454.650	Two way for wireline common carriers; mobile channels are spaced 5.0 MHz higher.

There are provisions in FCC *Rules* Part 22 for opening new paging channels in the 890 to 960-MHz band in the future.

21. Rural Radio Service

Where it is impracticable to extend wire communications service to remote areas, Subpart H of FCC *Rules* Part 22 provides for radio service between fixed stations on frequencies below 1000 MHz.

Rural central office and interoffice stations may be licensed to communications common carriers. Rural subscriber stations may be licensed to common carriers or to an individual user of the service.

Frequencies for this service are assigned within the bands listed below. See FCC *Rules* Sec. 22.601 for channeling details.

Central office frequency, MHz	Interoffice frequency, MHz	Rural subscriber frequency, MHz
152.51–152.8 1	152.51–152.81	157.77–158.07
454.375–454.650	157.77–158.07	158.49–158.67
454.375–454.650	459.025–459.650	
459.375–459.650		

22. Cellular Radio Systems

See Chap. 22 for a description of cellular radio systems. Existing and proposed communications common carriers are eligible to provide this service.

Frequencies are divided into Cellular System A bands for common carriers not also providing a public land-line message telephone service, and Cellular System B bands for others.

Cellular System A consists of 416 frequency pairs with 30-kHz spacing as follows:

Mobile frequencies

824.040 to 834.990 MHz

845.010 to 846.480 MHz

Base frequencies

869.040 to 879.990 MHz

890.010 to 891.480 MHz

Cellular System B consists of 416 frequency pairs with 30-kHz spacing as follows:

Mobile frequencies

835.020 to 844.980 MHz

846.510 to 848.970 MHz

Base frequencies

880.020 to 889.980 MHz

891.510 to 893.970 MHz

Twenty-one control-channel pairs are assigned to each cellular system. For System A, they are 834.390 to 834.990 MHz and 879.390 to 879.990 MHz. For System B, they are 835.020 to 835.620 MHz.

23. Offshore Radio Service (ORS)

ORS is a public mobile radio service for communication with stations in the offshore coastal waters of the Gulf of Mexico. Common carriers and private radio operators are eligible to provide service. Technical standards are the same for both.

Frequencies are in the 476- to 494-MHz band (shared with TV channels 15, 16, and 17); ORS channel centers are 25 kHz apart. Specific channel allocations, power and antenna restrictions, and permitted geographic areas of operation are shown in FCC *Rules* Sec. 22.1001 for common carrier systems, and 90.315 for private radio systems.

TERRESTRIAL MICROWAVE SERVICES

24. Private Operational-Fixed Microwave Radio Service (POFS)

POFS is for the use of the operator of the stations in a POFS system. With certain restrictions, excess channel capacities may be offered to others not directly associated with the operator as discussed below. These rules are undergoing review and the reader should review the latest FCC orders in *Rules* Sec. 94.9 in this regard.

Any person, or any governmental entity or agency, eligible under any of the private radio services is eligible to hold a license in this service. To apply for a license, complete FCC Form 402 and mail it with the proper filing fee and a document indicating that the frequency coordination steps detailed in FCC *Rules* Sec. 94.15 have been completed, to the FCC using certified U.S. mail, return receipt requested. If possible, watch the daily releases issued by the FCC during the following 2 to 4 weeks for notice of your application. If you are uncertain, call the FCC a month after filing at 717 337-1212 and ask for a status report on your application.

Some joint use of private systems is permitted; excess message capacity of a private system can be offered to the public at a profit to the operator under some conditions. This may be particularly useful to a system planner facing a shortage of common carrier frequencies in a desired area of operation. If private service frequencies are available, it might be that a slight change in business plan could make a communications system possible that would otherwise be blocked. Frequencies are assigned according to required bandwidth of operations as shown in Table 3.5. Channeling details are contained in FCC *Rules* Sec. 94.65.

25. Common Carrier Point-to-Point Microwave Radio Service

This is a communications service offered for hire to the public. Existing and proposed communications common carriers are eligible to provide the service. Construction permits are granted by the FCC upon receipt of the proper filing fee and

TABLE 3.5 Private Operational Fixed Service Frequencies

Frequency band, MHz	Allotted bandwidth, MHz	Frequency band, MHz	Allotted bandwidth, MHz
928.0125–928.9875	0.025	6730.0–6870.0	5.0
952.0125–952.8375	0.025	10,616.25–10,623.75	2.5
953.0–956.1	0.100	10,560.625–10,564.375	1.25
953.05–956.05	0.050	10,625.625–10,629.375	1.25
953.55–955.55	0.200	17,705.0–18,115.0	10.0
956.2625–956.4375	0.025	17,710.0–18,130.0	20.0
956.60–959.7	0.100	17,720.0–18,120.0	40.0
956.65–959.65	0.050	17,740.0–18,060.0	80
957.15–959.15	0.200	18,145.0–18,361.0	6.0
959.8625–959.9875	0.025	18,367.0–18,577.0	6.0
1855–1905	10.0	18,585.0–18,815.0	10.0
1860–1900	5.0	18,590.0–18,810.0	20.0
1940–1980	5.0	18,762.5–18,817.5	5.0
2130.8–2149.2	0.800	18,925.0–19,155.0	10.0
2180.8–2199.2	0.800	18,930.0–19,150.0	20.0
2451.9–2475.1	0.800	19,102.5–19,157.5	5.0
2476.3–2499.5	0.800	19,265.0–19,675.0	10.0
6545.0–6705.0	10.0	19,270.0–19,690.0	20.0
6550.0–6710.0	5.0	19,280.0–19,680.0	40.0
6715.0–6575.0	10.0	19,300.0–19,620.0	80.0

satisfactory review of a completed Form 494. Prior frequency coordination, as detailed in FCC *Rules* Sec. 21.100(d), is an essential part of the application.

Detailed regional channeling plans have evolved for this service; frequency coordinators will be familiar with them. The bands within which these systems operate are listed in Table 3.6.

26. Digital Electronic Message Service (DEMS or DTS)

In 1981, the FCC approved this new service, and, immediately, it became known both as DEMS and DTS (*digital termination system*). This is a two-way domestic end-to-end fixed radio service for the exchange of digital information. In each metropolitan area, a DEMS system would consist of a *nodal* station at the operator's central office that serves a number of *user* stations located on customer premises. Two bands of frequencies have been allocated for DEMS: 10.55 to 10.68 GHz and 17.7 to 19.7 GHz. Bandwidths assigned to *limited network operators* (i.e., fewer than 30 stations) and *extended network operators* (i.e., 30 or more stations) in the 17.7- to 19.7-GHz band are identical but, at 10.55 to 10.68 GHz, limited networks are assigned 2.5-MHz channels while extended networks are assigned 5.0-MHz channels. User stations and nodal stations in any network are assigned the same bandwidth. Frequency channeling details are shown in Table 3.7.

The application procedure is the same as for common-carrier point-to-point microwave stations, above. Existing or proposed communication common carriers are eligible.

27. Multipoint Distribution Service (MDS)

MDS is generally intended to provide a wideband one-way point-to-multipoint radio service within a metropolitan area. Eleven channels have been allocated; bandwidths in 10 of them are 6-MHz wide, thereby easily accommodating broadcast-quality television and ancillary signals. There is also a provision for low-power narrower-band *MDS response stations* to permit subscribers to an MDS service to transmit back to the central station. Frequencies for MDS stations are in the 2150- to 2162-MHz and 2596- to 2644-MHz bands, with response stations operating in the bands 2686 to 2690 MHz, 18,580 to 18,820 MHz, and

TABLE 3.6 Point-to-Point Common-Carrier Radio Service Frequencies

Frequency, MHz	Frequency, MHz
2110–2130	17700–18580
2160–2180	18820–19260
3700–4200*	21200–23600
5925–6425*	27500–29500
10550–10680	31000–31200
10700–11700*	38600–40000
13200–13250	

*Shared with the fixed satellite service.

TABLE 3.7   Digital Electronic Message Service Frequencies

Channel number	Channel Group A (nodal stations), MHz	Channel Group B (user stations), MHz
1	10,565–10,570	10,630–10,635
2	10,570–10,575	10,635–10,640
3	10,575–10,580	10,640–10,645
4	10,580–10,585	10,645–10,650
5	10,600–10,602.5	10,665–10,667.5
6	10,602.5–10,605	10,667.5–10,670
7	10,605–10,607.5	10,670–10,672.5
8	10,607.5–10,610	10,672.5–10,675
9	10,610–10,612.5	10,675–10,677.5
10	10,612.5–10,615	10,677.5–10,680
11	10,550–10,552.5	10,615–10,617.5
12	10,552.5–10,555	10,617.5–10,620
13	10,555–10,557.5	10,620–10,622.5
14	10,557.5–10,560	10,622.5–10,625
15	10,560–10,561.25	10,625–10,626.25
16	10,561.25–10,562.5	10,626.25–10,627.5
17	10,562.5–10,563.75	10,627.5–10,628.75
18	10,563.75–10,565	10,628.75–10,630
19	10,585–10,587.5	10,650–10,652.5
20	10,587.5–10,590	10,652.5–10,655
21	10,590–10,592.5	10,655–10,657.5
22	10,592.5–10,595	10,657.5–10,660
23	10,595–10,597.5	10,660–10,662.5
24	10,597.5–10,600	10,662.5–10,665
30	18,870–18,880	19,210–19,220
31	18,880–18,890	19,220–19,230
32	18,890–18,900	19,230–19,240
33	18,900–18,910	19,240–19,250
34	18,910–18,920	19,250–19,260

18,920 to 19,160 MHz. Response stations may transmit only voice or data communications. Applicants may be assigned a channel or channels as shown in Table 3.8.

Existing and proposed communications common carriers are eligible to provide the service, with the proviso that more than 50 percent of the service will be for subscribers not related to the carrier. The application procedure is the same as for common-carrier point-to-point microwave stations, above.

28.   Local Television Transmission Service

This is a common-carrier service (not a service for broadcasting to the public) intended for the relay of television material and related communications. The ap-

TABLE 3.8 Multipoint Distribution Service Frequencies

Channel	Frequency, MHz	Response channel frequency, MHz
1	2150–2156	
2	2156–2162	
2A	2156–2160	
E1	2596–2602	2686.5625
E2	2608–2614	2687.5625
E3	2620–2626	2688.5625
E4	2632–2638	2689.5625
F1	2602–2608	2686.6875
F2	2614–2620	2687.6875
F3	2626–2632	2688.6875
F4	2638–2644	2689.6875

plication procedure is the same as for common-carrier point-to-point microwave stations, above.

Frequency assignments (MHz) for television pickup and television nonbroadcast pickup common-carrier stations are

6425 to 6525

11,700 to 12,200

13,200 to 13,250

22,000 to 23,600

For television *studio transmitter link* (STL) common carrier stations, the frequency assignments in megahertz are

3700 to 4200

5925 to 6425

10,700 to 11,700

13,200 to 13,250

22,000 to 23,600

Channeling, power, and bandwidth restrictions are found in FCC *Rules,* Sec. 21.801.

SATELLITE RADIO SERVICES

29. Satellite Communications and Broadcasting Services

The FCC has defined the following satellite communications services:

Fixed satellite service

Land mobile satellite service

TABLE 3.9 Earth Exploration Satellite Frequencies

| Region 1 | | Region 2 | | Region 3 | | United States allocations | | | |
| | | | | | | Federal | | Private | |
1400.0000 2690.0000	1427.0000 2700.0000	1400.0000 2690.0000	1427.0000 2700.0000	1400.0000 2690.0000	1427.0000 2700.0000	1400.0000 2690.0000	1427.0000 2700.0000	1400.0000 2690.0000	1427.0000 2700.0000
Megahertz									
8025.0000	8400.0000	8025.0000	8400.0000	8025.0000	8400.0000	8025.0000	8400.0000		
Gigahertz									
10.6000	10.7000	10.6000	10.7000	10.6000	10.7000	10.6000	10.7000	10.6000	10.7000
15.3500	15.4000	15.3500	15.4000	15.3500	15.4000	15.3500	15.4000	15.3500	15.4000
21.2000	21.4000	18.6000	18.8000	21.2000	21.4000	18.6000	18.8000	18.6000	18.8000
22.2100	22.5000	21.2000	21.4000	22.2100	22.5000	21.2000	21.4000	21.2000	21.4000
23.6000	24.0000	22.2100	22.5000	23.6000	24.0000	22.2100	22.5000	22.2100	22.5000
31.3000	31.8000	23.6000	24.0000	31.3000	31.8000	23.6000	24.0000	23.6000	24.0000
36.0000	37.0000	31.3000	31.8000	36.0000	37.0000	31.3000	31.8000	31.3000	31.8000
50.2000	50.4000	36.0000	37.0000	50.2000	50.4000	36.0000	37.0000	36.0000	37.0000
51.4000	59.0000	50.2000	50.4000	51.4000	59.4000	50.2000	50.4000	50.2000	50.4000
64.0000	65.0000	51.4000	59.0000	64.0000	65.0000	51.4000	59.0000	51.4000	59.0000
65.0000	66.0000	64.0000	65.0000	65.0000	66.0000	64.0000	65.0000	64.0000	65.0000
86.0000	92.0000	65.0000	66.0000	86.0000	92.0000	65.0000	66.0000	65.0000	66.0000
100.0000	102.0000	86.0000	92.0000	100.0000	102.0000	86.0000	92.0000	86.0000	92.0000
105.0000	126.0000	100.0000	102.0000	105.0000	126.0000	100.0000	102.0000	100.0000	102.0000
150.0000	151.0000	105.0000	126.0000	150.0000	151.0000	105.0000	126.0000	105.0000	126.0000
164.0000	168.0000	150.0000	151.0000	164.0000	168.0000	150.0000	151.0000	150.0000	151.0000
174.5000	176.5000	164.0000	168.0000	174.5000	176.5000	164.0000	168.0000	164.0000	168.0000
182.0000	185.0000	174.5000	176.5000	182.0000	185.0000	174.5000	176.5000	174.5000	176.5000
200.0000	202.0000	182.0000	185.0000	200.0000	202.0000	182.0000	185.0000	182.0000	185.0000
217.0000	231.0000	200.0000	202.0000	217.0000	231.0000	200.0000	202.0000	200.0000	202.0000
235.0000	238.0000	217.0000	231.0000	235.0000	238.0000	217.0000	231.0000	217.0000	231.0000
250.0000	252.0000	235.0000	238.0000	250.0000	252.0000	235.0000	238.0000	235.0000	238.0000
		250.0000	252.0000			250.0000	252.0000	250.0000	252.0000

TABLE 3.10 Meteorological Satellite Service Frequencies, Megahertz

Region 1		Region 2		Region 3		United States allocations			
						Federal		Private	
137.0000	138.0000	137.0000	138.0000	137.0000	138.0000	137.0000	138.0000	137.0000	138.0000
400.1500	401.0000	400.1500	401.0000	400.1500	401.0000	400.1500	401.0000	1670.0000	1710.0000
1670.0000	1710.0000	1670.0000	1710.0000	1670.0000	1710.0000	1670.0000	1710.0000		
7450.0000	7550.0000	7450.0000	7550.0000	7450.0000	7550.0000	7450.0000	7550.0000		
8175.0000	8215.0000	8175.0000	8215.0000	8175.0000	8215.0000	8175.0000	8215.0000		

Maritime mobile satellite service

Aeronautical mobile satellite service

In addition, it has defined the broadcasting satellite service. The allocation of frequencies for these services, together with the policies and procedures for assignment, are described in Chap. 6.

30. Earth-Exploration Satellite Service

This is a radiocommunication service between earth stations and one or more space stations in which information relating to the characteristics of the earth and its natural phenomena is obtained from active or passive sensors on earth satellites. Frequency allocations for this service appear in Table 3.9.

31. Meteorological Satellite Service

This is a radiocommunication service between earth stations and one or more space stations in which meteorological information is obtained from active or passive sensors on earth satellites. Frequency allocations for this service appear in Table 3.10.

32. Radiodetermination Satellite Service (RDSS)

RDSS offers the promise of accurately locating any person or object on the earth's surface equipped with an RDSS transceiver. Simple alphanumeric messages can be accommodated. Allocated frequency bands include 1610 to 1626.5 MHz for uplinking, 2483.5 to 2500 MHz for downlinking, and 5117 to 5183 MHz for communications from RDSS satellites to a central ground control point to handle the computational functions for the RDSS system. Issues concerning technical standards, licensing, and procedures await further FCC action. Inquiries should be addressed to the Secretary, Federal Communications Commission, Washington, D.C. 20554.

CHAPTER 4
MICROWAVE TRANSMISSION SYSTEMS

Forrest F. Fulton, Jr.

Vice President, Advanced Engineering, Avantek, Inc.

CONTENTS

MICROWAVE TRANSMISSION OVERVIEW

1. Introduction

Microwave radio transmission is distinguished from other radio applications by its frequency range and by the use of highly directive antennas. These distinguishing features occur in both terrestrial systems and satellite systems, but the design problems are different enough to warrant separate engineering treatment. The terrestrial applications are primarily of a point-to-point nature and use the frequency range from about 1 GHz to 40 GHz.

2. Uses of Microwave Communication

The factors favoring radio over cable systems for transmitting voice, data, or video vary from case to case. Probably the most common factor is the requirement to communicate over rough or inaccessible terrain. Also, communication over water can frequently be accomplished more economically by radio in spite of the radio-reflecting characteristics of the water. Installation time pressures frequently favor radio solutions. Property-acquisition problems tend to be minimized, since radio systems require a small plot every 30 to 50 km compared to the need for a continuous right of way for cable. Radio systems can be reconfigured geographically to meet changing needs, whereas it is almost never practical to retrieve a buried cable. Occasionally, restoration of communications after a natural disaster is a significant mission for microwave radio.

After listing these positive factors, it is appropriate to point out that microwave radio has disadvantages as well. Radio transmission through the lower atmosphere is subject to propagation impairments which result in sporadic occurrences of transmission outage. As a radio wave skims along 30 m above a 13,000-m-diameter earth, it propagates through a boundary layer and is affected by conditions that human senses are not scaled to notice. The accumulated experience of many years, however, has provided statistical models to describe these conditions as they affect the radio system. The successful operation of tens of thousands of microwave radio hops attests to the satisfactory guidance which these models provide for system design. The third section of this chapter describes these models and their use in microwave system design.

For an additional discussion of the advantages and disadvantages of microwave transmission, see Chap. 9.

3. Frequencies Used in Microwave Transmission

Microwave frequencies are used for communications systems because it is practical to focus the radio energy into a beam and thus concentrate a higher percentage of that energy at a receiving location than would be the case for lower frequencies. For example, a 2-m-diameter parabolic antenna focuses a 2-GHz radio wave into a beamwidth of about 3° compared to a 30° beamwidth at a frequency of 200 MHz.

An additional reason for the use of microwave frequencies is *spectrum availability*. At the lower frequencies where nondirectional antennas are desired for broad coverage systems, the requests for spectrum allocations have been very great. As a result, only narrow slices of spectrum (tens of kilohertz) are normally licensed. In the microwave allocations, on the other hand, tens of megahertz are available; thus it is practical to build systems to handle large quantities of voice, video, and data traffic.

At present, the lowest frequency that is allocated for microwave transmission is 1.71 GHz; the highest is 40 GHz. See Chap. 3 for a complete listing of frequencies allocated to microwave systems.

Usage developed first at the lower frequencies in the vicinity of 2 GHz and 4 GHz because of the availability of hardware. The technical factor of better beam focusing and the regulatory factor of spectrum shortage both motivated an upward movement in frequencies utilized.

Systems operating in higher frequency bands, e.g., in the vicinity of 11 GHz, can utilize smaller antennas and do not require as much terrain clearance (see Eq. 4.12). Further, in densely populated metropolitan areas and on high-traffic intercity routes, low-frequency channels may not be available. On the other hand, atmospheric-caused propagation impairments, particularly rain attenuation, increase substantially at frequencies above 10 GHz.

In choosing the frequency band for a new system, the availability of a frequency assignment must first be investigated. A newly proposed system is required to coordinate the choice of operating frequencies with all other licensed systems within 125 mi of each location it expects to use. In some of the most commonly used bands, terrestrial microwave shares frequencies with satellite services, and satellite earth stations must be considered as well. Coordination is facilitated by companies specializing in maintaining a data base of frequencies and locations already licensed. These companies have developed skills in fitting new systems into the environment, but the use of a different frequency band or system rerouting may be required.

If spectrum availability is not a limiting factor, the choice can be made on the basis of cost and reliability. As a practical matter, frequencies in the vicinity of 2, 4, and 6 GHz are most commonly used for long-haul, multihop systems or in areas of extremely heavy rainfall. Frequencies in the vicinity of 11 GHz or higher often have satisfactory reliability for shorter single hops, and they may be the only ones available in urban areas. In some cases, a multihop intercity system operating in a lower band is terminated by a single higher-frequency hop to bring it into the city or a satellite earth station.

4. Propagation Impairments

Atmosphere-related *propagation impairments* cause the received signal strength to drop substantially below the calculated values for short periods of time. The easiest remedy is to design the radio link to provide a substantial "fade margin." The rule of thumb used before adequate statistical models were developed was to provide a 40-dB margin. With current knowledge, the outage time to be expected for a given fade margin is predictable with good confidence (though not with certainty), and an important part of the engineering task is to minimize the system cost (fade margin) while meeting the reliability objective for atmosphere-related propagation outage.

For very high reliability requirements, the approach of increasing the fade margin will not always be sufficient. For these cases, the solution is to use space diversity where the signal goes over two slightly different paths to separate receiving systems. If the paths are separated enough, the deep fades on the second path almost never occur at the same time as those on the first, and a "smart" receiving system can reduce the resulting outage time to very low numbers. In principle, the same result can be obtained by *frequency diversity,* i.e., by transmitting over the same path at sufficiently separated frequencies; however, in recent years the FCC has not permitted this approach because of the waste of spectrum space, and frequency diversity is not an available option in most circumstances.

PATH CALCULATIONS

Path calculations require a consideration of antenna gains, system losses, receiver performance, and radio-wave propagation. These elements are treated in detail in other chapters of this handbook. This section describes their relationship in calculating microwave path calculations.

5. Free-Space Transmission

The starting point for radio system path calculations is the transmission of a signal from a lossless isotropic transmitting antenna to a lossless isotropic receiving antenna in free space. The "isotropic antenna" is a mathematical model that transmits an equal signal in all directions with perfect spherical uniformity. Since it is used as the reference, its gain is 0 dB by definition. Its effective aperture area is $\lambda^2/4\pi$, where λ is the wavelength of the signal.

The ratio of transmitted power to received power for antennas separated by many wavelengths is the *free space loss* L_s and is

$$L_s = 20 \log (4\pi d/\lambda) \quad \text{dB} \tag{4.1}$$

where the distance d between the antennas is measured in the same units as λ.

With the frequency f in gigahertz and the distance in kilometers, this equation becomes

$$L_s = 20 \log d + 20 \log f + 92.45 \quad \text{dB} \tag{4.2}$$

In addition to the free-space loss, there are gains and losses in the passive components of a microwave link. These components are shown schematically in Fig. 4.1 with typical values in practical systems. The sum of these losses is known as the *fixed loss* L_f. The total loss in the system, from the transmitter output to receiver input is known as the *section loss*. It is given by the equation

$$\text{Section loss} = L_s + L_f - (G_t + G_r) \quad \text{dB} \tag{4.3}$$

where G_t and G_r are the gains of the transmitting and receiving antennas with respect to an isotropic antenna.

6. Fade Margin

Equations 4.1 and 4.2 give the space loss in free space, i.e., in the absence of atmospheric or terrain effects. It is essential to provide a *fade margin* in the design of the system so that it will continue to operate satisfactorily in the presence of atmospheric fading. The lowest signal power at which the receiver will operate properly is known as the *threshold*. If frequency modulation or amplitude modulation with digital information is employed, the threshold is quite sharp, and the quality of the signal deteriorates very rapidly at lower signal levels.

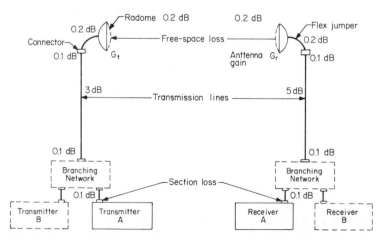

FIG. 4.1 Contributions to section loss.

The transmitter power minus the section loss gives the unfaded received signal power; the difference between this and the receiver threshold is the fade margin

$$\text{Fade margin} = P_t - P_r - \text{section loss} \qquad (4.4)$$

where P_t is the transmitter power in dBm and P_r is the receiver threshold level in dBm.

The selection of the appropriate fade margin is a critical decision for the system designer since it requires a tradeoff between cost and reliability. It is normally 20 dB or greater, and the practical maximum is about 50 dB.

Having selected the desired margin, additional tradeoff studies are required to make the most economical choice of antenna and transmission line sizes and costs (including the impact of antenna size on tower costs) and radio equipment parameters and costs. Figure 4.2 is a computer spreadsheet which can be used to calculate many path options rapidly.

The first step in calculating the necessary margin is to determine the acceptable outage time, usually in seconds or minutes per year. On the basis of this specification, the required margin can be calculated. This calculation is relatively complex because it must consider a variety of atmospheric and terrain propagation phenomena. Also it can be affected by the transmission path, particularly the terrain clearance. The relationship of the fade margin to these factors is described in the next two sections. Further information on this subject is found in Chap. 1.

		Aye		Bee
Site name				
Frequency, GHZ			2.145	
Path length	mi		17.9	
	km			
Free-space loss	dB		128.3	
Antenna height, above ground level	ft	275		105
	m			
Tower height	ft	280		110
	m			
Transmission-line length	ft	480		156
	m			
Transmission line part number		HJ7 - 50A		LDFS - 50A
Transmission line loss	db	5.3		3.1
External branching network loss	dB	0		0
Jumper cable or flex loss	dB	0.2		0.2
Radome Loss	dB	0		0
Connector loss	db	0.3		0.3
Total fixed losses	dB	5.8		3.6
Total losses (free space and fixed)	dB		137.7	
Antenna diameter	ft	12		10
	m			
Antenna model number		GP12F - 21A		HP10F-19C4
Antenna gain	dB	35.6		34
Total antenna gains	dB		69.6	
Net section loss	dB		68.1	
Radio equipment type and capacity		Vendor zee , 192 channel		
Minimum transmitter power	dBm	23		23
Receiver threshold sensitivity	dBm		-78	
Median received power	dBm	-45.1		-45.1
Fade margin	dB	32.9		32.9

FIG. 4.2 Path calculation worksheet.

ATMOSPHERIC EFFECTS

7. Atmospheric Refraction

The atmosphere is a dielectric medium having an index of refraction slightly greater than unity—normally about 1.0003—at sea level. The index ordinarily decreases with increasing altitude, and this causes the path of radio signals to bend slightly around the earth's curvature.

To analyze the effect of atmospheric refraction, a mathematical model can be used which consists of an atmosphere-free earth with a radius equal to k times the actual radius; see Fig. 4.3. Under "normal" conditions, k is approximately 4/3; i.e., the earth appears to have a larger radius, and radio waves are propagated beyond the geometric horizon.

Conditions may occur in which the rate of change of refractive index with elevation is abnormal. If k is smaller than 4/3, the atmosphere is said to be *subrefractive;* if larger than 4/3, it is *superrefractive*.

For analog microwave systems operating below 10 GHz, there are two abnormal atmospheric effects which must be considered in designing the link. These are *subrefractive conditions* and *atmospheric multipath*. For digital systems, atmospheric multipath not only causes signal-level fading but also a phenomenon known as *dispersive fading;* this is an increase in bit error rate resulting from phase and amplitude distortions across the transmission channel caused by the multipath condition. Finally, *rainfall attenuation* is a factor at frequencies above 10 GHz.

8. Subrefractive Conditions

In designing new systems, it is customary to consider how the terrain influences the system design under the normal $k = 4/3$ condition, and then to determine the effect of the terrain at lower k values. Figure 4.4 shows the probability of low k values in a continental temperate climate for a range of path lengths. Note that the probability is lower for longer paths; this is because they experience a better

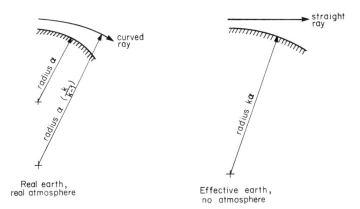

FIG. 4.3 Use of effective earth's radius (k factor) for propagation calculations.

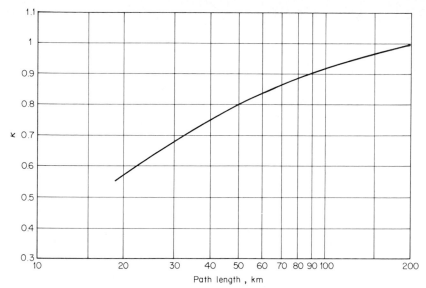

FIG. 4.4   Minimum effective value of k exceeded for approximately 99.9% of the time (continental temperature climate). (*From CCIR Report 338-4, "Propagation Data Required for Line-of-Sight Radio Relay Systems," Volume V,* Propagation Through Non-Ionized Media, *CCIR XVth Plenary, Geneva, 1982.*)

averaging of atmospheric conditions. A satisfactory engineering rule for typical paths in the 30- to 50-km range is to base the design on a minimum k value of 2/3 as the extreme condition.

9.   Atmospheric Multipath

So long as the dielectric constant decreases uniformly with height, there is only one transmission path from transmitting antenna to receiving antenna. If, however, the atmosphere develops an upper layer in which the constant decreases more rapidly, the path curvature will be greater than in the lower layer, and there will be more than one transmission path. This is shown in Fig. 4.5.

Since the upper path is longer than the lower, the phase relationship between the two rays is random, and they may either reinforce or cancel each other. Atmospheric conditions are not sufficiently stable to maintain a fixed phase relationship, and the signal amplitude fluctuates strongly.

Multipath propagation conditions may persist for many hours, most commonly before dawn in the summer. The signal amplitude as measured at a single frequency will fluctuate around a depressed level, with occasional dropouts to very low levels. The depressed levels are in the range of 15 to 30 dB below free-space levels, and if the system margin is 30 dB or more, they are not a serious problem. Extreme excursions, however, go to a fade depth of more than 40 dB for seconds at a time, and they cause system outages because of practical limits on the fade margin.

Sufficient statistical data have been obtained to develop empirical models which can be used to estimate the depth of multipath fading:[1]

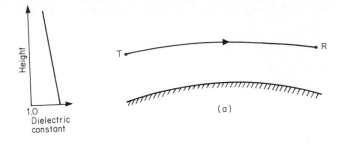

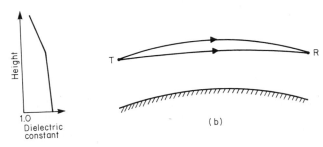

FIG. 4.5 Atmospheric dielectric constant and multipath propagation. (*a*) normal propagation; (*b*) multipath propagation.

The average time duration ($\langle t \rangle$ in seconds) of excursions below a relative signal level (voltage) V is

$$\langle t \rangle = 410V \qquad V \leq 0.1 \tag{4.5}$$

The *time fraction* of a month during the summer fading season during which the signal will be less than V is

$$T/T_o = rV^2 \qquad V \leq 0.1 \tag{4.6}$$

where T = total time the signal is below V during the observation period T_o
   r = scale factor based on terrain, climate, operating frequency, and path length

$r = a(f/4)D^3 10^{-5}$

where a = 4 over water and the Gulf Coast
   = 1 with average terrain and climate
   = 0.25 in mountains and dry climate
   f = frequency, GHz
   D = distance, mi

Finally, the *outage time* in seconds per year is

$$\text{Outage time} = 79abfD^3 \times 10^{-F/10} \tag{4.7}$$

where F = system fade margin, dB, and b accounts for a fading season of 1/2 year in hot humid areas, 1/4 year in temperate areas, and 1/8 year in dry or mountainous areas.

10. Multipath Signal Dispersion

In analog microwave systems, the only effect of multipath fading is increased noise. In digital systems operating at megabit rates, the distortions resulting from multiple paths can also cause excessive bit error rates, even when the signal-to-noise level is satisfactory. This effect is due to *dispersion,* and digital radio systems are said to have a *dispersive fade margin.* This is in addition to the "flat" fade margin which indicates the amount of reserve in the unfaded signal level.

Manufacturers of microwave radios have developed *adaptive equalizers* which adjust the amplitude and phase response of the radio channel on a dynamic basis to compensate for propagation distortions. The effectiveness of this circuit is indicated by specifying the dispersive fade margin F_d for the radio. This may be as great as 50 dB.

F_d is defined so that the outage results are obtained by the same climate and terrain scaling which has been accepted practice for analog radios which are not affected by dispersion. Thus, the total outage time for a digital radio system which is susceptible to dispersion is

$$\text{Outage time} = 79abfD^3 \, (10^{-F/10} + 10^{-F_d/10})\qquad(4.8)$$

F_d is calculated by reference to a model which was developed from a data base derived from measurements on a 26-mi path from Atlanta to Palmetto, Georgia, during June and July of 1977. Since the model is based on a single set of measurements, the numerical value of F_d can only approximate dispersive outages for other conditions. It provides a valid basis for comparing the performance of different receivers.

The most economical equipment option is usually one which provides a little better dispersive fade margin F_d, than the flat fade margin F.

11. Problem Paths

The calculations described in the previous section are based on empirical models, and the predicted performance results are not precise. The models predict performance well for most paths, but sometimes better and occasionally worse results are achieved after the system is built.

Better results than predicted by the model are usually obtained where the path has one terminal which is much higher than the other; an elevation difference of 500 m will practically ensure that no multipath outage will occur.

Worse results than predicted may be obtained where atmospheric stratification occurs more often than the fading hours per month included in the multipath model. This can be caused by local conditions which are conducive to substantial temperature inversions or by excessively high humidity resulting from irrigation or natural causes. In suspicious areas, it is good insurance to design the towers for economical expansion to space diversity but to make a nondiversity installation until a problem path is identified.

A path problem that is not easily cured is a condition known as *ducting.* Here the gradient is so strong that energy launched along the path from the transmitter is curved into the ground and misses the receiver; at the same time, the stratification changes abruptly just above the path, and energy launched slightly upward continues upward and also misses the receiver. These paths occur in high humidity locations with low, flat paths, particularly when enclosed by hills which hold cold air and strengthen temperature inversions. In susceptible locations, ducting can occur several percent of the time, and outages can be a number of hours' duration.

Curing the ducting problem requires substantial physical changes; increasing path elevation and slope is helpful, but adding a repeater may be the best solution. Changing to space diversity with the top antennas at 100 m and the bottom at a minimum height consistent with local terrain clearance has a high probability of success, but it may be more expensive than an additional repeater.

12. Space Diversity Improvement

If the calculated or actual outage time is greater than the design goals for the system, the use of space diversity can improve reliability substantially. Space diversity is achieved by using an additional antenna separated vertically from the primary antenna; under multipath propagation conditions, the vertical separation changes the relative phases of the paths so that phase cancellation conditions generally do not occur at both antennas simultaneously. The receiving system can be built to combine both antenna outputs; this achieves the most improvement. A more common arrangement is to utilize one antenna at a time, switching to the other when the signal fades near the outage-causing level.

The improvement in reliability can be estimated by dividing the outage time for the nondiversity path by an improvement factor I. The available experimental data show that I is given by[2, 3]

$$I = 7 \times 10^{-5} s^2 \, (f/D) \, 10^{F_{\text{div}}/10} \tag{4.9}$$

where s = vertical separation of antennas, ft
$\quad\quad f$ = frequency, GHz
$\quad\quad D$ = path length, mi
$\quad\quad F_{\text{div}}$ = fade margin of the system with the diversity antenna

This equation applies for F_{div} within 6 dB of F for the primary antenna and separations of 20 to 50 ft. Outside these ranges, for unusual distances, or for frequencies above 11 GHz, improvement can be expected but there is less confidence in the amount. Also, this improvement is achieved only for fade margins substantially higher than 20 dB; shallow fading of 15 to 20 dB tends to be nearly simultaneous at both antennas.

Space diversity is also useful on reflective paths where the antenna heights are sufficient for the reflected ray to arrive in phase cancellation with the direct ray at certain values of k. In these cases, the antenna separation is chosen such that one antenna is close to a signal reinforcement condition. See the section, *Terrain Effects,* below.

13. Rainfall Attenuation

At frequencies above 10 GHz, heavy rainfall can cause enough attenuation on a microwave path to require consideration in the system design. The physics of absorption and scattering by water droplets is well understood, but the system designer is faced with poor predictability of the amount of rainfall. As a result, it is standard practice to evaluate the system for peak rainfall rates which are expected to be exceeded 0.01 percent of the time or 53 minutes per year. Actual experience in any single year can easily differ from this by ± 25 percent.

Contours of rainfall intensity which are expected to be exceeded 0.01 percent of the time are shown in Fig. 4.6. The conversion of these rainfall levels to signal attenuation, γ_R, is given in Fig. 4.7.

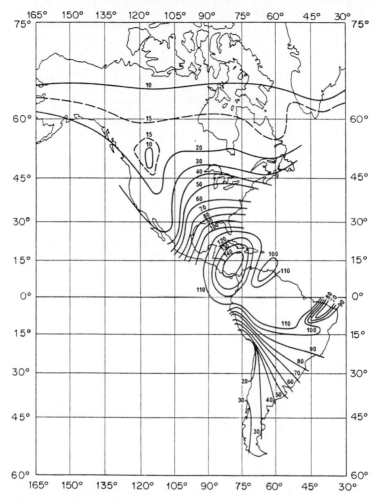

FIG. 4.6 Contours of rainfall intensity expected to be exceeded 0.01% of the time. (*From CCIR Report 563-2, "Radiometeorological Data," Volume V,* Propagation Through Non-Ionized Media, *CCIR XVth Plenary Assembly, Geneva, 1982.*)

Since it is unlikely that extremely heavy rainfall would exist along the entire path, an effective path length D_{eff} is used for calculating the attenuation

$$D_{\text{eff}} = 90D/(90 + 4D) \tag{4.10}$$

where D is the path length *in kilometers.* The attenuation which would be expected to be exceeded for 0.01 percent of the time or 53 minutes per, then, is

$$A = \gamma_R D_{\text{eff}} \tag{4.11}$$

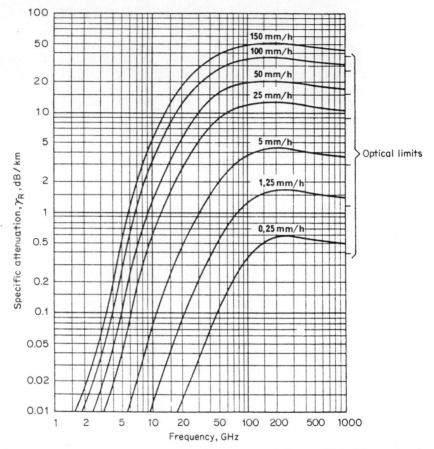

FIG. 4.7 Specific attenuation due to rain. (*From CCIR Report 721-1, "Attenuation by Hydrometeors," Volume V,* Propagation Through Non-Ionized Media, *CCIR XVth Plenary Assembly, Geneva, 1982.*)

TERRAIN EFFECTS

14. Diffraction

Opaque terrain obstructions in the path of a radio wave do not cast a sharp shadow, and the path of some of the energy is "bent" into the region which is within the geometrical shadow. This effect is called *diffraction*.

From a mathematical standpoint, the easiest example of diffraction to analyze is diffraction by a knife edge, Fig. 4.8. Although this situation seldom exists in nature, the analysis is useful because it gives a qualitative understanding of the characteristics and magnitude of the phenomenon as it occurs in real life.

The calculated knife-edge diffraction loss L_{dif} is plotted in Fig. 4.9 as a function of h/F_1 where F_1 is the radius of the first *Fresnel zone*

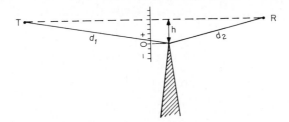

FIG. 4.8 Geometry for knife-edge diffraction calculations.

$$F_1 = (d_1 d_2 \lambda/d)^{1/2} \qquad (4.12)$$

where $d = d_1 + d_2$.

The shape of this loss function displays several aspects of diffraction. For large positive values of h, where the obstruction is well below the line of sight, the signal strength is approximately the same as in free space (diffraction loss, 0 dB). At $h/F_1 = 0.55$, the loss equals that in free space; as h is increased above this value, the signal strength oscillates around this level. At $h = F_1$, the signal is close to a maximum; this is called *first Fresnel zone* clearance. For $h/F_1 < 0.55$, the propagation is increasingly lossy; it is -6 dB relative to free space at $h = 0$ where the line of sight grazes the knife edge.

Real-world obstacles are not knife edges, but their effects are similar to those in Fig. 4.9. For paths with a line of sight well above grazing, propagation is approximately as in free space; for paths with line-of-sight approaching grazing, significant extra loss is added to the path. Hence the amount of clearance on a path

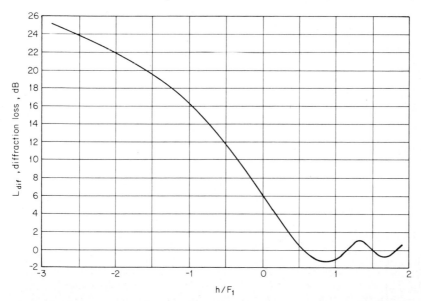

FIG. 4.9 Knife-edge diffraction loss, relative to free space, as a function of line-of-sight clearance above the edge.

depends on atmospheric conditions, and there is not a simple optimum design for tower heights.

A minimum tower-height—and thus minimum-cost—design is to provide for $0.6F_1$ clearance at the $k = 4/3$ condition. After the system is built, the received signal should equal the free-space value, thus providing a verification of the integrity of the installation. The problem with this design is that the signal levels will be depressed as the result of diffraction losses during time periods when k is less than 4/3. For the outage models described in Secs. 4.9, 4.10, and 4.12 to be valid, the path should show full signal strength almost all the time; i.e., diffraction losses should be essentially zero. For this reason, the path is normally designed with greater clearance; a full F_1 clearance at $k = 4/3$ is a common design; another common choice of clearance is $0.6F_1$ at $k = 1$. In either case, the clearance at $k = 2/3$ is also checked, and a clearance of $0.3F_1$ maintained.

The clearance rule is used to determine the required tower heights, and the fade margin is calculated from Eq. 4.4. The fade margin is then used to calculate the multipath propagation outage as discussed in Secs. 4.9, 4.10, and 4.12. An exception is transmission over water; this situation is treated in the next section.

The method of determining path clearance involves taking many terrain-height data points along the path line as drawn on a topographic map. These data are entered into a computer which has been programmed to include earth and atmospheric curvature effects together with vegetation heights in order to find the critical terrain feature and resulting clearance h for various combinations of tower heights at the opposite ends. Before a system design is implemented, it is important to conduct a field survey to look for recently constructed obstacles and verify the critical heights.

If local constraints such as zoning rules or airports make it impossible to achieve the desired clearance, the knife-edge model needs to be supplemented to provide a more accurate calculation of loss. A better model is to consider the obstacle to be a cylinder rather than a knife edge. This model has two additional components of loss: (1) a rounded edge produces weaker fields than the knife edge and (2) for paths at or below grazing, propagation along the surface reduces the signal strength. The additional attenuation resulting from these effects over average terrain is shown in Fig. 4.10. The accuracy of the results obtained by the use of this model is more affected by variations in the earth radius factor k than by the simplifications implicit in the model.

The additional loss due to the rounded edge is

$$L_{rd} = r(- 0.4r^3 + 2r^2 - 1.4r + 6) \qquad \text{dB} \qquad (4.13)$$

where

$$r = R^{1/3}\lambda^{2/3}/F_1 \qquad (4.14)$$

and R is the radius of the cylindrical surface. For estimating the magnitude of this effect on practical paths, R may be assumed to be 150 km. L_{rd} is typically 2 to 4 dB.

The additional loss in decibels due to propagation along the surface is[4]

for $0 \leq h/F_1 \qquad \leq 0.6 \qquad L_{sf} = -1.66L_{rd}\,(h/F_1)$ $\qquad (4.15)$
for $-2.73 \leq rh/F_1 \leq 0 \qquad L_{sf} = -18.3(rh/F_1)$
for $rh/F_1 \qquad \leq -2.73 \qquad L_{sf} = -24.9(rh/F_1) - 20 \log (-rhF_1) - 9.3$

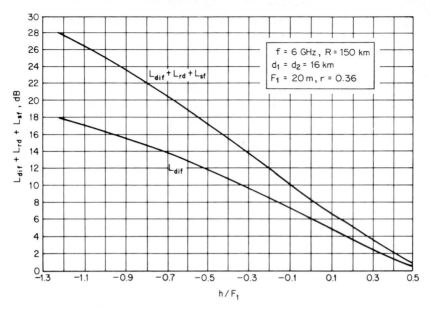

FIG. 4.10 Cylinder diffraction loss as a function of line-of-sight clearance above grazing.

L_{sf} changes sign at $h = 0$. Above grazing it is negative and reduces the effect of the rounding term, L_{rd}. Below grazing (where h is negative) it is positive and L_{sf} gives an additional loss.

For reduced clearances, the fade margin becomes:

$$\text{Fade margin} = P_t - P_r - \text{section loss} - L_{\text{dif}} - L_{rd} - L_{sf} \qquad (4.16)$$

See Fig. 4.10 for an example.

For additional treatment of diffraction effects, see Chap. 1, Sec. 7.

15. Terrain Reflections

For path clearances greater than $0.6F_1$, terrain reflections become significant, and a reflection model is substituted for the diffraction model.

The simplest model for terrain reflection is a flat surface of unlimited extent. For calculating the effect of the reflection from this surface, it is convenient to label the distances as shown in Fig. 4.11. The reflection point is where the grazing angle of incidence equals the grazing angle of reflection.

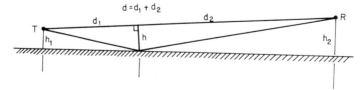

FIG. 4.11 Reflection from a plane surface.

The reflected signal travels a longer distance than the direct signal, and it may reinforce or tend to cancel it, depending on the distance differential and the phase shift at reflection. The differential distance is

$$\Delta = dh^2/2d_1d_2 \tag{4.17}$$

The phase shift in radians produced by the differential distance is $2\pi\Delta/\lambda$, and reflection from a perfectly conducting flat surface introduces a phase shift of 180°. The decibel loss resulting from reflection from such a surface is

$$L_{\text{ref}} = 20 \log |2 \sin [\pi\Delta/\lambda]| = 20 \log |2 \sin [(\pi/2)(h/F_1)^2]| \tag{4.18}$$

a flat area of seawater produces this kind of reflection for horizontal polarization at small grazing angles; the loss L_{ref} is plotted in Fig. 4.12. When the reflected signal combines constructively with the direct signal, there is a gain of 6 dB. Conversely, when the two signals arrive in opposite phase, the reflection cancels the direct signal.

In the real world, the reflected signal is seldom as strong as the direct signal with the result that the signal maxima are not as strong and the nulls are not as deep. The results in typical real-life situations are also shown on Fig. 4.12 for paths over rough terrain with an average reflectivity of 0.3 and over seawater.

Figure 4.12 is based on reflections from a plane surface. The convexity of the earth can be taken into account by multiplying the reflected signal field strength by a *divergence factor*

$$D = [1 + 2d_1d_2/ka(h'_1 + h'_2)]^{-1/2} \tag{4.19}$$

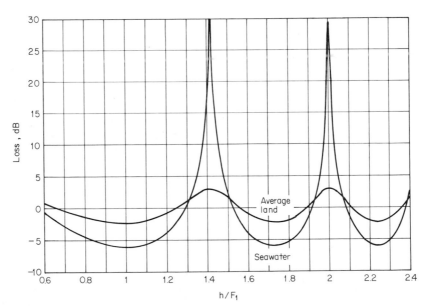

FIG. 4.12 Loss due to reflection of horizontally polarized signal from seawater and from average land with reflection coefficient 0.3.

where ka is the effective earth's radius. The heights h'_1 and h'_2 are measured from the plane that is tangent to the reflection point as indicated on Fig. 4.13.

On long hops, the divergence factor is several decibels, so that the possible signal reinforcement is reduced, but also the possible cancellation is less severe. Under extreme tropospheric conditions that produce negative k values, the apparent curvature becomes concave, enhancing the reflected signal; the same equation applies for calculating the effect.

Locating the reflection point when taking convexity into account requires solving a cubic equation; an efficient method of doing this is[5]

$$d_1 = d/2 - p \cos [(\Phi + \pi)/3] \qquad (4.20)$$
$$p^2 = [4ka(h_1 + h_2) + d^2]/3$$
$$\cos \Phi = 2ka(h_2 - h_1)d/p^3$$

In addition to the divergence factor of the spherical earth, there are other effects that reduce the magnitude of the reflected signal. The reflection coefficient for a vertically polarized microwave signal is less than unity by 2 or 3 dB for reflection from either land or water, even at grazing angles as shallow as 1°. Surface roughness reduces the reflected signal, and terrain features with dimensions of many wavelengths virtually eliminate the reflection. The terminal antennas will discriminate against the reflected signal, since it arrives at a different angle from the direct signal. A path that is partly over water may have a reflecting area that is much smaller than the projection of the first Fresnel zone onto the area, resulting in a lower reflection amplitude. Collectively, all of these points indicate that it is possible to have a successful microwave path even when the terminal heights are constrained to be much higher than $0.6F_1$, so that a reflection could tend to cancel the direct signal. Space diversity can provide satisfactory system performance even when the reflection is strong.

MICROWAVE TRANSMISSION METHODS

16. Analog Methods

The analog methods of microwave transmission are *frequency modulation* (FM), and *single sideband amplitude modulation* (AM). FM was used almost exclusively for thirty years, because it was easy to implement with early microwave power amplifier technology. Recently, the better spectral efficiency of AM combined with better power amplifiers has caused a major surge in AM systems.

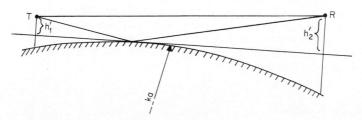

FIG. 4.13 Reflection from a curved surface.

The FM used in microwave point-to-point networks uses a low deviation of the carrier, in contrast to the high-deviation-ratio FM used for entertainment broadcasting. The low deviation causes the high-order modulation sidebands to be substantially below the first-order sidebands, so the FM system is effectively a double-sideband modulation.

The frequency-modulated signal has a constant envelope, which permits the use of nonlinear class C amplifiers. It is also possible to directly modulate the frequency of a microwave power oscillator. Both of these techniques are used by manufacturers of FM radio systems.

The modulation applied to the FM radio is most commonly a frequency-division-multiplexed set of voice channels, as described in Chap. 11. Alternatively, the modulation may be a video signal, for television distribution; see Chap. 17.

Most radio systems consist of both terminal points and also repeaters that are needed to extend the network for distances greater than about 50 km. For small networks using a few repeaters, the repeaters are operated as back-to-back terminals, with demodulation and remodulation at each repeater point. The radio design for these cases typically uses *direct modulation,* where the baseband signal modulates the frequency of a microwave oscillator at the desired transmitting frequency. The receiver downconverts the signal to an intermediate frequency, where it is filtered, then demodulated by a discriminator to recover the baseband modulation.

The alternative technique, when there are many repeaters in the system, is to use *heterodyne repeaters.* At heterodyne repeaters, the carrier frequency of the received signal is converted to an *intermediate frequency* (i-f), almost always 70 MHz. The i-f filter selects the desired signal which is then upconverted to a different carrier frequency to be retransmitted further along the system. The heterodyne repeaters are cheaper than back-to-back terminals, and introduce less distortion. For a system using heterodyne repeaters, the terminals are implemented by modulating a 70-MHz oscillator, rather than a microwave oscillator; this modulated oscillator is then upconverted as described for the heterodyne repeater.

The amplitude-modulated single-sideband radio cannot use nonlinear amplification; the effect of nonlinearities on the multiplexed voice channels is to create intermodulation products which add noise to the voice channels. In order to keep this noise low enough, the radio power amplifier is operated at a very small fraction of its maximum power output capability; typical "backed-off" operating levels are of the order of 20 dB below the maximum. In addition to this back-off, the distortion is further reduced by using "predistorter" circuits preceding the amplifier, which have a nonlinearity nearly opposite to that of the amplifier. The net result of these techniques is that single-sideband radios for 30-MHz allocations can amplify as many as 6000 multiplexed voice channels to usable power levels while maintaining low intermodulation noise.

The rest of the single-sideband radio is similar to the FM radio hardware, particularly in the use of heterodyne repeaters. Some refinements are needed in automatic gain control operation and in frequency control, but the cost of these and the linear power amplifiers are more than offset by the greater channel capacity of the single-sideband radio.

17. Digital Methods

Analog radio can be used for transmitting digital information by applying a voltage waveform representing digital values to the modulation port. However, much more effective radios are obtained by modulating the quadrature components of the carrier; for a carrier frequency of f_c, the voltage waveforms are $\cos(2\pi f_c t)$ and $\sin(2\pi f_c t)$. The importance of this is that a *coherent demodulator,* see Chap. 11, can separately recover the modulations applied to these two waveforms. Double-sideband suppressed-carrier amplitude modulation is applied to each of the quadrature components.

The digital radio interfaces with time-division-multiplexed voice channels and encodes the information into discrete-valued modulations. The modulation voltage waveforms are a sequence of pulse waveforms which are *band-limited* (filtered) to minimize the bandwidth of the signal. The modulator typically functions at 70 MHz, and the result is upconverted and amplified by a linear transmitter. At the receiver, the usual downconversion supplies an IF signal to a coherent demodulator. The values of the voltages at the output of the coherent demodulator are sampled at the appropriate time instants to determine the digital values represented by the *in-phase* (I) and *quadrature* (Q) modulations.

The repeaters in digital radio systems are almost always demodulating and regenerating repeaters rather than the heterodyne repeaters described for the analog methods. By regenerating the digital signal at repeater points, the noise contribution of each hop is removed, so that a multihop digital system is as noise-free as a one-hop system, in contrast to the analog situation where noise is accumulated throughout the system. However, when the signal to one of the digital regeneration points fades below the threshold, the perceived noise on a voice circuit increases more rapidly than for a corresponding analog system. The better noise performance of digital systems is particularly valuable when high-speed data traffic is being carried instead of voice channels.

For a conceptual description of digital modulation formats, the set of values that occur at the sampling instants is displayed as points on a two-dimensional plot and described as a signal constellation. The fundamental constellation is that for *quadrature phase-shift keying* (QPSK) and is shown in Fig. 4.14a. The QPSK name is historically applied to this constellation, but it is usually implemented by two-valued pulse waveforms amplitude modulating the quadrature carriers, producing four signal points, and can be called *4-state quadrature amplitude modulation* (4QAM). The logic values of 1 and 0 are coded into voltage values of $+1$ and -1.

Modern digital radios transmit more information bits per hertz of bandwidth by using more pulse levels and thus more points in the signal constellation. For example, 16QAM modulation results from using four levels on each axis, and 64

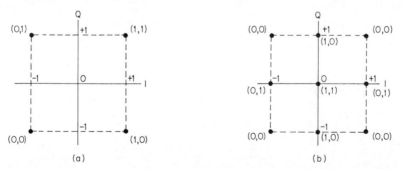

(a)   (b)

FIG. 4.14 Digital signal constellation display of voltage sample values, and digital logic value pairs: (*a*) *quadrature phase shift keying* (QPSK) and (*b*) *9-state partial response* (9QPR).

QAM from using eight levels. In these modulations, each signal point identifies four transmitted bits or six transmitted bits.

The filters which shape the pulses, and thus determine the bandwidth of the signal in QAM radios, are designed to reduce the signal spectrum by 6 dB at one-half the symbol rate, and down to nearly zero at frequencies about 125 percent greater. A different form of filter is used in some digital radios, one which attenuates the pulse spectrum to nominally zero at one-half the symbol rate. These filters are called *partial-response* filters, because at each sample time the demodulated voltage is partially due to the current pulse and partially due to the previous pulse. The overlapping pulses produce a systematic combining of information bits and voltage levels, with benefits in better spectrum occupancy; certain precoding and decoding techniques can also provide built-in error-detection capability. The partial-response signal-state constellation corresponding to the QPSK full-response modulation has nine points (9QPR) instead of four, and is illustrated in Fig. 4.14b. The four logic states (0, 0; 0, 1; 1, 0; and 1, 1) are constrained by the precoding rules to fall on the nine signal voltage states with prescribed patterns, and a violation of a pattern rule provides automatic error detection. Partial-response radios are also built with 25 states, 49 states, and 81 states.

SYSTEM CONFIGURATIONS

18. Radio System Elements

For any of the microwave transmission methods, the implementation of the system uses many common elements. As indicated in Fig. 4.15, the major distinguishing blocks perform the modulator-demodulator functions. The microwave oscillator source, frequency conversion, and amplifier functions are similar in the various radio implementations. It should not be inferred from Fig. 4.15, however, that the hardware could be interchanged from one implementation to another

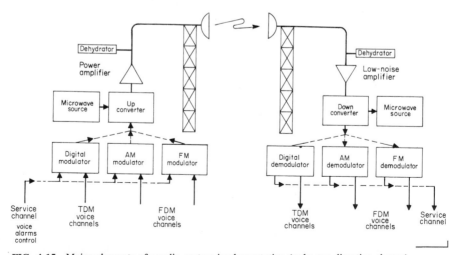

FIG. 4.15 Major elements of a radio system implementation (only one direction shown).

within the active electronic parts of radios. The more mechanical elements of towers, antennas, waveguide, or coaxial transmission line and dehydrators are not only interchangeable, but frequently are even shared by different methods as the network expands.

A necessary element to managing a radio system is a *service channel* or *auxiliary channel*. This provides for voice communication between maintenance people at different locations. It also carries an alarm and control system that does automatic reporting to a maintenance center of hardware failure indications, open door alarms, environmental conditions, and responses to remote commands.

Other cost elements of a radio system include land, buildings, and power sources, particularly at repeater locations, and hardware spares and trained maintenance personnel have to be included in the budget to assure system reliability.

REFERENCES

19. Cited References

1. Barnett, W. T., "Multipath Propagation at 4, 6, and 11 GHz," *Bell System Technical Journal,* vol. 51, no. 2, pp. 321–361, February 1972.

2. Vigants, A., "Space Diversity Engineering," *Bell System Technical Journal,* Vol. 54, no. 1, pp 103–143; January 1975.

3. White, R. F., "Engineering Considerations for Microwave Communications Systems," GTE Lenkurt, Inc., San Carlos, Calif., 1972.

4. Dougherty, H. T., and E. J. Dutton, "Quantifying the Effects of Terrain for VHF and Higher Frequency Application," National Telecommunications and Information Administration report no. 86–200, July 1986.

5. Fishback, W. T., in D. E. Kerr, "Propagation of Short Radio Waves," Vol. 13 of MIT Radiation Laboratory Series, McGraw-Hill, 1951, p. 113.

20. General References

Bean, B. R., and G. D. Thayer, "Models of the Atmospheric Radio Refractive Index," *Proc. IRE,* vol. 47, pp. 740–755, 1959.

Lundgren, C. W., and W. D. Rummler, "Digital Radio Outage due to Selective Fading— Observations and Prediction from Laboratory Simulation," *Bell System Tech. J.,* vol. 58, no. 5, pp. 1073–1100, May–June 1979.

Rummler, W. D., "More on the Multipath Fading Channel Model," *IEEE Transactions on Communications,* vol. Com-29, no. 3, pp. 346–352, March 1981.

CHAPTER 5

SATELLITE TRANSMISSION SYSTEMS

Jim Potts

Consultant; formerly Chief Engineer, COMSAT World Systems Division

CONTENTS

PREFACE

1. Introduction

The use of satellites for communications of all types is a technology which is now highly developed and in widespread use throughout the world. It has progressed in the past two decades from a technology of a quasi-experimental nature to one of routine provision of new services, taking advantage of the economies afforded by the unique characteristics of the *geostationary satellite orbit,* the GSO. Satellites are particularly useful for long-distance communications services, for services across oceans or difficult terrain, and for point-to-multipoint services such as television distribution.

The design of communications satellite systems is well understood, but the technology is still dynamic. The development of new modulation techniques and the use of higher regions of the radio-frequency spectrum continue to provide new opportunities and engineering problems. In particular, as the use of higher frequencies is heavily dependent upon meteorological phenomena which are not always amenable to engineering precision, the designs must be made with a good helping of prudent judgment.

This chapter deals with modern communications systems using satellites in the GSO, operating in the *fixed satellite service* (FSS). It provides an approach to transmission system design for such systems including the effects of the physical and interference environments. The approach to transmission system analysis is based on performance standards long established internationally for various transmission types, and it is applicable to any modulation technique and to nonstandard services.

2. Satellite Systems

The satellite is most commonly used to provide a radio-relay between widely separated points on the surface of the earth. It functions to provide line-of-sight conditions for each of the points and, in effect, produces line-of-sight path for points where it would not otherwise be possible. The design of the separate paths between these points on earth and the satellite and their combination are the principal elements of a systems transmission analysis. A special situation can occur where two satellites are used with a radio link between them. Although this type of link has not been used commercially to date, it will likely see use in the future.

Pictorially the general situation can be represented by Fig. 5.1 which shows the uplinks and downlinks as well as a satellite-to-satellite link. It can be seen that, geometrically, the up- and downlinks are the same, hence the design approaches are similar. The differences that do exist derive from the vastly different hardware problems involved in going from conventional operation in a terrestrial environment to that of space at about 36,000 km (22,300 mi) above the surface of the earth.

The radio transmission paths pass through the entire atmosphere and are influenced by a number of atmospheric effects which are discussed in detail in Chap. 1. Factors which affect satellite lifetime, satellite station-keeping, and orbital mechanics are not discussed except insofar as they affect the transmission performance.

As a final introductory note, the system design for satellite links uses the same basic equations as are used for terrestrial transmission systems; however, there are large differences in the magnitude of many of the resulting numbers compared to those that are encountered in conventional terrestrial systems. The path losses are considerably larger, and the received signals are considerably lower. Nonetheless, satellite links can be designed with relatively small margins as the propagation conditions are inherently stable.

SPACE SYSTEM GEOMETRY

3. Basic Geometrical Relationships

Figure 5.2 shows the geometrical relationships among the earth, the GSO, the satellite location on the GSO, and an earth station location at any point on the

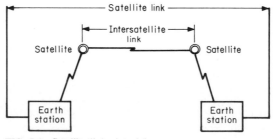

FIG. 5.1   Satellite link pictorial.

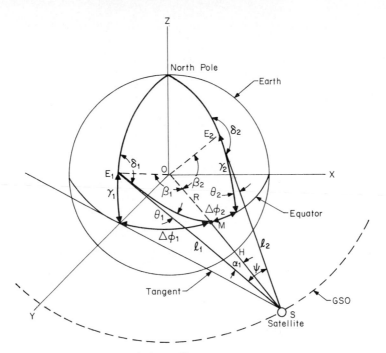

FIG. 5.2 GSO geometry: single satellite.

surface of the earth. Figure 5.3 provides additional geometrical parameters for the case of multiple satellites and multiple points on the surface of the earth. This section will define all of the geometrical parameters associated with satellites in the GSO, giving the equations for the calculation of those parameters necessary for transmission and interference analysis.

4. Geometric Term Definitions

The terms defined are those required for the calculation of transmission losses, system noise temperatures, margin requirements, and to characterize the interference environment. The terms are given in their most commonly used form or by their proper definition where possible.

Satellite Location **S.**   This is normally given as a longitudinal position, measured to the east from Greenwich, covering the range of 0 to 360°. Almost as common is the practice of using an east or west reference from Greenwich to the dateline (180°). The subsatellite point M is the point on the earth's surface directly below the satellite on the equator with the same longitude as the satellite.

Earth Station Location **E.**   This is given in earth surface coordinates of latitude γ and longitude and is the same for the northern or southern hemispheres by symmetry. For general analysis, the accuracy of the coordinates required is ade-

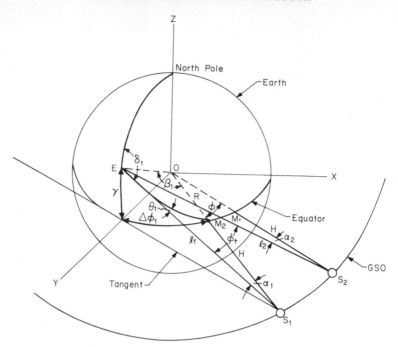

FIG. 5.3 GSO geometry: multiple satellites.

quately determined from a map, but for the purpose of proper earth station antenna pointing, coordinates should be determined by survey. The longitudinal coordinate used should be referenced to Greenwich in the same terms as the longitude of the subsatellite point of the satellite of interest.

Elevation Angle θ.   This is the angle above the earth's surface that an earth station antenna must be pointed to "see" the satellite at a particular orbital location. This angle is limited by international regulation to a minimum of 3°, while in the United States it is limited by the FCC *Rules and Regulations* to a minimum of 5° for transmitting earth station antennas.

Azimuth Angle δ.   This is the angle with respect to north that an earth station antenna must be pointed to "see" the satellite at a particular orbital location. For the entire visible orbit at an elevation angle greater than 5° from continental United States (CONUS) locations, this angle will range from about 95 to 265° using conventional compass notation.

Geocentric Angle ϕ.   This is the angle between two satellites on the GSO as measured from the center of the earth (see Fig. 5.3).

Topocentric Angle ϕ_t.   This is the angle between two satellites as seen from an earth station location (see Fig. 5.3) and is related to the geocentric angle above. The topocentric angle will be slightly larger than the geocentric angle and will be a maximum for earth stations on the equator near the limit of satellite visibility.

ϕ_r is often used to designate this angle; however, the use of ϕ_r here is consistent with the use of ϕ for the earth station sidelobe patterns.

Exocentric Angle ψ.   This is the angle between any two points on the earth's surface as seen from a point on the GSO.

The Angle α.   This angle is a particular case for the exocentric angle where one of the points is the subsatellite point.

The Angle β.   This is the angle between radii from the center of the earth to the subsatellite point and to a second point on the surface of the earth.

The Angle $\Delta\phi$.   This is the difference between the earth station longitude and the longitude of the subsatellite point.

Path Length (Range) l.   This is the line-of-sight distance from an earth station to a particular satellite and is a basic element of the path transmission loss. The range of values for satellites visible at elevation angles of 10° or greater is from 35,780 to 41,000 km (see Fig. 5.4).

5. Calculation of Geometric Parameters

The mathematical relationships among the various geometrical parameters are expressed in Eq. 5.1 through 5.7. Figures 5.4 and 5.5 provide graphical means for determining some of these parameters.

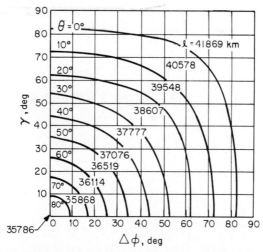

FIG. 5.4 Elevation angle and distance to satellite. θ: elevation angle at earth station location; l: distance to satellite from earth station; γ: earth station latitude; $\Delta\phi$: relative longitude, earth station to satellite.

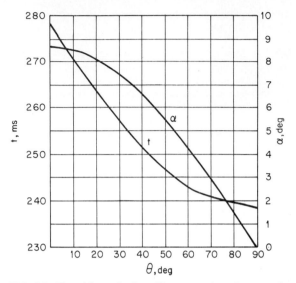

FIG. 5.5 Time delay and sub-satellite exocentric angle versus elevation angle. v: elevation angle at earth station location; α: angular separation of a point on earth and the subsatellite point from the GSO; t: Round-trip time delay to satellite from a point on the earth.

Path Length l $l = [R^2 + (R + H)^2 - 2R(R + H) \cos \beta]^{1/2}$ (5.1)

where β is given by Eq. 5.7, R = the earth's radius, 6378 km, and H = height of the GSO above the earth's surface at the equator, 35,786 km.

Elevation Angle θ

$$\cos v = \left(\frac{R + H}{l}\right) \sin \beta$$ (5.2)

Topocentric Angle ϕ_t

$$\cos \phi_t = \left[\frac{l_1^2 + l_2^2 - (84{,}328 \sin \phi/2)^2}{2l_1 l_2}\right]$$ (5.3)

where the subscripts 1 and 2 denote the path length to the two satellites respectively.

Azimuth Angle δ

$$\tan \delta = \frac{\tan \Delta\phi}{\sin \gamma}$$ (5.4)

For northern hemisphere, add 180° to the result; also observe sign convention of + for north latitude and − for south.

Exocentric Angle ψ

$$\cos \psi = \frac{l_1^2 + l_2^2 - 4R^2 \sin^2 \Gamma/2}{2l_1 l_2}$$ (5.5)

where

$$\cos \Gamma = \sin \gamma_1 \sin \gamma_2 + \cos \gamma_1 \cos \gamma_2 \cos \Delta\phi' \qquad (5.6)$$

The term $\Delta\phi'$ is the longitudinal difference of the two points, and the subscripts 1 and 2 denote the parameters associated with the two points. The $+/-$ convention for north/south latitudes should be observed.

For the special case of the exocentric angle when one point is the subsatellite point, Eq. 5.5 reduces to $\sin \alpha = (R/l) \sin \beta$.

The Angle β

$$\cos \beta = \cos \Delta\phi \cos \gamma \qquad (5.7)$$

SATELLITE DESIGN PRACTICES

6. Satellite Systems

Current practice in the design of satellites for commercial communications in the FSS derive from the coverage requirements and fall into the following groups:

- *Domestic satellites* which limit their coverage to national territories. There are many domestic satellites in orbit providing national services in a number of countries around the world. The United States alone had 50 FSS satellites in orbit or approved for launch as of mid-1985, while Brazil, Canada, China, Indonesia, Japan, and Russia all have satellite systems meeting some of their national requirements. Most of these systems operate in either the C band (6 and 4 GHz) or K_u band (14/11–12 GHz) frequency ranges (see Chap. 6 for allocation detail) with C band generally preferred because of superior propagation characteristics.

- *Regional systems* which limit their coverage to a group of countries for joint communications services. Examples of regional systems are the EUTELSAT regional network for European international traffic, which operates at K_u-band frequencies, and the Arabsat regional network, which operates at C-band frequencies.

- *International systems* which provide services on a global basis, i.e., to all countries visible from a single orbit location. Intelsat and Intersputnik are examples of this type of system, using both C-band and K_u-band frequencies.

- *The major difference among this group is the coverage provided, and this leads to major differences in the basic satellite designs used by each.*

7. FSS System Characteristics

C-Band (6/4-GHz) Satellites. C band was initially favored for communications satellites because of the favorable propagation characteristics for these frequencies; as a result, C-band technology is highly developed in terms of space-qualified components and earth-station subsystems. The specific bands in most common use are the 5925- to 6425-MHz (uplink) and the 3700- to 4200-MHz (downlink) band pair. Transponders on spacecraft using these bands have usually used 5-to-10 W *traveling-wave-tube amplifiers* (TWTAs). *Solid-state power amplifiers* (SSPAs) of up to 8.5 W are also available and are growing in popularity because of their superior linearity.

Another operational possibility at C band is the use of the 6425- to 6725-MHz band for the uplink paired with 4500 to 4800 MHz for the downlink. Allocation of these bands was made at the World Administrative Radio Conference of 1979 (WARC-79), but these bands have not been used to date. The same technology that is used in the conventional C band is applicable to this band pair. The use of these bands will be part of a worldwide allotment plan to be established in 1988 by the Second Session of WARC-ORE.*

Transponder bandwidths of 36 MHz are the most common in current use and are placed on 40-MHz centers. A satellite using a single polarization can provide 12 such transponders and twice that number using dual-polarization techniques. In addition, 72-MHz transponders are used, particularly where high-speed digital transmissions are needed.

Table 5.1 shows some of the main characteristics of the various classes of systems at 6 and 4 GHz.

C-band (6/4-GHz) Earth Stations.   Earth stations for these systems feature an extremely broad range of antenna sizes. Early stations used the largest antennas that were reasonable to build, and 32-m (105-ft) antennas are common in the INTELSAT system. As more satellite power has become available, antennas as small as 3 m (9.8 ft) have become common, particularly for receive-only applications. Table 5.2 shows the main characteristics of C-band earth stations.

K_u-band (14/11- or 14/12-GHz) Systems.   Systems in these bands are relatively new in the FSS, the first satellites being launched in 1976. The higher propagation losses characteristic of these frequencies require higher spacecraft *equivalent isotropically radiated power* (EIRP) to achieve the same transmission performance as C-band frequencies, and this is obtained from the use of greater spacecraft antenna gains. These are readily achievable at the higher frequencies.

---

*The First Session of WARC-ORB was held in 1985 and resulted in the decision to establish an allotment plan for the use of specific frequency bands.

TABLE 5.1 Typical C-Band Satellite Characteristics

Parameter	Type of coverage		
	Global	Regional	National
Satellite antenna gain, dBi			
Transmit	17–19	21–25	28–32
Receive	17–19	21–24	30–34
EIRP, dB	22–24	26–31	30–39*
Receiver noise temperature, K	800–2000	800–2000	800–2000
G/T, dB/K	− 17 to −14	− 12 to −5	− 3 to +5

*Associated with full-CONUS-coverage beams. Higher EIRPs are achieved with less than full CONUS coverage.

TABLE 5.2 Typical C-Band Earth-Station Characteristics

Parameter	Type of coverage		
	Global	Regional	National
Antenna size, m	4.5–32	4.5–25	3–30
Gain, dBi			
Transmit	47–64	47–62	43–63
Receive	43–61	43–59	40–60
Receive noise temperature, K	50–150	50–150	50–200
G/T, dB/K	23–41	23–38	17–41
Transmit power, kW	1–12	0.3–3	0.005–1
EIRP, dBW	46–95	46–74	45–84

In particular, as the 12-GHz band (11.7 to 12.2 GHz) is not shared with terrestrial systems, the *power flux density* (PFD) limitation is less stringent** and there is no requirement for coordination with terrestrial systems. The high powers that are feasible permit the use of earth stations with very small antennas, and the combination of these factors makes it possible to locate the earth stations at or near the user's premises. This results in an important economic advantage for many services and makes the use of the 12-GHz band very attractive. Even so, a good part of the higher power achievable is necessary to offset the additional attenuation that is experienced at these frequencies during heavy rain conditions.

Earth station antennas as small as 1 m are used, and earth station G/Ts vary over a very wide range, from as low as 14 dB/K to as high as 45 dB/K. Spacecraft transponder EIRPs vary from 35 to 50 dBW, while G/Ts (G/T is defined in Sec. 12 of this chapter) range from -3 to $+9$ dB/K. Transponder bandwidths are typically 36 to 72 MHz, although there is no fixed pattern of use.

8. Frequency Reuse

A common technique for increasing the capacity of a single satellite is to apply dual polarization, whereby the satellite capacity can be doubled. This requires the use of antenna feed systems for both spacecraft and earth stations that have polarization purity of the order of 30 dB of isolation between the orthogonal polarization senses. This is readily achievable with linear polarization as used by most systems. The INTELSAT system is the major exception; it uses circular polarization, although it also practices frequency reuse.

Table 5.3 illustrates a typical U. S. domestic system C-band transponder configuration providing for frequency reuse with dual polarization. Note that the frequencies of the odd and even transponders are interleaved so that the center of each transponder is located in the guard band between the transponders of the orthogonal polarization.

A second technique is to use more than one spacecraft antenna beam where the coverage requirements permit. In principle, two uses of the frequency band can be obtained from each beam by use of dual polarization, but this depends upon obtaining sufficient isolation between the beams from pattern characteristics. At 6/4 GHz, up to six total uses have been achieved; while at 14/12 GHz, designs for eight-fold reuse have been made.

**The PFD in the 11.7- to 12.2- GHz band is not limited by international agreements, but each country is authorized to estabish limits within its border. See Chap. 6 for limits established by the FCC for the United States.

TABLE 5.3 Typical C-Band Transponder Configuration

Horizontal polarization		Vertical polarization	
Transponder number	Transmit center frequency, MHz	Transponder number	Transmit center frequency, MHz
1	3720	2	3740
3	3760	4	3780
5	3800	6	3820
7	3840	8	3860
9	3880	10	3900
11	3920	12	3940
13	3960	14	3980
15	4000	16	4020
17	4040	18	4060
19	4080	20	4100
21	4120	22	4140
23	4160	24	4180

SATELLITE LINK CALCULATIONS

9. Transmission Performance

The performance of a communications channel is expressed by the signal-to-noise ratio after demodulation for analog transmission and by the *bit error ratio* (BER) for digital transmission. The performance of both types of transmission depends directly on the noise introduced by the transmission medium. Standards of performance established by the CCIR* and CCITT† are based on the concepts of a *hypothetical reference circuit* (HRC) for analog systems and a *hypothetical reference digital path* (HRDP) for digital systems. Figure 5.1, which depicts the satellite link, is also the schematic representation of the HRC as used by the CCIR to define the subsystems that make up such links and to provide a reference for the establishment of transmission performance standards which are in general worldwide use.

Chapter 10 provides a summary of the most important of these performance standards as contained in the most recent CCIR publications. It is noted that no performance standard exists for digital TV at the present time.

The standards for the FSS are under continuous review by Study Group 4 of the CCIR. A revised version of its Recommendations is published every four years as a result of the actions taken at CCIR Plenary Assemblies.

10. Link Budget Calculations

Transmission analysis for the satellite link is similar to that for other transmission systems. Figure 5.6 shows the component elements necessary to the analysis and

---

*Comité Consultatif International Rédiffusion, an organ of the International Telecommunications Union.

†Comité Consultatif International Telephonique and Telegraphique, an organ of the International Telecommunications Union.

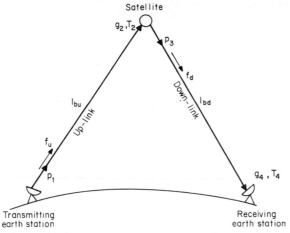

FIG. 5.6 Satellite transmission geometry and parameters. l_{bu}: basic transmission loss on the up-link; l_{bd}: basic transmission loss on the down-link; f_u: up-link frequency; f_d: down-link frequency; g_2: satellite receiving antenna gain; g_4: earth station receiving antenna gain; T_2: satellite receiving noise temperature; T_4: earth station receiving noise temperature; p_1: earth station EIRP ($p_e g_1$); p_3: satellite EIRP ($p_s g_3$).

indicates their relationship and the terms used in the equations. The objective of the analysis is to establish the parameters necessary to meet the performance objectives and to make tradeoffs in the choice of the implementation parameters in order to achieve a balance between cost and performance.

Transmission objectives consist of:

1. *Normal circuit quality,* i.e., signal-to-noise ratio or bit error ratio

2. *Circuit performance reliability,* the time percentage for which a minimum quality standard is met

3. *Circuit availability,* the time percentage a circuit is available, unaffected by outage from any cause; usually based on a minimum time interval (see Sec. 14, below)

The link budgets set up for the analysis are established to produce quality objectives expressed as a signal-to-noise ratio (or BER). Each noise power contribution is calculated as if it appeared at a reference point where it can be directly combined with other contributions for steady-state conditions. Statistical combination is used where time variations occur.

Each noise contribution can be calculated independently of the others and analyzed in accordance with its own budgeted contribution. A sample budget for a system for analog FM voice circuits which meet the CCIR-recommended performance is given in Table 5.4. The noise contributions are given in picowatts, psophometrically weighted, at the 0 reference level (pW0p). This unit is the international standard for voice channels and is based on the use of a psophometer for measuring the noise power in a channel. The frequency response of the psophometer is such that when an 800-Hz test signal is used to establish the voice channel reference condition, a 4-kHz band of white noise will measure 2.5 dB

TABLE 5.4   Sample Noise Budget

Source	Noise contribution, pW0p
Thermal noise	
Uplink	1,000
Downlink	3,000
Interference	
Terrestrial	1,000
Satellites	2,500
Intermodulation	2,000
Total	10,000

lower than its actual power level because of the shape of the filter of the psophometer.

The noise power values given in the table correspond to the long-term performance. The short-term performance values are calculated relative to this performance by taking into account the short-term attenuation conditions due to rain and other variable mechanisms.

To calculate the total link noise, it is usual to analyze the up- and downlinks separately, developing their cumulative distributions of noise. Then, as they are usually statistically independent, they are combined statistically. Note that the loss variations and the margins which characterize the uplink impact the total link performance in an entirely different way than those of the downlink.

11. Noise Components

Thermal Noise.   In determining the thermal noise component, a number of different expressions are in common use, each useful in its own way and easily converted, one to the other.

The noise performance of a communications channel is dependent upon the modulation technique employed and the noise of the system in the transmission link. $N = kTB$ expresses the total thermal noise in a system with a bandwidth B, and a noise temperature T (k is Boltzman's constant). The carrier-to-noise performance is expressed as

$$C/N = C/kTB \quad \text{dB} \tag{5.8}$$

A second noise expression in use, N_0, is independent of the bandwidth of the system and is expressed by

$$C/N_0 = C/kT \quad \text{dB} \bullet \text{Hz} \tag{5.9}$$

Finally, an expression dependent only on the noise temperature is

$$C/T = C/N_0 + 10 \log k \quad \text{dBW/K} \tag{5.10}$$

The latter two expressions are useful for making comparisons between carriers having different bandwidths, but the expression in Eq. 5.8 is probably the most convenient to use for general transmission analysis and will be followed here. However, all of the calculations that follow can be made using the other expressions with equal validity, including those pertaining to interference.

The general expression for a transmission link is then

$$C/N = P_t - L_b + G_r - 10 \log kT_s B \quad \text{dB} \quad (5.11)$$

where the terms are defined in Sec. 12.

Intermodulation Noise. A second important noise source to be accounted for is that due to *intermodulation*. The main source of this contribution is the satellite output amplifiers and the normal nonlinear operational mode. With single-carrier operation, this does not introduce noise of any significance; however, some applications require the simultaneous transmission of many carriers through the same amplifier resulting in a noise contribution arising from third-order distortion products. The calculation of the intermodulation noise is complicated, and it is usual to make measurements on the actual devices to be used to determine the values to use. For satellites that use TWTAs, values of the order of 2000 to 2500 pW0p are typical.

Interference Noise. As explained in Secs. 17 and 18, allowance for this noise source is made in the overall noise budget and is not actually determined. It consists of contributions from terrestrial systems and other satellite networks on the GSO. Depending upon the situation, there may be no actual contribution from either or both of these sources. It is important to identify for any particular source, whether it affects the uplink or the downlink, as it will affect the overall noise distribution differently.

Overall Link Noise. The overall satellite transmission system consists of the uplink and downlink in tandem as shown in Fig. 5.6. The total link noise expression is a combination of the up- and downlinks including the noise contributions on each due to interference plus intermodulation noise.

12. Basic Transmission Parameters

Basic Transmission Loss L_b. This consists of the free-space transmission loss plus atmospheric losses due to gases or rain (see Chap. 1). Two cases are of interest: (1) during a clear sky and (2) during meteorological events.

For satellites on the GSO, the free-space loss is a function of the elevation angle θ because of path length variation. L_{fs} is given in Fig. 5.7 for an elevation angle of 10° and for frequencies up to 50 GHz. For other elevation angles, the loss is calculated from

$$L_{fs}(\theta) = L_{fs}(10°) + 20 \log \frac{l(\theta)}{40,578} \quad \text{dB} \quad (5.12)$$

The path length as a function of the earth station location is obtained from Fig. 5.4. The basic transmission loss is then determined from

$$L_b = L_{fs} + A \quad \text{dB} \quad (5.13)$$

The *atmospheric absorption* factor A is given the appropriate value for the case of interest, i.e., A_a for clear sky or A_r for rain events.

A_a can be determined by multiplying the values obtained from the curves of Figs. 5.8 and 5.9 for the frequency and elevation angle of concern.

A_r must be determined for the climate, frequency, elevation angle, and the

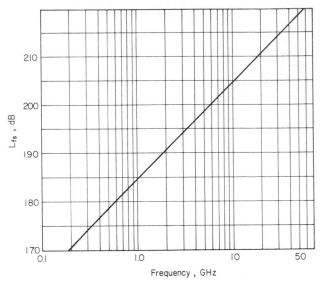

FIG. 5.7   Free-space loss for a 10° elevation angle. L_{fs}: free-space transmission loss; f: frequency, GHz.

percentages of time of interest for the uplink and downlink separately for each transmission direction.

The loss distribution for the uplink at station A is determined as well as the combined loss and noise-temperature distribution on the downlink at station B. These are combined statistically to determine the cumulative time distribution for one direction. The process is repeated in the reverse direction for the link from station B to station A. The distributions are usually determined for a period of a year or for the "worst month."

Chapter 1 defines the data necessary for this determination and also includes the most up-to-date procedure available to compute the statistical distribution of this component of the loss.

System Noise Temperature.   The diagram of Fig. 5.10 shows the basic components which contribute to the overall noise temperature of the earth-station receiving system. The basic equation for clear-sky conditions is

$$T_s = T_a + (L_f - 1)T_0 + L_f T_r \quad \text{K} \tag{5.14}$$

where $T_0 =$ ambient temperature of the transmission lines, K
$\quad L_f =$ loss of the transmission line
$\quad T_r =$ noise temperature of the receiver, K
$\quad T_a =$ contribution of the antenna including the sky noise, ground noise from unwanted antenna radiation, and noise temperature increase due to atmospheric absorption

This equation is general and is also applicable to the satellite receiving system; however, as the satellite antenna "sees" the warm earth instead of the cold sky,

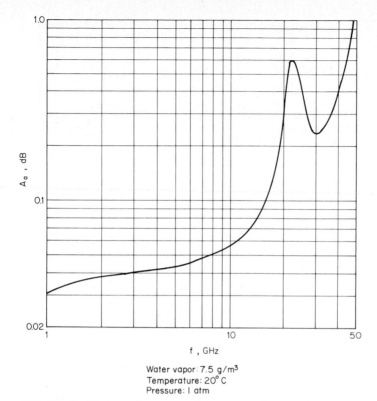

Water vapor: 7.5 g/m³
Temperature: 20° C
Pressure: I atm

FIG. 5.8 Zenith attenuation, gases and water vapor. A_a: atmospheric attenuation; f: frequency, GHz; water vapor: 7.5 g/m³; temperature: 20°C; pressure: 1 atm.

and as the satellite receiver usually has a much higher noise temperature than the earth station receiver, the transmission line losses for the satellite case have little effect on the total noise and can be ignored.

To calculate the clear-sky noise temperature, Figs. 5.8 and 5.9 are used to determine T_a from

$$T_a = 280\,[1 \,-\, 10^{A(\theta)/10}] \quad \text{K} \tag{5.15}$$

This equation is also used for the rain case by substitution of the values of A_r obtained for the rain rates of interest.

Gain-to-Noise Temperature Ratio, G/T.   A useful parameter for the calculation of link budgets is the ratio of the antenna gain to the noise temperature for receiving systems, either satellite or earth station. This is expressed in dB/K, by converting the noise temperature to its dB value relative to 1 Kelvin.

EIRP (P_r).   The *equivalent isotropically radiated power* is the final parameter needed for the link budget calculations. It is readily determined from the power

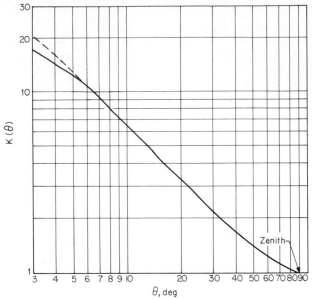

FIG. 5.9   Path length correction factor, $K(\theta)$.

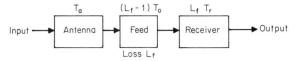

FIG. 5.10   Earth station receiving system.

input to the antenna expressed in dBw plus the antenna gain in dB.

$$P_1 \text{ (or } P_3) = P_e + G_1 \text{ (or } P_s + G_3) \qquad \text{dBW} \tag{5.16}$$

where P_e and P_s = earth station and satellite transmitter powers, respectively
P_1 and G_1 = earth station EIRP and antenna gains, respectively
P_3 and G_3 = satellite EIRP and antenna gains, respectively

13. Design Calculations

Uplink Design.   From Eq. 5.11 the equation for the uplink thermal noise budget is

$$c/n_{tu} = \frac{p_1 \, g_2}{l_{bu} k T_2 B} \tag{5.17}$$

(Note that the equations in this section are given in conventional mathematical form instead of decibels and that the subscripts are aligned with Fig. 5.6.)

Based on Eq. 5.17, the following equation takes into account the interference noise expressed as a carrier-to-noise ratio for the noise levels being budgeted for other space networks and for terrestrial systems:

$$c/n_u = \left[\frac{1}{c/n_{tu}} + \frac{1}{c/n_{iu}}\right]^{-1} \qquad (5.18)$$

This yields the total uplink noise for the long-term condition. If losses of short-term duration are a factor, then a margin is added which effectively increases the EIRP required. For example, at 14 GHz, a margin of more than 10 dB might be required. Also, the uplink fade will affect all sources of noise including interference, intermodulation, and the downlink, as the satellite output power is linearly related to the input power.

As the up- and downlinks are independent, it is usual to divide the time allowed for short-term noise conditions between them. A value of 0.005 percent might be allowed to each to achieve a 0.01 percent overall objective.

It should be noted that uplink power control is an alternative to adding a margin. While feasible, it is rarely practiced because of the cost of providing expensive facilities which would be required for only very short periods of time.

Downlink Design. The analysis follows the same general procedure as for the uplink:

$$c/n_{td} = \frac{p_3 g_4}{l_{bd} k T_4 B} \qquad (5.19)$$

and

$$c/n_d = \left[\frac{1}{c/n_{td}} + \frac{1}{c/n_{id}}\right]^{-1} \qquad (5.20)$$

The need for margin on the downlink is different from that on the uplink because only the downlink noise sources are affected by any losses. There is an added factor to consider on the downlink; not only can the carrier be attenuated, but the system noise temperature can increase above its normal value from the added losses due to rain (see Eq. 5.15). The increase is limited to a maximum value set by the finite rain temperature.

Total Link Performance. The total link expression is

$$c/n = \left[\frac{1}{c/n_u} + \frac{1}{c/n_d} + \frac{1}{c/n_{im}}\right]^{-1} \qquad (5.21)$$

where the intermodulation contribution is included.

While this expression is valid for most systems, a special case occurs when frequency reuse is practiced on a satellite. Additional sources of noise from self-interference have to be accounted for, one when the frequencies are used twice and three when the frequencies are used four times. The technique for including these sources in the analysis follows the same procedure outlined above for other sources. In particular, if polarization frequency reuse is employed, the effects of depolarization by rain are accounted for in a manner similar to that used for attenuation.

Overall Noise Distributions. Computing Eq. 5.21 for a satellite link will give the performance which is expected to prevail for a large percentage of the time and would correspond to the long-term CCIR performance. Margin to maintain this performance can be added to include losses due to antenna pointing errors and minor equipment performance variations, including degradations which occur gradually over time.

Sun Outage. A situation which occurs in two periods each year is the conjunction of the sun with a satellite for each earth-station antenna. The times when the elevation and azimuth of the sun equal that of the satellite can be determined from standard astronomical tables.

Sun outages cause a severe increase in the noise temperature of the receiving system and will cause an outage for most earth-station designs. The situation occurs daily during a period of up to 21 days depending on the earth-station location, either before or after the spring and fall equinoxes. The time duration of the outage can be calculated approximately from $480\phi_3$ s, where ϕ_3 is the antenna half-power beamwidth in degrees. The increase in noise temperature can be of the order of 20,000 K at 4 GHz when the antenna beamwidth is less than about 0.6°. The number of days that will have a conjunction can be approximated from $5\phi_3$. The times of occurrences and their duration are predictable, therefore operational procedures can be established to minimize the effects.

Any outage time resulting from sun conjunction is accounted for in making the circuit availability calculation, as such outages will normally exceed the 10-s limit (see Sec. 14).

Summary. The steps outlined above can be followed for any kind of system so long as any special factors unique to a particular situation are recognized. For systems operating around 6 and 4 GHz, little attention need be paid to rain attenuation as far as short-term criteria are concerned except for the case where polarization frequency reuse is involved.

Liberal reference to the documents of the CCIR has been made throughout and the list of references identifies the Reports and Recommendations most useful to various aspects of satellite system design.

14. Circuit Availability

Circuit availability is an important aspect of performance, as it describes the percentage of time that one can expect to be able to establish a connection using a particular facility. Periods of outage due to any cause which persist for more than a prescribed period, for example, 10 s, are defined as unavailable periods and are usually required to be a small percentage of the total time. This is commonly expressed in terms of the available time. CCIR Recommendation 579, which covers the FSS, specifies a value for the circuit availability of 99.8 percent of a year for outages due to all causes, with a specific allowance for propagation outages currently under study. A value of 99.96 percent of a year is being considered by the CCIR for this latter factor for digital paths.

An important aspect of propagation effects is the concept of *propagation availability* and its impact on transmission system design. When a propagation fading event causes a system to experience an outage which is longer than 10 s, the circuit is unavailable and this time is not charged to the short-term performance of the system. This concept has been recently introduced as a result of the particular performance requirements being developed by the CCITT for the ISDN where error-free intervals are of special concern. There is currently a body of data being developed on the duration of propagation fades at specific levels of attenuation with a view to establishing a relationship between propagation availability and the overall propagation attenuation data.

Generally, the need to give consideration to this aspect of system design applies only for systems which operate at frequencies above about 10 GHz. For example, at frequencies around 6 and 4 GHz, outages due to propagation would not be expected except in the most severe rain climates.

15. Time Delay

The use of the GSO has a particular characteristic which has been the source of some problems when used for telephony. The great distances involved lead to significant transit times, even at the speed of light. As can be seen from Fig. 5.5, the round-trip time for the satellite–earth-station path is of the order of 240 to 275 ms. For a complete link, earth-station–satellite–earth-station, the range is the same. As telephony systems require four-wire to two-wire hybrids at circuit ends, mismatches at these devices can cause *echoes* (reflections) and the round-trip time for these echoes can be over half a second. If the level of the echo is sufficiently high, the speaker hears his or her transmission delayed and will generally find this intolerable. Fortunately, the development of the echo canceller has all but eliminated this problem. The delay remains and while not intolerable for a single satellite hop, it generally limits the situations in which two satellite hops in tandem will be used.

For data or television or other one-way transmissions, the time delay is not a problem, even with multiple hops in tandem. However, some data systems require the use of special forward-error-correction techniques because of the time delay.

16. Solar Eclipses

A particular concern in the design of spacecraft power systems is the eclipse of the sun by the earth at the equinoxes. Since the primary power source is solar cells, either storage batteries must provide power to the spacecraft during eclipse or the satellite location is chosen so that the resulting outage occurs after local midnight. This is accomplished by placing the satellite to the west of the coverage area.

A total of 21 days on either side of the equinox will experience some eclipse. The duration of the eclipse, centered on midnight at the satellite longitude, increases as the sun approaches the equator to a maximum of 70 minutes and then decreases each day as the sun moves away from the equator.

It is usual to provide on-board storage batteries when the traffic requires continuous operation; however, the technique of establishing the outage occurrence at low traffic hours can be useful for some situations.

INTERFERENCE CONSIDERATIONS

17. Terrestrial System Interference

As most frequency bands allocated to the FSS are shared with the *fixed service* (FS, i.e., microwave), special attention must be paid to the various interference situations which can be present. Sharing criteria have been developed to establish a basis for the orderly use of the frequencies by the two services

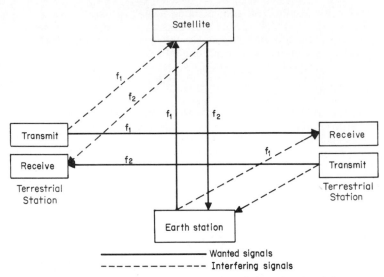

FIG. 5.11   Interference paths: FS and FSS sharing frequencies, f_1 and f_2.

without imposing unreasonable restrictions on either service. The various interference situations are pictured in Fig. 5.11, and the related sharing criteria are discussed below.

FS Transmitting Stations into Earth-Station Receivers.   This interference is controlled by placing limits on the EIRP of the terrestrial station transmitter and the amount of interference that can be caused to the earth-station receiver. A coordination procedure must be followed by earth-station planners if protection from terrestrial interference is required. This involves the establishment of a coordination area around the planned location of the earth station using the procedure contained in CCIR Report 382. This procedure and its application are outlined in the ITU *Radio Regulations*. Detailed interference calculations are made for all terrestrial stations within this area and tested against the allowable interference limits. After the successful coordination of an earth station using this procedure, any planning for a new terrestrial station within the coordination area must similarly include calculations to ensure that its operation will not exceed the allowable limit for interference into the earth station. The procedure for establishing the coordination area is also detailed in Part 25 of the FCC *Rules and Regulations,* while the appropriate interference criteria are given in CCIR Recommendation 356.

FS Transmitting Stations into Satellite Receivers.   This interference is controlled by limiting the EIRP radiated toward the GSO and effectively requires that the terrestrial station antenna be pointed to avoid intersections with the GSO by the main beam of the antenna. For specific details on these restrictions, see RR 2506 of the *Radio Regulations* and Part 21 of the FCC *Rules and Regulations.*

Satellite Transmitting Stations into FS Receivers.   This interference is controlled by imposing a *power flux density* (PFD) limit on satellite transmissions. The limits for various frequency bands are given in CCIR Recommendation 358 and by RR 2565 to 2583.

Earth Station Transmitters into FS Receivers. This interference is controlled by limiting the EIRP density in the horizontal plane. As in the case of the FS transmissions into the earth-station receivers, a coordination area around each earth station location is defined using procedures detailed in CCIR Report 382 and Appendix 28 of the *Radio Regulations.* Interference criteria for this case are covered by CCIR Recommendation 357.

Interface Budgets for FSS. For the purposes of transmission analysis, the actual interference noise powers are not usually calculated as operational data, for such calculations are not readily made, particularly in an early planning stage. The assumption is made that cochannel operation must be provided for, and this leads to the most severe interference situation. This requires making provision for the maximum permissible level of interference in the overall long-term noise budget. This amount is 1000 pW0p and is expected to occur on the uplink. The short-term interference is dominant on the downlink, and is related to the allowance of 50,000 pW0p for 0.3 percent of any month. For this case only a time portion is budgeted, i.e. 0.03 percent of any month.

18. Satellite System Interference

For the case of interference from other satellites of the FSS using the GSO, reference is made to Fig. 5.12, which illustrates the various interference modes. The principal modes are the interference from the earth-station transmitters on the uplink of one network into a satellite receiver of the second network, and the interference from the satellite transmitter of one network into an earth-station re-

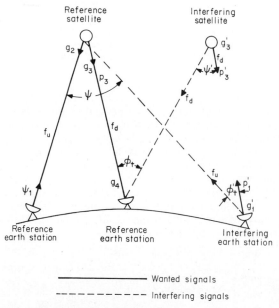

————————— Wanted signals

— — — — — — — — — Interfering signals

FIG. 5.12 Interference geometry: space networks operating on the same frequencies, f_u and f_d.

ceiver on the downlink of the second network. The equations applicable to each of these situations are

$$c/i_u = \frac{p_e g_1 g_2}{p'_e \, g'_1(\phi'_t) g_2(\psi)} \tag{5.22}$$

and

$$c/i_d = \frac{p_s g_3 g_4}{p'_s \, g'_3(\psi') g_4(\phi_t)} \tag{5.23}$$

The above interferences are given as *carrier-to-interference ratios* c/i and must be converted to their noise power equivalent. The noise power is dependent upon the modulation characteristics of the carriers involved and can be determined using Report 388 of the CCIR. The results are then combined by power addition to determine the total interference noise contribution. As for the terrestrial case, there are specific limits to the amount of interference that one network must accept from another. This is established by CCIR Recommendation 466 for analog FM systems and by Recommendation 523 for digital systems.

For satellites which provide coverage to the same service area, the only isolation which is available is derived from the earth-station antenna patterns and the spacing between the satellites on the GSO. To ensure a high level of orbit and spectrum utilization, reference patterns have been adopted by the CCIR in Recommendation 465 which serve as basis for coordination between networks of the FSS using the GSO. The current reference standard used by the CCIR for antennas with $D/\lambda > 100$ is

$$
\begin{aligned}
G(\phi) &= 32 - 25 \log \phi \quad \text{dBi} \quad & 1° \leqslant \phi \leqslant 48° \\
&= -10 \quad \text{dBi} & 48° \leqslant \phi
\end{aligned}
\tag{5.24}
$$

where ϕ is the angle from the antenna beam center in the orbital plane.

For D/λ less than or equal to 100

$$
\begin{aligned}
G(\phi) &= 52 - 10 \log(D/\lambda) - 25 \log \phi \;\; \text{dBi} \quad & (100\,\lambda/D)^2 \leqslant \phi < 48° \\
&= 10 - 10 \log(D/\lambda) \quad \text{dBi} & 48° \leqslant \phi \leqslant 180°
\end{aligned}
\tag{5.25}
$$

The resulting carrier-to-interference ratio is the difference between the main beam gain G_0 and the value determined from either Eq. 5.24 or 5.25, plus any difference in the respective carrier EIRPs. Spacing between cocoverage satellites is typically about 4° to achieve sufficiently high c/i ratios.

For the situation of U.S. domestic satellite systems, the FCC has adopted a more stringent earth-station antenna standard aimed at improving the orbit and spectrum utilization of the GSO by allowing satellites to be spaced as close as 2°. To accomplish this objective, the FCC requires for earth-station transmitting antennas a sidelobe envelope pattern in the plane of the orbit which is 3 dB better than that currently recommended by the CCIR in the near-in regions of the sidelobe pattern. The FCC pattern is given by

$$
\begin{aligned}
G(\phi) &= 29 - 25 \log \phi \quad \text{dBi} \quad & 1° \leqslant \phi \leqslant 7° \\
&= 8 \;\text{dBi} & 7° < \phi \leqslant 9.2° \\
&= 32 - 25 \log \phi \quad \text{dBi} & 9.2° < \phi \leqslant 48° \\
&= -10 \quad \text{dBi} & 48° < \phi
\end{aligned}
\tag{5.26}
$$

The additional 3 dB of isolation will permit a reduction in satellite spacing from 4 to 3°.

For the separate coverage case, i.e., where satellites serve different geographical areas, isolation can also be obtained from the spacecraft antenna pattern if it is designed to limit its radiation to the coverage area. Such designs have the additional advantage that more efficient use of spacecraft mass and power is made by confining the radiation to the area where it is most useful. Equations 5.22 and 5.23 contain a term of the form $g(\psi)$ which will account for this factor when it is present. The CCIR is currently developing a recommendation for reference spacecraft antenna patterns similar to those which exist for earth stations, to be used for coordination purposes and to increase the potential for access to the GSO by satellite networks of the FSS. Because of the wide variety of antenna types which can be used on spacecraft, several different references will probably be developed. The current references under consideration are the following.

Single-Feed Circular or Elliptical Beam Antennas

$$G(\psi) = G_0 - 3(\psi/\psi_0)^2 \text{ dBi} \qquad \text{for} \qquad \psi_0 \leqslant \psi \leqslant a\,\psi_0 \qquad (5.27)$$

$$= G_0 + L_s \text{ dBi} \qquad \text{for} \qquad a\,\psi_0 < \psi \leqslant b\,\psi_0 \qquad (5.28)$$

where L_s is the required near-in sidelobe level in decibels relative to the peak gain G_0 and ψ_0 is half the 3-dB beamwidth of the antenna. For $L_s = -25$ dB, $a = 2.88$ and $b = 6.32$. Values for a and b have yet to be determined for $L_s = -30$ dB, a desirable objective.

Multiple-Feed Shaped-Beam Antennas

$$G(\phi_c) = G_0 - [A + B\,(\Delta\psi_c + 1)^2] \text{ dBi} \qquad 0 \leqslant \Delta\psi_c \leqslant C \qquad (5.29)$$

$$= G_0 + L_s \text{ dBi} \qquad C < \Delta\psi_c\,0 \leqslant E \qquad (5.30)$$

where ϕ_c is the angular distance from the edge of coverage and is given by

$$\psi_c = \frac{\psi - \psi_0}{\psi_b} \qquad (5.31)$$

The term ψ is the angular distance from a reference point, ϕ_0 is the radial distance from the reference point to the edge of coverage in the direction of interest, and ψ_b is the angular radius of an elemental beamlet. A, B, C, and E are constants determined by the elemental beamlet size which depends upon the D/λ of the antenna and the design objective for L_s. For $L_s = -25$ dB, $B = 3.02$, $C = 1.88$, and $E = 3.85$. Also, $\psi_b = 36\,\lambda/D$.

Further developments on these reference patterns can be expected by the time of the Second Session of WARC-ORB.

FSS Coordination Requirements. At the earliest stages, satellite network planners must be aware of the current and planned level of occupancy of the GSO by other networks. The orbit and spectrum resource is finite, and in some arcs of the orbit, it can be very difficult to find a usable location for another satellite.

To place a satellite on orbit with proper international recognition of its frequencies and its interference characteristics relative to other systems requires that a coordination procedure be followed. Although similar to the procedure discussed in Sec. 17 regarding coordination with terrestrial systems, the satellite coordination procedure is more important because of the international impact which can be expected.

The coordination is carried out in accordance with Article 11 of the *Radio Regulations,* and eventually registration is performed under Article 13. A detailed discussion of these procedures is beyond the scope of this chapter; however, readers should be aware of the basic requirement. They should also be familiar with the procedures of Appendix 29 of the *Radio Regulations* which detail the calculations used to identify networks requiring coordination. This method examines the interference potential of satellite networks on the basis of an increase in noise temperature of a network from the transmissions of another network. Where the value calculated from the published parameters of the interfering network exceeds a prescribed limit, currently 4 percent, coordination is required.

As the burden of establishing the feasibility of the operation of a new satellite network rests with the newcomer, it is prudent to make such calculations with respect to existing networks before finalizing on designs of the transmission system, the spacecraft, and the operating orbit locations.

REFERENCES

19. Performance Requirements

CCIR Recommendation 353, Analog systems
CCIR Recommendation 522, Digital telephony
CCIR Recommendation 614, ISDN
CCIR Recommendation 567, Analog television
CCIR Recommendation 579, Availability
CCITT Recommendation G222, Analog systems
CCITT Recommendation G821, ISDN

20. Terrestrial Interference Limits

CCIR Recommendation 355, General sharing
CCIR Recommendation 358, PFD limits on the FSS
CCIR Recommendation 356, Interference from FS into FSS FM systems
CCIR Recommendation 357, Interference from FSS into FS analog angle-modulated systems
CCIR Recommendation 558, Interference from FS into FSS digital telephony systems
Recommendation 406, EIRP limits for FS systems

21. Sharing within the FSS

CCIR Recommendation 466, Interference into analog networks
CCIR Recommendation 523, Interference into digital networks
CCIR Recommendation 483, Interference into analog TV

CHAPTER 6
SATELLITES: REGULATORY SUMMARY

Carl J. Cangelosi

*Vice President and General Counsel, GE American
Communications, Inc.*

CONTENTS

FREQUENCY ALLOCATION AND ASSIGNMENT

1. Regulatory Authorities

Frequency allocations for United States satellites are a result of action by international and domestic regulatory organizations. The International Telecommunications Union (ITU) is the organization which on the international level governs radio services. The radio regulations adopted by the ITU contain specific frequency allocations to countries for radio services including those provided by satellites. The United States has ratified a treaty called the International Telecommunications Convention (and its various revisions). Except to the extent that the United States has taken reservations from these treaties, the United States is bound to enforce the Convention and the Radio Regulation adopted by the ITU.

The ITU meets infrequently, not more than once every five years. In the intervals, the policy making is conducted by the World Administrative Radio Conference (WARC) or Regional Administrative Radio Conferences (RARC).

2. FCC Frequency Allocations

The frequency allocations made by the ITU are frequently broader than those adopted by the Federal Communications Commission (FCC), the regulatory agency within the United States with the legal authority to make frequency allocations and assignments for nongovernment satellite usage. The allocations made by the ITU for satellites are shown in Chap. 3. The FCC's frequency assignments are shown below.

FCC FREQUENCY ALLOCATIONS, GHz

	Uplink	Downlink
Fixed satellite services		
C-Band	5.925–6.425	3.700–4.200
Ku-Band	14.0–14.5	11.7–12.2
Broadcast satellite service	17.3–17.8	12.2–12.7

3. International Frequency Registration Board

The International Frequency Registration Board (IFRB) is a permanent organization within the ITU which handles the activities of the ITU with regard to registration of frequencies and radio interference problems. The functions of the IFRB include:

- The processing of frequency assignment notices, including information about any associated orbital locations of geostationary satellites.

- The processing of information received from administrations in application of the advance publication, coordination, and other procedures relating to orbital assignments.

- The investigation of harmful interference and the formulation of recommendations to resolve such interference.

The ITU procedures for obtaining international rights to an orbital position are three-part: advance publication, coordination, and official notification and registration.

Advance publication is the procedure by which a country gives notice that it intends to establish a satellite at given frequencies and orbital location. The notice is to be given not earlier than five years and preferably not later than two years before the date of bringing the satellite into service. The IFRB publishes this information in its weekly circular. Any country that believes the new system will create unacceptable interference to its existing or planned satellites may file comments with the filing country. The two parties endeavor to resolve any difficulties and may seek the assistance of the IFRB in resolving the difficulties.

The *coordination* process is similar to advance publication except that more detailed information is required by the IFRB. A country effecting coordination must notify each country which has an assignment for a satellite or earth station that might be affected by the proposed satellite. Again a four-month period applies for comment, and the IFRB can supply assistance in coordination where difficulties arise.

The last step in the process is *official notification and registration*. The notification must be made not earlier than three years and not later than three months before the assignment is to be made. Upon the favorable finding of the IFRB, the frequency assignment is entered into the Master Register.

The rules relating to the IFRB are contained in the ITU *Radio Regulation*. Copies can be obtained from International Telecommunication Union, General Secreteriat, Sales Section, Place des Nations, CH-1211, Geneva 20, Switzerland.

SATELLITE LOCATION

4. The Orbital Arc

A communications satellite operates at a geosynchronous orbit which is 22,300 mi above the earth's equator. Satellites in each frequency band operate at the same frequency so that they must be spaced far enough apart to reduce interference to an acceptable level. At 22,300 mi, 1° of separation between satellites corresponds to 450 mi. For U.S. satellites, the FCC has established 2° spacing between satellites as its objective and is moving to a uniform 2° spacing environment at both 4/6 GHz (C band) and 12/14 GHz (K_u band).

5. Orbital Assignments

The present orbital assignments are shown on Table 6.1. Some of these satellites have spacings greater than 2° as a result of initial orbital assignments in the 1970s when greater separations were thought necessary. Figure 6.1 shows a map of the United States and various west longitude positions.

Orbital locations between approximately 60 and 135° west longitude (WL) provide an earth-station elevation angle of at least 5° throughout the contiguous United States. Generally 5° is the minimum acceptable elevation angle. Figure 6.2 shows how the angle of elevation is measured and Figure 6.3 shows the contour of the 5° elevation angle for satellites at various orbital locations. Orbital locations between approximately 60 and 135° WL provide a minimum 10° earth-station elevation angle throughout the same area. This higher elevation angle is desirable at 12/14 GHz to reduce the impact of increased propagation impairments at the higher frequencies.

The FCC has assigned orbital locations between 62 and 146° WL at 4/6 GHz and between 62 and 136° WL at 12/14 GHz. In the central portion of this arc, agreements have been reached among Canada, Mexico, and the United States so that orbital locations between 104.5 and 117.5° WL are used by Canadian and Mexican satellites in both the 4/6- and 12/14-GHz bands. However, the orbital positions at 105 and 103° WL are available for U.S. satellites in the 12/14-GHz bands. Generally the FCC will not assign new contiguous orbital locations to a single operator in the more desirable portions of the orbital arc. All orbital assignments by the FCC are temporary and are subject to change by summary order of the FCC on 30 days' notice.

LICENSING POLICIES AND PROCEDURES

6. Application of Authority

In order to submit an acceptable application for authority to construct and operate a domestic satellite, a would-be applicant must satisfy several requirements. First, the application must contain certain information. This information is described in detail in Appendix B to the FCC's Memorandum Opinion and Order, *Filing of Applications for New Space Stations in the Domestic Fixed-Satellite Service,* 93 FCC 2d 1260 (1983).

The required information includes a general description of the proposed satellite system including terrestrial facilities and a detailed description of the technical aspects of the planned satellite including frequency plan, reliability, redundancy, weight, power, energy dispersal, and the like. An applicant must also describe the types of services it plans to provide, the estimated demand for those services, and the technical transmission characteristics of each service. The applicant must state what launch arrangements it plans to make, describe its tracking, telemetry, and control proposals, and explain what orbital positions it seeks for its satellites. A detailed schedule of estimated costs must be shown as well as information on the financial qualifications of the applicant. Finally, the applicant should state the public interest considerations that support a grant of its application.

TABLE 6.1   Orbital Assignments

Location, west longitude	Satellite	Company*	Band, GHz
146°	Unassigned		4/6
144°	Unassigned		4/6
142°	Aurora-1	Alascom	4/6
140°	Galaxy-4	Hughes	4/6
138°	Satcom-1R	GE Americom	4/6
136°	Spacenet-4/GSTAR-3	GTE Spacenet	4/6/12/14
134°	Unassigned		4/6
134°	Unassigned		12/14
132°	Galaxy-1	Hughes	4/6
132°	Westar-B*	WU	12/14
130°	Satcom-3R	GE Americom	4/6
130°	Galaxy-B	Hughes	12/14
128°	ASC-2	ASC	4/6/12/14
126°	Telstar-1	AT&T	4/6
126°	Unassigned		12/14
124°	Westar-5	WU	4/6
124°	Fedex-B	Fedex	12/14
122°	Unassigned		4/6
122°	SBS-5	SBS	12/14
120°	Spacenet-1	GTE Spacenet	4/6/12/14
105°	GSTAR	GTE Spacenet	12/14
103°	GSTAR	GTE Spacenet	12/14
101°	Unassigned		4/6/12/14
99°	Westar-4	WU	4/6
99°	SBS-1	SBS	12/14
97°	Telstar-2	AT&T	4/6
97°	SBS-2	SBS	12/14
95°	Galaxy-3	Hughes	4/6
95°	SBS-3	SBS	12/14
93°	Unassigned		4/6/12/14
91°	Westar-65	WU	4/6
91°	SBS-4	SBS	12/14
89°	Unassigned		4/6
89°	Unassigned		12/14
87°	Spacenet-3	GTE Spacenet	4/6/12/14
85°	Telstar	AT&T	4/6
85°	GE-A	GE Americom	12/14
83°	ASC-2	ASC	4/6/12/14
81°	Satcom-4	GE Americom	4/6
81°	GE-B	GE Americom	12/14
79°	Westar-3	WU	4/6
79°	Unassigned		12/14
77°	Fedex-A	Fedex	12/14
76°	Comstar	Comsat Gen.	4/6
75°	Unassigned		12/14
74°	Galaxy-2	Hughes	4/6
73°	Westar-A*	WU	12/14
72°	Satcom-2R	GE Americom	4/6
71°	Galaxy-A	Hughes	12/14
69°	Spacenet-2	GTE Spacenet	4/6/12/14
67°	Unassigned		4/6
67°	GE-C	GE Americom	12/14

TABLE 6.1 Orbital Assignments (*Continued*)

Location, west longitude	Satellite	Company*	Band, GHz
64°	ASC-3/4	ASC	4/6/12/14
62°	Unassigned		4/6
62°	SBS-6	SBS	12/14

*Company: Alascom, Inc. (Alascom); American Satellite Corp. (ASC); American Telephone & Telegraph Company (AT&T); Comsat General (Comsat Gen.); Federal Express Corporation (Fedex); GTE Satellite Corporation (GTE Sat.); GTE Spacenet Corporation (GTE Spacenet); Hughes Communications Galaxy, Inc. (Hughes); GE American Communications, Inc. (GE Americom); Satellite Business Systems (SBS); Western Union Telegraph Company (WU).

7. Technical Standards

The applicant's proposed satellite system must also meet certain technical standards specified by the FCC. These include:

- The satellite must employ full frequency reuse of the 500 MHz of available spectrum in whatever band or bands the satellite will operate.
- The applicant must show that the operation of its proposed satellite is technically compatible with 2° spacing between adjacent satellites. This should be done through submission of an interference analysis.
- In C band, the power flux density on the ground produced by the satellite must not exceed certain values (-142 dBW/m^2 to -152 dBW/m^2 in any 4-kHz band) specified in §25.208 of the FCC *Rules*. The comparable limit for K_u band

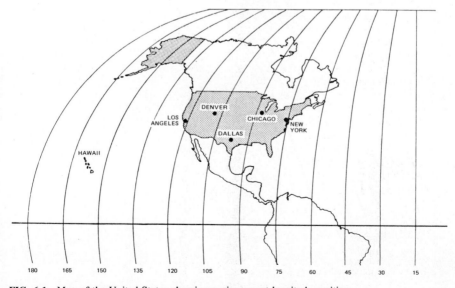

FIG. 6.1 Map of the United States showing various west longitude positions.

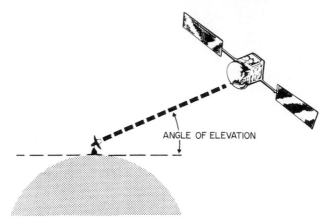

FIG. 6.2   Angle of elevation.

is -105 dBW/m^2. Additionally, there is a power (EIRP) limit of 53 dBw per television channel used for direct-broadcast-type services in the K$_u$ band.

There are also certain qualifications that the applicant itself must meet in order to be eligible for a satellite license. Thus, a station license cannot be granted to an alien, a foreign government, a corporation organized under the laws of a for-

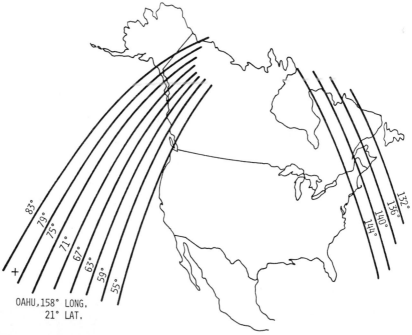

FIG. 6.3   Contours of 5° elevation angle.

eign government, or a corporation which has an alien as a director or officer. There are also restrictions on the amount of stock of a licensee corporation that can be owned or controlled by non-U.S. citizens (generally 20 to 25 percent). See §21.4 of the FCC *Rules*.

8. Hybrid Satellites

Most communications satellites operate in either the 4/6- or 12/14-GHz band. *Hybrid satellites* combine the ability to operate in both bands on a single satellite. Hybrid satellites may offer cost efficiencies to the satellite operator but pose significant difficulties in making orbital assignments. These difficulties were considerable when orbital spacings at 4/6 GHz were generally 4° and at 12/14 GHz were 3°. In this situation the two bands were coincident only every 12°. Today, with uniform 2° separations being implemented for both bands, potentially every satellite location is suitable for hybrid satellites. However, the FCC has identified as a problem that the orbital location assigned to a hybrid satellite must be compatible with the space-station polarization at 4/6 GHz. Since most existing hybrid satellites occupy the central portion of the arc and are designed with horizontal polarization at 4/6 GHz, new hybrid satellites will have to have vertical polarization if they are to be located in other than the edges of the orbital arc.

New hybrid satellites must have full frequency reuse in each band. This means that a hybrid satellite must be capable of providing at least the same transmission capacity as a 4/6-GHz satellite with 24 transponders with 36-MHz bandwidths and 8-W solid-state amplifiers and as a 12/14-GHz satellite with 20 transponders with 48-MHz bandwidth and 20-W amplifiers.

9. Common-Carrier versus Non-Common-Carrier Operation

Although the FCC's initial domestic satellite decision did not require domestic satellites to be operated solely on a common-carrier basis, all of the first-generation satellites were used only for common-carrier services. In 1981, several licensees proposed to sell transponders as opposed to providing them pursuant to tariff. The FCC has permitted such operations on the basis that they enable stable, long-term contractual relationships with individual customers which are provided with technically and operationally distinct portions of the satellite.

Applications to provide transponders on a non-common-carrier basis may be part of the original satellite application or part of a subsequent application for modification of license. These applications must contain the information specified by the FCC. The FCC has said that it will scrutinize the applications on a case-by-case basis to ensure that the grant of the applications does not artificially restrict the availability of domestic satellite communications services to the public.

SATELLITE AUTHORIZATIONS

10. Conditions

The FCC imposes a number of conditions when it grants authorization for the construction of new satellites. The two most significant conditions relate to timetables for construction and to maintenance of fill on in-orbit satellites.

As to the first item, the FCC requires that the construction of a satellite begin by a specified date, be completed by a specified date, and be launched and placed into service by a certain date. These dates have generally been those proposed by the applicant. Delays in commencement and completion of construction and launch activities beyond the specified dates may render the orbital assignment null and void unless FCC consent is obtained.

With respect to the second item, the FCC requires that in-orbit satellites maintain reasonable fill as represented in the applications. The availability of in-orbit spare capacity will be limited to reasonable amounts. The upper limit on in-orbit spare capacity for any licensee will be set at the equivalent of one spare satellite used for occasional or preemptable services within the system. If an operator fails at any time to demonstrate that its representations in its launch applications regarding traffic fill have been achieved, the FCC will consider the unfilled satellite as an excess spare, and this will result in the termination of the orbital assignment.

11. Launch Services Procurement

The FCC has eliminated its requirement for its approval prior to the procurement of launch services on the basis that such procurement generally involves refundable progress payments over an extended period of time and that there seems little likelihood that these expenditures would in any way influence the FCC's consideration of the underlying satellite application since unrecoverable expenditures tend to be a small percentage of the total cost.

12. Reporting Requirements

The authorizations for construction of domestic satellites contain a condition which requires the operator to file semiannual reports on the progress of construction and current satellite status and loading. The licensees must also submit measured spacecraft antenna-gain patterns and polarization vector directions (transmit and receive) obtained during construction and measured changes to space-station EIRP and sensitivity during its lifetime in orbit. Transponder-to-transponder variations are to be identified. Measured data will be required with respect to the capabilities of the satellite to operate at 2° orbital separations. Finally, a current rf carrier frequency plan for each satellite, providing the basic rf characteristics of each carrier, should be maintained on file with the FCC.

EARTH STATIONS

13. Receive-Only Stations

In 1979 the FCC instituted a policy of nonmandatory licensing of receive-only earth stations. This procedure permits the installation and operation of receive-only earth stations without any filings with the FCC. However, if a receive-only earth station is not licensed, no protection from interference due to existing or planned terrestrial systems is afforded.

14. Antenna Requirements

In 1983 the FCC amended its rules and regulations pertaining to the performance standards of licensed antennas. The thrust of these rule changes was to require receive antennas be built so as to better disregard unwanted signals from satellites adjacent to the satellite from which the signal was being received and for transmit antennas to better focus the signals on the intended satellite. The rules change provided that receive-only antennas did not have to conform with the higher performance standards generally made applicable. With respect to transmit-receive antennas, those antennas installed after July 1, 1985 were required to comply with the new technical standards. Antennas installed before that date were to be upgraded or replaced by January 1, 1987. The FCC has permitted the operation of nonconforming antennas existing prior to February 15, 1985 without modification upon a showing that the operation conforms with 2° spacing interference objectives or will not cause unacceptable interference with adjacent satellites.

15. FCC *Rules and Regulations*

The FCC's *Rules and Regulations* with respect to earth stations are contained in Part 25 of Chap. 1 of Title 47 of the Code of Federal Regulations.

CHAPTER 7

WIRE AND CABLE TRANSMISSION SYSTEMS

Thomas J. Aprille
Supervisor, AT&T Bell Laboratories

CONTENTS

INTRODUCTION

In this chapter, transmission systems that use the metallic wire medium will be discussed. Emphasis will be placed on data transmission and upon systems that are telephone network compatible. Facility termination electronics will be discussed to the extent needed to explain the associated transmission technique. Resulting network services available from the discussed systems are found in FCC tariffs for interstate (for example, AT&T FCC tariffs 9, 10, and 11) and state tariffs for intrastate.

The following sections (1 through 4) describe *voice-frequency* (VF) transmission over multipair sheathed metallic cables and voiceband data transmission over the same facilities. Following this, sections 5 through 13 treat baseband digital signal transmission, again using multipair metallic cables.

VOICE-FREQUENCY TRANSMISSION

1. Introduction

Even with the tremendous increase in digital data, analog VF and voiceband data transmission are still the predominant function of a telecommunications network. Likewise, even with the tremendous increase in radio and fiber, access to the network is still predominantly over metallic facilities.

A VF connection in the public network consists of a metallic connection starting at the carbon-block lightning protector network interface on the customer premises; customer-owned wiring completes the connections to the customer's telephone or PBX. The metallic connection then joins other such connections forming small pair-count distribution cable. Distribution cables join other such cables to become large pair-count feeder cable; this cable enters the *central office* (CO) and terminates on the main distribution frame. From the distribution frame, the metallic pair terminates on a switch (switched message service) or leaves the office for another customer premises (private line special service) or CO (foreign exchange or private line special service).

2. Metallic Cables

Telecommunication network distribution is realized with multiple twisted wire pair cable with final connections to individual subscribers implemented with drop-wire cable at splice connection points. Modern multiple metallic twisted wire pair cables are made with unitized cross sections as indicated in Fig. 7.1. Commonly, each unit will contain 12, 13, 25, 50, or 100 twisted wire pairs depending on cable size and number of units implemented. Pairs within a unit will stay in that unit the whole length of the cable; units within the cable will maintain their relative positions the whole length of the cable. Different pairs within the unit will have different, but constant, twist rates; this is done to minimize crosstalk between pairs. Some cables separate the units into two groupings by a metal shield; this screened cable greatly decreases crosstalk between pairs used for carrier systems. Most cables contain at least one small-sized unit called the service pair unit. This unit contains extra pairs for the purposes of substitution of defective pairs, order-wire circuit, pressure monitoring, and, in digital carrier systems, for fault-location pairs.

Cables are further classified by the gauge of wire used and by the method of insulating the wire pairs. The most commonly used gauges are 19, 22, 24, and 26. Common wire insulation is paper, pulp, and plastic. Plastic cables are often further delineated: PIC for *polyethylene-insulated conductor* cable, MAT for *metropolitan area trunk* cable, and LOCAP for *low-capacitance* cable.

Wire pairs in the cable have various transmission characteristics, with resistance and capacitance (between the wires in the pair) being the most important. Cable types are often characterized as standard capacitance cables (0.083 μF/mi) or low-capacitance cable (e.g., MAT and LOCAP cables). Further, all primary characteristics of the wire pair in the cable are a function of frequency and temperature.

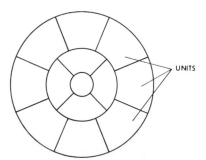

FIG. 7.1 Typical cable cross section.

3. Voice-Frequency Metallic Cable Engineering

In today's telecommunications plant, metallic interconnection is done on a two-
or four-wire basis. *Two-wire* is one twisted pair that carries transmission in both
directions; *four-wire* is two pair, one pair for each direction of transmission. Two-
wire operation is most often found in loop-plant applications (CO to customer
premises connections) and four-wire in trunking applications (CO-to-CO connections). An example of a metallic-only circuit connection is illustrated in Fig. 7.2.
There are several methods of engineering a two-wire metallic line to obtain
acceptable performance for VF application, e.g., resistance, unigauge, and long-
route designs. The resistance design has the simplest rules and is summarized as
follows. The loop resistance must not exceed 1500 Ω. If the total loop length exceeds 18,000 ft, the line must be loaded using the H88 loading scheme. Bridge tap
on nonloaded loops must not exceed 6000 ft. For loaded loops, no bridge tap
should exist between loading coils and any end bridge tap should not be loaded.
The *H* designates 6000-ft load-coil spacing and the *88* designates 88-mH load-coil
values; the first loading coil out of the CO is placed at 3000 ft and the last section
to the customer premises termination must be between 3000 and 12,000 ft. An
illustrative summary is given in Fig. 7.3.
Loading is necessary to correct excessive loss deviation with frequency that
results from long transmission lines. The loaded line has a loss in the VF band,
300 to 3000 Hz, that is reasonably flat. Since, with loading, the loading inductor
and cable capacitance dominate, the cable cutoff frequency becomes
$f_c = 1/\pi\sqrt{LC}$. With $L = 88$ mH and standard capacitance cable $C = 0.083$ μF,
then $f_c = 3724$ Hz.
Many telephone companies are now trying to avoid the installation of loaded
metallic loop lines. Where appropriate, loop carrier systems can be installed with
about the same amount of difficulty but with the advantage of multiple-channel
capability (see *T1 Carrier,* below).
For loops with over 1500 Ω resistance, special treatment, like range extension,
is required. This is needed, for example, because nominal office battery, -48 V,
is not sufficient to properly bias the carbon microphone in the station set. For
very long loaded lines, the insertion loss will become excessive and not allow
grade-of-service objectives to be met. In these circumstances, VF flat gain and
possibly further equalization must be added; this is usually done at the CO end of
the connection.

4. Voiceband Data Engineering

Data can be sent over voice-grade switched circuits at rates up to 4.8 kbs with
acceptable performance. Modems exist that convert digital data signals to VF

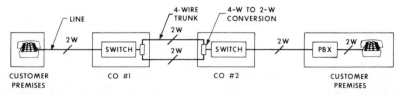

FIG. 7.2 Two-wire and four-wire circuits.

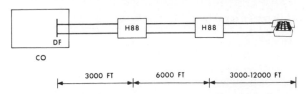

FIG. 7.3 H88 inductive loading.

band AM, quadrature AM, PSK, and FSK signals. Although new techniques may change the upper limit, anything above 4.8 kbs usually requires a private-line connection. In a switched connection, call-by-call facility variations make acceptable and predictable data performance difficult.

Switched data at rates up to 300 b/s requires little or no treatment of the metallic access facility. At higher rates, minimization of the line parameter variations with frequency is required to guarantee an acceptable call-by-call data performance. For voiceband data access, the metallic facility from the connecting block on the customer premises to the switch side connection at the CO main distribution frame must be engineered for a 9.0-dB insertion loss at 1 kHz, an envelope delay distortion of no more than 100 μs from 1.0 to 2.4 kHz, and an impulse noise requirement of no more than 15 counts above 59 dBrnc0 in any 15-min period. Line parameter treatment is administered with appropriate electronics placed at the CO appearance of the metallic circuit using a variety of available equipment for such purposes.

Nonswitched private line VF circuits can also be used for voiceband data applications. Interstate offerings exist under FCC tariffs and similar intrastate state tariffs also exist. Private-line facilities are cost-effective and should be used when there is high-usage point-to-point data transmission, minimal setup time is desired, or blocking is a major problem. Also, when operating at a high data rate, private-line facilities are a solution to switched circuit transmission variations and they allow for control of transmission characteristics via optional line-conditioning alternatives.

Services can be obtained that allow two-point or multipoint and simplex, half-duplex, or full-duplex operation. Circuits are engineered as two- or four-wire, depending upon the desired service, e.g., two-wire for simplex broadcast and four-wire for large multipoint circuits. Multipoint circuits can be two-wire by using conference bridges at junction nodes or four-wire by using hybrid splitters in master-slave arrangements. In general, four-wire circuits are implemented for two-point circuits when the terminating modems use the same transmit and receive frequency band and when turnaround time must be minimal. In multipoint circuits, four-wire must be implemented when two-wire performance does not meet objectives, e.g., sing margin.

An advantage of a private-line circuit is the ability to order optimal channel conditioning. Available options allow for 2.4-kbs, 4.8-kbs (type-B1 conditioning), and 9.6-kbs (type-B2 conditioning) limited-distance modem channels and for more general two-point and multipoint modem channels (types C1, C2, C3, C4, C5, C7, C8, D1, D2, D3, and D5 conditioning). A description of the major type C options is given in Table 7.1. Type D conditioning can be used with type C conditioning on interstate channels to place more stringent requirements on intermodulation distortion and signal-to-noise ratio. Needed conditioning is dependent upon the speed and equalization capabilities of the data sets that termi-

TABLE 7.1 Conditioning Options

Type	Frequency, Hz	Attenuation, dB*	Frequency	Delay distortion, μs
Basic	500 to 2500	− 2 to 8	800 to 2600	1750
	300 to 3000	− 3 to 12		
C1	1000 to 2400	− 1 to 3	100 to 2400	1000
	300 to 2700	− 2 to 6	800 to 2600	1750
	2700 to 3000	− 3 to 12		
C2	500 to 2800	− 1 to 3	1000 to 2600	500
	300 to 3000	− 2 to 6	600 to 2600	1500
			500 to 2800	3000
C4	500 to 3000	− 2 to 3	1000 to 2600	300
	300 to 3200	− 2 to 6	800 to 2800	500
			600 to 3000	1500
			500 to 3000	3000
C5	500 to 2800	− 0.5 to 1.5	1000 to 2600	100
	300 to 3000	− 3 to 3	600 to 2600	300
			500 to 2800	600

*Attenuation figures are with respect to 1000-Hz attenuation.

nate the ends of the private-line facility. Line conditioning is, when possible, administered at facility end or interconnect points. Various electronic circuits are available for providing the necessary gain, amplitude, and delay equalization.

Since the conditioning is administered with a fixed load impedance, it is important that modems have terminating impedances of $600 \pm 60 \ \Omega$ from 300 to 3000 Hz. Further, the transmit drive level should have a 0 dBm average and the receive level should have a −16 dBm ± 1 dB average.

DIGITAL DATA TRANSMISSION

5. Introduction

Baseband digital transmission services, on an absolute basis, are the most rapidly increasing services in the telephone network. Although digital transmission is presently only a small portion of the total telephone business, its percentage with respect to analog transmission is and will increase greatly, especially with the introduction of new digital-based services and systems. In fact, future telephone sets will supply digitally encoded voice on the loop.

Digital data transmission is desired because of its higher rates and improved error rate performance over voiceband data, cheaper and simpler terminals, simpler engineering, and easier testing. Presently, the most popular network services are offered with the *digital data system* (DDS) and the T1 carrier (discussed in *T1 Carrier,* below). Another recent transmission technique is *time-compression multiplexing* (TCM). Finally, an innovative new system is the *digital subscriber line* (DSL) that will be used as the principle access mechanism for the *Integrated Services Digital Network* (ISDN).

6. Digital Data System

DDS is a collection of hardware that allows for end-to-end digital connectivity. The most common application of the hardware is for Dataphone® digital service; it has also found applications in other tariffed services as well as in private networks.

A simplified, in both equipment and function, application of readily available DDS hardware is depicted in Fig. 7.4. Shown there is an end-to-end digital connection from customer premises 1 to customer premises 2 via three COs; CO2 acts as a tandem office. The system can work end to end at four different speeds: 2.4, 4.8, 9.6, and 56 kbs. The connection from the *data service unit* (DSU) or *channel service unit* (CSU) to the *office channel unit data port* (OCUDP) is via a four-wire loop connection; each transmission direction is on a separate two-wire loop. The CSU functions as a balanced line driver and terminator. Further, the receive side has a line equalizer and level slicer circuitry; the transmit side has a signal-shaping and interference band attenuation filter. A maintenance signal loopback relay also exists that is actuated by reversing, from the OCUDP, the balance split-pair splice sealing current. A DSU implements the CSU function and contains additional circuitry to convert the CSU signals to standard EIA interfaces. Transmit timing must be synchronized to the received signal.

The OCUDP provides the CSU function, sealing current drive, maintenance signal generation, and DDS line format to *D channel bank* (D bank) line format signal conversion. A D bank time-division-multiplexes 24 individual VF or data-port circuits onto one T1 carrier line (a 24-channel system running at 1.544 Mbs). At CO2, the terminating D bank has a *DS-0 data port* (DS-0 DP) for the purpose of providing a cross-office connection. The DS-0 signal is a 64-kbs nonreturn-to-zero bipolar signal implemented with a four-wire connection. An illustration of the nonreturn-to-zero bipolar format is given in Fig. 7.5. The DS-0 DP is a DS-0 to D-bank signal converter.

The loop-plant terminations are the DSU or CSU and OCUDP. Although up to two line repeaters are allowed for the 56-kbs rate case, they are not often used; essentially, the loop plant is without electronics. The loop rates are 2.4, 4.8, 9.6, and 56 kbs, and, with the optional secondary channel capability, 3.2, 6.4, 12.8, and 72 kbs. The line format is bipolar as illustrated in Fig. 7.5.

Loops used for the DDS application must be nonloaded and have limited

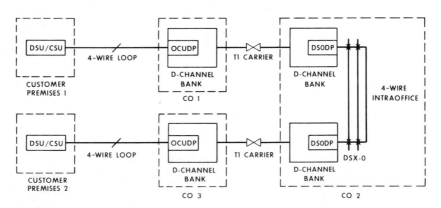

FIG. 7.4 Digital networking.

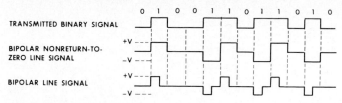

FIG. 7.5 Bipolar signal formats.

bridge tap. For the subrates (2.4, 4.8, 9.6, 3.2, 6.2, and 12.2 kbs) the sum of all bridge tap lengths must not exceed 6,000 ft. At the main rates (56 and 72 kbs), the sum of all bridge tap lengths must not exceed 2,500 ft and no single bridge tap can exceed 2,000 ft.

Line length is limited by the capabilities of the terminating circuits and by loop noise impairments. In general, the loop is engineered to a maximum line loss of 31.0 dB as measured at a frequency equal to one-half bit rate with 135-Ω terminations; the terminating circuits are designed to function a little beyond this range to allow for loop plant tolerance. For single-gauge loop plant, the maximum loop lengths are indicated in Table 7.2. The office connection from DS-0 DP to DS-0 DP is over office-grade cable, like ABAM, and is limited to 1500 ft.

7. Time-Compression Multiplexing

Loop digital transmission schemes that allow duplex operation on a two-wire medium are of great importance. Increased use of baseband digital transmission is forcing designers and suppliers of such systems to realize more economical digital access, i.e., two-wire systems rather than four-wire systems. Increased terminating electronics costs and difficult impairment issues limit how fast progress will take place in this two-wire direction.

One technique that is presently available in the *circuit switched digital capability* (CSDC) hardware is TCM. In this scheme, the desired digital signal at 56 kbs is upconverted to greater than twice its normal rate, i.e., greater than 112 kbs. This higher rate is transmitted on the two-wire line as a burst of bits. Thus, the line is utilized less than half the time. During the quiet time, the far-end 56-kbs signal is upconverted to twice the data rate and again transmitted in a burst on the same two-wire line. Overhead and control bits are needed with such a time-reversed channel scheme, and thus the actual line rate is greater than 2 × 56 kbs.

Figure 7.6 illustrates TCM in the CSDC system. The loop signal transmission is at 144 kbs. Engineering rules for a CSDC loop are dependent upon both VF and TCM constraints (this is because the CSDC system was designed to have an

TABLE 7.2 Maximum Loop Lengths

Wire gauge	Loop length, thousands of feet			
	2.4 kbs	4.8 kbs	9.6 kbs	56 kbs
19	114	86	67	41
22	73	56	43	24
24	56	42	32	17
26	42	32	25	13

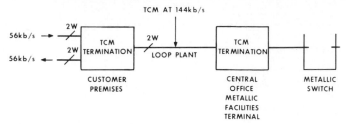

FIG. 7.6   CSDC using TCM transmission.

alternate voice-data transmission capability). Because of this, the loop loss must nominally be engineered for a maximum of 1100 Ω, be nonloaded and less than 15,000 ft in length, have a loss less than 40 dB when measured at 72 kbs, not have a bridge tap longer than 3000 ft, and not coexist in the same binder group as older carrier systems, e.g., SLC®-1, SLC®-8, or N carrier.

8.   Integrated Services Digital Network

ISDN will greatly accelerate digitization of the network. With ISDN, analog voice will be transmitted after conversion to a digital signal on the customer premises.

Access in ISDN is by a *digital subscriber loop* (DSL) that allows for two voice channels (or data channels) and one packet data and signaling channel. The DSL is realized on a two-wire loop plant as is indicated in Fig. 7.7.

The *network termination* (NT1) converts the two-wire loop plant signal, i.e., the U interface signal, to the four-wire customer premises signal, i.e., the T interface signal. The network terminates the DSL at the *loop termination* (LT); this termination will, in most cases, be integrated into an ISDN digital switch.

The bearer rate for DSL is 144 kbs, which allows for the transport of the 2B + D ISDN signal. Each B is a 64-kbs clear channel that can carry voice or data; the D is a 16-kbs channel that carries customer packet data and network signaling information. The excess line rate (160–144 kbs) is used for framing and line maintenance. For this application, nonloaded loops with losses in the 30- to 40-dB range will be possible.

T1 CARRIER

9.   Definitions and Applications

The 1.544 Mbs DS-1 and the 64 kbs DS-0 discussed in the previous section are components of the North American digital hierarchy. A full listing of these rates

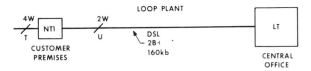

FIG. 7.7   ISDN DSL. B: 64 kbs; D: 16 bs.

TABLE 7.3 Digital Hierarchy

Level	North American designation	Line rate	CEPT line rate
0	DS-0	64 kbs	64 kbs
1	DS-1	1.544 mbs	2.048 mbs
1½	DS-1C	3.152 mbs	
2	DS-2	6.312 mbs	8.448 mbs
3	DS-3	44.736 mbs	34.368 mbs
4	DS-4	274.176 mbs	139.264 mbs
5			565.148 mbs

is given in Table 7.3 for both the North American and European (CEPT) hierarchies. DS-1 is by far the most important rate in public and private network use. The most ubiquitous bearer facility of DS-1 is the T1 carrier system.

The T1 digital carrier facility is a physical system used to transport DS-1 signals on two twisted pairs of metallic cable with interspersed repeaters, or digital regenerators, that compensate for cable loss. The DS-1 signal is electrically specified at the digital system crossconnect 1, DSX-1, delineated in Table 7.4. From this table, it is seen that the nominal T1 line rate is 1.544 Mbs. In a private network application, the full 1.544 Mbs can be used as the bearer rate with no limitations except those listed in Table 7.4. When interconnected with the public telecommunications network, two format options are allowed. The first is a nonchannelized option that is often used for end-customer point-to-point connections. This option allows only 1.536 Mbs throughput with the remaining 8 kbs constrained by public network equipment requirements. The second channelized option is selected when public network voice, data, or channelized services are desired. Each public network option is summarized in Fig. 7.8. For the channelized case each 64-kbs channel (8-bit × 8-kHz repetition rate) corresponds to PCM digitized voice or voiceband data or to digital data at rates 2.4, 4.8, 9.6, or 56 kbs.

Typical network applications of T1 digital carrier facilities are illustrated in Fig. 7.9. The initial and predominant use of T1 carriers is for trunking (the interconnection of VF circuits between switches) and is shown in Fig. 7.9 by the D channel bank and T-carrier connections to the trunk side of the office switches. Savings in copper pairs of 24 to 1 and improved transmission performance are possible with such arrangements. Similar cost savings in copper are also obtained with the application of T1 to the loop plant. This is illustrated with the subscriber loop carrier system in Fig. 7.9. Many private-line channelized services are also transported from office to office by D channel banks. For large-needs business customers, point-to-point high-capacity digital circuits can be provided by extending T1 facilities out to the customer premises.

10. T1 Components

A detailed application of the T1 carrier is depicted in Fig. 7.10. Terminating equipment can be primary multiplex (e.g., D channel banks), higher-order multiplex (e.g., M13 multiplex), or other non-T1 facilities (e.g., digital radio). The DSX-1 patch panel allows for DS-1 crossconnect capability and is the point of the standard DS-1 signal definition. The *office repeater* (OR) is the true start of theT1

TABLE 7.4   DSX-1 Signal Specification

Line rate	1.544 Mbs, $\pm$ 50 b/s
Line code	Bipolar with 12.5% average-ones density and no more than 15 consecutive zeros
Impedance	100-test loading
Isolated pulse shape	A normalized isolated pulse must fit within a template described by the following maximum and minimum straight-line curves:

Maximum curve		Minimum curve	
Time	Amplitude	Time	Amplitude
-0.77	0.05	-0.77	-0.05
-0.39	0.05	-0.23	-0.05
-0.27	0.8	-0.23	0.5
-0.27	1.15	-0.15	0.95
-0.12	1.15	0	0.95
0	1.05	0.15	0.9
0.27	1.05	0.23	0.5
0.35	-0.07	0.23	-0.45
0.93	0.05	0.46	-0.45
1.16	0.05	0.66	-0.2
		0.93	-0.05
		1.16	-0.05

	Time should be multiplied by the signal period in seconds of 1/1,544,000. The actual pulse center amplitude should be between 2.4 and 3.6 V before scaled to the above template.
Power constraints	Power P_{772} in the band 771 to 773 kHz should be greater than 12.6 dBm and less than 17.9 dBm for an all-ones signal. The power in the band 1543 to 1545 kHz should be 29 dB below P_{722} for an all-ones signal. There should be less than 0.5-dB difference between the power in the positive and negative pulses in an all-ones signal.

carrier facility; it contains the last receiving line repeater. Each line repeater is designed with two regenerators and is optimized for either outside aerial or buried-plant operation. The *distribution frame* (DF) is a hard-wired outside-plant cable-termination crossconnect.

A *span line* is defined as one CO-to-CO T1 repeated line, implementing both directions of transmission, starting and ending on the office-side input and output of the office repeaters. A collection of span lines between the same two central offices is defined as a span, or T1 span, between those offices. End-to-end connections formed by one or several tandem T1 span lines are described as T1 facilities.

Common office arrangement options are shown in Fig. 7.11 with equipment interconnection cable length limitations. As indicated, office repeaters are available with 3- and 6-V (peak voltage values) outputs.

The 6-V-output office repeaters are also equipped with an equalizer at their output that must be optioned, or selected, so that the cable (DSX-1 to OR) plus

ONE FRAME = 193 BITS AT 1.544 MB/S.

SEQUENTIAL FRAME RATE IS 8 kHZ.

BIT 1 OF EACH FRAME IS THE FRAMING BIT.

NONCHANNELIZED SIGNAL: BITS 2 THROUGH 193 ARE DATA.

CHANNELIZED SIGNAL: BITS 2 THROUGH 9, CHANNEL 1; BITS 10 THROUGH 17, CHANNEL 2; BITS 18 THROUGH 25, CHANNEL 3; ETC., TO BITS 186 THROUGH 193, CHANNEL 24.

REQUIRED FRAMING PATTERN:

    D4-TYPE, 12-FRAME, SUPERFRAME FORMAT
      FRAMING PATTERN = 1 0 0 0 1 1 0 1 1 1 0 0

      EXTENDED FRAMING (EF), 24-FRAME, SUPERFRAME FORMAT
      FRAMING PATTERN = D C D 0 D C D 0 D C D 1 D C D 0 D C
            D 1 D C D 1
            D = 4-kB/S DATA CHANNEL
            C = CRC-6 2-kB/S DATA VERIFICATION BITS

FURTHER DETAILS CAN BE OBTAINED IN BELL SYSTEM TECHNICAL REFERENCE PUB 62411.

FIG. 7.8 Bitstream formats.

equalizer has a flat 6-dB loss characteristic. Terminating equipment usually has a 6-V-output source with associated equalizer. The indicated 3-V office repeaters have an output signal that satisfies the Table 7.4 template and, thus, are allowed less interconnect distances. Selection of 6- or 3-V office repeaters and having or not having DSX-1 equipment is dependent upon office size; number of planned T1 systems; physical space availability; economics; operations, administration, and maintenance philosophies; and desired versatility. Note in Fig.

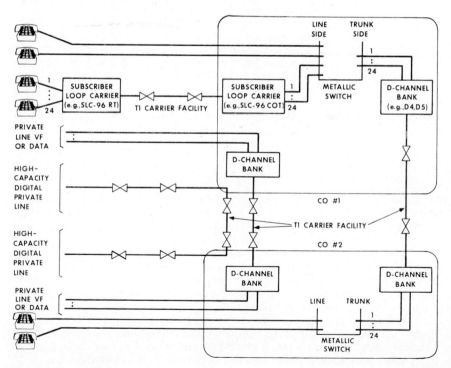

FIG. 7.9 T1 applications.

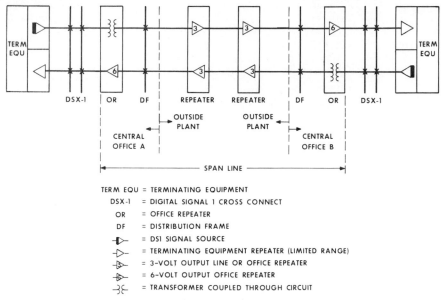

TERM EQU = TERMINATING EQUIPMENT
DSX-1   = DIGITAL SIGNAL 1 CROSS CONNECT
OR   = OFFICE REPEATER
DF   = DISTRIBUTION FRAME
-▷-   = DS1 SIGNAL SOURCE
-▷-   = TERMINATING EQUIPMENT REPEATER (LIMITED RANGE)
-▷-   = 3-VOLT OUTPUT LINE OR OFFICE REPEATER
-▷-   = 6-VOLT OUTPUT OFFICE REPEATER
-)(-   = TRANSFORMER COUPLED THROUGH CIRCUIT

FIG. 7.10   Typical T1 line.

7.11*b* that two span lines, 1 and 2, have been interconnected within one CO to result in a T1 through circuit in that office.

The arrangements indicated in Fig. 7.11 are commonly used in public network COs; they can also be implemented as private network arrangements. Where private and public networks interface with T1 facilities, the arrangement indicated in Fig. 7.12 is mandatory. In this figure, NCTEs are available with either 6- or 3-V terminating repeater capabilities and with circuitry required to satisfy FCC Part 68 network harms constraints and service-associated, tariff-defined capabilities.

11.   Digital Repeaters

A diagram showing the major signal processing blocks of typical line repeaters is given in Fig. 7.13. The 3-V repeater has a *line buildout* (LBO) network at its input, which is either fixed, and thus prescription selectable, or completely automatic (ALBO). The regenerator consists of a signal amplifier and level slicer with a buffering sample-and-hold output. Sampling times are obtained from the input signal via the timing recovery circuitry, usually consisting of a tuned LC network and buffer circuitry. The 6-V repeater has similar circuitry except that an additional prescription-selectable equalizer is required to generate a proper DS-1 signal at the DSX-1 crossconnect point.

Except for bridging applications, all line repeaters are equipped to handle two digital signals as indicated in Fig. 7.14. Bidirectional and unidirectional repeaters indicate opposite-direction signal processing and same-direction signal processing, respectively. Repeaters are available as bidirectional, unidirectional, or both. Outside plant repeaters are housed in apparatus cases; designs exist that allow for

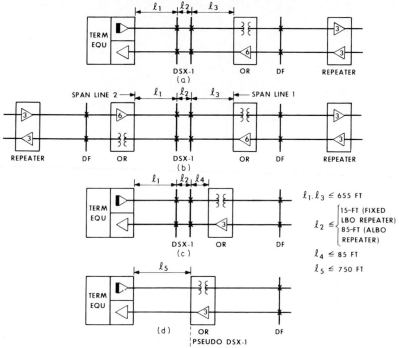

FIG. 7.11 Various office arrangements.

25 repeaters per case and for the less common pole-mounted 12 repeaters per case.

Office repeaters are different in that they only regenerate the receive signal as indicated in Fig. 7.15. The receive circuitry is very similar to the line regenerator circuitry with the equalizer implemented for 6-V ORs and not implemented for 3-V ORs. The transmit circuitry provides dc line isolation with no active gain or regeneration. Networks 1 and 2 (N1 and N2) are either pads or LBOs used to isolate the office repeater and terminating circuitry, respectively, from metallic line impedance discontinuities occurring up to 1000 ft from the CO. Such

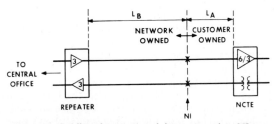

FIG. 7.12 Public-private network interconnection. NI: network interface; NCTE: network channel terminating equipment.

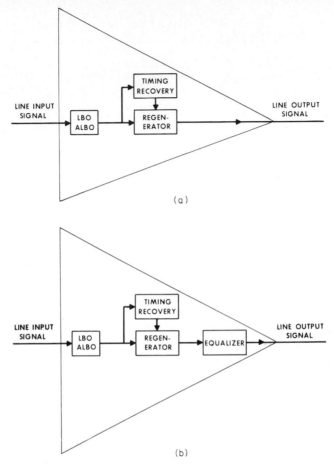

FIG. 7.13   Line regenerator block diagram. (*a*) 3-V regenerator; (*b*) 6-V regenerator.

discontinuities exist because of impedance mismatches created by office entrance cable or pressure plugs.

Networks N1 and N2 can be chosen as 3-dB pads to mitigate the impedance discontinuity problem. Additionally, for excessive office noise situations, N1 and N2 can be chosen as 7.5-dB LBO networks. When using a 6-V OR with 655-ft DSX-1 interconnect capability, the length to the first line repeater can be extended by choosing N1 as a 7.5-dB LBO and N2 as a 4.5-dB LBO. This can be done only when the first line repeater is an ALBO type.

Also shown in Fig. 7.15*b* is a bridging repeater. Such repeaters are office environment units used to bridge onto operating T1 lines to allow for spare line maintenance patching, signal monitoring, and testing. When used for monitoring, they must be isolated from the line by an external high-impedance pad.

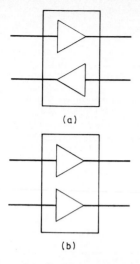

FIG. 7.14 Line repeater types. (*a*) bidirectional repeater; (*b*) unidirectional repeater.

12. Section Loss Engineering

Shown in Fig. 7.16 is a typical span line with repeater section losses explicitly noted. In the public network, T1 lines are engineered such that the end-to-end, or terminal-to-terminal, *bit error rate* (BER) does not exceed 1 bit error in 10^6 bits for more than 5 percent of all T1 systems. A 10^{-6} BER for T1 carrier carrying PCM VF and voiceband data channels is very acceptable; it also meets the objective for digital data channels. If the number of repeaters in a T1 system is N, then each repeater section loss is engineered such that the probability of the BER $\geq 10^{-6}$ is less than $0.05/N$. For network applications, the number N is nominally chosen at the design maximum of 50. The allowable loss value range between repeaters is a function of the repeater type, cable characteristics, and interfering noise characteristics.

Systems with greater than 50 repeaters are possible and do exist. But maintaining the end-to-end 10^{-6} BER requires more stringent engineering rules on repeater section loss and cable pair selection and also requires more complicated maintenance hardware.

There are a multitude of repeater types. One major type is the fixed LBO repeater which, from a loss point of view, will satisfy the BER objective when it operates with repeater section loss in the range of 31 ± 4 dB at 772 kHz. These repeaters will usually have a fixed LBO network before the regenerator input that is used to build out cable loss to the desired range. Another, and newer, type of repeater contains an ALBO that allows proper operation for signal losses ranging from 7.5 to 35 dB. To operate properly, the repeater must be further selected to match the cable being used.

Repeaters are also differentiated by application, as in one- or two-cable span lines and loop- or through-powering implementations. A one-cable span line has both directions of transmission in the same cable. A two-cable span separates the directions of transmission: one cable for one transmission direction and one cable for the opposite direction. Repeaters are further differentiated in their powering needs: 140- and 60-mA repeaters. For newer installations and customer-network interfaces (see Fig. 7.12), systems are being implemented with the low-power, 60-mA repeaters.

Finally, repeaters are still further differentiated as *protected* or *unprotected,* where *protected* implies enhanced surge and lightning immunity characteristics. Additional circuitry for protected repeaters results in cable loss values in a repeater section being reduced, additively, by 1.0 dB if driven by a protected repeater and 0.5 dB if terminated by a protected repeater.

The maximum cable loss that can be realized on a repeater section is determined by several factors. To be considered first is a limitation due to the 35-dB maximum capability of the repeater and to cable-type manufacturing tolerance and temperature characteristics. For a maximum operating temperature of 100°F, a representative list of maximum design losses and resulting maximum cable lengths is given in Table 7.5. These maximum cable loss values must be reduced by any fixed losses associated with repeater inputs, by office wiring losses asso-

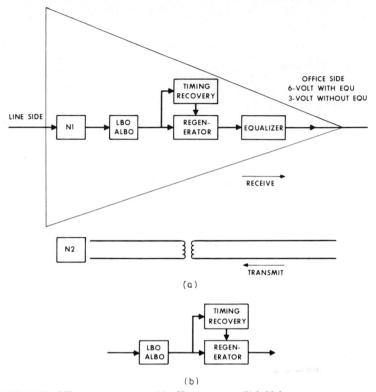

(a)

(b)

FIG. 7.15 Office repeater types. (a) office repeater; (b) bridging repeater.

ciated with end sections, by protected-repeater losses, and by multiple-cable junction losses.

Another maximum section cable loss figure is obtained when consideration is given to interfering noise effects. For all that follows, it is assumed that the repeaters of all span lines are collocated. For multiple span lines without collocated repeaters, the calculation of noise interference from adjacent systems becomes very complex. The first effect considered is *near-end crosstalk* (NEXT) interference due to bidirectional cable operation with a multiple line span. Equation 7.1 can be used to calculate, with 6-dB margin, the NEXT-limited maximum section loss value for a maximum cable operating temperature of 100°F:

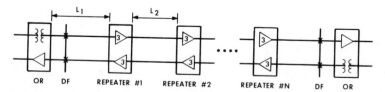

FIG. 7.16 T1 span line.

TABLE 7.5 Maximum Design Characteristics (100°F Maximum)

Cable type	Gauge	Capacitance, $\mu F/m$	Maximum design loss, dB	Maximum section length, thousands of feet
Pulp	19	0.083	31.4	8.1
Pulp	22	0.083	32.3	6.2
Pulp	24	0.083	32.2	5.1
Pulp	26	0.083	32.7	4.2
PIC	19	0.083	31.7	9.6
PIC	22	0.083	31.7	6.9
PIC	24	0.083	31.8	5.5
PIC	26	0.083	31.8	4.3
ICOT	24	0.060	31.5	8.0
MAT	25	0.064	31.3	6.1

$$L_{\text{NEXT}} = (m - \sigma - 32 - 10 \log n)/1.055 \qquad (7.1)$$

In Eq. 7.1, L is the NEXT-limited section loss in decibels, m is the mean, and σ is the standard deviation of the 772-kHz pair-to-pair coupling loss in the cable, and n is the total number of span lines.

It can be verified from this table that transmit and receive pairs are best separated as much as possible; the preferable order is nonadjacent units, adjacent units, and, lastly, same unit separation. The value of n should be chosen as the maximum number of lines including anticipated growth.

Representative values for $m - \sigma$ are found in Table 7.6. The NEXT crosstalk limit is mainly a problem for nonscreened one-cable spans. For equivalent performance from a 200-repeater T1 system, the numerical value 32 in Eq. 7.1 must be increased by 5 dB to 37.

For two-cable operation, NEXT is rarely a problem, but *far-end crosstalk* (FEXT) may be. The calculation of FEXT within-unit interference is a complicated function of cable, repeater-case, and repeater-splice type and positioning. In most cases, the nominal situation is for the maximum value from Table 7.5 to be limiting. When designing for a small number of T1 span lines, two-cable operation is not recommended. This is due to two-cable layout complexities, difficulty in FEXT evaluation, and engineering difficulties in future modifications.

One other noise source predominates: CO switching noise. This noise results from in-office relay switching transients and from within-cable dc loop current switching. For this reason, the final section loss L_1 of Fig. 7.16, is usually limited by this interference. Since there is up to 655 ft of CO wiring to the DSX-1 signal point and since existing noncarrier cable facilities have load coil splices and manholes at distances about equal to the half-spaced T-carrier loss for the first out-of-office load coil placement, L_1 can safely and conveniently be chosen to equal one-half of the nominal L_2 value of Fig. 7.11. In fact, since facilities usually exist for feeder cable load coil placement (3000 ft out of the office and 6000 ft thereafter), some telephone companies choose to place repeaters at existing load coil sites.

For private-public network interconnection, the engineering rules are summarized by tariff. In Fig. 7.12, L_A is the cable loss at 772 kHz from the network

TABLE 7.6   NEXT-Limited Design Values

Cable type	Gauge	Number of pairs	Pairs per unit	Use	$m - \sigma$, dB
Pulp	22	All	50	Same unit	65
Pulp	22	All	50	Adjacent unit	80
Pulp	22	≥ 200	50	Nonadjacent unit	94
Pulp	24	≥ 900	50	Same unit	66
Pulp	24	≥ 900	50	Adjacent unit	81
Pulp	24	≥ 900	50	Nonadjacent unit	95
PIC	19	≤ 50	12, 13	Same unit	58
PIC	19	≤ 50	12, 13	Adjacent unit	67
PIC	19	≤ 50	12, 13	Nonadjacent unit	76
PIC	19	300	12–25	Same unit	60
PIC	19	300	12–25	Adjacent unit	67
PIC	19	300	12–25	Nonadjacent unit	86
PIC	22	≤ 100	8–13	Same unit	57
PIC	22	≤ 100	8–13	Adjacent unit	68
PIC	22	≤ 100	8–13	Nonadjacent unit	75
PIC	22	300	12–25	Same unit	63
PIC	22	300	12–25	Adjacent unit	71
PIC	22	300	12–25	Nonadjacent unit	87
PIC	22	600	12–50	Same unit	66
PIC	22	600	12–50	Adjacent unit	78
PIC	22	600	17–50	Nonadjacent unit	98
MAT	25	≥ 400	25	Same unit	62
MAT	25	≥ 400	25	Adjacent unit	66

interface (NI) to the input of the NCTE and L_B is the cable loss from the last network-owned repeater to the NI. The public network entity must engineer its portion of the T1 facility such that $0.0 \leq L_B \leq 22.0$ dB and is responsible for informing the private network entity of the actual value of L_B. The private network entity must then engineer and supply the cable from NI to NCTE with the following constraint

$$15 \text{ dB} \leq L_A + L_B + \text{LBO} \leq 22.5 \text{ dB} \qquad (7.2)$$

where LBO is any additional LBO inserted at the line-side input to and output of the NCTE.

13. Line Powering

Power supplies used to power the T1 span lines are located in the *office repeater bay* (ORB); this is the same bay used to mount the line-terminating office repeaters. Supplies are available at 48 and 130 V and can be combined to allow line driving voltages of 48, 130, 178, and 260 V. The office powering repeater, called a *regulating repeater,* has a constant-current source regulator that drives the line at 60 mA or, for older equipment, 140 mA. A line must be engineered with all repeaters of the same current operating type.

A *bidirectional line powering setup* is illustrated in Fig. 7.17. Current is supplied to the line in a balanced simplex fashion starting at the regulating office repeater and terminating at the looping line repeater. Repeater power is obtained

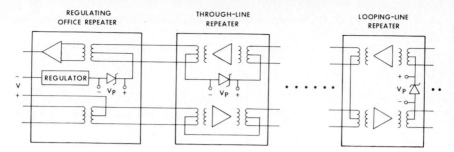

FIG. 7.17 T1 line powering.

from the dc balanced transformer isolation circuitry and is controlled by the zener diode. The zener diode breakdown voltage is 6.8 V for 60-mA repeaters and 10.9 V for 140-mA repeaters.

Beyond the looping line repeater could be another chain of through line repeaters that terminate at another regulating office repeater. This would imply that the span line is powered from both directions, as would be required when the number of repeaters and line resistance exceeded a single-office powering capability. Alternatively, the looping line repeater could be replaced by a terminating "looping" office repeater, thus allowing for the more desirable single-ended line powering.

The number of repeaters that can be powered in a span line is dependent upon the repeater power requirements, power loss due to metallic pair resistance, and to the maximum voltage available from the line power supplier. Using 60-mA, low-power repeaters and 260-V power sources at both ends of the span line, a maximum repeater count in the order of 33 is possible.

REFERENCES

14. General References

Aaron, M. R., "PCM Transmission in the Exchange Plant," *Bell System Tech. J.,* vol. 41, January 1962.

Ahamed, S. U., P. P. Bohn, and N. L. Gottfried, "A Tutorial on Two-Wire Digital Transmission in the Loop Plant," *IEEE Transactions on Communications,* vol. COM-29, no. 11, November 1981.

Aprille, T. J., S. Narayanan, P. G. St. Amand, and F. E. Weber, "The D4 Digital Channel Bank Family: Dataport Digital Access Through D4," *Bell System Tech. J.,* vol. 61, no. 9, part 3, November 1982.

Bell Telephone Laboratories Staff, *Engineering and Operations in the Bell System,* Western Electric Co., Inc., Winston-Salem, NC, 1977.

Bell Telephone Laboratories Staff, *Transmission System for Communication,* Western Electric Co., Inc., Winston-Salem, NC, 1982.

Bender, E. C., J. G. Kreuer, and W. J. Lawlers, "Digital Data System: Local Distribution Systems," *Bell System Tech. J.,* vol. 54, no. 5, May-June 1975.

Bosik, B. S., "Time Compression Multiplexing: Squeezing Digits Through Loops," *AT&T Bell Laboratories Record,* February 1984.

Brilliant, M. B., "Observations of Errors and Error Rates on T1 Digital Repeatered Lines," *Bell System Tech. J.,* vol. 57, no. 3, March 1978.

Bylanski, P., and D. G. W. Ingram, *Digital Transmission Systems,* Peter Peregrinus, Stevenage, England, 1976.

Cho, Y-S, J. W. Olson, and D. H. Williamson, "The D4 Digital Channel Bank Family: The SLC℠-96 System," *Bell System Tech. J.,* vol. 61, no. 9, part 3, November 1982.

Choratas, D. N., *The Handbook of Data Communications and Computer Networks,* Petrocelli Books, Princeton, N.J., 1985.

Clark, A. P., *Advanced Data-Transmission Systems,* John Wiley & Sons, New York, 1977.

Clark, A. P., *Principles of Digital Data Transmission,* John Wiley & Sons, New York, 1976.

Cravis, H., and T. V. Carter, "Engineering of T1 Carrier System Repeated Lines," *Bell System Tech. J.,* vol. 41, no. 1, March 1963.

Crue, C. R. et al., "The D4 Digital Channel Bank Family: The Channel Bank," *Bell System Tech. J.,* vol. 61, no. 9, Pt. 3, November 1982.

Decina, M., "Progress Towards User-Access Arrangements in Integrated Services Digital Network," *IEEE Transactions on Communications,* vol. COM-30, No. 9, September 1982.

Ebout, I. G., "The Evolution of Integrated Access Towards the ISDN," *IEEE Communications Magazine,* vol. 22, no. 4, April 1, 1984.

Horn, F. W., *Cables, Inside and Out,* Lee's ABC of the Telephone, Chicago, 1976.

Johnson, W. C., *Transmission Lines and Networks,* McGraw-Hill, New York, 1950.

Lathi, B. P., *Communications Systems,* John Wiley & Sons, New York, 1968.

Lucky, R. W., T. Salz, and E. J. Weldon, Jr., *Principles of Data Communication,* McGraw-Hill, New York, 1968.

Mahoney, J. J., Jr., J. J. Mansell, and R. G. Matlach, "Digital Data System: User's View of the Network," *Bell System Tech. J.,* vol. 54, no. 5, May-June 1975.

Manfred, M. T., G. A. Nelson, and C. H. Sharplers, "Loop Plant Electronics: Digital Loop Carrier Systems," *Bell System Tech. J.,* vol. 57, no. 4, April 1978.

Manhine, L. M., "Physical and Transmission Characteristics of Customer Loop Plant," *Bell System Tech. J.,* vol. 57, no. 1, January 1978.

McNamara, J. E., *Technical Aspects of Data Communication,* Digital Press, Bedford, Mass., 1982.

Murthy, B. R. N., "Crosstalk Loss Requirements for PCM Transmission," *IEEE Transactions on Communications,* vol. COM-24, no. 1, January 1976.

Nelson, K. C., *Understanding Station Carrier,* Lee's ABC of the Telephone, Chicago, 1975.

Riley, E. W., and V. E. Acuna, *Understanding Transmission,* Lee's ABC of the Telephone, Chicago, 1976.

Robin, G., "Customer Installations for the ISDN," *IEEE Communication Magazine,* vol. 22, no. 4, April 1984.

Sakrison, D. J., *Communication Theory,* John Wiley & Sons, New York, 1968.

Snow, N. E., and N. Knapp, Jr., "Digital Data System: System Overview," *Bell System Tech. J.,* vol. 54, no. 5, May-June 1975.

Viterbi, A. J. (ed.), *Advances in Communication Systems,* Academic Press, New York, 1975.

Connection of Equipment, System, and Protective Apparatus to the Telephone Network, Part 68 of FCC *Rules.*

"Data Communications Using Voiceband Private Line Channels," Bell System Technical Reference PUB 41004, AT&T, October 1973.

"High-Capacity Digital Service Channel Interface Specification," Bell System Technical Reference PUB 62411, AT&T, 1983.

Telecommunications Transmission Engineering: Volume 1, Principles, Western Electric, Winston-Salem, N.C., 1977.

Telecommunications Transmission Engineering: Volume 2, Facilities, Western Electric, Winston-Salem, N.C., 1977.

Telecommunications Transmission Engineering: Volume 3, Networks and Service, Western Electric, Winston-Salem, N.C., 1977.

CHAPTER 8
FIBER-OPTIC TRANSMISSION SYSTEMS

Ira Jacobs
Department of Electrical Engineering
Virginia Polytechnic Institute and State University
Formerly, AT&T Bell Laboratories

CONTENTS

HISTORICAL DEVELOPMENT

Light-guiding properties of *glass fibers* (filaments of glass with central core having higher index of refraction than surrounding cladding; light guided by total internal reflection at core-cladding boundary) have been known and used for many years. Typical absorption in glass (greater than 1 dB/m) and absence of suitable light sources restricted early applications to instrumentation using short fiber bundles. Key milestones in the evolution of *fiber-optic transmission systems* (transmission of information over extended distances) utilizing individual fibers are listed in Table 8.1.[1]

FIBER-OPTIC TECHNOLOGY

1. Transmission Media

Fiber. Principal fiber types used for transmission are *graded-index multimode* and *single-mode* fibers (Fig. 8.1). Principal fiber parameters, for transmission system design, are *attenuation* and *bandwidth*.

Progress in reducing attenuation is illustrated in Fig. 8.2. This progress is a result of improved fabrication and reductions in impurities.[10] Attenuations now being achieved (0.35 dB/km at 1.3 μm wavelength and 0.2 dB/km at 1.55 μm) are essentially at the theoretical limits achievable with silica-based glass.

Pulse spreading (dispersion) limits the maximum pulse rate or modulation

TABLE 8.1 Fiber-Optic System Evolution

1958	Invention of laser principle[2]
1960	First lasers
1962	First semiconductor lasers[3,4]
1966	Prediction of fiber telecommunications[5]
1970	First low-loss fiber[6]
	First semiconductor laser at room temperature[7]
1976-1977	System experiments and trials[8]
1979-1980	Beginning of commercial applications[9]

bandwidth that may be used with fibers. Dispersion τ (full-width half maximum) and bandwidth B for a gaussian pulse are related by

$$B = 0.44/\tau \tag{8.1}$$

Modal dispersion (in multimode fibers) is a result of different modes propagating with different delays. Bandwidth of multimode fibers, for a 1-km length, is typically 1 GHz or less. Bandwidth decreases with length to a power between 0.5 and 1.0. Modal dispersion bandwidth is wavelength-dependent (Fig. 8.3).

Chromatic dispersion (principal bandwidth limitation for single-mode fiber) results from the finite spectral width of the light source and the dependence of propagation delay on wavelength. It depends both on the material properties (material dispersion) and index of refraction profile of the fiber (waveguide disper-

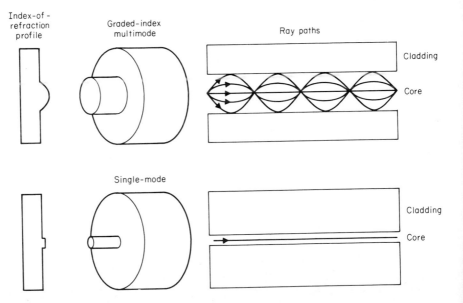

FIG. 8.1 Types of optical fiber. Typical single-mode fiber core diameters are 7 to 10 μm. Typical graded index multimode fiber core diameters are 50 to 85 μm. (A 50-μm core, 125 μm-o.d. is a common standard.) Larger core diameters are useful for short links to facilitate connectors and coupling from LED sources. (See Table 8.3.)

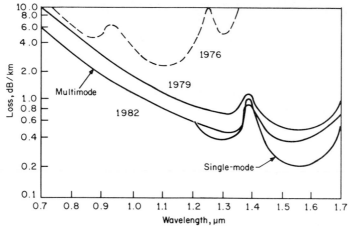

FIG. 8.2 Attenuation of an optical fiber as a function of wavelength. Loss decreases as $1/\lambda^4$ (Rayleigh Scattering). Lowest curve is single-mode fiber (single mode for wavelengths greater than 1.1 μm). Other curves are for multimode fiber. Loss peak at 1.4 μm is due to an OH-ion resonance. Increase in loss beyond 1.6 μm is due to silicon resonance absorption.

sion). Typical single-mode fibers have chromatic dispersion minimum at 1.3-μm wavelength. However, fibers may be designed (Fig. 8.4) having minimum dispersion at 1.55 μm (dispersion-shifted fiber) or having low dispersion over a wide wavelength region.

Note that modal dispersion is a property of the fiber. Chromatic dispersion depends not only on the fiber, but also on the optical source. (Limits on bit rate and repeater spacing caused by modal and chromatic dispersion are presented in Sec. 6.)

Other fiber parameters impacting system design are dimensional tolerances (important for splice loss and connectorization), strength (fibers are typically proof-tested at 50 to 100 kpsi*; high-strength fibers requiring special care in manufacture are made with strengths of 200 kpsi), and radiation effects (special composition fibers are required for application in high radiation environments).

Cable. The principal function of the cable structure is to provide mechanical protection to the fibers and to isolate them from external forces. Cable structures range from loose tube designs, with individually buffered fibers, to ribbon-structured cables in which multiple fibers are held between two strips of plastic. The latter configuration permits a higher packing density (larger number of fibers in a given diameter cable—e.g., a cable containing 144 fibers, 12 ribbons of 12 fibers each, is 12 mm in diameter) and facilitates multifiber splicing. Mechanical stresses may result in added attenuation (microbending loss), but with appropriate fiber and cable design, such loss may be held to less than 0.1 dB/km.

Splicing and Connectors. In addition to the attenuation of the fiber (and any added attenuation due to cabling), the attenuation due to splices and connectors

---

*The metric unit of strength is newtons per square meter, also known as pascals. Fiber strength is sometimes given in gigapascals (GPa); 1 GPa = 145 kpsi.

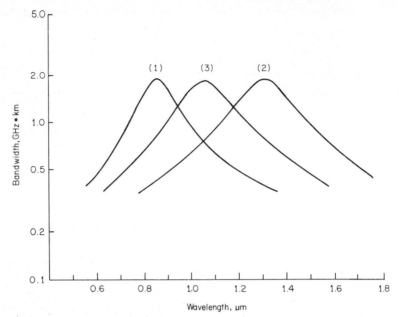

FIG. 8.3   Bandwidth of multimode fiber as a function of wavelength. (1) fiber profile optimized for 0.85 μm; (2) fiber profile optimized for 1.3 μm; (3) fiber profile to achieve wide bandwidth at 0.85 and 1.3 μm.

must be considered. This attenuation may be a significant part of the total link loss budget. Various factors affect splice loss, including misalignment of the fibers and different optical or dimensional properties of the fibers being spliced. Figure 8.5 indicates splice or connector loss as a function of lateral misalignment for typical multimode (50-μm core diameter) and single mode (8-μm core diameter) fiber.

Splices are made by fusing or bonding fibers together utilizing alignment fixtures. Demountable connectors utilize cylindrical ferrules or conic plugs in biconic receptacles (see Fig. 8.6) to achieve the appropriate centering. There is considerable variability in splice and connector losses associated with dimensional tolerances of the connector piece parts, dimensional tolerances of the fibers, fiber end preparation, cleanliness, etc. Connector losses somewhat below 1 dB are now being achieved with both multimode fiber and single-mode fiber. Splice losses below 0.2 dB are routinely achieved with multimode fiber. With appropriate care, similar results may be achieved with single-mode fiber. Indeed, single-mode adjustable splices used in conjunction with signal monitoring are now yielding losses of 0.05 dB.[11]

2. Transmitters

Principal system parameters for transmitters for fiber-optic systems are power coupled into the fibers, center wavelength, and spectral width. Typical parameters for semiconductor lasers and LEDs are summarized in Table 8.2.

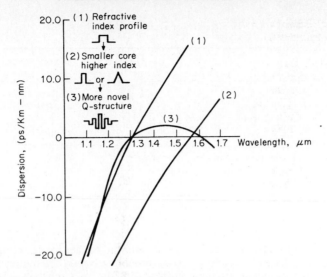

FIG. 8.4 Chromatic dispersion in single-mode fiber (picoseconds of dispersion per kilometer of length and per nanometer of source spectral width). (1) conventional single-mode fiber; (2) dispersion-shifted fiber; (3) wideband fiber (presently in research stage).

Lasers.   Light output versus current input for a typical semiconductor laser is shown in Fig. 8.7. Lasers are generally biased just below threshold with the modulation provided by a small drive current. Threshold varies with temperature and

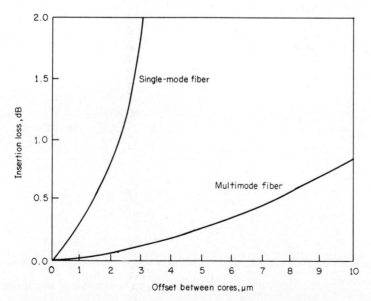

FIG. 8.5 Connector loss as function of lateral offset between fiber cores.

TABLE 8.2 Comparison of Lasers and LEDs
(Typical Values)

Parameter	Laser	High-radiance LED
Light output	6 dBm	6 dBm
Coupling loss	3 dB	20 dB*
Spectral width		
at 0.8 μm	2 nm	40 nm
at 1.3 μm	4 nm	100 nm
Temperature sensitivity	Strong	Weak
Feedback control	Yes	No
Failure mechanisms	Many	"Wear-out"

*Applicable to multimode fibers. Coupling loss from LEDs to single-mode fiber is considerably larger (see Table 8.3).

age of the laser. Laser output power is monitored (either by detecting the power from the back face of the laser, or by providing a tap on the output fiber) and a feedback control loop adjusts bias current to maintain a constant average output power from the laser.

Laser lifetime[12,13] has a log-normal distribution that is characterized by a median lifetime T_m and a variance σ. The *mean time before failure* (MTBF) is given by F_p

$$\text{MTBF} = T_m \exp(0.5\sigma^2) \tag{8.2}$$

For a log-normal distribution, the failure rate varies with time. Peak failure rate F_p is given by

$$\frac{\exp(\sigma^2)}{\sqrt{2\pi}\ \sigma}\left(\frac{1}{\text{MTBF}}\right) \tag{8.3}$$

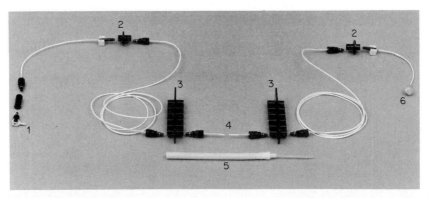

FIG. 8.6 Example of connectors in the optical path. (1) connectorized laser-transmitter package; (2) circuit-pack connector; (3) connectors at distribution panel; (4) patch cords used to provide interconnection; (5) ribbon-structured outside-plant cable (fan-out structure interfaces cable to distribution panel); (6) photodetector receiver with fiber pig-tail.

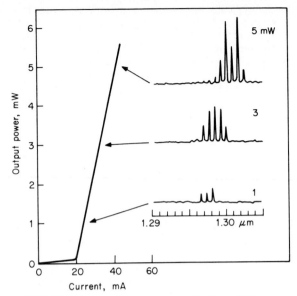

FIG. 8.7 Laser output power as a function of drive current (representative example). Spectra shown indicate that lasers are generally multifrequency and that the spectral distribution is a function of power.

and occurs at a time given by

$$T_p = (MTBF) \exp (- 3\sigma^2/2) \tag{8.4}$$

Typically $\sigma = 1$ to 2. T_m is a strong function of temperature, decreasing exponentially with increasing temperature. High-temperature burn-in and/or measurement of threshold current change with time is used to weed out potential early failures.[14] Thermoelectric coolers are often used to control the temperature of lasers.

Lasers are further characterized by their spatial and temporal modal characteristics. *Single-spatial-mode* (transverse-mode) lasers are used with single-mode fibers. Use of such lasers with multimode fibers can lead to modal noise resulting from spatial-mode conversions at discontinuities (such as connectors).

Semiconductor lasers typically emit a number of *discrete spectral lines* (temporal or longitudinal modes, Fig. 8.7). Mode partition noise may result from the dynamic redistribution of power among the laser modes coupling with the chromatic dispersion of the fiber[15] (see Sec. 5 for the effect of this noise).

To obtain a single-wavelength output [*single longitudinal mode* (SLM)] laser, either an external cavity or special resonant structures within the laser (e.g., distributed feedback or cleaved coupled-cavity structures) may be employed. SLM lasers are of particular interest at a wavelength of 1.55 μm, where conventional single-mode fibers do not have minimum dispersion, and narrower spectral width sources are required to avoid limitations imposed by chromatic dispersion and mode partition noise.

LEDs. Light-emitting diodes have a linear relation between light output and drive current, and are simpler and more reliable than lasers. They couple less power into the fiber and typically may not be modulated at as high rates as lasers. (Lasers may be modulated at rates of several gigabits per second. LEDs are typically limited to several hundred megabits per second.) Edge-emitting LEDs are intermediate in performance and complexity to surface-emitting LEDs and to lasers. Their utility in system applications is not yet clear. Representative powers coupled from LEDs into various fiber structures are summarized in Table 8.3.

3. Receivers

The key receiver parameter is *sensitivity*—viz., the minimum received power P_r required to achieve a given error probability P_e at a given bit rate B. Alternatively, sensitivity may be expressed as the minimum average number of photons $\bar{n}$ that must be received per pulse:

$$P_r = \bar{n}B\,hc/\lambda \tag{8.5}$$

where hc/λ is the photon energy (h is Planck's constant, 6.6252×10^{-34} J $\cdot$ s, c the velocity of light in vacuum, 3×10^8 m/s; and λ the wavelength of light).

Theoretical Sensitivity.   There is an inherent statistical fluctuation (Poisson statistics) in the number of photons within a pulse, and this results in

$$P_e = \tfrac{1}{2} \exp\left(-\bar{n}\right) \tag{8.6}$$

For example, $\bar{n} = 20$ to achieve $P_e = 10^{-9}$. For a 100-Mbs system (with binary on-off modulation), this corresponds to a minimum of 10^9 photons per second. If each photon generated one photoelectron, the resultant electrical current would be 1.6×10^{-10} amperes. Practical receivers require considerable amplification following the detector. (The electrical energy at the output of the detector is considerably less than the optical energy at the input.)

The *equivalent minimum noise spectral density* of receivers is given by

$$N_o = \frac{hc/\lambda}{1 - \exp\left(-\dfrac{hc}{\lambda}\dfrac{1}{kT}\right)} \tag{8.7}$$

where $k = 1.3804 \times 10^{-23}$ J/K is Boltzmann's constant and T is the noise temperature in degrees Kelvin. For conventional electronic systems, $hc/\lambda \ll kT$ and Eq. 8.7

TABLE 8.3   Representative Average Powers Coupled from LEDs into Fiber

	Single-mode fiber	Multimode fiber	
LED		50-μm core, NA* = 0.23	62.5-μm core, NA = 0.29
Surface emitter	− 34 dBm	− 16 dBm	− 12 dBm
Edge emitter	− 26 dBm	− 12 dBm	− 8 dBm

*NA (numerical aperture) is a measure of the range of angles over which light is accepted by the fiber.

reduces to the thermal noise limit $N_o = kT$. However, at optical wavelengths, $hc/\lambda \gg kT$ (e.g., at $\lambda = 1.3$ μm, $hc/k \approx 11{,}000$ K, which is much larger than the noise temperature of good microwave receivers) and Eq. 8.7 reduces to $N_o = hc/\lambda$ (quantum limit).

The quantum noise limit at optical wavelengths is much greater than the thermal noise limit of standard electronic systems. Thus, as a result of the large photon energy at optical wavelengths, fiber-optic systems require considerably more received optical power than the received electrical power required for low-noise microwave or baseband electronic systems.

PIN and APD Receivers. Amplification following the photodetector introduces additional noise. In PIN photodetectors all of the amplification is by external electronic amplifiers, although there is considerable research on integrating the detection and amplification functions (integrated optoelectronics). Typical PIN receivers require 20 to 30 dB more received signal power than the ideal quantum limit.

In an *avalanche photodetector* (APD), there is gain internal to the detector (photoelectrons and holes, accelerated by a high electric field, generate additional electrons and holes). Present APDs at 1.3-μm wavelength result in receivers 6 to 10 dB more sensitive than PIN receivers.

Sensitivities of APD and PIN receivers are compared with the quantum limit in Fig. 8.8.

Coherent Receivers. If the incoming optical signal is mixed with a coherent local oscillator then the output photocurrent is increased (relative to direct detection) by the ratio of the local oscillator to the incoming signal field strengths. The theoretical performance achievable with *coherent* (homodyne or heterodyne) detection is summarized in Table 8.4.[16] It is contemplated that these theoretical limits may be more closely approached with coherent than with direct-detection noncoherent systems, giving perhaps a 10-dB sensitivity advantage for coherent systems. However, this requires extremely stable sources (source spectral line width less than about one-thousandth of the modulation rate).

4. Passive Components

Processing of signals in optical form is still at a very early stage relative to electronic signal processing. This is an active area of current research.

An important subclass of optical processing is provided by passive components such as optical attenuators, splitters, and couplers, and *wavelength-division-multiplexing* (WDM) filters.

Optical Attenuators. Dynamic range limitations of optical receivers may necessitate use of optical attenuators in some systems. A variety of fixed or adjustable attenuators are available. These are often implemented within connector structures utilizing separation of the fiber ends or the addition of absorbing material between the two fibers to achieve the desired attenuation.

Couplers. Optical signals may be split between two (or more) paths utilizing Y junctions in fibers. These may be implemented by fusing fibers together, or, alternatively, utilizing the scattering loss occurring in the vicinity of a controlled bend in the fiber. These are illustrated schematically in Fig. 8.9. Such

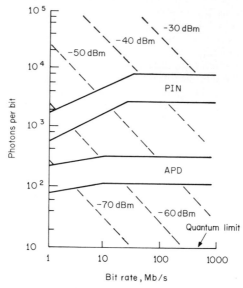

FIG. 8.8   Receiver Sensitivity. Photons per bit (required at the receiver) to achieve 10^{-9} bit error probability. Translation to received power level (in dBm) shown by dashed lines. APD performance band applies to silicon APDs at 0.85 μm wavelength. Present APDs at 1.3 or 1.55 μm do not achieve this performance and result in sensitivities about midway between the APD and PIN bands in this figure.

techniques are suitable for both single-mode and multimode structures. In the single-mode case, optical waveguides etched into planar substrates (monolithic integrated optics) provide considerably greater functionality without need for the precise positioning of discrete fiber piece parts. Electrooptic effects may be used to vary the coupling, thereby obtaining electronically controlled switches.

Wavelength-Division-Multiplexing (WDM) Filters.   Optical techniques may be used to combine (multiplex) and separate (demultiplex) different wavelengths of light so that several wavelengths may be used simultaneously on a single fiber. WDM filters suitable for multimode or single-mode fibers typically utilize thin-

TABLE 8.4   Average Number of Photons per Bit to Achieve BER $< 10^{-9}$ with Ideal Binary Receiver

	Receiver		
Modulation	Heterodyne	Homodyne	Direct detection
Phase-shift keying (PSK)	18	9	
Frequency-shift keying (FSK)	36		
On-off keying (ASK)	36	18	10

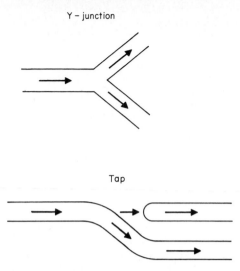

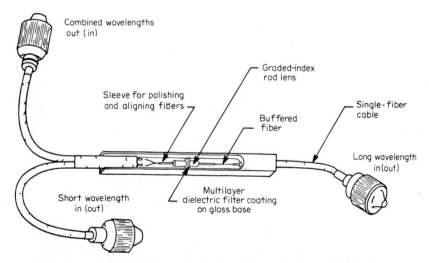

FIG. 8.9 Fiber couplers providing equal distribution of power between two paths (Y junction) and tap of small amount of power from primary path.

film dielectric filters which transmit one wavelength and reflect the second wavelength (see Fig. 8.10).[17] The insertion loss of such filters may be held to about 3 dB. Integrated optics techniques may be used to achieve greater wavelength selectivity and (using electrooptic materials) to achieve tunable structures.

FIG. 8.10 Schematic of filter used for wavelength division multiplexing and demultiplexing.[17] Such filters may be connected at fiber distribution panels to extend the communication capacity of the fiber cable.

FIBER-OPTIC SYSTEM DESIGN

5. Digital Systems

Fiber-optic systems are typically power-limited (i.e., low-power sources, high-noise receivers), but bandwidth is plentiful. In such cases, wideband modulation techniques which trade signal-to-noise ratio for bandwidth are desirable. *Pulse-code modulation* (PCM) is a particularly efficient way of making this trade, and thus fiber-optic systems are compatible and, indeed, are a further spur to the growing trend towards digital transmission of voice and other information.

Hierarchical Bit Rates. The three internationally established digital hierarchies are summarized in Table 8.5. Digital fiber-optic systems typically interface at a standard hierarchical level. Additional stages of multiplexing may be contained in the fiber-optic system terminal so that the transmission line capacity is an integral multiple of the interface capacity. Most systems in the United States interface at the third level of the North American digital hierarchy (45 Mbs). Systems presently exist or are planned with transmission capacities corresponding to the following multiples of the basic DS-3 (45 Mbs) capacity: 1, 2, 3, 4, 6, 9, 12, 18, 24, 36. (This corresponds to bit rates between 45 Mbs and 1.7 Gbs.) Modular system architectures allow achieving a variety of line rates. Standardization is at the DS-3 interface, not at the line rate. A general block diagram of a digital fiber-optic transmission system is provided in Fig. 8.11.

Modulation. Fiber-optic systems generally utilize simple *binary* (on-off) intensity modulation. Variants on this modulation method, to achieve easier clock ex-

TABLE 8.5 Digital Hierarchies

	System		
Characteristic	North American	European (CEPT)	Japanese
Level 1			
Bit rate, Mbs	1.544	2.048	1.544
Number of voice circuits	24	30	24
Level 2			
Bit rate, Mbs	6.312	8.448	6.312
Number of voice circuits	96	120	96
Level 3			
Bit rate, Mbs	44.736	34.368	32.064
Number of voice circuits	672	480	480
Level 4			
Bit rate, Mbs	274.176	139.264	97.728
Number of voice circuits	4032	1920	1440
Level 5			
Bit rate, Mbs	Not defined	565.148	397.20
Number of voice circuits		7680	5760

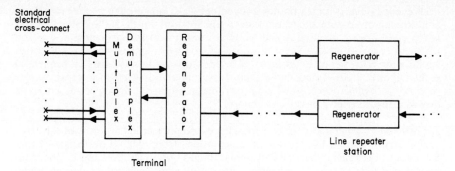

FIG. 8.11 Fiber optics digital transmission system. Regenerator contains photodetector, amplification, timing recovery, decision circuit, and optical transmitter. Additional circuits may also be in the regenerator to allow remote fault location from the terminals. Terminals also generally contain maintenance circuits (performance monitoring, protection switching, and telemetry access).

traction, performance monitoring capability, or improved power efficiency are illustrated in Fig. 8.12.

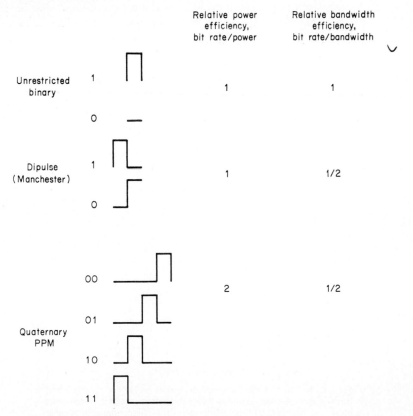

FIG. 8.12 Examples of pulse codes for fiber optic transmission.

Present optical sources do not conveniently permit frequency-shift or phase-shift keying of an optical carrier. Indeed the spectral width of the optical carrier is typically large compared to the modulation bandwidth. For example, even an SLM laser might have a spectral width of about 0.1 nm in wavelength corresponding to a spread of 1 part in 10^4. The spectral width (measured in frequency) is then about 20 GHz, much larger than most modulation rates of interest. Active research is in progress on stabilizing semiconductor lasers, and spectral widths as low as a few kilohertz have been reported.

Performance. Digital system performance is often specified in terms of the error probability. Figure 8.13 illustrates the deviations from ideal in performance that can occur in practical fiber-optic systems. In addition to additive noise, fiber systems (as indeed all digital systems) can suffer signal-dependent degradations. Intersymbol interference, improper setting of decision thresholds, degradation of laser on-off ratios and a host of other phenomena lead to "eye degradations" which necessitate a larger received power to achieve a given error probability. Additional eye margin allowance is generally required for aging effects. The relation between eye margin and power margin is provided in Fig. 8.14.

In addition to eye degradations, *signal-dependent noise* can occur (modal noise, mode partition noise, laser oscillations) which result in a lower bound to the error probability. It is important to assure that there is no appreciable bending of the error rate characteristic over the error probability range of interest.

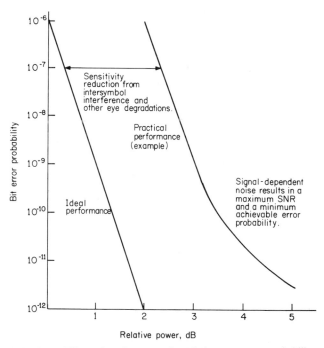

FIG. 8.13 Effect of performance degradations on error probability. Intersymbol interference and threshold variations result in a sensitivity reduction (see Fig. 8.8 for sensitivity, dBm). Signal-dependent noise (modal noise and mode partition noise) may result in a noise floor.

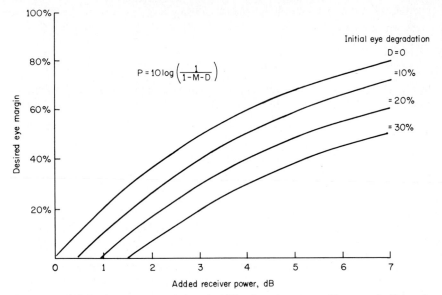

FIG. 8.14 Relation between eye margin and additional receiver power. For example, if there is a 30 percent initial eye degradation (from ideal) and allowance is for an additional 20 percent degradation from electronic component aging, then 3 dB of additional receiver power (relative to the ideal sensitivity) is required.

Regenerator Spacing Calculation. The loss budget for a regenerator span is calculated by taking the difference between the transmitter power P_T and receiver sensitivity P_R, subtracting for connector losses L_C at the transmitter and receiver locations (there are typically connectors not only in the regenerators, but also at patch panels that provide flexible access to the fiber cable), and allowing for various degradations L_D and typically some unallocated margin M. The resultant outside plant loss allowance L_1

$$L_1 = P_T - P_R - L_C - L_D - M \qquad (8.8)$$

must then be allocated to the installed fiber cable, including splice loss (an allowance for a repair splice may be included in M).

There are statistical variations in cabled fiber loss, splice loss, and temperature effects. A worst-case calculation would be overly pessimistic. Having the means and standard deviations of the individual loss components, the 2σ outside plant loss is calculated as follows[18]

$$L_2 = (\mu_f + \mu_T)l + (\mu_S + \mu_{ST})N \qquad (8.9)$$
$$+ 2[(\sigma_f^2 + \sigma_T^2)l^2/(N - 1) + (\sigma_S^2 + \sigma_{ST}^2)N]^{1/2}$$

where μ = mean value
σ = standard deviation
l = length of regenerator section
N = number of splices

and the subscripts have the following meanings:

f is fiber in cable

T is temperature

S is splice

The outside plant loss L_2 must be less than or equal to the loss allowance L_1. A simplified calculation for a 400 Mbs single-mode system at 1.3 μm is as follows:

$$P_T = -3 \text{ dBm}$$
$$P_R = -33 \text{ dBm (for a bit error rate of } 10^{-9})$$
$$L_C + L_D = 8 \text{ dB (combined in a statistical analysis)}$$
$$M = 3 \text{ dB}$$
$$L_1 = 19 \text{ dB}$$
$$l = 38 \text{ km}$$
$$N = 18$$
$$\mu_f = 0.4 \text{ dB/km} \qquad \sigma_f = 0.1 \text{ dB/km}$$
$$\mu_S = 0.1 \text{ dB} \qquad \sigma_S = 0.1 \text{ dB}$$

All other means and standard deviations are assumed to be zero.

$$L_2 = 19 \text{ dB} \qquad \text{(15.2-dB mean cabled fiber,}$$
$$\text{1.8-dB mean splice loss,}$$
$$\text{2-dB } 2\sigma \text{ variation)}$$

Bandwidth.   The above calculation assumes spacing is determined by loss and that bandwidth is not limiting. If fiber bandwidth is not large compared to the bit rate, then a dispersion penalty should be included in the degradation allowance. Dispersion penalty is shown as a function of ratio of bandwidth B to bit rate R in Fig. 8.15. Equalization techniques (including decision feedback equalization) may be used to reduce this penalty and to operate with $B/R - 0.5$. Such techniques are useful to extend the capability of multimode systems, but are typically not required for single-mode systems.

In single-mode systems, mode partition noise, more so than chromatic dispersion limits bit-rate–distance product (see Fig. 8.16).

Bit Rate–Repeater Spacing Technology Boundaries.   Figure 8.17 illustrates representative repeater spacing of practical terrestrial systems with most of the various technologies being used or currently planned. At low bit rates, repeater spacing is limited by receiver noise (loss limited). At high bit rates it may be limited by pulse dispersion. Figure 8.17 does not represent hard technology limits. Indeed, system experiments have been performed at 1.5 μm with *single-longitudinal-mode* (SLM) lasers achieving performance well above these boundaries, e.g., 203 km at 420 Mbs and 117 km at 4 Gbs.

6. Analog Systems

Analog transmission is used on fibers for some *point-to-point* (no intermediate repeater) applications in which the economics do not presently warrant A/D conversion.

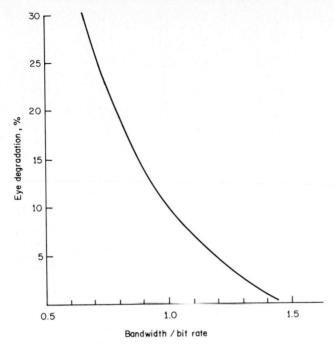

FIG. 8.15 Percent eye degradation caused by dispersion for typical lightwave regenerator without equalization.

Modulation

With direct detection receivers, only intensity modulation of the optical source is practical. The information signal may be used directly to intensity-modulate the light source. Alternatively, the information signal may modulate a subcarrier which in turn intensity-modulates the light source. Frequency modulation of the subcarrier provides a constant-amplitude signal for modulation of the light source. Relative to direct intensity modulation, this alleviates linearity requirements on the light source and also provides signal-to-noise-ratio enhancement.

The subcarrier frequency should be chosen large compared to the modulation bandwidth, and the fiber bandwidth should be sufficiently high to support the subcarrier frequency.

Performance.   The electrical signal at the output of the optical detector is the modulated subcarrier with a signal-to-noise ratio given by[19]

$$\text{SNR} = \left(\frac{\gamma\eta e\lambda}{hc}\right)^2 \frac{P_o^2\,\overline{M^2}}{\left(\frac{2P_o\,\lambda\eta e^2\,\overline{M^2}}{hc}B + N_{th}\right)} \tag{8.10}$$

where γ = intensity modulation index

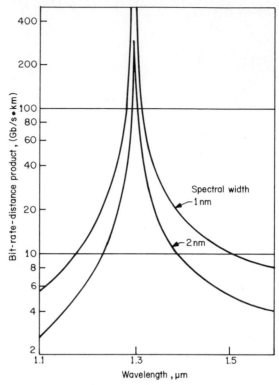

FIG. 8.16 Limitation on bit rate distance product caused by
mode partition noise for a single-mode fiber with minimum
dispersion at 1.3 μm (Reference 15).

η = detector quantum efficiency
e = electron charge $1.6021(10)^{-19}$ coul
P_o = received optical power at the input to the detector
M = detector avalanche gain
B = bandwidth of the amplifier following the detector
N_{th} = square of the equivalent input thermal noise current of the amplifier

For good APDs, the mean square and average avalanche gain are related by

$$\overline{M^2} = \overline{M}^2\left[\kappa\overline{M} + \left(2 - \frac{1}{\overline{M}}\right)(1 - \kappa)\right] \tag{8.11}$$

where κ is the ratio of ionization probabilities for holes and electrons.

At high SNR (required for simple intensity modulation), Eq. 8.10 reduces to
the quantum limit

$$P_o(\text{min}) = 2\,(\text{SNR})\,\frac{hc}{\lambda\eta}\,\frac{B}{\gamma^2} \tag{8.12}$$

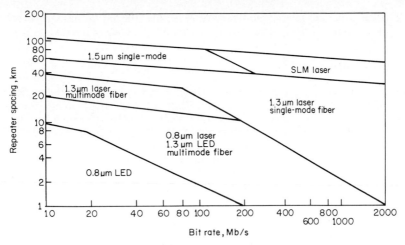

FIG. 8.17   Approximate (and somewhat conservative) fiber optic system repeater spacing technology boundaries for practical systems with present technology.

For example, for SNR $= 10^5$ (50 dB), $\eta = 0.8$, $\lambda = 1.3\ \mu m$, $B = 4$ MHz (corresponding to a single video channel), and $\gamma = 0.2$, it follows from Eq. 8.12 that

$$P_o(\text{min}) = 3.82 \times 10^{-6}\ W = -24\ \text{dBm}$$

This is a considerably higher received power than is required for digital modulation.

The use of wideband subcarrier modulation results in an S/N improvement and requires less received power. Approximate calculations are based on using Eq. 8.10 to determine an intermediate (i-f) S/N. Baseband S/N is then calculated using standard communication system formulas relating i-f to baseband S/N for the appropriate subcarrier modulation.

Nonlinear distortion and intermodulation effects further limit the application of analog modulation to fiber-optic systems.

TELECOMMUNICATION APPLICATIONS

7. Scope of Applications

Telecommunication applications of fiber-optic systems range from short data links within a building to long transcontinental and intercontinental systems. Principal technology choices and factors impacting these applications are summarized below.

8. Building and Campus

Fiber data links are being used to provide interconnection within telephone switching equipment and to interconnect computer peripherals. LED sources,

PIN detectors, and multimode fiber are generally utilized in these applications. Key advantages of fiber optics are noise and interference immunity.

9. Subscriber Loops

Fibers are used in conjunction with subscriber carrier systems in the feeder portion of the loop plant (from the central office to an interface point). For many applications to business subscribers, the interface point is located on the subscriber's premises. Principal advantages of fiber in these applications are that it eliminates the intermediate repeaters required in metallic systems and it has the capacity for growth. Initial applications utilized LED sources, PIN detectors, and multimode fiber, but there is growing interest in single-mode fibers for these applications.

10. CATV

Fiber systems have been used for point-to-point *community antenna television* (CATV) supertrunks (e.g., to connect a satellite earth station with a cable head end). There have been many trial and demonstration systems utilizing fiber for CATV distribution (e.g., HI-OVIS in Japan and in Biarritz, France; Elie, Canada; and Milton-Keynes, England), but the economics do not yet appear generally favorable. Coaxial cable distribution utilizes a tapped-bus T architecture ideally suited for one-way broadcast service. Fiber technology is more suited for star architectures. Such architectures are capable of providing two-way switched services. It remains to be seen whether a single architecture may be used economically to provide a combination of switched and broadcast services.

11. Intracity Trunking

Fiber systems are extensively used for interoffice trunking in metropolitan areas. Economics are generally favorable, since a fiber replaces many copper pairs, and intermediate repeaters are not generally required between central offices. The technology has gone largely to single-mode fiber and to higher-capacity transmission systems.

12. Intercity Trunking

The economic prove-in of fiber-optic systems for intercity trunking was enhanced by terminal cost savings in combining digital transmission with digital switching and the utilization of existing right-of-way for fiber cable placement. All new applications are with single-mode fiber, with the trend toward higher-capacity transmission and longer repeater spacings. Installed systems may be upgraded in capacity by either higher-bit-rate transmission or by wavelength-division multiplexing.

13. Undersea

Fiber has supplanted coaxial cable as the cable medium of choice for undersea applications. Trans-Atlantic and trans-Pacific systems are planned for installation in 1988. Reliability is critical for undersea communications systems, with 25-year operation as the system requirement. To achieve such performance extremely strong fiber and ultrareliable optical components had to be developed. Presently committed systems utilize single-mode fiber at 1.3 μm with repeater spacings of about 50 km. There is interest in the future use of 1.55 μm to achieve longer repeater spacings.

14. End-to-End Capability

Relative to coaxial cable, fiber-optic systems are more cost-effective because of their larger capacities, smaller-diameter cable, and longer distances between repeaters. Also, unlike coaxial cable where maximum voice circuit transmission capability is achieved with analog transmission, fiber optic systems are most efficiently used with digital transmission. Thus, fiber systems in the long-haul terrestrial and undersea plants are a key step in the evolution to an end-to-end digital transmission capability spanning the globe.

ACKNOWLEDGMENTS

In addition to the cited references, the author has benefited considerably from input from innumerable colleagues throughout AT&T.

REFERENCES

15. Cited References

1. Li, T. "Advances in Optical Fiber Communication: An Historical Perspective," *IEEE J. Sel. Areas in Comm.*, vol. SAC-1, no. 3, pp. 356–372, April 1983.
2. Schawlow, A. L. and C. H. Townes, "Masers and Maser Communication System," U. S. Patent 2,929,922, filed July 30, 1958.
3. Hall, R. N., et al., "Coherent Light Emission from GaAs Junctions," *Phys. Rev. Letters*, vol. 9, p. 366, November 1962.
4. Nathan, M. I., et al., "Stimulated Emission of Radiation from GaAs p-n Junctions," *Appl. Phys. Letters*, vol. 1, p. 62, November 1962.
5. Kao, C. K., and G. A. Hockham, "Dielectric-Fibre Surface Waveguides for Optical Frequencies," *Proc. IEE*, vol. 113, pp. 1151–1158, July 1966.
6. Kapron, F. P., et al., "Radiation Losses in Glass Optical Waveguides," *Appl. Phys. Letters*, vol. 17, p. 423, Nov. 15, 1970.
7. Panish, M. B., et al., "Double-Heterostructure Injection Lasers with Room Temperature Thresholds as Low as 2300 A/cm^2," *Appl. Phys. Letters*, vol. 16, p. 326, April 1970.
8. Jacobs, I., et al., "Atlanta Fiber System Experiment," *Bell System Tech. J.*, vol. 57, pt. 1, July-August 1978.
9. Cook, J. S., and O. I. Szentisi, "North American Field Trials and Early Applications in Telephony," *IEEE J. Sel. Areas in Comm.*, vol. SAC-1, no. 3, pp. 393–397, April 1983.
10. Li, T., (ed.), *Optical Fiber Communications*, vol 1, *Fiber Fabrication*, Academic Press, 1985.

11. Miller, C. M., and G. F. DeVeau, "Simple High-Performance Mechanical Splice for Single-Mode Fibers," *1985 Optical Fiber Conference Proceedings.*

12. Hartman, R. L., R. W. Dixon, "Reliability of DH GaAs Lasers at Elevated Temperatures," *Appl. Phys. Letters,* vol. 26, no. 5, pp. 239–242, 1975.

13. Cheng, S. S., "Optimal Replacement Rate of Device with Lognormal Failure Distributions," *IEEE Trans. on Reliability,* vol. R-26, no. 3, pp. 174–178, 1977.

14. Easton, R. L., et al., "Assuring High Reliability of Lasers and Photodetectors for Submarine Lightwave Cable Systems," *AT&T Tech. J.,* vol. 64, no. 3, March 1985 (special issue).

15. Ogawa, K., "Considerations for Single-Mode Fiber Systems," *Bell System Tech. J.,* vol. 61, no. 8, pp. 1919–1931, October 1982.

16. Henry, P. S., "Lightwave Primer," *J. Quantum Electronics,* December 1985.

17. Stakelon, T. S., and F. S. Welsh, "Optical Frequency Control for Wavelength Multiplex Systems," *Proc. Frequency Control Symposium,* 1985.

18. Meskell, D. J., Jr., "Fiber Optic System Engineering," *Proceedings of the National Communications Forum,* vol. 38, pp. 142–148, 1984.

19. Personick, S. D., "Receiver Design," Chap. 19 in *Optical Fiber Telecommunications,* Academic Press, New York, 1979.

16. General References

Miller, S. E., and A. G. Chynoweth (eds.), *Optical Fiber Telecommunications,* Academic Press, New York, 1979.

Kao, C. K., *Optical Fiber Systems: Technology Design, and Applications,* McGraw-Hill, New York, 1982.

Midwinter, J. E., *Optical Fibers for Transmission,* John Wiley & Sons, New York, 1979.

Personick, S. D., *Optical Fiber Transmission Systems,* Plenum Press, New York, 1981.

Personick, S. D., *Fiber Optics Technology and Applications,* Plenum Press, New York, 1985.

CHAPTER 9

SELECTION OF TRANSMISSION MEDIA

Mark S. Whitty
Advisory Engineer

Arthur R. Roberts
Vice President, Microwave Engineering and Construction,
MCI Telecommunications Corporation

CONTENTS

INTRODUCTION

1. Scope of Chapter

This chapter describes telecommunications transmission requirements and the principal media available to meet them. Typical system models are used to compare costs of different media when applied to the models.

The material in the chapter is devoted primarily to voice and data requirements in networks operated by public carriers and major private systems; other traffic, such as video and other types of signals, is covered briefly.

The transmission media covered are microwave, fiber optics, and satellites with emphasis on their application to long-distance and metropolitan trunking requirements. There is no discussion of subscriber loop application or metallic wire media. See Chap. 7 for coverage of metallic wire systems.

The principal criterion used to guide the choice of transmission media is first cost; this is established as cost per channel mile for the range of models covering long haul, medium haul, and short haul for both high and low capacities. Construction carrying costs as well as operation and maintenance costs are also discussed.

NETWORK TRANSMISSION REQUIREMENTS

2. Intercity Circuits

The intercity portion of a network is the *long-distance* facility which interconnects metropolitan areas.

Principal users of intercity circuits may include:

1. *Interexchange carriers* (ICs).
2. Private business networks.
3. Private operational fixed systems. These are systems which are operated mostly by regulated utilities which are eligible for *operational fixed* licenses.
4. Television broadcasters and networks.

Typical cross sections for these intercity circuits are tabulated in Table 9.1.

Typical circuit lengths and the normal and limit noise specifications for intercity analog circuits are shown in Table 9.2.

The most important specification for a digital circuit is its *bit error ratio* (*BER*); however, the way in which errors occur is also important and so measures such as *error-free seconds* (*EFS*) are also included. Typical distance and performance requirements for intercity digital circuits are tabulated in Table 9.3.

3. Intracity Trunks

Figure 9.1 illustrates a simplified city area network in a configuration commonly used in the United States. At the present time, the major part of this network is operated by the franchised local telephone company—the *exchange carrier* (*EC*). However, private operators and interexchange carriers do install some of their own intracity circuits. Subject to regulatory approval, these may include *bypass,* i.e., a direct connection from the user's premises to the IC's *point of presence* (*POP*).

The principal trunking facilities included in the intracity network are

- Intra-EC: interoffice; toll connecting; access lines
- IC/EC: access trunks; terminal to large user (where permitted)
- Intra-IC: interterminal

TABLE 9.1 Intercity Cross Sections

User	Voice	Data	Video
Interexchange carriers	20,000–50,000	DS-1–DS-3*	4-6
Private networks	50–500	20–50% 1–2 × DS-1	
Operational fixed utility	100–1000	20–40% 1–4 × DS-1	

*DS1, 1.544 Mbs; DS3, 44.736 Mbs.

TABLE 9.2 Analog Performance

Designation	Distance, mi (km)		Normal noise	Limit noise
United States	4000	(6500)	43 dbrnC0	55 dbrnC0
CCIR	2500	(4000)	7500 pwp0	47,500 pwp0
United States (industrial)	1000	(1610)	38 dbrnC0	58 dbrnC0

TABLE 9.3 Digital Performance

Designation	Distance, mi (km)	Normal BER or EFS	Limit
United States		99.5% EFS (56 kbs) 95% EFS (DS-1)	
Canada	4000 (6500)	10^{-6}/95% 99.5% EFS	10^{-3}/0.2%
CCITT	17,000 (27,500)	10^{-6}/90% 92% EFS	10^{-3}/0.2%

The cross-section requirements of intracity trunks vary widely from a few circuits in small towns to thousands of circuits between large metropolitan exchanges.

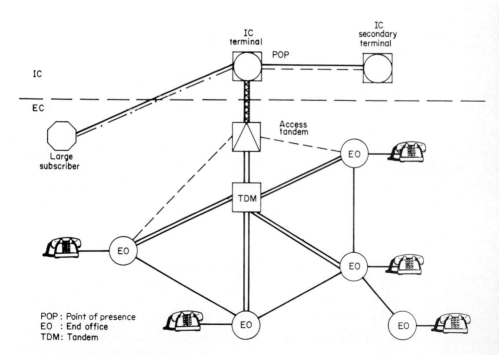

POP : Point of presence
EO : End office
TDM: Tandem

FIG. 9.1 Elements of an intracity network.

This chapter is concerned primarily with trunk routes having large cross sections; typical requirements for these circuits are shown in Table 9.4.

The length of intracity trunks will vary from less than one mile to several tens of miles.

The specification of performance of a communication system requires an allocation of noise between its short-haul and long-haul portions. The short-haul noise objective* generally used in the United States is 28 dBrnC0 for a 60-mi (97-km) trunk with the allowed noise adjusted in proportion to distance (3 dB for doubling).

The CCIR and CCITT have also published recommendations for the short-distance portions of the network. Refer to CCIR Recommendations 395-2, 594, and Report 930 and CCITT Recommendations G215, G821. This general subject is the subject of further study by both CCIR and CCITT.

TRANSMISSION MEDIA

4. Transmission Media—Introduction

The available transmission media include:

- Terrestrial line-of-sight microwave
- Fiber-optic cable
- Satellite
- Coaxial cable
- Paired copper cable

Within each of these general categories there are many different systems, having different bandwidths, modulations, frequency bands, etc., which would affect the choice for a given application.

Terrestrial microwave and *fiber-optics transmission systems* (*FOTS*) can apply to both long-haul and short-haul requirements. Satellite systems are practical only for long-haul.

Coaxial cable systems have been used principally in long-distance intercity applications. Some of these systems are now being converted to digital transmission, but no new installations are being built since they are not competitive with FOTS. Coaxial cable is used as a medium for distribution of *community antenna*

---

*Notes on the Network, AT&T Network Planning Division, Sec. 7, p. 43, 1980.

TABLE 9.4   Intracity Cross Sections

Type of interconnection	Cross-section voice circuits
EC/EC interoffice trunks	1000–10,000
EC/end user access	50–200
IC/EC access	2000–10,000
IC/end user access	50–500
IC/IC interterminal	1000–5000

television (CATV) signals, and these systems are also being used for local distribution of data and video as part of carrier and business networks.

The principal systems available within the three major media are now summarized.

5. Terrestrial Microwave

The operating frequency band is an important factor in the application of microwave.

Within the various frequency bands a large number of equipments, both analog and digital, are available from manufacturers. Equipments in some cases are aimed at a specific class of user particularly in the United States, where most of the frequency bands are allocated by user category.

Table 9.5 shows some of the principal bands and typical available system capabilities in the United States. Except for the 2-GHz bands, those shown also apply internationally. Some additional bands with wide band allocations are available below 10 GHz internationally.

The cost-effectiveness of long-haul, high-capacity systems is largely governed by total capacity of the route (all rf channels) and repeater spacing. As Table 9.5 shows, the 4- and 6-GHz bands are in the optimum region for long-haul.

Performance and Reliability. Most microwave equipments operating below 10 GHz are capable of meeting either the long-haul or short-haul performance objectives discussed in Sec. 4. With the high reliability of modern solid-state equipment, path engineering and propagation outage are the controlling factors in overall system reliability.

TABLE 9.5 U.S. Microwave Systems

Band, MHz	User category*	No. of 2-way rf channels	Typical capacity per rf channel	Typical path length, mi (km)
1850–1990	Ind.	6	600 VF 8 DS-1	27–32 (43–52)
2110–2130 2160–2180	CC	6	252 VF 4 DS-1	27–32 (43–52)
3700–4200	CC	12	1500 VF (A) 2 × DS-3	25–30 (40–48)
5925–6425	CC	8	5400 VF (A) (SSB) 2700 VF (A) (FM) 3 × DS-3	20–27 (32–43)
6425–6875	Ind.	8	600 VF 16 × DS-1	20–27 (32–43)
10,700–11,700	CC	12	2700 VF (A) 3 × DS-3	8–12 (13–19)
17,700–19,700	Ind. CC Pvt.	Various 6 × DS3	1 × DS-3	1–3 (1.6–4.8)

*CC = common carrier; Ind. = industrial; Pvt. = private.

Systems in the 11-GHz bands can meet the short-haul objectives and could probably meet long-haul requirements over areas of relatively light rainfall. In areas which experience high rainfall rates, the system availability becomes harder to predict, and in many instances outages in bad seasons would be likely to exceed the objectives.

For bands higher than 11 GHz, such as the 18-GHz band, the ability to meet outage objectives becomes even more dependent on rainfall rates.

6. Fiber Optics

Like all cable or *guided-wave* systems, fiber optic frequency bands do not need to be regulated. In these cases the wavelength used is determined by the technology, capabilities, and the application.

Table 9.6 shows some key characteristics of the range of fibers currently used in telecommunications. They are described in more detail in Chap. 8.

Until recently, single-mode fiber with high-speed capability was significantly more expensive than multimode. However, the heavy demand for long-haul, high-capacity systems in the United States has resulted in a rapid reduction in single-mode fiber cost, with the effect that it is now being used in short-haul and lower-capacity applications.

Table 9.7 summarizes line rates, repeater spacings, and applications currently practicable with FOTS cable systems. The relationship between repeater spacing and bandwidth is shown in Fig. 8.15, Chap. 8.

Performance and Reliability. Fiber-optic cable systems have relatively predictable characteristics, although some uncertainty exists as to the quality and uniformity of the fibers along their length. Also, the precision of installing and splicing can affect the end-to-end performance. Nevertheless, the performance of these systems is within the control of the system designer and the long-haul objectives as discussed in Sec. 4 can be met.

With current single-mode systems, repeater spacings of about 20 mi (32 km) are common, and outages resulting from electronic equipment failures are rare.

To a large extent, the reliability of fiber-cable systems depends on the system layout, care of installation and the choice of right of way. As with all cable systems, the greatest danger is the threat of a cable cut.

TABLE 9.6   Fiber Characteristics

Fiber type	Core diameter, μ	Attenuation wavelength, db/km	Bandwidth, MHz·km	Dispersion, ps/nm·km
Step-index, multimode	100	2.8/850	20	
Graded-index, multimode	50–100	0.7/1310	1000	
Single-mode	8	0.4/1310		1
		0.3/1550		10

TABLE 9.7 Fiber Applications

Fiber type	Line Rate, Mbs	Span length, mi (km)	Application
Step-index, multimode		0.6 (1)	Local building distribution
Graded-index, multimode	45	9 (15)	Short-haul telephony
Single-mode	565	19 (30)	Long-haul telephony

7. Satellite Systems

The frequency bands and bandwidths available for various satellite services throughout the world are described in detail in Chap. 6 and are summarized in Table 9.8. Virtually all current telecommunications systems occupy the C or K_u bands. The utilization of the 500-MHz bandwidth available for uplinks and downlinks is described in Chap. 5.

For long-distance transmission of telephony, the trend in all media has been to develop increased cross-section capacity. This can be simply visualized in terrestrial microwave and cable systems, where it is clear that increased capacity will reduce costs per circuit mile. In the case of satellite systems, means of increasing the capacity through the satellite are equally desirable but the nature of the satellite network may be more complex, with differing goals making it desirable, for example, to trade off throughput capacity for flexibility.

Access and Modulation. In a point-to-point configuration or in point-to-multipoint service, a carrier from a single uplink occupies a complete transponder 100 percent of the time. This is *single access*.

For some applications, such as low-density telephone service to smaller towns, it may be desirable to allow signals from different points to make use of one transponder. This is *multiple access*.

Multiple access may employ *frequency-division multiple-access* (*FDMA*) or *time-division multiple-access* (*TDMA*) commercial systems.

The most commonly used modulation methods for satellite systems are *frequency modulation* (*FM*), single sideband (*SSB*), and *quadrature phase-shift keying* (*QPSK*). Spread-spectrum techniques are also used in some data communications systems where it is desired to obtain a low earth-station cost by reducing the capacity of the transponder.

Table 9.9 summarizes transponder capacities for various types of access and modulation.

TABLE 9.8 Satellite Frequency Bands

Frequency band, GHz	Uplink, GHz	Downlink, GHz	Bandwidth, MHz
4/6 (C band)	5.925–6.425	3.7–4.2	500
12/14 (Ku band)			
Domestic	14.0–14.5	11.7–12.2	500
International	14.0–14.5	10.95–11.2	250
	14.0–14.5	11.45–11.7	250
19/29	27.5–31.0	17.7–21.2	3500

TABLE 9.9 Transponder Capacity*

Modulation	Single access	Multiple access
Companded FM	2892	1500
SCPC/FM	800/1300†	800/1300†
SCPC/QPSK	800/1600‡	800/1600‡
SCPC/QPSK/T1	576/1152‡	576/1152‡
TDMA/QPSK	900/1800‡	900/1800‡
TDMA/DSI/QPSK	900/1800‡	900/1800‡
TDMA/DSI/QPSK	1800/3600‡	1800/3600‡
SSB	7200	7200

*Maximum achievable capacity in a 36-MHz transponder (half-circuits).
†Companded FM.
‡32 kbs.

Performance and Reliability.   Performance criteria as stated in Sec. 4 are very similar for satellite and terrestrial circuits, and in general the transmission system designer would want to achieve comparable performance over the two types of media. Since the satellite link is essentially independent of distance, its performance is usually expected to be comparable to the transcontinental distance performance of a terrestrial system.

The noise performance and reliability of satellite systems can be expected to be good. The greatest concern for telephony applications is the delay due to propagation time.

Adequately designed satellite links can generally meet the long-haul objectives for data and video transmission. Since much data transmission requires interactive acknowledgment, the transmission delay can be a problem, but systems have been developed which perform a buffering type of function and which largely eliminate the delay problem in packet systems for lower data speeds.

COSTS OF SYSTEM MODELS

8. Transmission Costs—General

In this section, the approximate 1985 first costs of various system models are developed for purposes of comparison. Except for fiber optic *cable*, the cost elements of the various media are changing with time at approximately the same rate. Thus the *cost comparisons* for the various media (although not the *absolute* costs) continue to be valid except for fiber optic systems. The cost of fiber optic cable, which represents 20 to 30 percent of an installed system, is decreasing rapidly, thus somewhat reducing the comparative costs of fiber optic systems.

Figure 9.2 shows the basic models for digital and analog including a repeatered line segment, the cost of which is distance sensitive (except for satellite systems), plus multiplex equipment whose cost is fixed.

In most systems a large portion of the cost is in the initial construction phase, including fixed plant such as buildings, towers, the cable itself, land, and right-

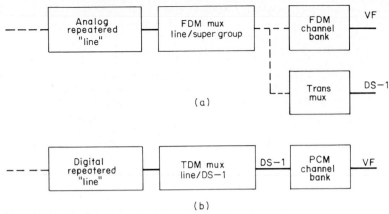

FIG. 9.2 Major cost elements: (*a*) analog system; (*b*) digital system.

of-way. In such cases the cost per circuit mile decreases rapidly as electronics are added and the potential capacity is utilized.

For the system models which follow, estimates of first cost are developed for the line segment of the system. This is expressed as a cost per *voice-frequency* (*VF*) circuit mile.

In order to obtain an estimate of the cost for a given circuit, multiplex costs are added for both ends as follows:*

Analog circuit cost to VF interface   = cost/circuit-mile × miles + \$355 × 2
Analog circuit cost to DS-1 interface = cost/circuit-mile × miles + \$270 × 2
Digital circuit cost to VF interface = cost/circuit-mile × miles + \$140 × 2
Digital circuit cost to DS-1 interface = cost/circuit-mile × miles + \$20 × 2

9. Terrestrial Microwave

Cost Elements.   The major items which make up the cost of a microwave system are

Microwave equipment electronics

Towers

Building and facilities

Power plant

Site acquisition

Engineering

These elements will vary quite widely depending on the application. At the high end of the scale, the typical common-carrier backbone system is designed for maximum usage of frequencies over intercity and transcontinental distances. The lowest-cost application is low-capacity short-distance equipment, for in-

---

*All fixed costs (the dollar values at the right side of the equations) are based on 1985 dollars.

stance the connection from a carrier's terminal to a subscriber. Such a system would normally not involve any repeaters. In between these examples, a lower capacity short-haul system, either private or a connection from a carrier's backbone, to a small city would be representative.

High-Capacity Long Haul.   The model for this application is an eight-hop modulation section equipped for ultimate full expansion of both the 6-GHz and 4-GHz bands. The 6-GHz band is equipped with SSB equipment capable of carrying 5400 voice channels per rf channel, and the 4-GHz is equipped with digital 90-Mbs equipment capable of 1344 standard PCM voice channels per rf channel. This configuration is typical of major carriers now being built in the United States. The system has dual-band antennas, includes automatic protection equipment on a 1:N basis, and is built to the highest standards to meet long-haul performance and availability objectives. The 1:N protection utilizes one rf channel to protect any one of the N-equipped traffic-carrying channels.

Table 9.10 shows the approximate costs for such a system, when it is equipped at 1:1 SSB only, equipped at 1:3 (both 4 and 6 GHz), and when fully equipped at 1:7 (6 GHz) and 1:11 (4 GHz). These costs are capital (first) costs only. As can be seen, the fixed-plant cost is a significant proportion of the initial cost.

The eight-hop modulation section gives a good indication of costs per mile for a high-performance common-carrier system and would apply for most distances except for very short systems such as single-hop spurs from junction stations to downtown terminals.

Medium-Capacity Short Haul.   This model is based on a 20-mi (32-km) no-repeater connection such as might exist between a junction station on the outskirts of a city to a downtown terminal. The model assumes the 6-GHz band only, equipped with digital microwave having three DS-3 capacity. Also it assumes that buildings and facilities exist at both ends and no major tower construction is required.

Table 9.11 lists the first costs of this model.

Low-Capacity Medium Haul.   This model is a 100 mi (161 km) eight-hop, hot-standby section of 6-GHz microwave such as might be used to connect from a carrier's backbone to a small city. The microwave is digital with 3 × DS-3 capacity.

For this model it is assumed that no expansion is allowed and therefore fixed-plant costs are lower than the backbone case.

TABLE 9.10   Long-Haul First Costs*

Item	1:1 (SSB)	1:3 + 1:3	1:7 + 1:11
RF equipment, thousands of dollars	1,440	5,120	11,360
Fixed plant, thousands of dollars	5,400	5,400	5,400
Engineering, thousands of dollars	1,000	1,000	1,000
	7,840	11,520	17,760
Cross-section capacity, circuits	5,400	20,232	52,584
Cost per circuit-mile	$7.30	$2.85	$1.69
Base cost, thousands of dollars	6,400		

*Long-haul, high-capacity backbone microwave 4-GHz digital (2 × DS-3) plus 6-GHz SSB (5400-circuit) eight-hop, seven-repeater, 200-mi (320-km) section. Based on 1985 costs.

TABLE 9.11   Short-Haul First Costs*

Item	1:3	1:7
RF equipment, thousands of dollars	320	640
Fixed plant, thousands of dollars	300	300
Engineering, thousands of dollars	70	80
	690	1,020
Cross-section capacity, circuits	6,048	14,112
Cost per circuit-mile	$5.70	$3.60
Base cost	370	

*Medium-capacity short-haul microwave; one-hop, 20-mi (32 km), 6-GHz; 3 × DS-3system. Based on 1985 costs.

Table 9.12 lists the first costs. Costs per circuit mile are given for the stage when only one DS-3 is utilized and when all three are used. At the single DS-3–equipped level, this model is reasonably representative of an industrial system also.

Very Short Local Connection.   This model is a single-hop, low-capacity system such as may be used to interconnect a large business subscriber to a carrier's terminal. The model assumes an 18-GHz hop using equipment capable of carrying four DS-1 signals which equates to 96 standard PCM voice channels. It is assumed that the system is between two existing buildings and only minimum costs are involved to house it, provide power, and support antennas.

The distance is assumed to be 4 mi (6.5 km), which is practical at 18 GHz in average- to low-rainfall areas. The equipment includes hot standby protection.

Total first cost for this model is calculated to be $120,000 (1985 dollars).

When converted to cost per circuit mile this becomes:

$$\frac{\$120,000}{4 \times 96} = \$313/\text{channel-mile}$$

This system is vastly more costly per channel-mile than the high-capacity backbone system, but it still may be effective compared to other means of making this type of local area connection. Also, such figures will vary widely for different situations.

TABLE 9.12   Low-capacity costs*

Item	Value
RF equipment, thousands of dollars	800
Fixed plant, thousands of dollars	1250
Engineering, thousands of dollars	200
	2250
Cross-section capacity, circuits	2016
Cost per circuit-mile	
1 × DS-3	$33.50
3 × DS-3	$11.50
Base cost, thousands of dollars	1450

*Low-capacity, medium-haul microwave; hot-standby, 100-mi (160-km), five-hop, 6-GHz, 3 × DS-3 system. Based on 1985 costs.

10. Fiber Optics

Cost Elements.   The major cost elements in a fiber-optic cable system are

Cable
Electronics (regenerators and terminals)
Repeater facility including power
Installation/construction
Right-of-way
Engineering

Several models are considered to illustrate first-cost magnitudes and relationships among different applications.

High-Capacity Long Haul.   This model is a 22-fiber cable of single-mode fibers operating at 405 Mbs line rate. Thus each fiber is capable of carrying 6048 voice channels of standard PCM. An average repeater spacing of 22 mi (35 km) is assumed so that a 200-mi (320-km) section requires eight repeaters plus two end terminals. For the model purposes it is assumed that all fibers are ultimately equipped at 405 Mbs although it is clear that higher rates will be available for fibers in subsequent years. Table 9.13 shows the first costs for this model when equipped 1:1, 1:3, and 1:10.* A cost of right-of-way of $8000/mi is typical of current costs. This figure could vary greatly in practice. Installation costs are based on average long-haul conditions requiring a mix of direct burial and duct installation, plus road and bridge crossings.

Intermediate-Capacity Short Haul.   This model is a one-hop (no-repeater) system for metro area exchange trunking or for connection from an IC junction to a downtown terminal. The line rate is 405 Mbs and an eight-fiber single-mode cable is included. Table 9.14 shows first costs at the 1:1 and 1:3 states. This model assumes existing buildings at both ends and installation into existing ducts.

---

*One equipped fiber is used as the "protection" channel for the N installed working fibers.

TABLE 9.13   High-Capacity FOTS*

Item	1:1	1:3	1:10
Cable, thousands of dollars	4,710	4,710	4,710
Splicing and installation, thousands of dollars	5,400	5,400	5,400
Buildings facilities, thousands of dollars	2,490	2,490	3,050
Right-of-way, thousands of dollars	1,600	1,600	1,600
Electronics, thousands of dollars	550	1,090	3,000
	14,750	15,290	17,760
Cross-section capacity, circuits	6000	18,000	60,000
Cost per circuit-mile	$12.30	$4.25	$1.48
Base cost, thousands of dollars	14,200		

*High-capacity, long-haul FOTS, 405 Mbs, 200-mi (320-km), eight-repeater, 22-fiber, single-mode system. Based on 1985 costs.

TABLE 9.14 Short-Haul FOTS*

Item	1:1	1:3
Cable, thousands of dollars	210	210
Splicing and installation, thousands of dollars	200	200
Buildings and facilities, thousands of dollars	100	100
Right-of-way, thousands of dollars	160	160
Electronics, thousands of dollars	300	600
	970	1270
Cross-section capacity, circuits	6000	18,000
Cost per circuit-mile	$8.10	$3.52
Base cost, thousands of dollars	670	

*Short-haul intermediate-capacity FOTS; 20-mi (32-km), 405-Mbs, single-mode, eight-fiber system. Based on 1985 costs.

Low-Capacity Medium Haul.   This model is a relatively low capacity system 100 mi in length, as might be used to interconnect a small city to a major backbone. The model assumes no major expansion and only four-fiber cable is included. Installation costs are based on easy conditions requiring mostly direct burial.

Table 9.15 shows the first costs for this model.

11. Satellite Systems

Cost Elements.   The major cost elements in a satellite system are

The space equipment—transponders

Ground-station antenna

Ground-station rf equipment

Ground-station modem and terminal equipment

Ground-station site and facilities

Terrestrial interconnection to terminal

TABLE 9.15 Low-Capacity FOTS*

Item	1:1 system
Cable, thousands of dollars	540
Splicing and installation, thousands of dollars	1250
Buildings and facilities, thousands of dollars	640
Right-of-way, thousands of dollars	800
Electronics, thousands of dollars	240
	3,470
Cross-section capacity, circuits	2000
Cost per circuit-mile	$17.35
Base cost	3230

*Low-capacity, medium-haul FOTS; 100-mi (160-km), four-repeater, 135-Mbs, four-fiber, single-mode system. Based on 1985 costs.

The telephony type of applications for which a satellite system may be considered range from a highly concentrated large-cross-section trunking system between two major nodes to a system which provides low-capacity service to a large number of locations.

In order to see the magnitude of costs involved in satellite systems, two models are discussed.

In all systems, a fixed cost per transponder is used for the space-segment first cost. This cost will vary considerably depending on the approach taken to acquisition of transponders; this could range from construction and launch of a complete satellite to leasing of individual transponders from another party. Also, depending on this approach, the timing of expenditure with need will vary over a wide range. Based on the availability of transponders for lease today, the capital cost used in these models is $5 million. This figure includes *tracking, telemetry, and command (TT&C)* facilities.

Large-Cross-Section, Two-Node System. This model is an FM/FDM system using compandored FDM voice channels. With a 30-m antenna, an HPA of 3 kW and a receiver G/T of 42 dB/K, this system can transmit 2900 voice channels through a transponder. Table 9.16 shows the magnitude of first costs for the earth station for such a system and Table 9.17 develops system costs for a two-node system including the space segment.

As Table 9.17 shows, the cost per two-way voice circuit for this model is $7800; for a 1000-mi (1620-km) circuit, this would equate to $7.80/circuit-mile. With the expanded utilization the cost per channel-mile decreases to $5.90 for the 1000-mi (1620-km) circuit. (As the table indicates, costs are in 1985 dollars.)

Small-Cross-Section Multinode System. This model is based on a TDMA system designed to interconnect a number of small cities to several gateway earth stations. The system uses a 60 Mbs-burst modem with a maximum capacity of 35 DS-1s—equivalent to 840 standard *pulse-code modulation (PCM)* voice channels. In this configuration, the system can use 11-m antennas, HPAs of 3 kW and 60° LNAs. Table 9.18 shows the major cost elements for an earth station with traffic from up to 35 DS-1s plus a gateway station that could handle four times this load. ADPCM (Adaptive Differential Pulse Code Modulation) is included, providing a 2:1 compression.

Table 9.19 derives the system cost including the space segment for a model made up of 10 earth stations, two of which are gateways. The total traffic load from the eight nongateway stations sums to the equivalent of 8×35 DS-1s (840 voice circuits in standard PCM), all of which is split between the two gateways.

TABLE 9.16 Earth-Station Costs*

Item	First costs, thousands of dollars	
	1:1 system	1:2 system
Antenna and rf equipment	2,150	300
Modem and ancillary equipment	800	400
Building and facilities	1,900	
Microwave interconnect	1,500	200
	6,350	900

*30-m FM/FDM earth station. Based on 1985 costs.

TABLE 9.17   Two-Node Satellite System*

Item	1:1 system	1:2 system
Two earth stations, thousands of dollars	12,700	14,500
Two transponders, thousands of dollars	10,000	
Four transponders	—	20,000
	22,700	34,500
Number of two-way circuits	2900	5800
Cost per two-way cct, thousands of dollars	$7.80	$5.95
Cost per circuit-mile (1000 mi)	$7.80	$5.95

*Based on 1985 costs.

TABLE 9.18   TDMA Earth-Station First Costs*

	Cost, thousands of dollars	
Item	Outlying earth station	Gateway earth station
Antenna and rf equipment	450	500
Modems and ancillary electronics	640	1270
Buildings and facilities	400	550
Microwave interconnect	600	700
	2090	3020

*12-m TDMA earth station. Based on 1985 costs.

For this load, eight transponders are needed; the costs are high but may be acceptable, depending on distances and other alternatives.

COMPARISONS AND CHOICES

12.   Bases for Choice of Medium

The factors affecting choice of transmission medium will vary greatly from one situation to another. A large established carrier with a lot of existing plant will have different influences from a business deciding to build a private system where none exists.

Most telephone carriers are rapidly moving to digital switching plant, and thus transmission media which are most economical for digital transmission are favored. Nevertheless, in the United States the ability to overbuild existing microwave routes with SSB and thereby more than double the capacity has prolonged the use of analog techniques.

Assuming that several media can satisfy the capacity needs (present and future) and meet the performance standards, factors such as ready availability of right-of-way may govern the decision.

Obviously the goal is to achieve maximum cost effectiveness considering the many complex factors for each case.

TABLE 9.19   Ten-Node System First Cost*

Item	Quantity	Cost, thousands of dollars
Outlying earth stations	8	16,720
Gateway earth stations	2	6,040
Control center	1	500
Transponders	8	40,000
Two-way circuits (8 × 840)	6700	63,260
Cost per two-way circuit		9.44

*Small cross-section multinode system; 12-m ten-node TDMA system, two gateways, ADPCM 2:1. Based on 1985 costs.

13. Comparisons of First Costs

High-Capacity Long Haul.   Figure 9.3 shows cost per circuit-mile for microwave, fiber optics, and satellite line systems. The satellite system cost (based on 1000 miles) levels off since each increase in capacity is dominated by a high incremental transponder cost. The higher start-up costs for the FOTS system show up at lower utilized capacities; however, the FOTS system costs descend rapidly and

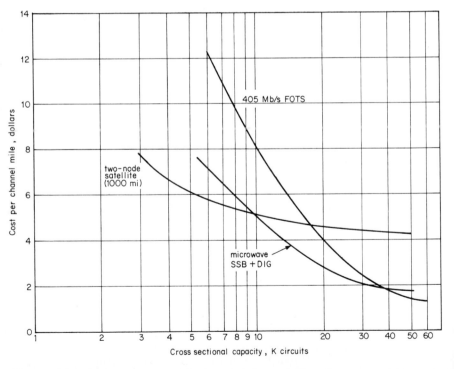

FIG. 9.3 Comparative costs versus utilized cross sections: satellite, microwave, and fiber optic systems (based on 1985 costs).

ultimately are lower than microwave. In view of its potential for wider bandwidth, the FOTS system has cost advantages where cross-section capacities above 30,000 to 40,000 circuits will be utilized.

Figure 9.4 shows total cost per circuit versus distance when cross-section capacity utilized is 20,000 circuits (1:3 SSB + 1:3 DIG) and for 50,000 circuits (1:7 SSB + 1:11 DIG). For this comparison all circuits have multiplex costs added to bring them to the DS-1 interface. This is typical of a modern network using all-digital switching.

The satellite system is cost-effective for very long coast-to-coast spans or for transoceanic applications. However, at shorter distances, and, as the utilized cross-section capacity increases to tens of thousands of circuits, the terrestrial systems have the cost advantage.

Intermediate-Capacity, Short Haul. For the short-haul intracity system with fairly high cross-section requirements, the single-span microwave and FOTS systems may be compared in Tables 9.11 and 9.14. As the tables show, the FOTS system has higher start-up costs but becomes more cost-effective as higher cross sections are utilized.

These models are highly sensitive to the conditions of any given case. If line of sight is not available, a repeater would rule out the microwave; also the 20-mi (320-km) distance makes most efficient use of the fixed-cost microwave hop. For shorter distances the FOTS system would become more attractive since most of its costs remain distance-sensitive. Figure 9.5 shows the cost per circuit versus distance for the two systems with 6000-circuit cross-section utilization. Under about 10 mi (16-km) the FOTS system is favored. Other factors, such as existing right-of-way owned by the user, could change the crossover significantly. Costs for digital multiplex to the DS-1 interface are included.

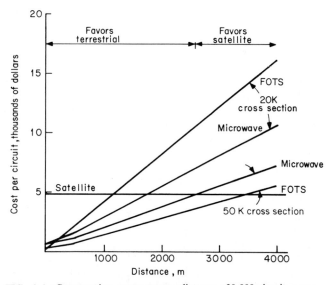

FIG. 9.4 Comparative costs versus distance—20,000-circuit cross section: satellite, microwave, and fiber optic systems (based on 1985 costs).

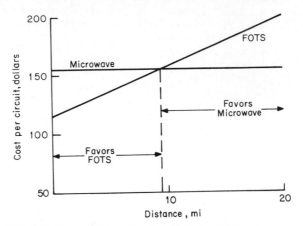

FIG. 9.5 Comparative costs: short haul circuits (based on 1985 costs).

Low-Capacity, Medium Haul.   For this application where the cross section is not expected to exceed about 2000 circuits, the *light-route* microwave system is very cost-effective. A direct cost comparison may be made between microwave, Table 9.12, and the low-capacity FOTS system, Table 9.15. Both models are 100 mi long and, as chosen, have equal capacities so the total costs can be compared directly as shown in Fig. 9.6. The high fixed costs of the fiber-cable system penalize this approach. While factors such as existing right-of-way could alter the comparison, a fiber system would not generally be chosen for this application unless cross-section requirements were projected to be considerably higher.

Another comparison may be made in this category between the low-capacity

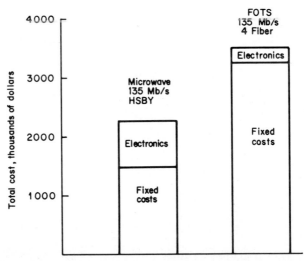

FIG. 9.6 Comparative costs: medium haul circuits (based on 1985 costs).

microwave and the multinode satellite system (Table 9.17). As can be seen from these tables, the low-capacity microwave is considerably more cost-effective unless large distances are involved; the 10 cities must be at an average distance of more than 300 mi from the backbone before the satellite system would be competitive with the microwave.

14. Construction Carrying Costs

An important factor in comparing the costs of different transmission media is the rate of expenditures versus the rate at which capacity becomes available. The construction carrying costs, usually considered to be the interest on costs as they are incurred during construction, are part of the capital costs of the system and must be considered as such.

The importance of this factor is shown in Fig. 9.7, which compares the first-cost expenditures for 200-mi microwave and FOTS transmission systems as a function of system capacity. While the FOTS system appears cheaper for cross sections in excess of 50,000 circuits, the cost of the higher investment in the early years when the utilization may be lower is a significant factor; the importance of this factor depends on the rate at which the full capacity is utilized. However, if all the analog circuits on the SSB microwave are converted to DS-1 interface by the use of transmultiplexers, the microwave system cost increases dramatically compared to negligible increase in FOTS and the crossover point with FOTS moves to a lower cross-section value. For all digital requirements an all-digital

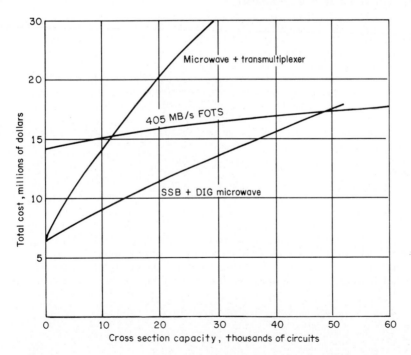

FIG. 9.7 Total system cost comparison: 200-mile system (based on 1985 costs).

(4-GHz and 6-GHz) microwave would not incur these multiplex costs but might not satisfy the ultimate capacity requirement. Each case must be studied on a *present worth of annual charges (PWAC)* or similar basis.

Satellite systems may also require high up-front expenditures, depending on the method and timing of the transponder procurement.

The same consideration will affect the choice of size (number of fibers) of a FOTS cable. The high cost of installation and construction tend to favor sizing the cable to satisfy very long term projections. Offsetting that is the fact that a higher investment will remain unused in the early years. In view of the speed at which line rates are being increased on FOTS systems, the choice of cable size for a large-capacity application requires some degree of clairvoyance.

Construction carrying costs are also a function of construction times.

The time to complete a new route will vary depending on the given circumstances plus effort applied; however, in comparing different media certain generalizations can be made:

- In a start-from-scratch situation, the obtaining of continuous right-of-way for FOTS will usually take longer than obtaining microwave sites.

- Environmental difficulties are increasingly slowing the rate of obtaining microwave sites.

- Large earth-station sites are likely to move slowly because of their major environmental impact.

15. Operation and Maintenance

Costs for operation and maintenance of the system will also be included in a rigorous PWAC study. As noted earlier, however, they are usually small in comparison with capital costs.

Annual maintenance costs for large microwave networks average 1 to 2 percent of the capital cost. Similar costs are projected for a single-mode FOTS system with comparable repeater spacings except that cable cuts add significant cost for restoration compared to microwave. Satellite systems maintenance costs are in the same percentage range as the earth-station costs. Since fewer sites and smaller dollars are deployed on the ground these costs are lower than for the two terrestrial technologies.

16. Security

Both physical and communication security are increasingly important. Satellite systems, since they involve the least number of sites, have the advantage in physical security. Cable systems must be considered most vulnerable due to the physical presence of the cable over the entire length of the system.

On the other hand, cable systems are virtually invulnerable to eavesdropping; microwave and satellite systems require encryption to protect against this risk.

CHAPTER 10

COMMUNICATION SIGNAL INFORMATION AND SPECIFICATIONS

Kerns H. Powers

*Staff Vice President, Communications Research, RCA
Laboratories, David Sarnoff Research Center*

CONTENTS

INTRODUCTION

1. Definition of Terms

This chapter contains information (or references to information) about baseband signal formats that are in common use, or for which standard specifications have been developed by those national or international organizations that concern themselves with standards and engineering practice. It will begin with introductory material that will lead to a better understanding of the terms that are used in communication practice to help the reader comprehend the definitive specifications in the chapter, whether applied to systems for audio, video, or data transmission.

An example of the need for careful definition is the confusion in the lexicon between *digital* transmission and *data* transmission. The term *digital,* as opposed to *analog,* refers to the *form* in which the information is communicated, whereas the term *data* refers to the *type* of information being communicated, as opposed to *audio, video,* or *facsimile,* for example. Data can come from an inherently digital source, e.g., written English or computer data, or it can come from an analog source, e.g., telemetry. Another example is the confusion between *sampled* signals that are defined only at discrete instants of time, *quantized* signals that have values only at discrete amplitude levels, and *digital* signals that are manifestations of a sequence of symbols, or digits. A digital signal may have been derived from both the sampling and quantizing of an analog signal, but the quantized samples must have been further processed by encoding each sample into a sequence of *digits,* before such a signal qualifies for the adjective *digital.*

2. Communication Channel Capacity

The capacity of a communication channel to convey information depends primarily on two properties: (1) its bandwidth and (2) its signal power to overcome noise in the channel. From the early work of Hartley[1] and Shannon,[2] this capacity can, in its simplest form, be expressed as

$$C = W \log_2 (1 + S/N) \tag{10.1}$$

where C = channel capacity, bits per second
W = channel bandwidth, Hz
S = signal power
N = noise power

This expression, although most often used in discussions of digital transmission, applies specifically to an analog channel that is linear, single-path, and nondispersive with additive gaussian noise. The channel capacity expression represents, not the information rate itself, but the maximum rate that can be achieved *error-free* under conditions of optimum encoding and decoding, usually implying infinite delay in the coding process.

Equation 10.1 does not apply in its simple form to more complex forms of noise or interference, e.g., multiplicative noise (fading), doppler, and multipath. The noise power N of Eq. 10.1 usually increases with bandwidth W and, in the case of *white noise,* can be expressed as $N = N_0 W$, where N_0 is the constant power density of the noise. It can be stated generally that the capacity of any communication channel increases monotonically with its bandwidth.

This chapter is concerned only with the bandwidth parameter of Eq. 10.1 or, more specifically, with the baseband spectral characteristics and other specifications of the waveforms that are commonly used as information-bearing signals in the transmission of audio, video, and data services. The effects on link performance of the signal-to-noise power ratio in the channel are discussed in Chaps. 4, 5, 11, and 17.

INFORMATION CONTENT AND POWER DENSITY SPECTRA

3. The Modulation Process

The transmission of a waveform through an allocated radio-frequency or cable channel generally involves the modulation of a carrier (typically, a single-frequency sinusoidal carrier) so that the transmitted signal can be represented by the mathematical expression

$$s(t) = a(t) \cos \left[2\pi f_c t + p(t)\right] \tag{10.2}$$

where $a(t)$ = instantaneous amplitude
$p(t)$ = instantaneous phase
f_c = carrier frequency of the modulated wave

For the case of $a(t)$ = constant, Eq. 10.2 is that for *phase* or *frequency modulation* (PM or FM), whereas $p(t)$ = constant implies standard *amplitude modulation* (AM). The functions $a(t)$ and $p(t)$ are usually referred to as the modulating functions, and are the conveyors of the information content in the transmitted signal.

Hybrid AM-FM systems may have independent modulating functions, conveying two separate information entities, or they may be dependent as, e.g., in single-sideband transmission, wherein neither $a(t)$ nor $p(t)$ is the independent modulating function. From trigonometric identities, Eq. 10.2 can be expressed in the form

$$s(t) = \sigma\,(t) \cos 2\pi f_c t + \psi(t) \sin 2\pi f_c t \qquad (10.3)$$

expressing a hybrid AM-FM signal as amplitude modulation of quadrature carriers. If $\sigma(t)$ is an information-conveying modulating function of finite energy and if

$$\psi(t) = \hat{\sigma}(t) = \frac{1}{\pi} \int_{-\infty}^{\infty} \frac{\sigma(\tau)}{t - \tau}\, d\tau \qquad (10.4)$$

the Hilbert transform of $\sigma(t)$, the function $s(t) = \sigma(t) \cos 2\pi f_c t \pm \hat{\sigma}(t) \sin 2\pi f_c t$ represents a *single-sideband* (SSB) signal[3] whose power spectral density is confined to a single sideband on only one side of the carrier frequency, the upper or lower as determined by the choice of sign in the equation. The information-conveying modulating function of an SSB wave is thus the "in-phase" component of the wave, the quadrature component being a dependent function.

If $f(t)$ is a waveform of finite energy, i.e., of finite power and of essentially finite duration, the spectrum of the waveform or, more precisely, its amplitude and phase distribution in frequency, is given by the Fourier transform

$$F(\omega) = \int_{-\infty}^{\infty} f(t)\, e^{-i\omega t}\, dt = |F(\omega)|\, e^{iP(\omega)} \qquad (10.5)$$

which is a complex function of the angular frequency $\omega = 2\pi f$, wherein $|f)\omega)|$ and $P(\omega)$ are, respectively, the amplitude and phase components of the spectrum.

4. Baseband and Modulated-Carrier Amplitude Spectra

The amplitude spectrum of most waveforms used as modulating functions in communication (e.g., a segment of voice, video, or an analog signal embodying a symbol from a digital data stream) will have, typically, a characteristic similar to that shown in Fig. 10.1, in which maximum spectral density occurs at a low frequency and the spectrum decays to near zero at some upper frequency B, called the *bandwidth*, or more appropriate to this chapter, the *basebandwidth* of the signal. If the waveform or segment has a dc component, or a nonzero average value over its duration, the spectrum characteristic will have a nonzero value at zero frequency equal to that average times the duration T; but if the dc component persists over all time, the spectrum will exhibit in addition an impulse

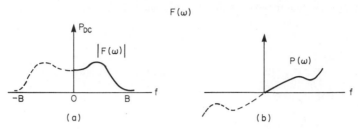

FIG. 10.1 Amplitude and phase spectra of a typical transient waveform.

function at zero frequency of weight (power) proportional to the square of the dc value. The Fourier transform function of Eq. 10.5 is defined for both positive or negative values of frequency ω, but for real functions $f(t)$, $F(-\omega) = F^*(\omega)$, the complex conjugate of $F(\omega)$. Thus the amplitude spectral characteristic $|f)\omega)|$ of Fig. 10.1 is symmetric (mirror-imaged about zero) and the phase component $P(\omega)$ is skew-symmetric, $P(-\omega) = -P(\omega)$.

If the waveform $f(t)$ is used to amplitude-modulate a sine-wave carrier, the spectrum $F_M(\omega)$ of the modulated wave is given by

$$F_M(\omega) = \int_{-\infty}^{\infty} f(t) \cos 2\pi f_c t \; e^{-i\omega t} = \tfrac{1}{2}F(\omega + 2\pi f_c) + \tfrac{1}{2}F(\omega - 2\pi f_c) \qquad (10.6)$$

and is illustrated in Fig. 10.2. The dc component has become an impulse function at carrier frequency, and the baseband component $F(\omega)$ has been translated as upper and lower sidebands about the carrier frequency having identical shapes as the (positive- and negative-frequency) baseband spectrum of Fig. 10.1. It is seen that the bandwidth of the modulated wave is exactly twice the basebandwidth in the case of AM.

Further aspects of modulation theory for both analog and digital transmission systems are found in Chap. 11 and in Refs. 3 and 4. Frequency modulation, in particular, requires more extensive treatment; the reader should see in addition Ref. 5.

The previous discussion of modulation applies to both analog and digital transmission systems, as the latter typically employ *analog* waveforms in the commu-

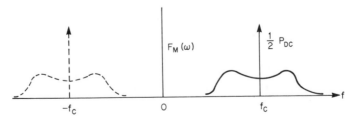

FIG. 10.2 Spectrum of modulated wave.

nication channel itself. The key distinguishing characteristic of digital transmission is that the waveforms constitute a finite set of possible waveforms selected discretely in accordance with a sequence of digital symbols, or equivalently, a digital data stream. That digital stream, in turn, might well have been derived from an analog signal as, e.g., in the transmission of digital voice or digital video. Or, alternatively, the digital stream might come from a source that is by nature data as, e.g., in teletypewriter or computer data.

Figure 10.3 tabulates several simple waveforms $f(t)$ defined over the time interval $(0, T)$ together with their spectra as defined in Eq. 10.5. These examples will be useful in deriving power density spectra of the baseband signals commonly used in digital transmission of a random bit stream (see Sec. 6).

5. The Autocorrelation Function

Information theory[6] states that a transmission system conveys information only if the waveform transmitted is to some degree unexpected by the receiver and there implicitly exists a set of probability distributions that governs its (random) occurrence. In other words, to convey information, the signal waveform (or segment of a waveform) $f(t)$ of duration T must be one random sample out of an *ensemble* of such possible waveforms having defined statistical properties. If, for example, the segment $f(t)$ represents the sound-pressure waveform of a given word or syllable spoken by a given speaker, the amplitude and phase spectra of that specific segment are really of no special interest in communications except, perhaps, to

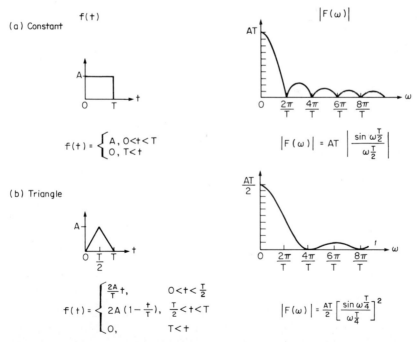

FIG. 10.3 Spectra of waveform segments. (*a*) constant; (*b*) triangle; (*c*) half-sine; (*d*) sine-square.

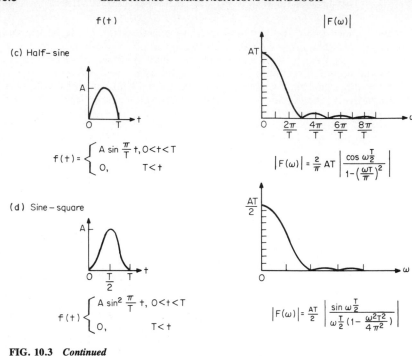

FIG. 10.3 *Continued*

evaluate a system's transient response. Of more concern is the *power spectral density* of *continuous* speech averaged over all possible words, phrases, and sentences, further averaged over the *ensemble* of all possible languages and speakers. Of course, both the average value and the deviation from the average of any spectral parameter of interest must be known to determine the required channel capacity.

To deal with an information-conveying modulating function $g(t)$ of finite power but of *infinite* duration, the power spectral density is derived through the so-called *autocorrelation function*

$$\phi(\tau) = \lim_{T \to \infty} \frac{1}{2T} \int_{-T}^{T} g(t)g(t + \tau) \, dt \tag{10.7}$$

which is a measure of how well the waveform $g(t)$ correlates with itself under translation by a time interval τ. The function $\phi(\tau)$ generally goes to zero as $\tau \to \infty$ if the signal $g(t)$ is random and has no dc or periodic components. It should be clear that the wider the basebandwidth of $g(t)$, the finer is its structure in time, and the faster will be the decay of $\phi(\tau)$ with the time shift τ. It should also be apparent from Eq. 10.7 that the *average power* of $g(t)$ is given by $\phi(0)$, the value at $\tau = 0$. The autocorrelation function is also symmetric, i.e., $\phi(-\tau) = \phi(\tau)$, if the statistical properties are time-invariant.

The autocorrelation function $\phi(\tau)$ at any displacement τ may be interpreted as the average value over all time of the product of $g(t)$ and its displaced self $g(t + \tau)$. If the statistical properties of the random variable governing $g(t)$ do not change with time (i.e., the random variable is *stationary*), the averaging can be performed over the ensemble of possible waveforms $g(t)$, and the autocorrelation function is sometimes written as the *expected* value

$$\phi(\tau) = E\left[g(t)g(t + \tau)\right] = \overline{g(t)g(t + \tau)} \tag{10.8}$$

where the bar is read *average value of*. An ensemble of random variables whose governing statistics are stationary in time and unvarying over the ensemble is often referred to as an *ergodic process* or an *ergodic ensemble,* having the property that time and ensemble averages are equal.[4]

The power density spectrum of the function $g(t)$ is defined by the Fourier transform of the autocorrelation function of $g(t)$,

$$\Phi(\omega) = \int_{-\infty}^{\infty} \phi(\tau)e^{-i\omega\tau}\, d\tau \tag{10.9}$$

Equations 10.7 through 10.9 will be used to derive the spectra of some commonly used digital baseband waveforms under the assumption that they convey the information contained in a *purely random* bit stream. This will demonstrate that the power density spectrum of a probabilistically repeating waveform segment is related to the amplitude and phase spectrum of that waveform, as described in Sec. 3.

6. The Power Density Spectrum of a Random Bit Stream

The calculation of the power density spectrum of a random bit stream employs the function $f(t)$ of Eq. 10.5, which is assumed to have finite energy and finite duration T. Assume that time is divided into intervals of duration T seconds, and in each successive interval the function $f(t)$ (defined over 0, T) will occur with probability $\frac{1}{2}$ and with the same probability the interval will be occupied by the negative function $-f(t)$. Assume that the occurrence of $\pm f(t)$ in a given interval will be statistically independent of all previous and future occurrences (*purely random*). The resulting random waveform $g(t)$ is shown at a particular point in time in Fig. 10.4.

The random waveform of successive positive or negative segments $f(t)$ is now of infinite duration but still is of finite power. The autocorrelation function $\phi(\tau)$ may be obtained from Eq. 10.7:

$$\phi(\tau) = \lim_{T \to \infty} \frac{1}{2T} \int_{-T}^{T} g(t)g(t + \tau)\, dt = \overline{g(t)g(t + \tau)} \tag{10.10}$$

Figure 10.5 shows the set of possible conditions in which $g(t)$ in the interval (0, T) is one of the segments $\pm f(t)$; and the translated function $g(t + \tau)$ is shown

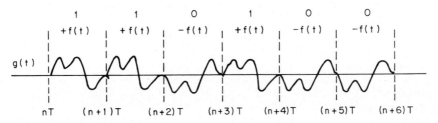

FIG. 10.4   A digital wavetrain.

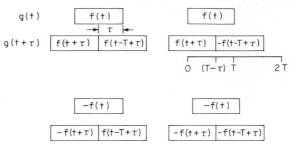

FIG. 10.5 Ensemble of autocorrelation conditions for $0 < |\tau| < T$

over the two intervals $(0, T)$ and $(T, 2T)$ for which the signal in the latter interval is the segment $\pm f(t - T + \tau)$, the displaced adjacent segment. As Fig. 10.5 constitutes the complete *ensemble* of all possible conditions for the case $0 \le \tau \le T$, the autocorrelation function $\phi(\tau)$ is obtained by averaging in time over the interval $(0, T)$ and further averaging the ensemble over all possible conditions, each weighted by the probability of occurrence of that condition. Thus, as each condition in Fig. 10.5 occurs with equal probability ¼,

$$\phi(\tau) = \overline{g(t)g(t + \tau)} \qquad 0 \le \tau \le T$$

$$= \frac{1}{4}\left\{\frac{1}{T}\int_0^{T-\tau} f(t)f(t + \tau)\, dt + \frac{1}{T}\int_{T-\tau}^T f(t)f(t - T + \tau)\, dt\right\}$$

$$+ \frac{1}{4}\left\{\frac{1}{T}\int_0^{T-\tau} f(t)f(t + \tau)\, dt + \frac{1}{T}\int_{T-\tau}^T f(t)[-f(t - T + \tau)]\, dt\right\}$$

$$+ \frac{1}{4}\left\{\frac{1}{T}\int_0^{T-\tau} [-f(t)][-f(t+\tau)]\, dt + \frac{1}{T}\int_{T-\tau}^T [-f(t)]f(t - T + \tau)\, dt\right\}$$

$$+ \frac{1}{4}\left\{\frac{1}{T}\int_0^{T-\tau} [-f(t)][-f(t+\tau)]\, dt + \frac{1}{T}\int_{T-\tau}^T [-f(t)][-f(t - T + \tau)]\, dt\right\}$$

$$= \frac{1}{T}\int_0^{T-\tau} f(t)\,f(t + \tau)\, dt \tag{10.11}$$

all other terms canceling.

The case for $|\tau| > T$ is shown in Fig. 10.6, wherein the translated function $g(t + \tau)$ under these larger translations overlaps segments of $g(t) = \pm f(t)$ with statistical independence, and the self-term occurs only probabilistically. There are eight equiprobable conditions, and for every positive term, there is an equal corresponding negative term; thus $\phi(\tau) = 0$ for $|\tau| > T$. From the property that $\phi(-\tau) = \sigma(\tau)$,

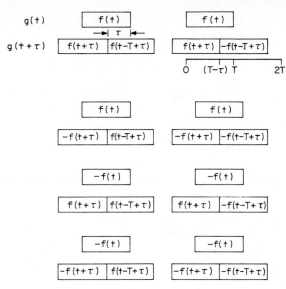

FIG. 10.6   Ensemble of autocorrelation conditions for $T < |\tau|$

$$\phi(\tau) = \frac{1}{T} \int_0^{T-|\tau|} f(t)f(t + |\tau|)\, dt \qquad |\tau| \leq T$$
$$= 0 \qquad\qquad\qquad\qquad\qquad\quad |\tau| > T \tag{10.12}$$

As the product $f(t)f(t + |\tau|)$ is zero outside the interval $(0, T - |\tau|)$, the integration limits may be extended without loss of generality and the autocorrelation function (for positive τ) becomes

$$\phi(\tau) = \frac{1}{T} \int_{-\infty}^{\infty} f(t)f(t + \tau)\, dt$$

$$= \frac{1}{T} \int_{-\infty}^{\infty} f(t) \left[\frac{1}{2\pi} \int_{-\infty}^{\infty} F(\omega)e^{i\omega(t+\tau)}\, d\omega \right] dt \tag{10.13}$$

Interchanging the order of integration gives

$$\phi(\tau) = \frac{1}{2\pi T} \int_{-\infty}^{\infty} F(\omega)e^{i\omega\tau} \int_{-\infty}^{\infty} f(t)e^{i\omega t}\, dt\, d\omega$$

$$= \frac{1}{2\pi T} \int_{-\infty}^{\infty} F(\omega)F^*(\omega)e^{i\omega\tau}\, d\omega \tag{10.14}$$

Thus from Eq. 10.9, the power density spectrum $\Phi(\omega)$ of the signal $g(t)$ can be seen to be

$$\Phi(\omega) = \frac{1}{T} \, |F(\omega)|^2 \qquad (10.15)$$

which is proportional to the squared magnitude of the amplitude spectrum of the finite segment $f(t)$.

ANALOG SIGNAL FORMATS—VOICE AND RADIO

Basebandwidths of 3, 5, and 15 kHz are used in the transmission of speech and high-fidelity music. Telephone circuits are designed to have bandwidths of approximately 3 to 4 kHz. In broadcasting, radio stations utilizing amplitude modulation and frequency modulation limit the baseband audio signal to 5 and 15 kHz, respectively. These limits are selected on the basis of tradeoffs between technical performance characteristics, spectrum conservation, and economic channel costs.

Frequency and intensity ranges of speech and orchestral music are shown in Fig. 10.7. The solid curve in the figure shows the minimum audible threshold and the threshold of feeling for an average listener. To reproduce music faithfully, a baseband extending from approximately 45 Hz to over 12 kHz with 70-dB dynamic range capability is required. To reproduce speech the requirements are more modest.

The formats for analog video services are described in Sec. 17 and Chap. 17.

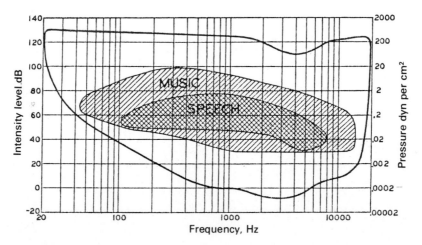

FIG. 10.7 Frequency and intensity ranges of speech and music (from *Bell Laboratories Record,* June, 1934).

7. Voice Services

Mathematical models have been developed[7] that relate the process of speech production to physical characteristics of the human vocal system. Direct experimental measurements have also been performed that give power spectra for typical conversational speech signals averaged over many talkers. Figure 10.8 shows data obtained by Dunn and White[8] for both male and female speakers.

Probability density for speech signals have also been estimated by determining histograms of signal amplitude for a large number of speech samples. Davenport[9] made extensive measurements of this kind. Paez and Glisson[10] have shown that the experimental data is well approximated by a gamma distribution of the form

$$p(\chi) = \left[\frac{\sqrt{3}}{8\pi\sigma_\chi |\chi|}\right]^{1/2} e^{-\sqrt{3}|\chi|/2\sigma_\chi} \tag{10.16}$$

or by the simpler Laplacian density of the form

$$p(\chi) = \frac{1}{\sqrt{2}\,\sigma_\chi} e^{-\sqrt{2}|\chi|/\sigma_\chi} \tag{10.17}$$

Figure 10.9 shows plots of experimental data and the gamma and Laplacian approximations.

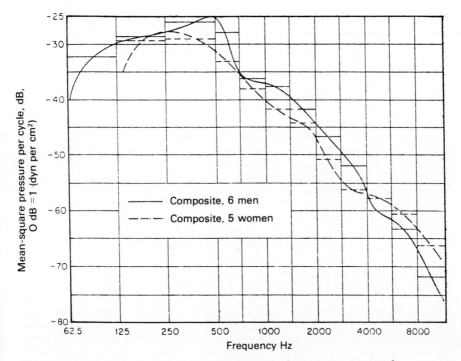

FIG. 10.8 Power spectra for typical conversational speech (After Dunn and White[8]).

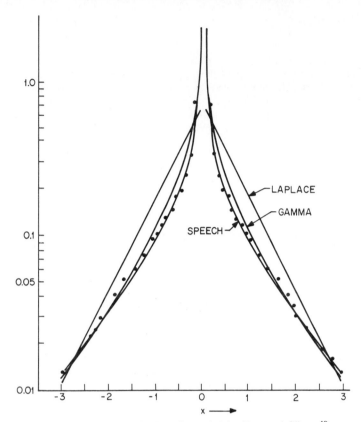

FIG. 10.9   Probability distribution of speech (after Paez and Glisson[10]).

Effects of band limiting and reduced dynamic range on quality of speech have been extensively investigated. The primary tool used in these investigations is to measure percentage of words or sounds uttered by a talker and correctly perceived by a listener. This percentage is referred to as percent articulation.

Figure 10.10 shows loss in articulation and change in signal energy caused by elimination of all speech frequencies above or below a given frequency.[11] For example, passing all frequencies above 1 kHz reduces articulation to 86 percent and signal energy to 17 percent, while low-pass filtering with a 4-kHz cutoff frequency reduces articulation to 88 percent and signal energy to 98 percent. Additional information and extensive references on effects of channel impairments on articulation and speech quality are given by Meeker.[12]

Recommended specifications for attenuation distortion of a telephone voice channel are given in CCITT Vol. III, Recommendations G.132 and G.151A. CCITT specifies reference frequency at 800 Hz, whereas 1000 Hz is more commonly used in North America. Figure 10.11 shows a typical attenuation distortion curve for a voice channel (1000-Hz reference frequency). Figure 10.12 shows limits to attenuation distortion recommended by CCITT for the worst case normally encountered in telephone transmission, namely 12 circuits of a four-wire chain in tandem.

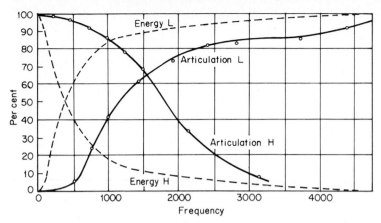

FIG. 10.10 Loss in articulation and signal energy from restriction of bandwidth (from Fletcher[11]).

Some telephone circuits are designed with a 3-kHz nominal bandwidth, i.e., a band that is narrower than that implied by the attenuation distortion curve of Fig. 10.11. For such equipment CCITT recommends that limits on attenuation distortion be as shown in Fig. 10.12.

In addition to attenuation distortion, envelope delay (derivative of phase shift with respect to angular frequency) variation across a baseband contributes to quality of transmitted speech. Envelope delay distortion of a voice channel is a function of the transmission medium used. Figure 10.13 shows some typical envelope delay data for a variety of transmission media. Delay data are usually expressed in milliseconds about a reference frequency.

CCITT in Recommendation G.133 recommends that envelope delay distortion (group-delay distortion) not exceed limits given in Table 10.1. These limits are for the worst case of 12 circuits in tandem. Delay values specified in Table 10.1 are

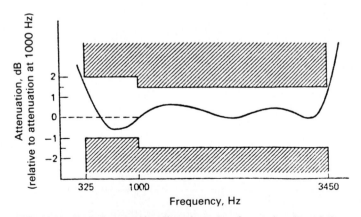

FIG. 10.11 Typical attenuation distortion curve for a voice channel (from Freeman, R. L., *Telecommunication Transmission Handbook*, 2nd Ed., John Wiley & Sons, 1981).

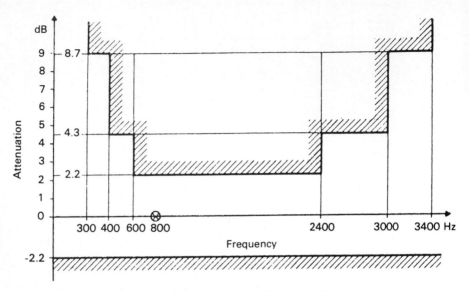

FIG. 10.12   Limits on attenuation distortion (from CCITT Rec. 132).

at band edges which are in the range of 260 to 320 Hz for the low edge and 3150 to 3400 Hz for the high edge.

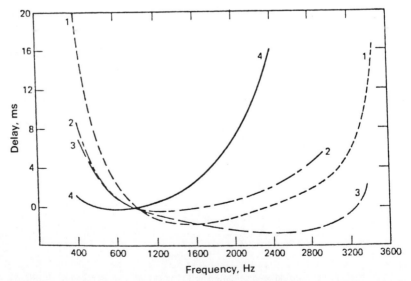

FIG. 10.13   Typical envelope delay characteristics (from Freeman, R. L., *Telecommunication Transmission Handbook,* 2nd Ed., John Wiley & Sons, 1981).

TABLE 10.1 Limits on Group Delay Distortion

Communication link	Delay at lower limit of frequency band, ms	Delay at upper limit of frequency band, ms
International chain	30	15
Each of the national four-wire extensions	15	7.5
On the whole four-wire chain	60	30

Source: Ref. 15

8. Radio Broadcast Services

For radio broadcast services, the basebandwidth represents a compromise among many factors. AM radio service occupies the frequency range from 535 to 1605 kHz. This range is divided into 107 channels, with assigned carrier frequencies 10 kHz apart.[13] At the transmitter, baseband frequency response is typically well within ± 1 dB from 50 Hz to 10 kHz. This bandwidth is adequate for high-quality transmission of speech and orchestral music with spectral properties as shown in Fig. 10.7.

Two carriers separated by 10 kHz and amplitude modulated with a baseband signal of 10 kHz width will produce strongly interfering signals in a receiver. Approaches adopted to reduce this interference include band limiting the receiver response to 5 kHz or less, band limiting the modulating signal bandwidth at the transmitter, assigning carrier frequencies on a geographical basis to separate adjacent carriers physically, and tailoring the radiation pattern and radiated power from a transmitting antenna to limit coverage. The effective baseband response of most consumer-type AM receivers is from (at −3 dB) a high-end frequency of between 2.5 and 5 kHz, with 2.5 kHz being more common, to a low-end frequency of between 100 and 300 Hz, with 200 Hz being more common.[13]

FM radio service occupies the frequency range from 88 to 108 MHz. This range is divided into 100 channels of 200 kHz width. Baseband frequency response extends from 50 Hz to 15 kHz. Considerable improvement[13] in received *signal-to-noise ratio* (*S/N*) is realized by boosting baseband high-frequency response in the transmitter (*preemphasis*), and reducing it in the receiver (*deemphasis*). The standard preemphasis curve is defined as an ideal *RC* network with a time constant of 75 μs and a frequency response as shown in Fig. 10.14. Improvement in received *S/N* resulting from deemphasis is shown in Fig. 10.15.

DIGITAL SIGNAL FORMATS

9. Digital Audio

Voice Signals. A number of approaches have been developed for digital representation of analog signal waveforms. The simplest of these, and the one most extensively used, is *pulse-code modulation* (PCM). In PCM the peak-to-peak range of an analog signal is divided into a set of uniform ranges. The number of ranges is normally chosen to be a binary number of the form 2^N. A code word of N bits is assigned to each range. The analog signal is sampled at, say, the Nyquist rate, and each sample is represented by a code word corresponding to the range

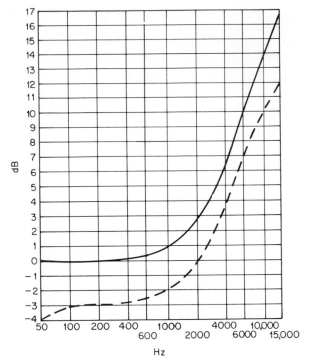

FIG. 10.14   The FCC standard preemphasis curve (from NAB Engineering Handbook[13]).

in which the sample level falls. The sequence of code words may be transmitted as a serial bit stream at a rate of $F \times N$ b/s, where F is the sampling rate and N is the number of bits per sample (code word).

In reconstructing an analog signal from a digital bit stream, the amplitude of each reconstructed sample is the midrange value of the range corresponding to the received code word. Thus the reconstructed signal differs from the original signal by an error determined by the quantization step size. For random analog signals and small quantization step sizes, the error takes on characteristics of noise and may be described by a simple statistical model.[7] Under such conditions, a signal-to-quantizing-noise ratio is an objective measure of quality of digital representation of an analog waveform. The S/N in decibels is given by[7]

$$S/N = 10 \log (\text{signal variance/noise variance}) = 6N + 4.77 - \quad (10.18)$$
$$20 \log (\text{signal peak/signal standard deviation})$$

The S/N ratio is seen to be independent of variance of quantizing noise, but dependent on signal characteristics. This is a result of uniform quantization which leads to a constant noise variance. For both speech and music, quality of a decoded signal will thus vary substantially during soft and loud passages.

For band-limited signals of a telephone voice channel, an 8-kHz sampling rate is adequate for converting input analog waveforms to digital bit streams. Subjec-

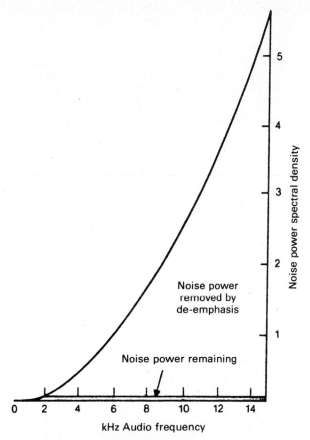

FIG. 10.15 Improvement in SNR from de-emphasis (from NAB Engineering Handbook[13]).

tive testing indicates that for linear PCM coding, i.e., uniform quantizing, speech quality is equivalent to that of an analog channel when the digitizing rate is a minimum of 14 bits per sample, equivalent to a 112-kbs data rate.

Figure 10.16 shows schematically the technique used to reduce the digitizing rate for telephone voice-frequency signals. The input analog signal is first compressed by a network with a nonlinear transfer function, and is then converted

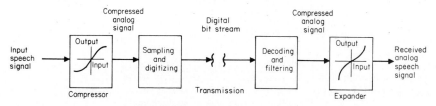

FIG. 10.16 Companding for voice-frequency signals.

into 8-kHz samples at eight bits per sample, i.e., a 64-kbs data rate. Nonlinear signal compression effectively converts uniform quantizing to nonuniform quantizing. At the receiver the incoming bit stream is converted to an analog signal which is expanded with a network having a transfer function that is inverse to that of the compressor. The expanded signal is the desired output. The operation of compression-expansion is referred to as companding. Subjective quality of speech is approximately the same for linear PCM coding at 14 bits per sample and companded PCM coding at eight bits per sample.

CCITT Recommendation G.711 advises use of one of two alternative companding laws, μ law and A law. For a normalized coding range of ± 1, the companding laws, shown graphically in Figs. 10.17 and 10.18, are given by

$$\mu \text{ law:} \quad F_\mu(\xi) = \text{sgn}\ (\xi)\ \frac{\ln\ (1 + \mu|\xi|)}{\ln\ (1 + \mu)} \qquad -1 \le \xi \le 1 \qquad (10.19)$$

$$A \text{ law:} \quad F_A(\xi) = \text{sgn}\ (\xi)\ \frac{1 + \ln A|\xi|)}{1 + \ln A} \qquad \frac{1}{A} \le \xi \le 1 \qquad (10.20)$$

$$= \text{sgn}\ (\xi)\ \frac{A|\xi|}{1 + \ln A} \qquad 0 \le |\xi| \le \frac{1}{A} \qquad (10.21)$$

In actual PCM systems the companding law is approximated by a set of linear segments. With the advent of large-scale integrated circuits, companding is performed in the digital domain by converting the linear PCM code at 14 bits per sample to a companded PCM code at 8 bits per sample.

Digital transmission of high-fidelity audio signals is discussed in CCIR Report 647. For a signal bandwidth extending from 40 Hz to 15 kHz, the CCIR[14] proposes A-law companding of the signal, 32-kHz sampling frequency, and 10 bits per sample digitizing, resulting in a 320 kbs transmission rate. CCITT Recommendation J.41 suggests a 384-kbs transmission rate, resulting from digitizing the A-law companded signal at 11 bits per sample and adding a 32 kbs error protection code.[15] CCIR Recommendation 606 and Report 648-2 suggest use of a 32-kHz sampling rate and discuss a number of alternative companding and digitizing strategies along with error control redundancies.[16]

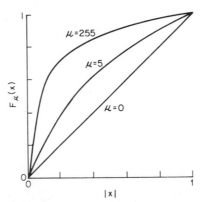

FIG. 10.17 μ-Law compression characteristics.

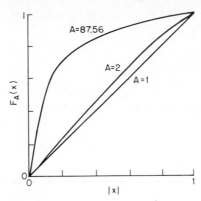

FIG. 10.18 *A*-law compression characteristics.

Radio Broadcast Signals. An operational system, known as the *digital audio transmission system* (DATS), is used by all the major radio networks in the United States to transmit radio programming from the networks' Manhattan studios to affiliate radio stations via satellite. The system uses 15-bit linear PCM coding at a sampling rate of 32 kHz. The digital signal is converted to 11-bit μ law companded ($\mu = 255$) plus one bit for parity, i.e., 12 bits per sample, for a transmission rate of 384 kbs.

Alternative sampling and bit rates are in use for digitizing high-fidelity analog signals for specialized applications. For example, compact disk digital audio systems use a 44.1-kHz sampling rate and 16 bit per sample linear PCM coding[17] to accommodate a 20-kHz signal baseband. Similarly the Sony DVR/DVPC-1000 television studio digital videotape recorder uses a sampling rate of 48 kHz and 16-bit linear PCM to record a 20-kHz audio signal.

10. Integrated Services

A digital transmission system might convey signals derived from a digital data source, such as a sequence of alphanumeric characters in a reservation system, or signals derived from the sampling, quantizing, and digitizing (encoding) of an analog source, such as voice, video, or telemetry. An *integrated services digital network* (ISDN) might thus convey voice, data, graphics, facsimile, or other such signals derived from either analog or digital sources. In either case, digital encoding permits the signal to be transmitted to the receiver as noisefree as desired (even for a fixed transmitter power) if the transmission channel has sufficient bandwidth.

The generic digital transmission system is shown in the block diagram of Fig. 10.19. If the signal source is analog, the first process is generally *antialias prefiltering* to limit the spectrum of the signal (and also the source noise) to be within the Nyquist limits of the sampler.

Perhaps the most important design parameter in a digital transmission system that will carry analog source signals is the choice of a high enough *sampling rate* to control *aliasing,* a particularly annoying type of noise or impairment that can occur in sampled systems. Figure 10.20*a* illustrates the principle of sampling a

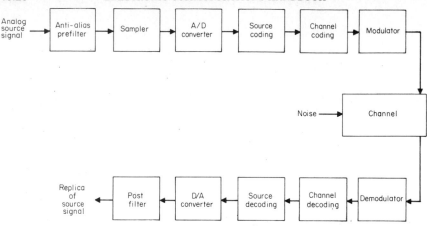

FIG. 10.19   The digital communication system.

continuous signal to derive a sequence of discrete samples. The sampling process is tantamount to multiplying the continuous signal by a periodic train of impulses each of constant energy. The pulse train in effect becomes a *carrier* with fundamental frequency equal to the sample rate, but a carrier that is very rich in harmonics. The narrower the sampling pulse, the slower the harmonics decay with order. For all practical purposes, the level of the harmonics can be considered to be flat to infinity. The sampled signal thus becomes an amplitude-modulated wave with a spectrum constituting upper and lower sidebands displaced about each harmonic, as illustrated in Fig. 10.20*b*. It is clear from the figure that if the sampling frequency is less than twice the basebandwidth of the wave being sampled, the lower sideband of the first (fundamental) carrier component overlaps the baseband spectrum. This overlap region produces spurious components called "aliases" that become inextricably imbedded in the sampled wave.

Figure 10.21 illustrates a similar sampling process with a sampling rate that is comfortably above $2W$ Hz, the so-called Nyquist rate. (The *Nyquist criterion* for sampling states simply that the sampling frequency should provide at least two samples per cycle of the highest frequency component in the spectrum of the signal being sampled.) An ideal low-pass filter of bandwidth $2W$ can be applied to the

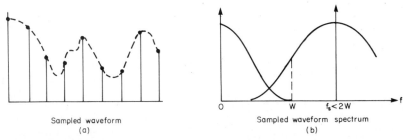

Sampled waveform
(a)

Sampled waveform spectrum
(b)

FIG. 10.20   The alias effect of subsampling. (a) sampled waveform; (b) sampled waveform spectrum.

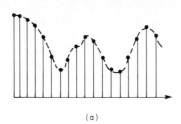 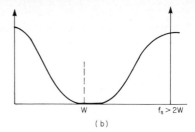

(a)   (b)

FIG. 10.21 Sampling above the Nyquist rate.

sampled signal to recover the original wave. In practice, because the original signal must be low-pass-filtered ahead of the sampler to remove any noise or other components above the basebandwidth, the headroom required by practical low-pass filters generally demands that the sampling frequency be chosen to be well above the Nyquist rate, say about $2.5W$, giving 2½ samples per cycle in the sampled signal.

In Fig. 10.22, the effect of sub-Nyquist sampling of a sinusoidal wave is illustrated. The samples (the envelope of which is indicated by the dashed line) are taken at slightly greater than 1 sample per cycle of the wave. Low-pass filtering of the samples would yield the dashed envelope rather than the original wave, which could never be recovered from the value of the samples. Thus the alias is manifest by a spurious frequency of a much lower frequency. An additional sample taken midway between each of those shown would provide greater than two samples per cycle, and the original wave would be essentially traced by the envelope of these samples.

One of the major advantages of digital transmission is that noise does not accumulate over cascade links, as long as each link has sufficient capacity to avoid errors in the transmission of the digits. At each link receiver in a cascade of links, a decision is made as to which digit or symbol was transmitted and the symbol waveform is regenerated for transmission over the next link, thereby "erasing" the noise. When the received signal at the terminal link of a cascade is converted back to its analog form, the residual noise is essentially limited (except for the occasional transmission error) to the quantization noise that occurred at the analog-to-digital conversion point ahead of the transmission path. Thus the desired S/N ratio at the end of the complete transmission path can be established by the *dynamic range* (or number of quantization levels, expressed in decibels) of the *analog-to-digital* (A/D) converter shown in Fig. 10.19.

The dynamic range (the ratio of the peak signal level to the rms quantization noise) of the A/D converter is given approximately by the expression $20 \log_{10} L$ decibels, where L is the number of quantizing levels. As A/D converters are usually designed for a binary output, L is typically a power of 2 and the dynamic

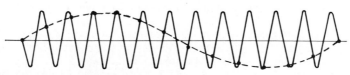

FIG. 10.22 Subsampling of a sine wave.

range becomes approximately $6 \log_2 L$ decibels. For a quantizer of 1024 levels (10 bits), both expressions yield about 60 dB. These approximations make use of the simplifying assumptions (1) that the rms quantization noise is equal to one quantizing level, and (2) that $20 \log_{10} 2 = 6$, neither of which is exact. Both assumptions are useful as rules of thumb, however, and the latter assumption yields 6 dB per bit.

If the digitized signal is to be further processed before transmission, the A/D converter output is usually in parallel format, with each bit on a separate line.

The box labeled *source coding* in Fig. 10.19 comprises those processes that are applied to the source signal to *alter* its statistical or spectral properties. These processes might involve further (digital) filtering, bit-rate reduction to remove excess redundancy in the signal, and possibly sub-Nyquist sampling (or subsampling), a particular bit-rate-reduction technique that is often applied to video signals.[18] Subsampling creates aliases in the signal, but if applied in a controlled way, transmission can be achieved over a reduced-rate channel and postprocessing at the receiver can be applied to ameliorate the alias effects.

The box labeled *channel coding* contains those processes that take into account the statistical characteristics of the *channel*. The most common such process is *error-control,* or *error-detection and/or error-correction,* encoding, in which redundancy is deliberately introduced into the signal (in a controlled way) to reduce the probability of errors in the *channel decoder,* where the inverse error-control decoding is applied. The *source decoder* similarly applies the appropriate inverse decoding processes to restore the signal to (as nearly as possible) its original (digital) form at the source.

The *digital-to-analog* (D/A) converter and the postfilter [or $(\sin x)/x$ equalizing filter] provide a replica of the input analog source signal. Differences between the replica and the original source signal constitute the residual transmission noise that, as discussed previously, can be made as small as desired (down to the quantizing-noise limit at the A/D converter) by the channel encoding-decoding process. This assumes that the channel has sufficient marginal capacity over the source information rate, and the delay and complexity of the encoding-decoding process can be tolerated by the communication application.

The block diagram of Fig. 10.19 shows only one channel, but cascades of channels (with separate and possibly distinct channel encoding-decoding processes) usually occur. As pointed out earlier, cascades of *digital links* have the advantage over analog links that channel noise does not accumulate as long as each link has sufficient capacity to limit the error probability to a specified small value.

If the source signal originates in a digital form as, e.g., from a data source, the prefiltering, A/D, D/A, and postfiltering boxes of Fig. 10.19 are eliminated. Both source coding and channel coding can be and usually are applied to signals from a data source. The *modulator* and *demodulator* on either side of the channel in Fig. 10.19 comprise, respectively, the selection at the transmitter of a waveform in accordance with the symbol (or bit) to be transmitted, and the decision (after filtering or other processing) at the receiver as to which of the possible symbols (or bits) was probably transmitted (a maximum-likelihood detection). As most digital transmission systems tend to be implemented in full duplex, the modulator and demodulator are generally combined into a common box of hardware, called a *modem.* (Unfortunately, the term *modem* is sometimes incorrectly applied in simplex systems to mean either modulator or demodulator.)

As most digital transmission systems are based on *binary* selection and decision, the digital symbols are usually *binary digits* or bits, and this chapter deals only with binary transmissions. However, there is an important and growing class of digital transmission systems that are *nonbinary* (*M*-ary) in nature, and the selection and decision processes in the modems involve nonbinary symbols (sometimes derived from multilevel signals or samples) or may involve the transmission of several successive bits at a time for each waveform selected, e.g., an eight-bit alphanumeric character transmitted as a 64-ary waveform. See Refs. 6 and 19 for theory and Ref. 20 for some recent practice in nonbinary transmission.

The *baseband* signal of a digital transmission system appears directly at the output of the modulator only in the case of a *baseband transmission channel,* e.g., a twisted-pair telephone line, a coaxial cable, or a tape or disk recording. A *baseband channel* is defined as a channel whose upper and lower frequency limits have values the ratio of which is much larger than one ($\gg$ 1). Such a channel often does not involve a carrier. Otherwise, the baseband signal occurs only inside the modulator, or perhaps, implicitly, in a modulator that selects a *radiofrequency* (rf) modulated-carrier waveform in accordance with the input digital symbol. In such a case, there exists implicitly a modulating function whose waveform constitutes the baseband signal.

To determine the spectra of various digital baseband waveforms, assume that the input to the modulator consists of a time sequence of the binary symbols 0 or 1 occurring in each time slot with statistical independence of their occurrence in other time slots (a purely random bit stream). Figure 10.23 illustrates such a bit stream (*a*) and some of the commonly used digital baseband signal waveforms that would be generated in accordance with the symbols from the bit stream. Some of these waveforms are useful primarily for illustration purposes while others are important because of their widespread use. In the former category are (*b*) on-off keying and (*c*) RZ (return-to-zero); in the latter category are (*d*) NRZ (nonreturn-to-zero), (*e*) Miller, and (*f*) Manchester. All of these examples are applicable to baseband channels explicitly (*d, e,* and *f* are commonly used in digital recording). Others are the implicit baseband waveforms in rf channels. For example, NRZ is the implicit baseband waveform in binary PSK, which can be regarded as amplitude modulation of a sinusoidal carrier by the NRZ waveform.

DATA SERVICES

11. History and Characteristics

Spurred by the rapid growth of the computer industry just after World War II, data communications has become the most rapidly growing communication segment. The first electrical medium, Morse's telegraph, was in concept not very different from today's digital systems. The history of modern data communications can thus be traced in the evolution of the terminals from the telegraph key and sounder to the teletypewriter to the modern computer data terminal, over the span of a century.

The major sophistication added in this century of evolution has been not the technology of the terminals nor the transmission media, but rather the overlay of the *network* to the earlier point-to-point links, and in the complex processing that is applied in the network to the signals and to the data itself. Thus a modern-day data communication system is typically a network of processors and peripherals

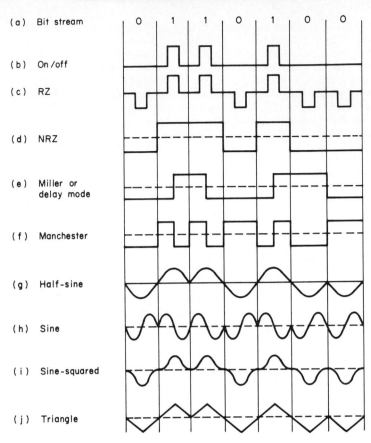

FIG. 10.23 Digital wavetrains.

all interconnected to move data between sources and users that are spread over a wide geographical area. In fact, the terms data communications and computer communications have become virtually synonymous.

The architecture of a modern data network involves many design tradeoffs that might be specific to the application, but the complete specifications for the system will include:

- Transaction classification
- Network topology
- Network operating system
- Source codes
- Communication protocols
- Transmission links
- Software
- Hardware

12.   Transaction Classification

A *transaction* is a business event, such as placing an order, sending an invoice, making a payment, or simply asking or answering a query. The term *data services* has generally been applied to the restricted class of information services, i.e., providing a data base and access to it, but more and more, the term is being applied more broadly to include transaction services, e.g., placing a reservation or making an electronic funds transfer. Thus a data network might be classified as to the type of transaction it is designed to implement. However, in a data communication network, a *transaction* is manifest simply by the transmission and *acknowledgment* (ACK) or *nonacknowledgment* (NAK) of a message. The *transaction classification* of a network is a design factor, as it is representative of the average length and frequency of a message in each direction, or the average time for a transaction. These classifications include conversational, inquiry/response (I/R), record update (R/U) (data entry or collection), and batch or remote job entry (RJE).

Conversational transactions are characterized by short messages of approximately balanced or equal length in both directions. *Inquiry/response* transactions tend to be short messages for the inquiry and somewhat longer messages in the response direction. *Record update* transactions, which can be either data entry or collection, will typically have very much longer messages in one direction than the other. Finally, *batch* transactions might involve a very long message of data and program from a terminal for, say, overnight processing at a computer center with another very long message back the next day containing the processed results.

13.   Network Topology

The *topology* of a network is a pattern or diagram of its interconnectivity as illustrated by the classes of topology in Fig. 10.24. The earliest computer time-sharing networks were connected in a simple star configuration, which is appropriate when the terminals communicate only with the central computer. However, for applications in which terminals communicate with each other (e.g., electronic mail), and processors communicate with other processors, more complex configurations such as meshes of trees can be appropriate. Here some processors serve only to control communications (front-end processors, circuit and message switches, concentrators). The ring topology is a common configuration used in some *local-area networks* (LANs).

14.   Network Operating System

Since one of the chief benefits to be gained from a complex multicomputer network is resource sharing, the concept of a *network operating system* was developed to permit terminals on the network to access any computer on the network

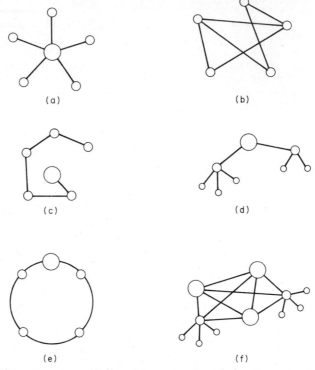

FIG. 10.24 The topologies of data networks. (a) star; (b) mesh; (c) multidrop line; (d) tree; (e) loop; (f) mesh of trees.

in a common operating system, even though each computer may operate in its own (different) operating system.[21]

15. Source Codes

The *source codes* of a data communication system are the alphabets of the source signals and the data bases. Thus the source codes of the telegraph were the Morse codes,[22] those of the teletypewriter and telex were Baudot and Moore,[23] and today's transaction networks are likely to use ASCII or EBCDIC, the alphabets of computers (see Fig. 10.25).

16. Communication Protocols

The *protocols* of a communication network are, as the term implies, a set of rules whereby the communicating terminals establish a connection or "handshake," and take specific actions during the communication process. The earliest time-sharing computer systems used a simple polling protocol, whereby each terminal in the network is polled in sequence and responds with a message that incorporates an error detection code. If no errors are detected by the com-

b4 b3 b2 b1	ROW	000 (0)	001 (1)	010 (2)	011 (3)	100 (4)	101 (5)	110 (6)	111 (7)
0 0 0 0	0	NUL	DLE	SP	0	@	P	`	p
0 0 0 1	1	SOH	DC1	!	1	A	Q	a	q
0 0 1 0	2	STX	DC2	"	2	B	R	b	r
0 0 1 1	3	ETX	DC3	#	3	C	S	c	s
0 1 0 0	4	EOT	DC4	$	4	D	T	d	t
0 1 0 1	5	ENQ	NAK	%	5	E	U	e	u
0 1 1 0	6	ACK	SYN	&	6	F	V	f	v
0 1 1 1	7	BEL	ETB	'	7	G	W	g	w
1 0 0 0	8	BS	CAN	(	8	H	X	h	x
1 0 0 1	9	HT	EM	)	9	I	Y	i	y
1 0 1 0	10	LF	SUB	*	:	J	Z	j	z
1 0 1 1	11	VT	ESC	+	;	K	[	k	{
1 1 0 0	12	FF	FS	,	<	L	\	l	|
1 1 0 1	13	CR	GS	-	=	M	]	m	}
1 1 1 0	14	SO	RS	.	>	N	^	n	~
1 1 1 1	15	SI	US	/	?	O	_	o	DEL

(a)

Most significant bits (bit 8 transmitted last)

bit positions 8,7,6,5

Least significant bits (bit 1 transmitted first)

4 3 2 1		0	1	2	3	4	5	6	7	8	9	10	11	12	13	14	15
0000	0	NUL	DLE	DS		SP	&	-						(	)	\	0
0001	1	SOH	DCI	SOS						a	j	~		A	J		1
0010	2	STX	DC2	FS	SYN					b	k	s		B	K	S	2
0011	3	ETX	DC3							c	l	t		C	L	T	3
0100	4	PF	RES	BYP	PN					d	m	u		D	M	U	4
0101	5	HT	NL	LF	RS					e	n	v		E	N	V	5
0110	6	LC	BS	ETB	UC					f	o	w		F	O	W	6
0111	7	DEL	IL	ESC	EOT					g	p	x		G	P	X	7
1000	8		CAN							h	q	y		H	Q	Y	8
1001	9	RLF	EM						\	i	r	z		I	R	Z	9
1010	10	SMM	CC	SM		¢	!	|	:								
1011	11	VT				.	$	,	#								
1100	12	FF	IFS		DC4	<	•	%	@								
1101	13	CR	IGS	ENQ	NAK	(	)	_	'								
1110	14	SO	IRS	ACK		+	;	>	=								
1111	15	SI	IUS	BEL	SUB	|	¬	?	"								

(b)

FIG. 10.25 (a) ASCII code set: communications control characters are outlined; (b) EBCDIC code set.

puter, an ACK message is transmitted back and the central computer polls the next terminal for an exchange. In the event of a channel error in either direction, a NAK response requests a repeat of the failed message. This ACK/NAK protocol originated in the *automatic repeat request* (ARQ) technique of early teletype systems and is still used extensively today.

To minimize the response time in computer time sharing, the polling protocol was originally carried out on a character-by-character basis with the ACK and NAK messages being control characters in the alphabet of the source code. As data networks became more complex and as data channel capacities have become higher, the higher efficiency (lower overhead) of longer messages has displaced character-by-character polling except in local networks. In extensive resource-sharing computer networks such as ARPANET,[24] the use of packetized messages of variable lengths has proven to be the most flexible signal format for data networks that are designed for a wide range of applications.

The transmission formats of the links of a network using a polling protocol are *time-division multiplex* (TDM) or *time-division multiple access* (TDMA), as the messages appear in the links in nonoverlapping sequences with guard times between messages being determined by the propagation delay. Obviously, in a star network with a large number of terminals, each link is used very inefficiently because it is unoccupied during most of the polling cycle. This inefficiency (and its high line cost) led to the use of multidrop lines as in configuration *b* of Fig. 10.24. These shared lines work well in the polling protocol because only one terminal is ever transmitting at any time.

Even with multidrop lines, however, a polling protocol can represent inefficient use of the lines, especially for terminals that operate with low duty cycle. Even though the computer controls the polling sequence, in most data networks the communication is actually initiated by the terminal. Thus much time is wasted in the polling of a terminal that has at that instant no message ready to transmit. To address these low-duty-factor applications, the so-called *contention protocols* were developed.

The best early example of contention signaling was the ALOHA system[25] at the University of Hawaii, a time-shared computer system that interconnects terminal points spread over the Hawaiian Islands through a single UHF radio channel. Operating much like a multidrop line, the radio channel is shared by all terminals. However, polling is not used in the ALOHA system. Instead, on the assumption that most terminals are idle at a given instant of time, a terminal that has a message packet ready for transmission sends it in the hope that the channel is clear, then waits for an ACK from the intended recipient. If the transmitted packet collides (overlaps in time) with a packet from another terminal, no acknowledgments will be forthcoming and each terminal pauses for a random delay and repeats the transmission. The probability of a second collision is small if the average delay is large compared with the packet duration. For a small number of light-duty-factor terminals, the contention protocol uses the channel more efficiently than polling, and the system can accommodate a larger total number of terminals, as long as their duty cycle of activity is small. On the other hand, it should be clear that if too many terminals become active at the same time, communication breaks down abruptly and the response time of the system becomes intolerable. For a good discussion of the advantages and limitations of various link protocols, see Ref. 26.

The previous paragraphs discussed only the simplest forms of multiaccess link-control protocols. A complete discussion of the very complex protocol science that has evolved simultaneously with the complex data communication networks is beyond the scope of this chapter. The ISO and CCITT have begun to develop standards for seven layers of protocol that one might encounter in the most sophisticated network. At the bottom of the hierarchy, the physical layer refers to standards for the physical interface between pieces of communication hardware (connector specification, pinout functions, control leads, etc.). The link protocols constitute the second layer. As one moves up the hierarchy, the language of protocols shifts from electrical and communication engineering to the highly specialized nomenclature of computer science. Table 10.2 lists the standards that have been completed or are under active work by the various standards bodies that devote effort to data communication. See Refs. 24, 26, and 27 for more details.

VIDEO SIGNAL FORMATS

17. Analog Video Signal Formats

Television broadcasting on the 6-MHz bandwidth VHF and UHF channels allocated in the United States employs *vestigial-sideband amplitude modulation* (VSB-AM) of the 4+ -MHz-basebandwidth video. Most point-to-point microwave and satellite distributions use wide-deviation FM to achieve a high degree of immunity to noise, cochannel interference, and multipath. Video recording typically employs a narrow-deviation FM on a relatively low frequency carrier.

NTSC, PAL, and SECAM.   Video signals are inherently analog in nature, and, until recently, analog formats have been employed almost exclusively for video broadcasting, transmission, and recording. The three formats commonly used for color video signals are NTSC (National Television Systems Committee), PAL (phase-alternating line), and SECAM (sequential color avec memoire). NTSC is used in the United States, Canada, Mexico, Japan, and most Central and South American countries. PAL is used in western European countries, Australia, Argentina, Israel, and most former members of the British commonwealth. SECAM is used in France, the Soviet Union, eastern European countries, and most Communist-bloc countries.

The signal formats for NTSC and PAL are described in Chap. 17.

Multiplexed Analog Component Standards.   Both the NTSC and PAL formats transmit chroma information by a subcarrier which is superimposed on the luminance signal. The presence of a high-energy subcarrier at the upper end of the video baseband creates a number of problems, particularly in recording and FM transmission. To reduce or eliminate these problems and to establish a common format for video recording in Europe, the line-plex system of Schoenfelder[28] was proposed in 1980.

This system employs time-division multiplexing of chrominance and luminance in a single video signal in which the luminance component is compressed in time by a factor of 3:2 to occupy two-thirds of the active line time. One color-difference signal, e.g., $R - Y$, is time-compressed by a factor of 3:1 to occupy the remaining one-third of the active line. On the next succeeding line, the other color-difference signal, $B - Y$, is substituted for $R - Y$. As the luminance Y in the digital component standard has twice the horizontal bandwidth of the

TABLE 10.2   Data Communications Standards

Organization	Title description
CCITT	
V.3	International Alphabet No. 5
V.4	General structure of signals of International Alphabet No. 5 code for data transmission over public telephone networks
V.5	Standardization of data signalling rates for synchronous data transmission in the general switched telephone network
V.6	Standardization of data signalling rates for synchronous data transmission of leased telephone-type circuits
V.10	Electrical characteristics for unbalanced double-current interchange circuits for general use with integrated circuit equipment in the field of data communications
V.11	Electrical characteristics for balanced double-current interchange circuits for general use with intergrated circuit equipment in the field of data communications
V.21	300 bps duplex modem standardized for use on the general switched network
V.22	1200 bps duplex modem standardized for use on the general switched telephone network and on leased circuits
V.23	600/1200 bps modem standardized for use on the general switched network
V.24	List of definitions for interchange circuits between data terminal equipment and data circuit-terminating equipment
V.25	Automatic calling and/or answering equipment on the general switched telephone network, including disabling of echo suppressors on manually established calls
V.26	2400 bps modem standardized for use on 4-wire leased circuits
V.26bis	2400/1200 bits per second modem standardized for use in the general switched telephone network
V.27	4800 bps modem with manual equalizer standardized for use on leased telephone-type circuits
V.27bis	4800/2400 bits per second modem with automatic equalizer standardized for use on leased telephone-type circuits
V.27ter	4800/2400 bits per second modem standarized for use in the generalized switched telephone network
V.28	Electrical characteristics for unbalanced double-current interchange circuits
V.29	9600 bits per second modem standardized for use on point-to-point leased telephone-type circuits
V.31	Electrical characteristics for single-current interchange circuits controlled by contact closure
V.35	Data transmission at 48,000 bps using 60-180 kHz group band circuits
V.36	Modems for synchronous data transmission using 60-108 kHz group band circuits
V.54	Loop test devices for modems
X.1	International user classes of service in public data networks
X.2	International user services and facilities in public data networks
X.3	Packet assembly/disassembly facility (PAD) in a public data network
X.20	Interface between DTE and DCE equipment for start-stop transmission on public data networks

TABLE 10.2 Data Communications Standards (*Continued*)

Organization	Title description
X .21	Interface between DTE and DCE equipment for synchronous operation on public data networks
X.22	Multiplex DTE/DCE interface for user classes 3-6
X.24	List of definitions for interchange circuits between DTE/DCE on public data networks
X.28	DTE/DCE interface for a start-stop mode data terminal equipment accessing the PAD in a public data network situated in the same country
X.29	Procedures for the exchange of control information and user data between a PAD and a packet mode DTE or another PAD
X.75	Terminal and transmit call control procedures and data transfer system on international circuits between packet-switched data networks
X.87	Principles and procedures for realization of international user facilities and network utilities in public data networks
X.92	Hypothetical reference connections for public synchronous data networks
X.121	International numbering plan for public data networks
X.150	DTE and DCE test loops in public data networks
ANSI	
X3.44	Determination of the performance of data communications systems
X3.57	Structure for formatting message headings for information exchange using ASCII
ISO	
ISO 2593	Connector pin allocations for use with high speed data terminal equipment
ISO 4903	Data communication—15 pin DTE/DCE interface connector and pin assignments
ISO 7498 (draft)	Open Systems Interconnection—Basic Reference Model
EIA	
RS-363	Standard for specifying signal quality for transmitting and receiving data processing terminal equipments using serial data transmisions at the interface with nonsynchronous data comm. equipment
RS-404	Standard for start-stop signal quality between data terminal equipment and nonsynchronous data comm. equipment
RS-410	Standard for the electrical characteristics of class A closure interchange circuits
FTSC	
FED-STD-1002	Time and frequency reference information in telecommunication systems
INTERIM FED STD-001033	Telecommunications: digital communication performance parameters
FED-STD-1037	Glossary of telecommunication terms
ECMA	
ECMA-72	Transport protocol

SOURCE: Ref. 26.

chrominance signals, the multiplexed-component signal in line-plex has 50 percent larger bandwidth than Y in both its Y and C segments. The alternate-line-chroma encoding reduces the vertical chroma resolution by a factor of 2, which results in equalizing the horizontal and vertical resolution of chroma. This reduction is consistent with SECAM encoding, and PAL typically employs a 1H (one horizontal line) delay line in decoding which also reduces the vertical chroma resolution for display.

In 1981, the Independent Broadcasting Authority (IBA) of the United Kingdom proposed[29] a similar system as a common European standard for satellite broadcasting and further proposed to transmit sound and data digitally during horizontal blanking. These time-multiplexed analog component (or MAC) systems were investigated by the EBU and were characterized as A-MAC, B-MAC, and C-MAC. The A designates an *analog* subcarrier for sound, B denotes digital encoding of sound and data time-multiplexed with the video at *baseband,* and C denotes digital sound and data encoded directly at the rf *carrier* (interrupted time multiplexing between PCM and FM).

The EBU (European Broadcast Union) has developed standards for C-MAC and for a system called D2-MAC, a baseband multiplexed (B-type) signal that incorporates provisions for a wider image aspect ratio (5:3) rather than the normal 4:3 of television. In the United States, the B-MAC standard has been developed by the Advanced Television Systems Committee (ATSC). Details of these satellite proposals are to be found in the CCIR "Green Books" of the 1986 Plenary Session in Volume 12.

In the Society of Motion Picture and Television Engineers (SMPTE), a standard is being developed for the use of multiplexed analog components in the television *studio* (S-MAC) as a possible bridge between the analog composite studio of today and the digital component studio of the future. S-MAC differs from the alternate-line chroma systems discussed above by retaining both $R - Y$ and $B - Y$ chroma components on every line. The higher vertical resolution of chroma is deemed necessary to achieve good chroma key (color matting) and other special effects in post-production processing. To make room for the additional chroma signals, Y is compressed 2:1 and each C component by 4:1 to yield a multiplexed signal of about 11-MHz bandwidth. S-MAC is essentially an analog equivalent of CCIR Recommendation 601 (see Sec. 18) for digital encoding for the television studio.

18. Digital Video Standards

As the telephone common carriers of the world began in the 1970s to contemplate the transition from analog to digital transmission in segments of their plant, the need to distribute video in digital form became a priority item. As all video signals delivered to the common-carriers for transmission are in component NTSC, PAL, or SECAM format, sampling of the video was generally assumed to require coherence with a harmonic of the color subcarrier to avoid the possibility of interference from incoherent beats between the sampling frequency and such subcarrier harmonics. At about the same time, the broadcasters were beginning to use digital video equipment for time-base correction of video recorder instabilities and for the implementation of digital special effects (fades, wipes, zoombacks, and image manipulations).

For digital transmission, conservation of bandwidth suggested a sampling frequency of $3f_{sc}$, or about 10 MHz for NTSC and 13 MHz for PAL/SECAM. On the other hand, a sampling frequency of $4f_{sc}$ (or 14.318 MHz for NTSC) provides greater flexibility in signal processing because $4f_{sc}$ sampling along the I and Q axes gives samples at the peaks of one component simultaneously with zero crossings of the other (quadrature) component.

The baseband NTSC color television signal can be represented (during the active-picture portion on each scan line) by the expression:

$$Y(t) + I(t) \cos (2\pi f_{sc}t + 33°) + Q(t) \sin (2\pi f_{sc}t + 33°) \qquad (10.22)$$

From an examination of Eq. 10.22, it can be seen that samples at a rate of $4f_{sc}$, coherent with the I and Q axes (33° from burst), will yield for the cosine term the sequence $+I$, 0, $-I$, 0 etc., whereas the sine term yields the sequence 0, Q, 0, $-Q$, etc. Thus successive samples of the composite signal yield the sequence $Y + I$, $Y + Q$, $Y - I$, $Y - Q$, etc., for the periodic set of four successive samples (pixels). It should be clear that, since the phase of the subcarrier inverts 180° at a given pixel position on the next scan line, the corresponding set of adjacent pixels on the next succeeding scan line will be $Y - I$, $Y - Q$, $Y + I$, $Y + Q$, etc. Thus sampling an NTSC signal at a rate of $4f_{sc}$ provides an array of pixel values that can be processed to aid separation of a composite signal into its component values. For example, by adding and subtracting the pixel values on adjacent lines, the Y, I, and Q signals can be derived simply (this process is the digital equivalent of a line-comb filter[30]). In the processing of an image for a special effect (such as image rotation) processing on the separate component signals is necessary to avoid the complex effects that would occur with subcarrier phase if such processing were applied directly to the composite signal. After such image manipulation is applied, the analog composite signal is simply restored by creating the new sequence $Y + I$, $Y + Q$, $Y - I$, $Y - Q$ at the $4f_{sc}$ rate which is then processed by a D/A converter and postfilter.

Precise coherence of the sampling frequency with the color subcarrier has less advantage in PAL or SECAM because of the 25-Hz offset in PAL and the FM nature of the subcarrier in SECAM. Thus European broadcasters recognized that the transition to digital could be advantageously used to remove some of the impediments of multiple standards, if studios would maintain video signals in component rather than composite form for intrastudio distribution as well as for common-carrier transmission. Thus, about 1980, standardizing bodies (the SMPTE in North America and the EBU in Europe) began to concentrate on the possibility of a common worldwide component digital studio standard that could maximize the commonality between studio equipments designed for the two markets. This very effective collaboration between the SMPTE and the EBU (with Japanese cooperation in the SMPTE) led to the adoption[31] in 1982 by the CCIR of Recommendation 601, Digital Encoding for the Television Studio.

It was deemed desirable to establish an international standard for digital video that would use a common sampling frequency for both 525/60 and 625/50 scanning systems, since both systems have nearly the same video bandwidth (about 6 MHz) in the studio and approximately the same line frequencies. But the desire to have sampling coherent with the color subcarriers in both NTSC and PAL made a single frequency choice difficult. The Europeans, however, decided that the advantages of a *common* digital standard for both the PAL and SECAM countries overwhelmed the advantages of subcarrier-coherent processing in the interim period between the analog-composite studio and the all-digital-component studio of the future. To reach agreement with the NTSC world, on the other hand, was not quite so simple.

Then a fortuitous discovery was made by S. Baron of Thompson-CSF Labo-

TABLE 10.3   Encoding Parameter Values for the 4:2:2 Member of the Digital Family

Parameters	525-line, 60 field/s systems	625-line, 50 field/s systems
1. Coded signals: Y, C_R, C_B	These signals are obtained from gamma precorrected signals, namely: E'_Y, $E'_R - E'_Y$, $E'_B - E'_Y$ (Annex II, Sec. 2)	
2. Number of samples per total line:		
Luminance signal (Y)	858	864
Each color-difference signal (C_R, C_B)	429	432
3. Sampling structure	Orthogonal; line, field, and picture repetitive. C_R and C_B samples co-sited with odd (1st, 3d, 5th, etc.) Y samples in each line.	
4. Sampling frequency:		
Luminance signal	13.5 MHz	
Each color-difference signal	6.75 MHz	
	The tolerance for the sampling frequencies should coincide with the tolerance for the line frequency of the relevant color television standard.	
5. Form of coding	Uniformly quantized PCM, eight bits per sample, for the luminance signal and each color-difference signal	
6. Number of samples per digital active line:		
Luminance signal	720	
Each color-difference signal	360	
7. Analog-to-digital horizontal timing relationship: from end of digital active line to O_H	16 luminance clock periods	12 luminance clock periods
8. Correspondence between video signal levels and quantization levels:		
Scale	0 to 255	
Luminance signal	220 quantization levels with the black level corresponding to level 16 and the peak white level corresponding to level 235	
Each color-difference signal	225 quantization levels in the center part of the quantization scale with zero signal corresponding to level 128	
9. Code-word usage	Code words corresponding to quantization levels 0 and 255 are used exclusively for synchronization. Levels 1 to 254 are available for video.	

SOURCE: CCIR Recommendation 601, 1986

ratories in Stamford, Connecticut. The discovery was that the ratio of the line frequencies of 525/60 and 625/50 is a rational number:

$$\frac{15,625}{15,734.266} = \frac{143}{144} \tag{10.23}$$

Equivalently, both line frequencies are integral submultiples of 2.25 MHz, thus any sampling frequency that is an integral multiple of 2.25 MHz would be line-locked (an integral number of samples per line) in both 525/60 and 625/50. Thus 13.5 MHz was proposed as the common sampling frequency for an international digital video studio standard, and after extensive tests by both EBU and SMPTE the bargain was struck. A second advantageous feature was a further agreement that the active-picture line duration for both 525/60 and 625/50 would be pegged at 720 sample periods or 53.33 μs in both systems. This agreement, which actually extends the active line duration for both systems into the black-level horizontal blanking interval, permits equipments that perform line-related signal processing (dropout compensation, vertical filtering, etc.) to employ identical active-line selector and processing circuits while accommodating the tolerances of both systems for active line duration. The detail specifications for CCIR Recommendation 601 are given in Table 10.3 (see opposite).

19. High-Definition Television (HDTV) Standards

The Japan Broadcasting Corporation (NHK) proposed[32] in 1974 a new *higher-definition television (HDTV)* system based on 1125-line scanning. The HDTV system has a sufficient bandwidth to accommodate twice the horizontal resolution of today's television and also a widescreen (5:3) aspect ratio.

HDTV has been under study by the CCIR, EBU, SMPTE, and others for over a decade, but at the time of this writing, no definitive standards have been adopted. Further information on high- and extended-definition television can be found in Refs. 33–34.

REFERENCES

1. Hartley, R. V. L., "Transmission of Information," *Bell Syst. Tech. J.,* vol. 7, pp. 535–563, July 1928.

2. Shannon, C. E., "A Mathematical Theory of Communication," *Bell Syst. Tech. J.,* vol. 27, pp. 379–423, 623–656, July–October 1948.

3. Panter, P. F., *Modulation, Noise, and Spectral Analysis,* McGraw-Hill, New York, 1965.

4. Schwartz, M., *Information Transmission, Modulation, and Noise,* McGraw-Hill, New York, 1970.

5. Baghdady, E. J., *Lectures on Communication System Theory,* Chap. 19, McGraw-Hill, New York, 1961.

6. Fano, R. M., *Transmission of Information,* MIT Press, John Wiley & Sons, New York, 1961.

7. Rabiner, L. R., and R. W. Schafer, *Digital Processing of Speech Signals,* Prentice-Hall, Englewood-Cliffs, N.J., 1978.

8. Dunn, H. J., and S. D. White, "Statistical measurements on Conversational Speech," *J. Acoust. Soc. Am.,* vol. 11, pp. 278–288, January 1940.

9. Davenport, W. B., "An experimental study of speech-wave probability distributions," *J. Acoust. Soc. Am.,* vol. 24, pp. 390–399, July 1952.

10. Paez, M. D., and J. H. Glisson, "Minimum mean squared-error quantization in speech," *IEEE Trans. Comm.,* vol. WM-20, pp. 225–230, April 1972.

11. Fletcher, H., *Speech and Hearing in Communications,* Van Nostrand, New York, 1953.

12. Meeker, W. F., "Speech Characteristics and Acoustic Effects," Chap. 3, *Communication Systems Engineering Handbook,* D. H. Hamsher, ed., McGraw-Hill, New York, 1967.

13. *NAB Engineering Handbook,* 7th ed., National Association of Broadcasters, Washington, D.C.

14. CCIR XIII Plenary Assembly, "Green Books," vol. 12, p. 210, Geneva, 1974.

15. CCITT, "Red Books," Fascicle III.4, p. 161, Geneva, 1985.

16. CCIR XV Plenary Assembly, "Green Books," vol. 12, p. 193, Geneva, 1982.

17. Peck, J. B. H., "Communication Aspects of the Compact Disc Digital Audio System," *Communications Magazine,* vol. 23, no. 2, IEEE Communications Society, February 1985.

18. Stott, J. H., and G. J. Phillips, "Digital Video: Multiple Sub-Nyquist Coding, Report No. 1977/21, BBC Research Department, Kingswood Warren, Tadworth, Surry, 1977.

19. Viterbi, A. J., "Performance of an *M*-ary Orthogonal Communication System using Stationary Stochastic Signals," *IEEE Trans. Info. Theory,* vol. IT-13, no. 3, pp. 414–22, July 1967.

20. Kretzmer, E. R., "The design of modems for digital data communication," 1976 Zurich Conference.

21. Forsdick, H. C., R. E. Schantz, and R. H. Thomas, "Operating Systems for Computer Networks," *Computer,* January 1978.

22. Jordan, E. C., (ed.), *Reference Data for Engineers: Radio, Electronics, Computer, and Communications,* 7th ed., Howard Sams, 1985.

23. Freeman, R. L., *Telecommunication Transmission Handbook,* 2d ed., John Wiley & Sons, New York, 1981.

24. Green, P. E., Jr., *Computer Network Architectures and Protocols,* Plenum Press, New York, 1982.

25. Abramson, N., "The Aloha System," in *Computer Communication Networks,* N. Abramson and Kuo, eds., Prentice-Hall, Englewood Cliffs, N.J., 1973.

26. *Datapro Communications Solutions,* vol. 1, Datapro Research Corp., Delran, N.J., December 1982.

27. CCITT, "Red Books," Recommendation I.320, vol. III, Fiscicle III.5, Geneva, 1985.

28. Brand, G., et al., "TIMEPLEX—ein serielles Farbcodiervarfahren fur Heim-Video-recorder," Fernseh- und Kinotechnik (34), Heft 12, 1980, p. 451–458.

29. Lucas, K., and M. D. Windram, "Direct Television Broadcasts by Satellite—Desirability of a New Transmission Standard," E&D Report 116/81, Independent Broadcasting Authority, Winchester, Hants, U.K., 1981.

30. Pritchard, D. H., "A CCD Comb Filter for Color TV Receiver Picture Enhancement," *RCA Review,* December 1949.

31. CCIR "Green Books," vol. XI, Part I, Recommendation 601, p. 271.

32. Fujio, T., "High-Definition Wide-Screen Television System for the Future," *IEEE Trans. on Broadcasting,* vol. BC-26, no. 4, p. 113, December 1980.

33. Jackson, R. N., and M. J. J. C. Annegarne, "Compatible Systems for High-Quality Television," *Jour. SMPTE,* vol. 92, no. 7, July 1983.

34. Powers, K. H., "Techniques for Increasing the Picture Quality of NTSC Transmissions in Direct Satellite Broadcasting," *IEEE J. Sel. Areas in Comm.,* vol. SAC-3, no. 1, pp. 57–64, January 1985.

CHAPTER 11
SIGNAL TRANSMISSION MODES

Alauddin Javed
Director, Radio Systems, Bell Northern Research

CONTENTS

INTRODUCTION

1. Modulation and Multiplexing

All information-carrying signals must ultimately be transmitted over a medium separating the transmitter and the receiver. Efficiency of transmission requires that the information be processed in some manner before being transmitted. The step of processing the signal for more efficient transmission is called *modulation*.

Modulated signals are mainly of two types:

1. *Continuous-wave* (cw) or analog modulation in which the amplitude, phase, or frequency of a sine wave is altered in accordance with the information being transmitted.

2. Pulse modulation in which the height, width, or position of a set of pulses is altered in a definite pattern corresponding to the information being transmitted.

The modulation process enables signals to be transmitted at frequencies much higher than the signal frequency components. Thus the entire electromagnetic frequency spectrum can be utilized for the transmission of signals.

In practical communications systems, it is often necessary to convey multiple

11.1

messages simultaneously from a number of information sources. This is accomplished by *multiplexing,* which combines multiple message sources for transmission as a group over a single transmission facility.

There are two generic types of multiplexing, *frequency-division multiplexing (FDM)* and *time-division multiplexing (TDM).*

FDM is directly applicable to continuous-wave or analog sources. It is accomplished by "stacking" modulated signals from several information sources side by side in frequency to form a composite signal. This composite signal is then used to modulate a carrier in a conventional manner. The individual messages are recovered after reception by bandpass filtering and frequency selection of the channel.

TDM, a logical extension of pulse modulation, involves interleaving the pulses of several pulse-modulated signals in time to form a composite pulse transmission system. Separation of time-multiplexed signals at the receiver is accomplished by gating the appropriate pulses into individual channel filters.

ANALOG MODULATION

2. Conventional Amplitude Modulation (AM-DSB)

In conventional *amplitude modulation with double sidebands (AM-DSB),* the carrier envelope varies about a mean value in a manner which is linearly related to the modulating signal. Conceptually, an AM waveform can be generated by adding a constant value to the modulating signal and multiplying the sum by a sinusoidal carrier:

$$m(t) = [1 + s(t)] \cos \omega_c t \qquad (11.1)$$

where $m(t)$ = the instantaneous value of the modulated wave
$\quad \omega_c = 2\pi f_c$
$\quad f_c$ = carrier frequency
$\quad s(t)$ = the instantaneous value of the modulating signal

Usually the equation for a modulated wave is written as

$$m(t) = [1 + m_a s(t)] \cos \omega_c t \qquad (11.2)$$

where m_a is the *modulation index* and m_a and $s(t)$ are constrained such that

$$|s(t)_{\max}| \leq 1 \qquad \text{and} \qquad 0 < m_a < 1 \qquad (11.3)$$

The frequency components of $s(t)$ are spectrally confined to the region $|f| f_{\max}$. The maximum modulating frequency component $f_{\max}$ is usually much less than f_c, and the average value of $s(t)$ is 0. The degree of modulation m_a is often stated as the percentage modulation.

Figure 11.1 illustrates how the modulated carrier envelope follows the modulation.

The frequency spectrum of the AM signal is the Fourier transform of Eq. 11.2,

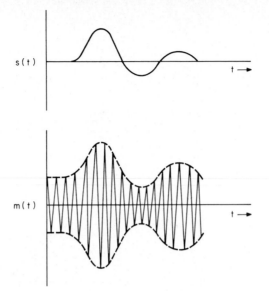

FIG. 11.1 Baseband   and   amplitude-modulated waveforms.

$$M(f) = \underbrace{\tfrac{1}{2}[\delta(f_c - f) + \delta(f_c + f)]}_{\text{Carrier}} + \underbrace{m_a[S(f_c - f) + S(f_c + f)]}_{\text{Sidebands}} \qquad (11.4)$$

where $\delta(f)$ = unit impulse in frequency and $S(f)$ = the baseband as a function of frequency.

Figure 11.2 shows $S(f)$ for a typical baseband and the resulting AM signal spectrum. The signal spectrum contains a carrier component in addition to the translation of the baseband spectrum to sidebands on both sides of the carrier. Thus the AM transmission bandwidth is $2f_{max}$.

In the simplest case where the baseband is a sine wave, $s(t) = \cos \omega_m t$, the AM signal is

$$m(t) = A_c (1 + m_a \cos \omega_m t) \cos \omega_c t = A_c \cos \omega_c t + m_a A_c/2[\cos 2 \pi(f_c - f_m)t$$
$$+ \cos 2 \pi(f_c + f_m)t] \qquad (11.5)$$

where A_c = unmodulated carrier amplitude.

The relative power relationships in the modulated carrier can be determined from Eq. 11.5. If the AM carrier voltage is applied across a 1- Ω resistor, the carrier power is

$$P_c = A_c^2/2 \qquad (11.6)$$

The power in each sideband is

$$P_{sbl} = P_{sbu} = m_a^2 A_c^2/8 \qquad (11.7)$$

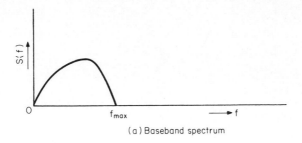

(a) Baseband spectrum

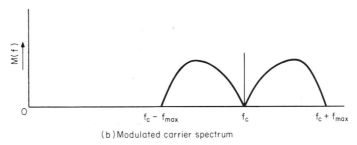

(b) Modulated carrier spectrum

FIG. 11.2   Amplitude-modulation double-sideband spectra.

while the total sideband power is

$$P_{sb} = P_{sbl} + P_{sbu} = m_a^2 A_c^2/4 \qquad (11.8)$$

The maximum power in the sidebands, the information-bearing portion of the AM signal, occurs when $m_a = 1$:

$$P_{sb} = A_c^2/4 = P_c/2 \qquad (11.9)$$

Since the maximum possible sideband power is one-half the carrier power, at best only one-third of the total available power is utilized in the information-carrying sidebands. Improved power utilization can be achieved by the use of suppressed-carrier double-sideband or single-sideband modulations as described in the next sections.

3.   Suppressed-Carrier Double-Sideband (SC-DSB) Modulation

An SC-DSB signal is an AM signal with the carrier removed or suppressed. It can be expressed mathematically as the product of the modulating and carrier time functions:

$$m_{dsb}(t) = s(t) \cos \omega_c t \qquad f_c \gg f_m \qquad (11.10)$$

Suppressed-carrier double-sideband modulation can be produced by a balanced modulator as shown in Fig. 11.3. The message or modulating function $S(t)$ is a

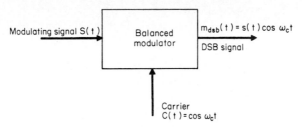

FIG. 11.3   Suppressed-carrier double-sideband modulation.

band-limited signal whose spectrum is limited to the region $|f| \leq f_m$. There is no carrier frequency component in the DSB spectrum because $S(t)$ contains no zero-frequency component.

Figure 11.4 illustrates a typical modulation waveform and the resultant DSB signal.

The spectrum of the DSB signal is given by

$$M_{dsb}(f) = m_a[S(f_c - f) + S(f_c + f)]/2 \qquad (11.11)$$

which represents a simple frequency translation to sidebands on either side of the carrier. Figure 11.5 shows the spectrum of a suppressed-carrier double-sideband signal.

The detection of a DSB signal requires a special receiver which can reinsert or reconstitute the carrier. The frequency and phase of the reinserted carrier must be controlled within narrow tolerances. This can be accomplished with a phase-locked receiver or by the transmission of a low-amplitude carrier.

4.   Single-Sideband (SSB) Modulation

Although SC-DSB modulation reduces the required transmitter power, it occupies the same bandwidth as conventional AM and transmits both sidebands, either of which supplies the full information. *Single-sideband (SSB)* modulation halves the transmission bandwidth by eliminating the carrier and one sideband.

The most commonly used technique for generating SSB signals is to form the DSB signal with a balanced modulator and then to reject one sideband by filtering as shown in Fig. 11.6. This method of generating SSB signals requires the use of filters with very tight tolerances. To avoid these rather difficult filter requirements, other techniques have been developed. The alternate techniques all involve generation of two DSB signals with carriers in phase quadrature. With suitable phasing arrangements, the unwanted sideband is balanced out when the two signals are combined. Very careful control of the amplitude and phase of these signals is required to achieve a high degree of suppression of the unwanted sideband.

SSB demodulation is usually accomplished in a product detector in which the received signal is multiplied by a locally generated carrier. This provides translation of the sideband information down to baseband. Low-pass filtering then removes the undesired double frequency term. Some SSB systems transmit an appreciably reduced *vestigial carrier* or *pilot tone* to facilitate proper frequency and phase control at the receiver. The carrier is recovered by placing a very sharply

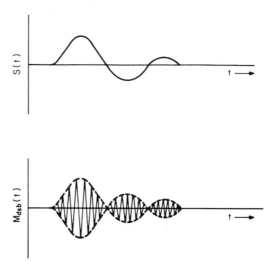

FIG. 11.4   Baseband and suppressed-carrier waveforms.

tuned filter at the carrier frequency and greatly amplifying the pilot tone. The recovered carrier can then be used directly or indirectly to synchronize the local oscillator.

5. Implementation Complexity of AM Systems

SSB is widely used in frequency-division multiplexing systems and finds some application in the crowded HF and VHF bands to conserve spectrum space. It is

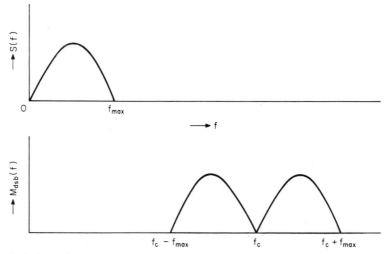

FIG. 11.5   Suppressed-carrier double-sideband spectra.

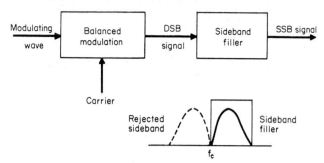

FIG. 11.6   Generation of a single-sideband signal.

not used more extensively because of equipment complexity and the cost of both transmitters and receivers. Since generation of SSB signals necessarily occurs at low power levels, SSB transmitters require linear power amplifiers to prevent distortion of the transmission. By contrast, convenient methods exist for generating AM and DSB signals directly at high power levels, thus employing more efficient nonlinear power amplifiers.

6. Vestigial Sideband Modulation

A form of amplitude modulation known as *vestigial sideband* is widely used throughout the world for the transmission of television broadcast signals. This combines double-sideband AM transmission of the lower-frequency components of the signal waveform and single-sideband for the higher. The amplitude of the higher-frequency components can be restored at the receiver by shaping the response of the i-f amplifier.

The purpose of vestigial sideband is to reduce the bandwidth requirements of the radio-frequency channel. In the United States and Canada, 6-MHz channels are employed for television broadcasting. With double-sideband AM, a channel width of 8 to 9 MHz would be required to obtain the same performance.

The spectrum of a U.S. TV broadcast channel as specified by the FCC is shown in Fig. 11.7 (Sec. 73.699, Fig. 5 of the FCC *Rules and Regulations*). (Low-power UHF stations are allowed to broadcast near-double-sideband signals.) Worldwide, the CCIR has designated 13 different standards for TV broadcast channels (CCIR Recommendation 470-1, Report 642-2).

7. Signal-to-Noise Ratios in Amplitude-Modulated Systems

The signal-to-noise ratio S/N is defined as:

$$S/N = \text{average signal power/average noise power} \qquad (11.12)$$

In AM-DSB systems:

$$S/N = m_a{}^2 A_c{}^2/2N \qquad (11.13)$$

where A_c = carrier amplitude

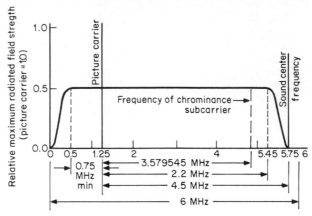

FIG. 11.7 The FCC channel specification for monochrome and color television transmission.

N = average noise power = $2n_of_{max}$
n_o = noise spectral density
f_{max} = maximum modulating frequency = ½ channel bandwidth

In a DSB system, the signal-to-noise is

$$S/N = A_c^2/2N \tag{11.14}$$

For the same transmitted power, the greater efficiency of DSB resulting from the suppression of the carrier results in the relationship

$$(S/N)_{dsb} \geq 3(S/N)_{am} \tag{11.15}$$

with equality applying for $m_a = 1$ or 100 percent modulation.

Since the bandwidth of an SSB signal is one-half that of AM and DSB signals, the noise admitted by the SSB predetection filter is one-half that of DSB and AM systems. Since AM and DSB experience an improvement in detection gain by a factor of 2,

$$(S/N)_{SSB} = (S/N)_{AM \text{ or } DSB} \tag{11.16}$$

8. Frequency Modulation

Frequency modulation is one type of a general class known as *angle modulation.* In angle modulation, the amplitude A of a carrier signal $A \cos (\omega_c t + v)$ is maintained constant while the phase angle v of the carrier is varied in some manner according to the message waveform $s(t)$. The phase angle can be varied by phase or frequency changes. With frequency modulation, the instantaneous frequency is varied linearly in accordance with $s(t)$.

With FM, the time-varying phase function $v(t)$ is the integral of the modulating signal $f(t)$:

$$(t) = k_f \int_{-\infty}^{t} S(t)\, dt \tag{11.17}$$

where k_f is a system constant.

The FM waveform then is

$$m_{\text{FM}}(t) = \cos \left[\omega_c\, t + k_f \int_{-\infty}^{t} S(t)\, dt \right] \tag{11.18}$$

The peak frequency deviation Δf, is defined as

$$\Delta f = [k_f/2\ \pi]|S(t)_{\text{max}}|\ \ [k_f/2\pi]|S(t_{\text{max}})| \tag{11.19}$$

and indicates the maximum excursion of frequency from the carrier frequency.

The modulation index β is defined as

$$\beta = f/f_m = \text{frequency deviation/highest modulating frequency}$$

The average signal power of an FM system is a constant value, $A_c{}^2/2$.

Two types of FM are in common use, *narrowband* and *wideband*. As their names imply, they differ in the amount of frequency deviation.

Narrowband FM. In narrowband FM, the modulation index is constrained to small values (β — 1). Narrowband FM is an important operational mode for digital communications in a form known as frequency-shift keying; see Sec. 17. With analog communications, however, relatively high signal-to-noise ratios are usually desired, and this dictates the use of wideband FM with β in the order of unity or greater.

With small values of β and sinusoidal modulation, Eq. 11.18 becomes approximately

$$m_{\text{FM}}(t) = \cos \omega_c t - [\beta/2]\,[\cos 2\pi(f_c - f_m)t - \cos 2\pi\,(f_c + f_m)t] \tag{11.20}$$

Note the similarity of Eq. 11.20 to Eq. 11.5 for AM. Equation 11.20 differs primarily in the phase of the lower sideband. The spectra for AM and narrowband FM are compared in Fig. 11.8.

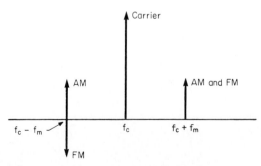

FIG. 11.8　AM and narrow-band FM spectra.

Wideband FM.   Wideband FM is used for radio broadcasting, high-quality voice communications, and video transmission via microwave and satellite. Further information on these applications is found in Chaps. 4, 5, and 17. As compared with AM and narrowband FM, wideband FM provides significant improvements in S/N and cochannel interference rejection. These improvements are achieved, of course, at the cost of greater bandwidth requirements.

The simplification of Eq. 11.5 embodied in Eq. 11.20 for narrowband systems is not appropriate for values of β in the vicinity of 1 or larger, and a much more complex expression must be used. For single-frequency sinusoidal modulation, the instantaneous value of the modulated wave can be expressed by an infinite series of Bessel functions:

$$
\begin{aligned}
m_{\mathrm{FM}}(t) = {}& A_c \, J_o(\beta) \cos \omega_c t \\
& - A_c \sum_{n=0}^{n=\infty} J_{2n+1}(\beta)\{\cos 2\pi[f_c - (2n + 1)f_m]t \\
& - \cos 2\pi[f_c + (2n + 1)f_m]t\} \\
& + A_c \sum_{n=0}^{n=\infty} J_{2n}(\beta)[\cos 2\pi(f_c - 2nf_m)]t \\
& + \cos 2\pi(f_c + 2\pi f_m)t
\end{aligned}
\tag{11.21}
$$

Thus, a wideband FM signal with sinusoidal modulation of frequency f_m consists of a carrier frequency component f_c and an infinite number of sideband components spaced $\pm f_m$ Hz from the carrier and from each other. The amplitude of each term depends on the modulation index and the appropriate Bessel function, as shown in Fig. 11.9.

The complexity of the expressions for frequency modulation signals makes rigorous mathematical analysis difficult, and much of the engineering development has resulted from approximations and empirical studies.

Although the number of sidebands surrounding the carrier of an FM signal is infinite, the number of components with significant energy is finite. An approximate expression for the bandwidth of a wideband FM signal is

$$
B = 2(\Delta f + f_m)
\tag{11.22}
$$

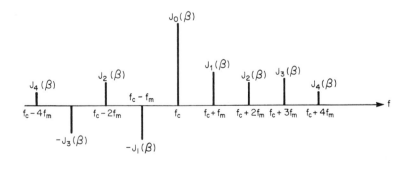

FIG. 11.9   FM spectrum for sinusoidal modulation.

9. Signal-to-Noise Improvement in FM Systems

Preemphasis-Deemphasis. Most FM systems employ preemphasis-deemphasis systems to improve the signal-to-noise ratio. Preemphasis is accomplished by increasing the frequency deviation Δf for the higher-frequency components in the signal in accordance with a standardized response curve. Deemphasis is achieved by using a receiver with a complementary response (less amplification at the higher frequencies). Since the noise output of an uncompensated FM receiver increases with baseband frequency, deemphasis improves the S/N.

In FM systems having values of β in the vicinity of unity, e.g., land mobile systems which employ a frequency deviation of ± 5 kHz, preemphasis is often achieved automatically by the use of phase modulation where the phase function $v(t)$ rather than the instantaneous frequency is varied in accordance with $f(t)$:

$$m_{\text{PM}}(t) = \cos \left[\omega_c t + k_p s(t) \right] \tag{11.23}$$

where k_p is a system constant with the dimensions of radians per second. The result of phase modulation is to increase the deviation Δf linearly with increasing modulation frequency, thus maintaining the modulation index at a constant value at all modulating frequencies and achieving the desired preemphasis.

For FM systems with larger values of β, e.g., FM broadcast, standard preemphasis curves are specified.

FM Improvement Factor. Wideband FM has the inherent property of producing a higher S/N in the demodulated signal than in the undemodulated carrier. The ratio of the S/N in the demodulated signal to the carrier-to-noise ratio C/N is known as the FM improvement factor:

$$\text{Improvement factor} = (S/N)/(C/N) \tag{11.24}$$

The improvement factor is usually expressed in decibels and is approximated by the expression:

$$\text{Improvement factor} = 20 \log (\Delta f/B_m) + 10 \log (B_{\text{if}}/B_m) + 7 \qquad \text{db} \tag{11.25}$$

where B_m = the baseband signal bandwidth and B_{if} = the bandwidth of the receiver i-f amplifier.

Equation 11.25 is valid only for relatively large values of Δf and for C/N values above the receiver threshold, typically 8 dB.

In practical systems, the improvement factor can be quite large and, when combined with preemphasis-deemphasis, can be in the range of 20 to 50 dB. It provides a useful tradeoff of bandwidth for power. See Chaps. 4 and 17 for further discussion.

PULSE MODULATION

---

10. Introduction

In a pulse-modulation system, the message information is transmitted intermittently rather than continuously. A series of discrete pulses (unquantized code elements) rather than a continuously changing wave carries the specification of the message. When the message is not already made up of values arranged in a discrete time sequence, it must be translated to this form before transmission.

Sampling affords a means of representing a continuously varying signal by a series of single values. The sampling principle states: if a signal that is a magnitude-time function is sampled instantaneously at regular intervals and at a rate slightly higher than twice the highest significant signal frequency, the samples contain all of the information of the original signal.

Familiar examples of pulse modulation are *pulse-amplitude modulation* (PAM), *pulse-duration modulation* (PDM), and *pulse-position modulation* (PPM).

11. Pulse-Amplitude Modulation

The essentials of the pulse-amplitude modulation process are shown in Fig. 11.10. The signal waveform is sampled by a unit sampling function at intervals T for a duration of t_o. The resulting single-polarity amplitude-modulated pulses shown in Figure 11.10a can then be transmitted.

The unit sampling function $U(t)$ shown in Fig. 11.10b can be expressed as

$$U(t) = k \sum_{-\infty}^{+\infty} [\sin mk\pi/mk\pi] \cos m\omega_c t \qquad (11.26)$$

where $k = t_o/T$ and $\omega_c = 2\pi/T$.

The spectrum of the amplitude-modulated pulses can be obtained by multiplying the unit sampling function by the message, $S(t) = A_v \cos (\omega_v t + \theta_v)$, and adding UK:

$$m(t) = kA_v \cos (\omega_v t + \theta_v)$$

$$+ kA_v \sum_{1}^{\infty} [\sin mk\pi/mk\pi] \cos [(m\omega_c \pm \omega_v)t \pm \theta_v]$$

$$+ Kk \sum_{1}^{\infty} [\sin mk\pi/mk\pi] \cos m\omega_c t \qquad (11.27)$$

where K = constant.

Almost all types of amplitude-modulated pulses are easily demodulated by a low-pass filter having appropriate frequency characteristics. The loss-frequency characteristics of the filter should be flat in the passband. In addition, an equalization network to compensate for the aperture effect, namely [sin $(k\pi\omega_v/\omega_c)]/[k\pi\omega_v/\omega_c]$, is required. Above the passband, the filter must introduce enough loss for adequate suppression of other spectrum components.

12. Pulse-Duration Modulation

In *pulse-duration modulation* of a pulse carrier, the value of each sample of a continuously varying modulating wave produces a pulse of proportional length. The modulating wave may vary the time of occurrence of the leading edge, the trailing edge, or both edges of the pulse.

The precision with which the demodulator detects the time location of the

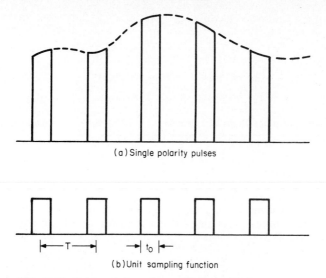

(a) Single polarity pulses

(b) Unit sampling function

FIG. 11.10 Pulse train for pulse-amplitude modulation.

edge of the pulse determines the signal-to-noise ratio of the output signal. In-creasing the bandwidth increases the sharpness of these edges and minimizes time distortions caused by noise in the system. Thus pulse-duration modulation has an advantage analogous to that of frequency modulation. Noise and interfer-ence can be reduced at the expense of increased bandwidth, provided the extra-neous disturbances are below the threshold.

The three types of pulse-duration modulation are shown in Fig. 11.11.

The spectrum of duration-modulated pulses with the trailing edges modulated

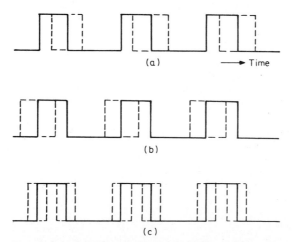

FIG. 11.11 Pulse-duration modulation. (a) trailing edge modulated, leading edge fixed; (b) leading edge modulated, trailing edge fixed; (c) both edges modulated.

by samples of a sinusoidal modulating wave, $M \cos \omega_v t$, is given by

$$m(t) = k + (M/2) \cos \omega_c t + \sum_{m=1}^{\infty} \sin (m\omega_c t/m\pi)$$

$$- (J_o m\pi M/m\pi) \sin (m\omega_c t - 2m\pi k)$$

$$- (J_n m\pi M/m) \sin (m\omega_c t + n\omega_v t - 2m\pi k - n/2) \qquad (11.28)$$

Pulses with their leading edges modulated become pulses with trailing edges modulated when the time scale is reversed. Thus

$$f_2(t) = f_1(-t) \qquad (11.29)$$

where $f_2(t)$ is the series representing a train of duration-modulated pulses with trailing-edge modulated.

Pulses with both edges modulated can be considered as a combination of two pulse trains, one with the leading edge and one with the trailing edge modulated:

$$f_3(t) = f_1(t) + f_2(t) = f_1(t) + f_1(-t) \qquad (11.30)$$

13. Pulse-Position Modulation

In *pulse-position modulation*, the value of each instantaneous sample of a modulating wave is caused to vary the position of pulses in time relative to their unmodulated time of occurrence. The variation in relative position may be related to the modulated wave in any predetermined manner. As in pulse-duration modulation, the maximum modulating signal must not cause a pulse to enter adjacent allotted time intervals. This requirement leads to a wasteful use of time space when used for multiplexing many signal channels.

Spectra of PPM. An unmodulated pulse train can be represented by the unit sampling function expressed as an even function of time:

$$U = k [(\sin mk\pi)/mk\pi] \cos m\omega_c t \qquad (11.31)$$

where $k = t_o/T$
    t_o = duration of a pulse
    $T = 2\pi/\omega_c$ = unmodulated pulse repetition period

The unmodulated wave is represented by:

$$S(t) = A_v \sin \omega_v t \qquad (11.32)$$

Let

$$k_1 S(t) = M \sin \omega_v t \qquad (11.33)$$

where $\pm MT/2$ = the maximum excursion of a pulse from its unmodulated time of occurrence.

If m_{PPM} denotes the representation of the unit pulse's position modulated by $A_v \sin \omega_v t$, then

$$m_{PPM} = \Sigma\Sigma a_{mn} \cos (m\omega_c + n\omega_v)t \qquad (11.34)$$

For uniform sampling, it can be shown that:

$$a_{mn} = k(-1)^n\{[\sin k\pi (m + n \omega_v/\omega_c)/k\pi (m + n\omega_v/\omega_c)]\} \qquad (11.35)$$
$$\times\{j_n [M\pi(m + n\omega_v/\omega_c)] \}$$

where J_n [ ] = Bessel function of first kind and nth order.

Position-modulated pulses may be produced by deriving them from duration-modulated pulses. The modulating wave is sampled and converted to duration-modulated pulses as shown in the previous section. The duration-modulated pulses are caused to generate a rectangular pulse of short duration each time the modulated edge of the duration-modulated pulse passes through a specified value. The position-modulated pulses thus produced are all of equal duration.

When the maximum excursion of a pulse from its unmodulated time of recurrence is moderately small, a delayed copy of the original signal with little distortion can be obtained by passing the PPM signal through a low-pass filter.

14. Comparison of Pulse-Modulation Systems

Pulse-position modulation and pulse-duration modulation systems require much greater bandwidth than pulse-amplitude modulation systems. The increased bandwidth of PPM and PDM signals results in improved signal-to-noise ratio, just as wideband FM provides an improvement in analog transmission. The ratio of average signal power to average noise power at the output of each multiplexed channel of PPM and PDM systems is proportional to the square of the ratio of the prorated bandwidth gainfully utilized by the channel to the bandwidth of the channel itself. There is no signal-to-noise improvement in PAM systems, just as AM provides no improvement in analog transmission.

The modulated pulses of PPM and PDM are of uniform amplitude so that they can be reshaped periodically to avoid accumulation of excessive distortion. PAM systems cannot be treated similarly since the pulse amplitudes are modulated by the transmitted signal.

In both PDM and PPM, the maximum modulating signal must not cause a pulse to enter adjacent allocated time intervals. In time-division-multiplexed systems, this requirement leads to a very wasteful use of time space. Thus PDM and PPM both fall short of the theoretical ideal when used for multiplexing a large number of channels. PAM does not have this limitation.

DIGITAL MODULATION

15. Introduction to Digital Modulation

Digital modulation, like pulse modulation, employs a train of pulses for the transmission of signal information. The pulses are of equal amplitude and duration, and the information is transmitted by encoding the spacing between pulses as they are sent in sequence. A description of pulse coding is given in Chap. 10. The discussion which follows relates to the transmission of pulse trains where the presence or absence of a pulse is described as a bit.

All of the digital modulation systems discussed in this chapter can be described in terms of a radio-frequency sinusoid (carrier) modulated by a low-frequency (baseband) signal that conveys the digital information. These baseband signals may be filtered, weighted, or otherwise shaped prior to modulating the carrier in order to achieve the desired results. At the receiver, the baseband information is recovered by a detection process. Coherent detection requires a sinusoidal reference signal perfectly matched in both frequency and phase to the carrier. The phase reference can be obtained either from a transmitted pilot tone or the modulated signal itself. Noncoherent detection, which is based on waveform characteristics independent of phase, does not require a phase reference.

Detection is usually followed by a decision process which converts the recorded baseband modulation signal into a sequence of digital bits. This process requires bit synchronization which is generally extracted from the received waveform. With most modulation systems, decisions can be made on a bit-by-bit basis with no loss of performance. With others, however, an advantage can be gained by examining the signal over several-bit intervals before making each bit decision.

16. Amplitude-Shift Keying (ASK)

Amplitude-shift keying is a form of amplitude modulation, and, like analog modulation, it can employ *amplitude modulation* (AM-DSB), *suppressed-carrier double-sideband* (SC-DSB), or *single-sideband* (SSB).

AM-DSB.   The simplest digital AM technique is AM-DSB with the carrier modulated by a binary signal. The waveform is

$$m_{AM}(t) = [A/2] [1 + m(t)] \cos \omega_c t \tag{11.36}$$

If $m(t) = \pm 1$, the result is a *nonreturn-to-zero* (NRZ) binary data waveform which produces *on-off keying* (OOK) modulation. An OOK waveform can be detected either coherently or noncoherently, but the advantage of coherent detection is so small that it does not justify the added complexity.

SC-DBS.   Since the carrier conveys no information, the transmission efficiency can be improved by the use of suppressed-carrier double-sideband modulation.
The expression for SC-DSB is

$$m_{sc} = Am(t) \cos \omega_c t \tag{11.37}$$

When $m(t)$ is either 1 or zero, the result is OOK. When $m(t)$ equals ± 1, the result is binary shift keying.

SSB.   Both DSB techniques involve the transmission of a redundant sideband. For applications in which spectral efficiency is important, the bandwidth can be reduced by a factor of 2 by the use of SSB. The SSB waveform can be written as

$$m_{SSB}(t) = A[m(t) \cos \omega_c t + m(t) \sin \omega_c t] \tag{11.38}$$

where $m(t)$ = Hilbert transform of $m(t)$.

SSB signals are usually generated by the use of a bandpass filter to suppress the upper or lower sideband. The sharp cutoff characteristics required for the filter present problems, and a smooth roll-off is often used. This results in a vestigial sideband.

17. Frequency-Shift Keying (FSK)

Frequency-shift keying is a form of frequency modulation in which binary signaling is accomplished by the use of two frequencies separated by Δf Hz. The frequency deviation Δf is small compared with the carrier frequency f_c. It is common practice in FSK systems to specify the frequency spacing in terms of a modulation index d, defined as $d = ft_o$, where t_o is the symbol duration.

FSK can be detected either coherently or noncoherently. Noncoherent detection can be effected by two bandpass filters followed by envelope detectors and a decision device. With this approach, the frequency spacing must be at least $1/t_o$ to prevent significant overlap of the passbands of the two filters. Alternatively, a discriminator can be used to convert the frequency variations to amplitude variations. This eliminates the constraint on the frequency spacing.

Recently there has been considerable interest in modified versions of FSK including some coherent systems. They are based on the concept of *continuous-phase FSK* (CP-FSK) which avoids the abrupt phase changes characteristic of other FSK systems at the bit-transition instants. This version of FSK results in rapid spectral roll-off and improved efficiency. The improvement is attained by the use of observation intervals greater than one bit. With coherent detection, values of d in the neighborhood of 0.7 have been shown to provide optimal performance for any observation interval.

Another FM technique that has received considerable interest in recent years is *minimum phase-shift keying* (MSK), also called fast frequency-shift keying. MSK is a special case of CP-FSK for which $d = 0.5$ and coherent detection is used. This technique achieves performance equal to coherent PSK and exhibits the superior spectral properties of CP-FSK. MSK has the additional advantage of relatively simple self-synchronization, an advantage that coherent CP-FSK with $d = 0.7$ does not share.

18. Phase-Shift Keying (PSK)

Most digital phase-modulation schemes require coherent detection. The most straightforward approach is coherent *binary phase-shift keying* (BPSK) in which the carrier is shifted by 0° or 180°. Detection requires a precise phase reference which is normally obtained by performing a nonlinear operation on the received waveform.

Since some phase-extraction techniques exhibit 180° phase ambiguities, a modified form of PSK called *differentially encoded PSK* (DE-PSK) is often used. With DE-PSK, the information is conveyed via transitions in carrier phase. Since a bit decision error on the current bit will induce another error in the subsequent bit, the performance of DE-PSK is slightly inferior to that of coherent PSK.

The third version of binary PSK is *differential PSK* (DPSK) in which, as with DE-PSK, the information is differentially encoded. The difference between DPSK and DE-PSK is in the detector. With DPSK, no attempt is made to extract a coherent phase reference. Rather, the signal from the previous bit interval is used as a phase reference for the current bit interval. Since the phase reference is not smoothed over many bit intervals, the performance of DPSK is somewhat worse than that of DE-PSK.

Coherent quaternary PSK (QPSK) schemes involve encoding two bits at a time with one of four possible carrier phases spaced by 90°. As in the binary case, the data can be differentially encoded and differentially detected with a concomitant loss in performance. This also applies to *m*-array PSK schemes.

A modification of QPSK called *offset-keyed QPSK* (OK-QPSK) has come into use. OK-QPSK can be considered to consist of in-phase and quadrature components. With normal QPSK, during each two-bit time interval of *T* seconds, the I carrier is binary-PSK modulated by one bit and the Q carrier is modulated by the other. The resulting signal can take any one of the four possible phases, and abrupt transitions of 0°, 90°, and 180° can occur. With OK-QPSK, the Q channel is shifted by *T*/2 seconds with respect to the I channel. The transition rules are designed so that when the I and Q channels are added together, the resulting signal is shifted abruptly by less than 90°.

19. Hybrid AM/PM Techniques

The need for increased bandwidth and power conservation has led to the use of a class of hybrid AM/PM techniques called *quadrature amplitude modulation* (QAM). The resulting information signals can best be represented in phase-amplitude signal space as shown in Fig. 11.12. This diagram shows the location of each of 16 possible transmitted signals for QAM-16. This scheme transmits four bits of information during each signaling interval and thus requires less bandwidth for the same information rate. Modulations up to QAM-256 are being used in digital radio systems.

20. Partial-Response Systems

According to the Nyquist criterion, the maximum theoretical rate at which symbols can be sent in a bandwidth of *W* Hz is 2*W*. However, by introducing dependencies or correlations between the amplitude of transmitted pulses and by changing the detection procedure, a maximum packing rate equal to or greater than 2 symbols/(s·Hz) can be obtained using realizable filters. The schematic of a generalized *partial-response system* is shown in Fig. 11.13. By choosing various combinations of integer values for the coefficients f_k, one can obtain different

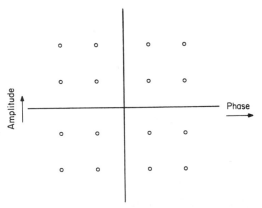

FIG. 11.12 Sixteen-ary quadrature amplitude modulation.

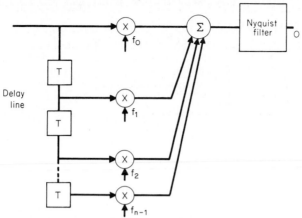

FIG. 11.13   Generalized partial response system.

partial response systems having a controlled amount of intersymbol interference over a span of a few bits.

21.   Comparison of Digital Modulation Methods

Tables 11.1 to 11.4 in this section compare the performance of the various modulation methods described in previous sections. A performance measurement utilized throughout this section is E_b/N_o, the ratio of the average signal energy per bit to the noise spectral density (as measured at the input of the receiver) required to achieve a bit error rate of 10^{-4}. The results were taken from a variety of sources, and there are slight inconsistencies resulting from small differences in the filters used. These tables, however, are indicative of the results which can be expected and provide a basis for determining the tradeoffs which are required for the design of a practical system.

Ideal Performance.   Table 11.1 presents the ideal performance of representative modulation methods in the presence of additive white gaussian noise but in the absence of bandwidth limitations and continuous-wave interference.

Bandwidth Requirements.   The bandwidth required to transmit a specified information rate is an important performance criterion. The figure of merit is the "speed" of a modulation method, R/W, where R is the maximum data rate and W is the i-f bandwidth. Table 11.2 presents the speed of each method and the required E_b/N_o. A comparison of this parameter with the corresponding values in Table 11.1 shows the degrading effect of finite bandwidth.

Effects of Interference.   Table 11.3 shows the effect of cochannel cw interference at two levels, -10 dB and -15 dB, with respect to the desired channel.

Effects of Fading.   Fading is a problem often encountered on digital radio links. Table 11.4 presents the performance results for a channel with Rayleigh fading.

TABLE 11.1   Ideal Performance of Representative Modulation Methods

Modulation method	E_b/N_o, dB*
Amplitude-shift keying	
OOK—coherent detection	11.4
OOK—envelope detection	11.9
Frequency-shift keying	
FSK—noncoherent detection ($d = 1$)	12.5
CP-FSK—coherent detection ($d = 0.7$)	7.4†
CP-FSK—noncoherent detection ($d = 0.7$)	9.2†
MSK ($d = 0.5$)	8.4
MSK—differential encoding ($d = 0.5$)	9.4
Phase-shift keying	
BPSK coherent detection	8.4
DE-BPSK	8.9
DPSK	9.3
QPSK	8.4
DQPSK	10.7
OK-QPSK	8.4
8-ary PSK coherent detection	11.8
16-aryPSK coherent detection	16.2
QAM	
16-ary QAM	12.4

*Required for a bit error rate of 10^{-4}.
†For a three-bit observation interval.
(Source: Oetting, J.D., "A Comparison of Modulation Techniques for Digital Radio," *IEEE Transactions on Communications*, vol. 23, no. 12, December 1979.)

Because of the severe effects of this degree of fading, a required bit rate of 10^{-2} is assumed in this table. This error rate is high for most digital radio applications, but error-control coding could be used to achieve a satisfactory result. Since the values in Table 11.4 are a weighted average of the ideal performance figures, the relative performance of the various methods does not differ markedly from that indicated in Table 11.1.

Cost and Complexity.   The relative costs of the various modulation methods for a specific communication system cannot be evaluated accurately without conducting a full-scale investigation of the tradeoffs involved for the various options. Nevertheless, the modulation methods can be ranked according to their relative complexity, and this provides the basis for an initial estimate of their relative costs. This ranking is shown in Fig. 11.14.

SPREAD-SPECTRUM MODULATION TECHNIQUES

22.   Introduction

Spread-spectrum systems are characterized by transmission bandwidths that are much wider than the minimum required to transmit the information. The large

TABLE 11.2   Signal Speed of Representative Modulation Methods

Modulation method	Speed, (bs/Hz)	E_b/N_o, dB*
Amplitude-shift keying		
OOK—coherent detection	0.8	12.5
Frequency-shift keying		
FSK—noncoherent detection ($d = 1$)	0.8	11.8†
CP-FSK—noncoherent detection ($d = 0.7$)	1.0	10.7
MSK ($d = 0.5$)	1.9	9.4
MSK—differential encoding ($d = 0.5$)	1.9	10.4
Phase-shift keying		
BPSK—coherent detection	0.8	9.4
DE-BPSK	0.8	9.9
DPSK	0.8	10.6
QPSK	1.9	9.9
DQPSK	1.8	11.8
8-ary PSK—coherent detection	2.6	12.8
16-ary PSK—coherent detection	2.9	17.2
QAM		
16-ary QAM	3.1	13.4

*Required for a bit error rate of 10^{-4}.
†Discriminator detection.
(Source: Oetting, J.D., "A Comparision of Modulation Techniques for Digital Radio," *IEEE Transactions on Communications*, vol. 23, no. 12, December 1979.)

bandwidth redundancy gives spread-spectrum systems the ability to overcome severe levels of interference.

A second important characteristic of spread spectrum signals is *pseudorandomness*. This makes the signals appear similar to random noise and difficult to demodulate by receivers other than the intended ones. This provides protection against unauthorized reception by casual listeners (privacy) or, in its more complex forms, by sophisticated listeners with analytic equipment (security).

Spread-spectrum modulated signals are used for:

1. Combating or suppressing interference due to jamming, other users of the channel, signals from adjacent satellites, and the like.

2. Hiding the signal by transmitting it at low power and thus making it difficult for unwanted listeners to detect.

3. Achieving message privacy or security in the presence of other listeners.

23. Spread-Spectrum Modulation Systems

Spread-spectrum systems employ three general types of modulation:

TABLE 11.3   Performance of Representative Modulation Methods in the Presence of CW Interference

Modulation method	E_b/N_o, dB*	
	$S/I = 10$ dB	$S/I = 15$ dB
Amplitude-shift keying		
OOK—envelope detection	> 20.0	14.5
Frequency-shift keying		
FSK—noncoherent detection ($d = 1$)	14.7	13.3
Phase-shift keying		
BPSK—coherent detection	10.5	9.2
DE-BPSK	11.0	9.7
DPSK	12.0	10.3
QPSK	12.2	9.8
DQPSK	> 20.0	14.0
8-ary PSK—coherent detection	> 20.0	15.8
16-ary PSK—coherent detection		> 24.0

*Required for a bit error rate of 10^{-4}.
(Source: Oetting, J.D., "A Comparision of Modulation Techniques for Digital Radio," *IEEE Transactions on Communications*, vol. 23, no. 12, December 1979.)

1. Modulation of a carrier by a digital code sequence with a bit rate much larger than the information bit rate. Such systems are called *direct sequence* (DS) or *pseudo-noise* (PN) modulated systems.

2. Shifting of the carrier frequency in discrete increments in a pattern determined by a code sequence. The resulting signal is called a *frequency-hopped* (FH) spread-spectrum signal.

3. "Chirp" modulation in which the carrier is swept over a wide band during each pulse interval.

A technique related to frequency hopping is time hopping in which the time of transmission is governed by a code sequence.

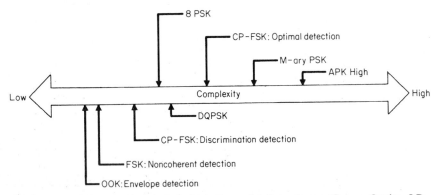

FIG. 11.14   Relative complexity of representative modulation schemes. (Source: Oetting, J.D., "A Comparision of Modulation Techniques for Digital Radio," *IEEE Transactions on Communications*, vol. 23, no. 12, December 1979.)

TABLE 11.4   Performance of Representative Modulation Methods on a Channel with Rayleigh Fading

Modulation method	E_b/N_o, dB*
Amplitude-shift keying	
OOK—coherent detection	17
OOK—envelope detection	19
Frequency-shift keying	
FSK—noncoherent detection ($d = 1$)	20
CP-FSK—coherent detection ($d = 0.7$)	13†
CP-FSK—noncoherent detection ($d = 0.7$)	18†
MSK ($d = 0.5$)	14
MSK—differential encoding ($d = 0.5$)	17
Phase-shift keying	
BPSK—coherent detection	14
DE-BPSK	17
DPSK	17
QPSK	13.5
DQPSK	20
OK-QPSK	13.5
8-ary PSK—coherent detection	16.5
16-ary PSK—coherent detection	21
QAM	
16-ary QAM	18

*Required for a bit error rate of 10^{-2}.
†For a three-bit observation interval.
(Source: Oetting, J.D., "A Comparision of Modulation Techniques for Digital Radio," *IEEE Transactions on Communications*, vol. 23, no. 12, December 1979.)

Direct Sequence Modulation Systems.   A DS waveform is produced by digital modulation of a carrier with a pseudo-random code sequence. Pseudo-noise waveforms are produced by filtering of DS signals, producing signals which resemble noise.

A common modulation format is 180° biphase-shift keying. The energy spectrum typical of this signal is shown in Fig. 11.15. The mainlobe bandwidth (from null to null) is twice the clock rate of the code sequence used as a modulating signal and contains 90 percent of the total power.

The receiver bandwidth is chosen judiciously by a tradeoff which rejects more noise power than signal, hence obtaining a net gain. Narrowing the receiver bandwidth to 60 percent of the mainlobe bandwidth produces a net signal-to-noise ratio gain of approximately 1.7 dB.

Typically, the direct sequence biphase modulator uses a balanced mixer whose inputs are a code sequence and an rf carrier. The output has the following advantages over other modulation forms:

1. The suppressed carrier produced by the balanced mixer is difficult to detect without resorting to complicated techniques.

2. More power is available for sending useful information.

3. It is an extremely simple device.

The *process gain* of a modulation system represents the difference in output and input signal-to-noise ratios and is an important characteristic of spread-

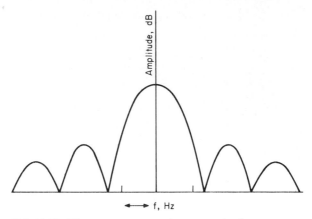

FIG. 11.15   Direct sequence spread spectrum signal.

spectrum systems. For DS modulation, it is given by the equation:

$$\text{Process gain} = G_p = \text{BW}_{rf}/R_{inf} \qquad (11.39)$$

where BW_{rf} = bandwidth of the transmitted spectrum and R_{inf} is the data rate of the baseband channel.

Direct sequence process gains typical for practical systems are of the order of 40 dB.

The *jamming margin* expresses the capability of the system to perform in a hostile environment where interfering signals have power levels larger than the desired signals. It can be calculated from

$$\text{Jamming margin} = M_j = G_p - [L_{sys} + (S/N)_{out}] \qquad (11.40)$$

where L_{sys} is the system implementation loss and $(S/N)_{out}$ is the signal-to-noise ratio at the information output.

Frequency-Hopping Systems.   Frequency hopping is FSK except that the set of available frequencies is large, often measured in the thousands.

A frequency-hopping system consists of a code generator and a frequency synthesizer capable of responding to the outputs from the code generator. The transmitter should be designed to transmit the same amount of power in every channel. The received frequency-hopping signal is mixed with a locally generated replica which is offset in frequency by a fixed amount, thus producing a constant difference frequency when the transmitter and receiver sequences are in synchronism.

The process gain for a frequency-hopping system is equal to the number of available frequency choices. The minimum-frequency switching rate usable in frequency hopping is determined by the information rate, the redundancy used, and the strength of the interfering signal.

Figure 11.16 displays the error rate versus the fraction of channels jammed (J/N) for various frequency (chirp) decision criteria in multifrequency (chirp) transmission.

Time-Hopping Systems.   In time-hopping systems, the code sequence is used to key the transmitter on and off at pseudo-random times. The simplicity of this

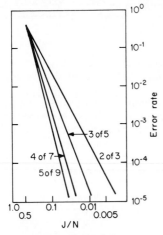

modulator is evident, and any pulse-modulatable signal source capable of following code waveform is an eligible time-hopping modulator. Simple time-hopping modulation offers little in the way of interference rejection because a continuous carrier at the signal center frequency can block communications effectively. Because of its relative vulnerability to interference, simple time hopping is not used for antijamming unless combined with frequency hopping to prevent single-frequency interferers from causing significant losses.

FIG. 11.16 Error rate versus fraction of channel jammed (J/N) for various frequencies (chirps) decision criteria in multifrequency (chirp) transmission

Pulsed FM (Chirp) Systems. Chirp transmissions are pulsed rf signals whose frequency varies in a known way during each pulse period. The receiver used for the chirp signals is a filter matched to the angular rate of change of the transmitted frequency-swept signal. Coding is not normally used with this type of matched filter. Most chirp systems use a linear sweep pattern, but any pattern which is compatible with the requirement for a matching receiver is suitable.

The dispersive filter is the heart of the chirp receiver. It is a storage and summing device that accumulates the energy received over an interval, assembles it, and releases it in one coherent burst. Thus a received signal is stored until the entire ensemble of frequencies in one sweep arrives and the summed power of all frequencies is released at one time. The chirp matched filter compresses a frequency sweep (usually linear) which provides an improvement in output signal-to-noise ratio D equal to:

$$D = (\Delta T) (\Delta f) \tag{11.41}$$

where Δf = frequency sweep and ΔT = time occupied by sweep in frequency.

Typical bandwidth parameters available with delay lines of the dispersive type are shown in Fig. 11.17.

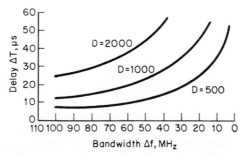

FIG. 11.17 Chirp filter TW product curve (Source: Dixon, R.C., *Spread Spectrum*, John Wiley, New York, 1976.)

MULTIPLEXING

24. Frequency-Division Multiplexing

The advent of broadband analog transmission facilities (cable and microwave radio) led to the development of FDM systems, which have provided efficient bandwidth utilization of the transmission media.

A basic building-block group of twelve 4-kHz channels in the frequency band 60 to 108 kHz was established for early voice carrier systems (J- and K-type carriers) in 1930. Since then, the multiplexing hierarchy has been broadened to encompass a 13,200-channel spectrum. Each step in the evolution of the FDM hierarchy was built upon and utilized existing equipment designs in order to achieve economical terminal equipment for new transmission systems.

The FDM hierarchy most widely used in North America is derived by means of five separate frequency translations, each of which places signals at higher frequencies and in larger groupings. The basic groupings of 4-kHz channels in the hierarchy are defined as follows:

Designation	Number of channels	Frequency range, kHz
Group	12	60–108
Super group	60	312–552
Master group	600	564–3084
Jumbo group	3600	564–17,548

The basic North American hierarchical plan is shown in Fig. 11.18.

The A-6 channel bank shown in Fig. 11.19 is an example of FDM implementation. The group bank is formed by two steps of modulation.

The first step places each channel at a preassigned frequency near 8 MHz. The resulting double-sideband signal at the output of the modulator is passed through a high-pass filter to suppress residual baseband energy and a bandpass filter to suppress the upper sidebands and residual carrier energy from the balanced modulator. Only the lower sideband components are retained to be combined with 11 other signals, each in a different portion of the group spectrum.

The second step translates the entire group down to the group frequency band.

This two-step process was adopted to permit the use of a new channel filter design using monolithic quartz crystals optimized at frequencies close to 8 MHz.

A switching office with FDM equipment contains one or more reliable *primary frequency supplies* (PFS) and/or a *jumbo frequency supply* (JFS) with redundant circuits which are automatically switched in the event of failure. The operating frequencies of these units are controlled by synchronization with standard sources such as the Bell System reference frequency standard. This reference is currently supplied nationwide through regional frequency supplies. In the event of failure of the reference frequency standard, synchronization is maintained by the regional supplies.

25. Time-Division Multiplexing

Interleaving of pulse trains representing different channel signals into a single

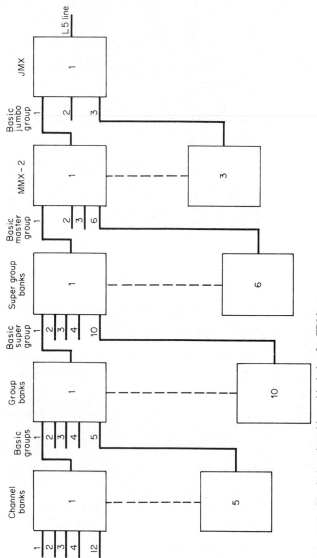

FIG. 11.18 North American hierarchical plan for FDM.

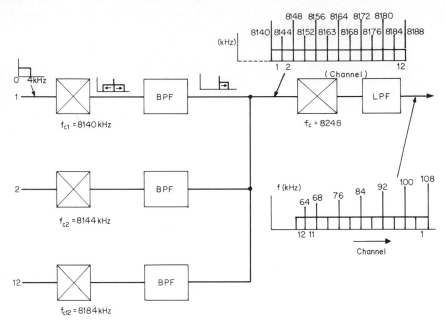

FIG. 11.19 Block diagram of A-6 transmitting channel bank

continuous stream is called *time-division multiplexing* (TDM). This process involves three major functions—timing, framing, and synchronization.

Timing. All the major logical processes in a digital system terminal depend upon accurate timing of pulses associated with each signal and with the multiplexing of the coded signal into a single pulse stream. In most terminals, the timing is provided by or derived from a single clock circuit that distributes a stream of pulses with a highly precise repetition rate. The sampling and coding functions are controlled by this timing signal so that all the pulses are properly related with respect to the repetition rate, width, and position in a time sequence.

Framing. Pulses associated with each specific signal need to be identified and separated from all other signals in the stream. This is accomplished by organizing the pulse stream into frames. The separation of pulses of one signal from others is then accomplished by counting pulse position relative to the beginning of the frame.

Synchronization. For satisfactory performance, terminal circuits must be properly synchronized with each other. In the originating terminal this function is fulfilled by the previously discussed timing signal. If the equipment must multiplex digital signals from different sources, it is necessary that all sources be in synchronism or that their rates be adjusted before multiplexing. Because of the necessity of synchronism, the clock circuits must be extremely reliable and must produce timing signals with great precision.

Transmission-Rate Hierarchies. A hierarchy of transmission rates or levels has evolved in digital networks similar to that of analog. It will undoubtedly grow as digital networks expand in flexibility and size. These levels are designated by digital signal numbers with increasing rates from DS-0 to DS-4 as shown in Fig. 11.20. The rates are not integral multiples of lower rates because, at each level, additional bits are "stuffed" to facilitate multiplexing (framing, synchronization, etc.) and to provide other service functions (parity bits, etc.).

The 24-channel T1 carrier system has been established as the basic building block in the TDM hierarchy. The transmission rate is 1.544 Mbs, a rate suitable for time-division multiplexing of 24 digitally encoded voice signals.

The digital multiplex units used to translate from one digital level into another are designated by a prefix M (for multiplex) followed by two digits that designate the steps in the translation process.

DS-1 Signal. DS-1 is a 50 percent duty-cycle signal consisting of 1.544 million time slots per second, each of which can transmit one bit, a 0 (no pulse) or 1 (pulse). It accommodates 24 voice-frequency channels (sometimes called digroup) digitized by pulse-code modulation (8-kHz sampling rate and 8 bits/sample). Each frame contains 24 channel samples with 8 bits/sample or 192 bits. At the end of each frame, an extra bit is added to the signal to identify the frame sequence and synchronize the terminal equipment. When the frame rate is combined with the sampling rate of 8000 samples per second, the DS-1 signaling rate is $193 \times 8000 = 1.544$ Mbs.

The framing bits that follow each 192-bit sequence are transmitted as alternate 1s and 0s, and thus provide a coded sequence of framing pulses $(101010 \cdots)$ that can be recognized by the receiving terminals.

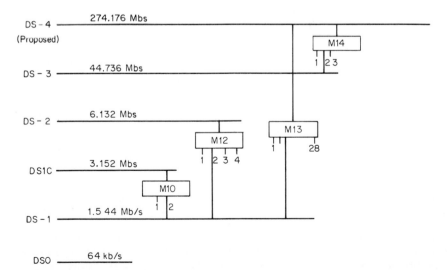

FIG. 11.20 Existing North American hierarchical plan for TDM.

DS-1C Signal.   The transmission rate for the DS-1C signal is 3.152 Mbs. It contains two DS-1 signals plus a 64-kbs signal used for synchronization and framing. The synchronization of the two DS-1 signals is achieved by pulse stuffing. These signals, received from the crossconnect frame in bipolar form, are first converted to unipolar signals. Before multiplexing, the second signal is inverted logically to control the signal statistics of the transmitted pulse stream. After inversion and stuffing of this signal, the two are multiplexed by interleaving them bit by bit according to the input numbering sequence.

The multiplexed bit stream is scrambled and multiplexed with a control bit sequence that permits demultiplexing and deletion of the stuffed bits from the signals at the receiving end. Each control bit sequence precedes a block of 52 bits from the multiplexed DS-1 signals. The control bits form a repetitive sequence 26 bits long which, with the information bits associated with each control bit, defines a 1272-bit block; see Fig. 11.21. Each 26-bit control sequence (or word) is made up of three subsequences designated M, F, and C. The symbol 0 with subscripts denotes how the information bits from two DS-1 signals are interleaved.

The M sequence consists of four bits designated M_1, M_2, M_3, and M_4. They are the first, seventh, thirteenth, and nineteenth bits in the 1272-bit M frame. The first three bits, 011, are used to identify the M frame format and the fourth is used as a signaling channel to indicate alarm conditions at the transmitting terminal. A 1 indicates no alarm while a 0 indicates the presence of an alarm.

The F sequence is made up of alternate 1s and 0s ($F_1 = 1$ and $F_0 = 0$) that appear at the beginning of every third 52-bit information sequence. This code is used to identify the scrambled input signals and control bit time slots.

The C-bit sequence is used to identify the presence or absence of stuff pulses in the information bit portion of each frame. There is a sequence of three C bits in each subframe. If a stuff pulse is to be inserted during the subframe, the C bits are all 1s, otherwise they are 0s.

DS-2 Signal.   The DS-2 signal, 6.312 Mbs, consists of four DS-1 signals and a number of control, framing, and stuff bits. The frame format is shown in Fig. 11.22. Synchronization of the four DS-1 signals is necessary because they may originate from different sources with unsynchronized timing clocks. Synchronization is accomplished by adding stuff pulses to each signal so that all four are timed to the same gate which is controlled by a common clock at the multiplex equipment.

DS-2 frames contain 1176 bits which are divided into four 294-bit subframes. The control-bit word begins with an M bit as shown in Fig. 11.22. The four M bits are transmitted as 011X where the fourth bit, which may be a 1 or a 0, may be used as an alarm indication. When a 0 is transmitted, an alarm is present. The 011 sequence is used at the receiving circuits to identify the frame. Within each subframe, two other sequences are used for control purposes. Each control bit is followed by a 48-bit block of information with 12 bits taken from each of the four DS-1 signals. The first bit in the third and sixth blocks is designated as an F bit. The F bits are in a 0101 ⋯ sequence and are used to identify the location of the control-bit sequence and the start of each block of information bits. The stuff-control bits are transmitted at the beginning of each of the 48-bit blocks which are numbered 2, 4, and 5 within each subframe. When these control bits, designated C, are 000, no stuff pulse is present; when C bits are 111, a stuff pulse is added in the stuff position.

Prior to multiplexing, input signals 2 and 4 (DS-1s) are logically inverted. At the output of the multiplex equipment, the multiplex signal is unipolar and must

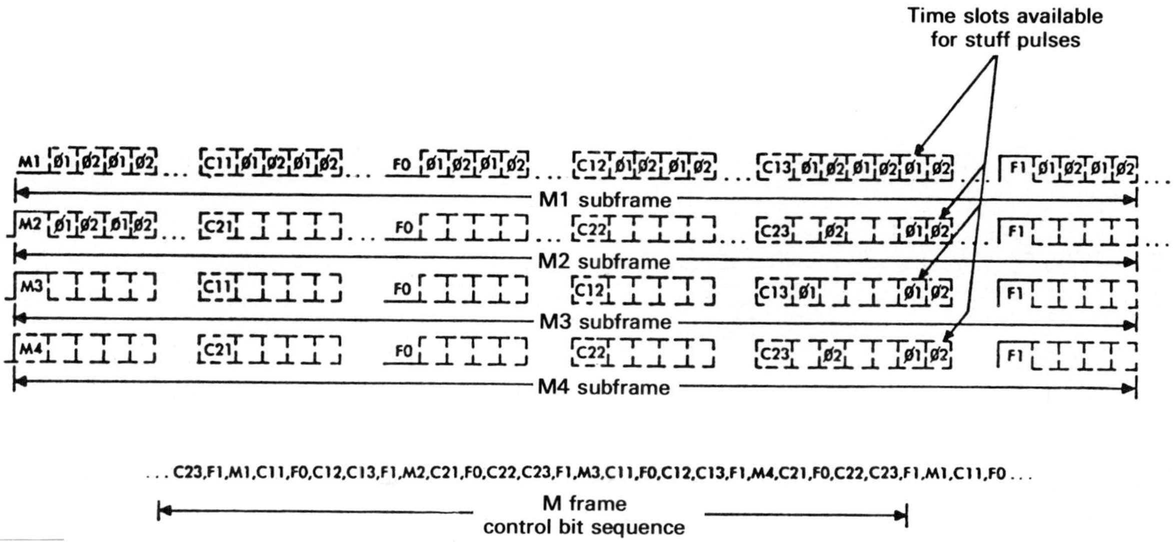

FIG. 11.21 DS1C frame format. (Source: *Telecommunication System Engineering*, vol. 2, Bell System Center for Technical Education, 1977.)

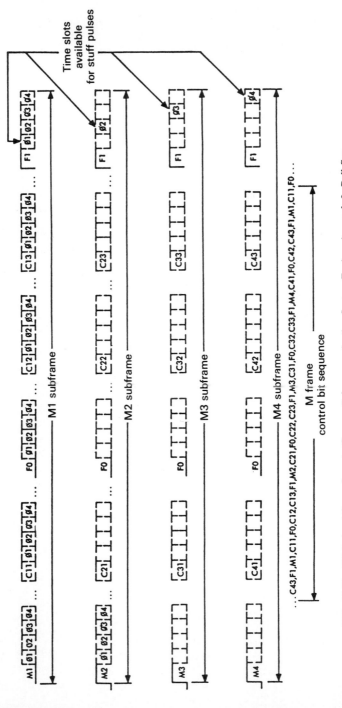

FIG. 11.22 DS-2 signal frame format. (Source: *Telecommunication System Engineering*, vol. 2, Bell System Center for Technical Education, 1977.)

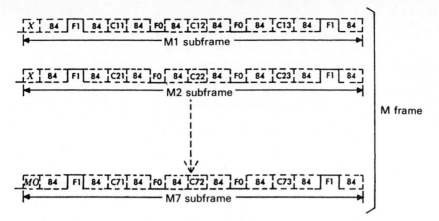

Fig. 11.23 DS-3 signal frame format. (Source: *Telecommunication System Engineering*, vol. 2, Bell System Center for Technical Education, 1977.)

be converted into a bipolar format with a 50 percent duty cycle for transmission. The format used at the DS-2 signal level is bipolar with 6-zero substitution.

DS-3 Signal.   The DS-3 format is shown in Fig. 11.23. It consists of four DS-1 signals and seven DS-2 signals multiplexed. The bit rate is 44.734 Mbs. It employs a modified bipolar format with three zeros substitution, B3ZS, and a 50 percent duty cycle.

The DS-3 signal is partitioned into frames of 4760 bits which are subdivided into seven subframes of 680 bits. Each subframe, in turn, is divided into eight blocks of 85 bits with the first bit in each block used as a control bit. The initial bits in successive subframes are X, X, P, P, M0, M_1, and M_0. The first time slot in each of the first two subframes (called X bit) is used for alarm or other maintenance or operational purpose. The two X bits in a frame must be the same.

The first time slots in the third and fourth subframes are designated as P bits. These are used to convey parity information relating to 4704 information time slots following the first X bit in the previous frame. If the sum of all information bits is 1, PP = 11 and if the sum is 0, PP = 00.

The first time slots in subframes 5, 6, and 7 are designated M bits. These three bits always carry the code, 010, which is used as a multiframe alignment signal.

In each subframe, blocks 2, 4, 6, and 8 carry F bits. These are transmitted in the first time slot of each of these blocks as a 1001 code that is used as a frame alignment signal to identify all control-bit time slots.

The first time slot in subframes 3, 5, and 7 carry bits to indicate the presence or absence of a stuff pulse in the subframe. In the C-bit positions, a 111 code indicates that a stuff pulse has been added and a 000 code indicates that no pulse has been added.

FURTHER READING

Black, H. S., *Modulation Theory,* D. Van Nostrand, New York, 1953.

Dixon, R. C., *Spread Spectrum,* John Wiley & Sons, New York, 1976.

Holmes, J. K., *Coherent Spread Spectrum Systems,* John Wiley & Sons, New York, 1982.

Lucky, R. W., J. Salz, and E. J. Weldon, *Principles of Data Communications,* McGraw-Hill, New York, 1968.

Oetting, J. D., "A Comparison of Modulation Techniques for Digital Radio," *IEEE Transactions on Communications,* vol. Com-23, no. 12, December 1979.

Proakis, J. G., *Digital Communications,* McGraw-Hill, New York, 1983.

Schwartz, M., *Information Transmission, Modulation, and Noise,* McGraw-Hill, New York, 1959.

Stein, S., and J. J. Jones, *Modern Communication Principles,* McGraw-Hill, New York, 1967.

Taub, H., and D. L. Schilling, *Principles of Communication Systems,* McGraw-Hill, New York, 1971.

Telecommunication System Engineering, vol. 2, Bell System Center for Technical Education, 1977.

CHAPTER 12
CIRCUIT SWITCHING

Amos E. Joel, Jr.
Consultant, formerly AT&T Bell Laboratories

CONTENTS

INTRODUCTION

1. Functions of Switching Systems

In telecommunication systems the most important element is the message. If there are only two stations to originate and terminate messages, no switching is required. Unless messages are broadcast, switching is required when there are more than two stations, to provide *selective* communication among the stations. Signaling is usually required to alert a station that a message is to be sent.

Generally privacy is required so that a message will be received only by the station for which it is intended. Communication privacy requires that *contention* be included in the selection process to ensure that the message reaches only the selected station and not also be received by some other station that might be attempting to transmit a message at the same time. Switching is, therefore, both selection and contention.

TYPES OF SWITCHING

2. Definitions

A message may be directed only from one station to another or there may be interactive or two-way messages. For two-way messages that occur in real time, switching systems establish a circuit between the two or more stations exchanging messages. These systems are given the name *circuit switch.*

To obtain efficiencies, the switching system may remove or disconnect the circuit during silent periods in the message and reestablish the circuit when the message information resumes. This is a technique that is sometimes referred to as *virtual circuit switching.* Virtual circuit switching assumes that for the application, there is no perceptible degradation of the message. This perception depends upon the service quality expectations of the user.

Establishment of the circuit by the switch is sometimes known as a *connection.* The process of requesting and establishing a connection through the switch is known as a *call.*

A combination of transmission, switching, and stations forms a *telecommunications network.* Where the nature of the messages may not require real-time interaction, other forms of networks with different economic criteria for service quality and message delay may be employed. They may also be used where the network serves more than one type of telecommunications, e.g., voice and data.

These networks may utilize different forms of switching known as *message switching* and *packet switching,* which are described in Chaps. 13 and 14.

These switches and the networks in which they are employed allow for the whole message or segments of the message to be delayed. They are generally used where one desires a different economic balance of the telecommunication network costs and service quality factors.

Message switching is employed where whole one-way messages may be delayed until the transmission and/or station is available for its receipt. Packet switching is generally used where the short messages are interactive and transmission is segmented.

SWITCHING SYSTEM ARCHITECTURES

3. Noncentral Switching

Transmission facilities must connect each station with a switch. When the switch is located near the station, the function has been given a variety of names, such as station or noncentral switching. See Fig. 12.1. A particular form of implementation in circuit switching is known as a *key system.* In packet switching a current popular form of implementation is known as a *local-area network* (LAN).

4. Central Switching

By moving the switching for a number of stations to a centralized location, cost reduction efficiencies are possible. See Fig. 12.2. Additional signaling is then required to remotely control the selection and contention functions of the switch. Also required is signaling to indicate when the station is requesting service, or originating a call, sometimes known as *attending,* and when the station has answered a call after being *alerted.*

5. Hierarchical Switching

In large networks, a single centralized switching point or node may not represent the optimum economic cost between transmission and switching. The average

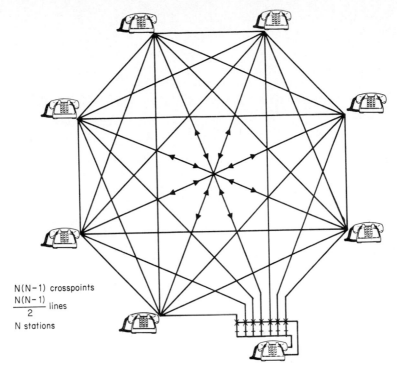

$N(N-1)$ crosspoints
$\dfrac{N(N-1)}{2}$ lines
N stations

FIG. 12.1 Noncentral switching. Each of the N stations reaches the remaining stations through individual station, or noncentral, switches with N-1 switches, known as "crosspoints." The lines interconnecting the stations are used in both directions.

length of lines that serve the stations is longer with centralized switching. Therefore in large networks it is generally more economical to provide a number of nodes arranged in hierarchies and to interconnect them with common arteries for traffic, known as *trunks*. Nodes that only interconnect other nodes are known as *tandem* switches or nodes. See Fig. 12.3.

A switching node may serve lines as a "local" or "end" office or as a combination local and tandem office. It then switches calls line-to-line and trunk-to-trunk and traffic between lines and trunks.

One of the efficiencies possible with switching is the introduction of the concept of *concentration*. With concentration the switch provides for fewer simultaneous connections than would be required if each station were simultaneously busy on a connection.

The use of concentration makes it necessary to introduce the inverse selection concept of *expansion* into the switch; this causes the caller to be connected to the called line. In modern switching systems, the concentration and expansion functions are combined, and the resulting subsystem is usually called a *line concentrator*.

To obtain an even greater cost optimization between transmission and switching, the line concentrator function may be located close to groups of stations. In this case, the subsystem is located nearer to the stations and at a distance from the remainder of the switching system. The concentrator is known as a *remote line concentrator*. The remainder of the switching system is known as the *host*.

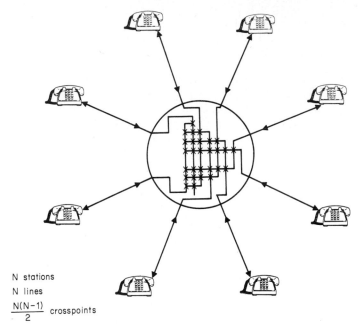

N stations

N lines

$\dfrac{N(N-1)}{2}$ crosspoints

FIG. 12.2 Central switching. Each station has one 2-way line to a central node. There is one central remotely controlled switching network which, with nonblocking, requires half the number of crosspoints as for noncentral switching (Fig. 12.1). The number of crosspoints may be considerably reduced by introducing blocking, which is possible with central switching. The central also needs to recognize an "attending" or request for service.

With modern technology it has become possible to include additional capability in the remote switching node besides concentration and expansion. In particular it may be able to process calls within the concentrator should the trunks between it and the host be temporarily disabled. When the remote concentrator is given greater intelligence, it is known as a *remote switching system, unit,* or *module.*

In circuit switching, remote concentrators and switching units are part of a *distributed network.* They may also reflect the application of a particular form of system architecture as covered in this section.

CLASSIFICATION OF CIRCUIT SWITCHING

Circuit switches may be classified in many ways. The following cover the technology and techniques of circuit switching.

6. Circuit Switching Technologies

The basic switching element is known as a *crosspoint.* Any device that opens and closes a circuit or changes from a high to low electrical impedance is a crosspoint.

Manual Devices—Plugs and Jacks. The plugs are usually connected together through flexible cords, several of which may operate over the same field of jacks that in effect constitute the crosspoints.

Gross-Motion Electromechanical. Gross-motion electromechanical switches include two-motion (up and around) step-by-step switches and those that move in only one direction—such as rotary and panel selectors that move over a vertical plane. In gross-motion switches, wipers or brushes move over fixed banks of contacts. The combination of wiper and bank constitutes crosspoints.

Fine-Motion Electromechanical. Fine-motion electromechanical switches are coordinate matrix devices such as crossbar switches where sets of contacts are operated at the matrix intersections. The contact sets are crosspoints. They may be individual electromagnetic relays or contacts operated by cooperatively moving actuators.

Electronic. Electronic switches employ PNPN bistable transistor devices or any other electronic components that are inherently bistable or are used to create a bistable condition, including gas and vacuum tubes. Also any gating device, such as an AND gate may be used as a crosspoint. Electronic crosspoints should possess linear characteristics over the range of use if the signals are of an analog nature, e.g., for 4kHz speech. Nonlinear crosspoints may pass digital signals.

Photonic. Photonic switches employ optical beam switches or crosspoints in the optical medium, e.g., *directional decouplers*.

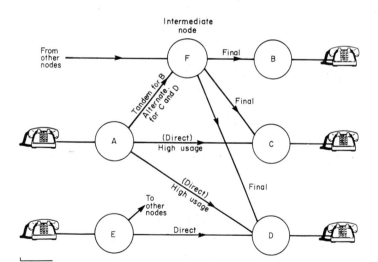

FIG. 12.3 Intermediate switching. One-way trunk groups interconnect nodes (E–D). Some direct trunk groups provide the only route between particular nodes (E–D). Intermediate or tandem nodes (F) form a hierarchical network of offices. Some offices may be reached only through the intermediate offices (A–B), thereby eliminating the need for direct trunk groups between all offices. Traffic between nodes (A–C) (A–D) may be routed over direct (high usage) trunk groups as well as tandem trunk groups for the alternate routing of traffic when all direct trunks are busy.

7. Circuit Switching Techniques

While crosspoints may be assembled in various ways to form a switch, the switching art contains the analog of transmission multiplex techniques, viz. multiplexing in the frequency and time domains. In addition, when looked at broadly, transmission channels packaged into multipair cables might be considered to be multiplexed in the *space* domain. In switching, these domains have been called *divisions*.

Space Division.   In space-division switching, crosspoints are assembled into arrays such that each call or connection uses a distinct crosspoint or set of crosspoints in series to establish a separate path in "space."

Time Division.   In time-division switching, as in transmission, channels are established periodically as time slots. The time slot rate relates to the bandwidth of the message channel. For speech a common value is 8000 time slots per second to faithfully transmit speech signals.

   In time-division switching, time slots assigned to calls may be moved in time and multiplexed among different transmission lines. This movement in time is known as *time-slot interchange* and the multiplexing between different lines is known as *time-multiplex switching*. The moving of time slots implies delaying the transmission. When used for interactive speech, these delays must be held to levels that are imperceptible to the users.

Frequency Division.   Frequency division uses the same transmission medium within the switch to carry a plurality of calls simultaneously by assigning different frequencies to each. Frequency division requires an equivalent space-division *network-control access network* (NCAN) (see below) and therefore has no economic advantage.

TRAFFIC PRINCIPLES

8. Measuring Traffic

Traffic is measured for the load offered to a switch, a transmission facility, or circuit, or an operator force. The principal measure of traffic is *usage*. The dimension of usage is time. If 100 traffic sources, such as lines or trunks, are busy during a particular hour for an average of 3 minutes each, then the total usage is 300 call-minutes, 18,000 call-seconds (ccs), 180 hundred call-seconds, 5 call-hours, or 5 erlangs. (The erlang is named after a famous traffic theorist.) One erlang is a call-hour. Five erlangs means that on the average the system is offered and/or carries five simultaneous calls for an hour.

   The busy period for each call is known as *holding time*. Usage therefore is equal to the sum of holding times in a given period of time. This period is usually chosen as the anticipated or measured busiest hour of the busiest day of the busiest season of the year.

9. Blocking

It was early realized that the number of circuits or operators needed to provide service for the transmission of messages was not proportional to the offered load.

The greater the load, the fewer the circuits in proportion that were required to provide the same quality or grade of service. This assumes that calls originate at random and those that cannot be served are abandoned. There are many theories that provide curves such as shown in Fig. 12.4 that show the carried load versus offered load for two grades of service. Grades of service are described as a measure of "blocking" and are quoted as a probability of blocking, p, such as 1 in 100 or 1 percent. The greater the blocking, the fewer the number of circuits provided to serve the same offered traffic.

Also, as shown in Fig. 12.5, for a given grade of service or blocking loss, the greater the carried load, the higher the average occupancy of the circuits. This demonstrates the principle that the more the offered traffic, the smoother the peaks and valleys of the random offered traffic. Average circuit occupancy is greater as the blocking is increased. In the extreme, with only a few circuits to carry a load, very high blocking will occur. When one-way messages may be stored in the switch, high circuit occupancy and blocking are utilized to obtain a more economical balance between expensive transmission facilities and memory within the switch or terminal.

Delay is another important attribute of teletraffic, both in message switching and in the control portion of switching systems. The subject is treated in more detail in Chap. 13.

10. Blocking and Nonblocking Systems

Lines are the transmission facilities that bring call attempts and subsequent messages to switching systems. Line circuits or ports are the line terminations on switching systems. (The term *loop* is used outside of the switch to describe a particular two-wire, two-way baseband voice transmission line.)

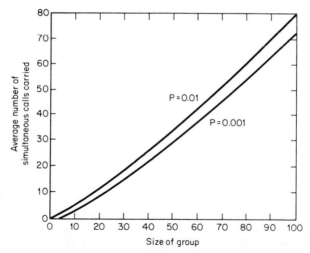

FIG. 12.4 Carried load vs. offered load. P is the probability of blocking, or "grade of service." The size of the group of equally available links or trunks. The equally available links or trunks form a group (abscissa). Less blocking for the same offered load occurs as the size of the group is increased. The relationship is nonlinear.

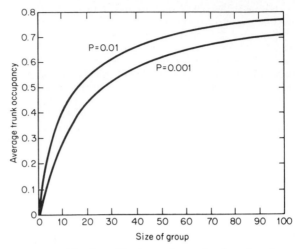

FIG. 12.5   Blocking loss. The lower the grade of service, the greater the average occupancy of the serving group.

Within a circuit switching system, the connections are made by a function known as the *switching center network* (SCN), or simply the switching network. If the switching network enables any desired connection to be established between idle ports independently of the offered traffic, the network is known as *nonblocking*.

Each input of the switching network must be able to reach every output, or all other ports. This is known as *availability*. However, when the traffic offered cannot be carried, then the network is called *blocking*.

When the ports or inputs are two-way they may have a single appearance on the switching network, which is called *single-sided*. The traffic flows bidirectionally through such networks. There are also two-sided networks with ports on each side. Each connection has an input and an output. Such networks may be unidirectional or bidirectional.

11.   Concentration and Distribution

If the number of inputs is greater than the number of outputs in a two-sided unidirectional network, the switching network functions as a *concentrator* of traffic. Conversely if the number of inputs is less than the number of outputs the network is known as an *expandor*. For bidirectional networks the two attributes are combined, concentrating in one direction and expanding in the other. Generally, the single term *concentration* is used. These networks are always blocking and are generally the major contributors to the blocking in switching systems. The purpose of concentrators is to implement the chosen grade of service. When the number of inputs on each side of a two-sided network is the same, the network is called a *distribution* switching network. This network may be blocking or nonblocking.

12. Multistage Systems

All space-division and parts of time-division switching networks are implemented with crosspoints. Therefore, a nonblocking distribution network with n inputs is a square of $n \times n$ crosspoints. Similarly, for concentration or expansion the number of crosspoints is equal to $n \times m$, where n does not equal m. When n and m are large, the number of crosspoints required in a single matrix becomes very large. To reduce the number of crosspoints per input from m or n to a small number requires the introduction of the concept of additional stages of switching or *multistage* switching networks. The paths between matrices are known as *links*.

One interesting innovation in switching networks was the invention of C. Clos,[1] who postulated a multistage nonblocking network that has fewer crosspoints than the obvious single square matrix. He obtained this capability by dividing the inputs into a number of smaller switch matrices. Each of these matrices introduces expansion into the first stage and concentration into the third stage of a three-stage network. The expansion and concentration is almost 2-to-1. The center stage consists of a greater number of smaller distribution (square) matrices. With this configuration, shown in Fig. 12.6, if all inputs of a given matrix were to be connected to the corresponding outputs of a third-stage concentration matrix, there would be sufficient middle-stage matrices through which connections could be made between any other input and output. Three-stage nonblocking switching may be substituted for each of the middle-stage square switches. Therefore a nonblocking switching network of any size may be achieved by employing multistage networks with an odd number of stages. The same principles may also be applied to one-sided networks.

By gradually reducing the expansion and concentration ratios of the first-stage switches, thereby reducing the total number of crosspoints, blocking may be introduced into these odd-stage networks. As would be expected, blocking networks require fewer crosspoints than nonblocking networks.

Networks with an even number of stages may also be used to achieve blocking networks. Generally these networks use many small-square matrices of

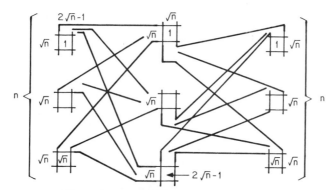

FIG. 12.6 Clos switching network. Two-sided multistage switching networks are made nonblocking by providing $2 < n - 1$ center stage switches, where n is the number of inputs and outputs. Center-stage switches are single-stage nonblocking matrices that may further employ the Clos principle. For large numbers of inputs, the center stages may be made into additional Clos networks.

crosspoints in each stage. The link interconnections are known as *grids,* with each matrix connected with each matrix in the next stage by one or more links. With a single link between each switch in two adjacent stages, there is *full access;* that is, each input can reach every output, but the blocking is very high. By adding pairs of stages, parallel paths are introduced between end switch matrices. In this way, blocking is gradually reduced as more pairs of stages are employed. From the above, it is obvious that multimatrix, multistage techniques may be used to obtain networks of large capacity with fewer crosspoints at the desired blocking levels. One other principle used to reduce the crosspoints by introducing blocking in multistage networks is known as *graded multipling.* Here access is reduced to less than the required availability; that is, each input cannot reach all outputs of a given stage.

BASIC SWITCHING FUNCTIONS

13. Switching Function Descriptions

There are three functions included in all circuit switches. These are switching, control, and signaling. (See Fig. 12.7.) The switching networks are the major function of switching systems, even those, such as data switching systems, that do not employ switching-center networks. There are other networks within switching systems that have been broadly classified as *access networks.*

The second general function is the controls of switching systems. This function is called *call information processing.* It includes not only the hardware but also the software dimension.

Finally, there is the *signaling* function. Signaling is essential to switching. The term is a holdover from the days of electromechanical switching when special protocols were required for each system. Today standard protocols have taken the place of the many earlier signaling techniques. While the techniques have changed, *call signal processing* (CSP) is still a major function of a switching system.

The remainder of this chapter is devoted to a description of these basic functions: switching networks, control, and signaling.

SWITCHING NETWORKS

14. Access Networks

Switching networks provide connective functions within switching systems. Many of these connectives have already been described in connection with other functions, and their characteristics will be recapitulated here.

The principal network in a circuit switching system is the *switching center network* (SCN), but there are many other minor or access connectives. These are the *signal access* (SAN), *control access* (CAN), *intracontrol access* (ICAN), and *network control access* (NCAN) networks. The actual names given to these networks are different for each switching system. Some popular names are *scanners, signal distributors,* and *connectors.* They frequently require a high degree of concentration or expansion, where n or m may be as small as 1.

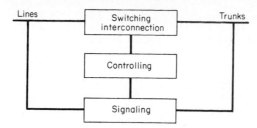

FIG. 12.7 Functional design: switching, control, and signaling. Interrelation among the basic functions of a circuit switching system.

These networks may employ either of the practical switching network techniques of space and time division. Space-division networks require a choice of technology. Time-division networks always use electronic technology, and the principles for these networks are given below. Frequency division, while important for transmission multiplexing, is not practical in switching since a space-division network-control access network equivalent in size to a frequency-division network is required for its implementation.

15. Time-Division Switching Elements

Time division requires a sampling of the analog message signals at a rate equal to at least twice the maximum bandwidth to be transmitted. These signal samples are usually digitized for advantageous use in transmission systems. Techniques are also available to the switching system designer for using the samples in analog form without digitizing. This is known as *pulse-amplitude switching* (*PAS*). (See Fig. 12.8.) High-speed gates connect analog pulse sample sources with sinks. In one technique known as *resonant transfer* there is no loss of signal power in this process. There is, however, a limit to how far the pulse samples may be faithfully transmitted within a switch. Therefore PAS is used only in small systems where the common medium may serve a maximum of a few hundred gates. Another PAS technique uses two common media, one for transmitting and one for receiving, with a pulse amplifier between them. This is an "amplified bus" PAS.

Most newer time-division switching systems employ digital techniques. Switching digital signals permits the switching of digitized voice or video message signals as well as signals from inherently digital sources such as computers. These systems are known as *time-division digital switching systems,* but for many the misappelation "digital switching" has become widely utilized.

16. Time-Multiplex Switching (TMS)

There are two major switching actions that are needed to switch messages from one channel, or time slot, on a digital transmission line to a channel on another line. These actions are known as *time-multiplex switching* and *time-slot interchange.* (See Figs. 12.9 and 12.10.) In time-multiplex switching, a number of frame-synchronized digital lines are connected as required to pass information in

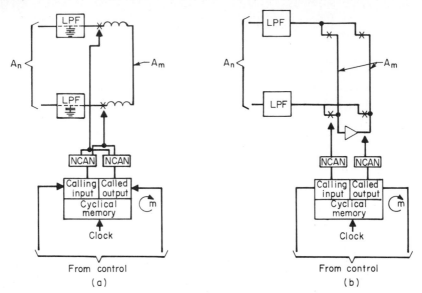

FIG. 12.8 Pulse amplitude switching. (*a*) Pulse-amplitude resonant-transfer switching. Cyclical memory of *m* time slots are loaded with the calling input and called output addresses. The network control access networks (NCAN) simultaneously operate the gates of the calling and called addresses and the instantaneous voltages are interchanged. The gating rates are twice the bandwidth of the analog signals (An). (*b*) Amplified-bus pulse-amplitude switching. The input and output gates of the terminations are gated separately. An amplifier is inserted between the calling and called gating buses.

the same time slot from one line to another. In each time slot the association of input with output lines may be different.

TMS is performed by high-speed electronic gate circuits that, for the duration of one time slot, connect one multiplexed transmission line with another. The gates are arranged topologically as two-sided networks. The simplest such network is a single ungraded, or nonblocking, matrix stage. Since such matrices are used in space-division switching, the TMS stages are also called *space* stages; however, here the switching takes place at time-slot rates, e.g., for 24 voice channels, $24 \times 8000 = 192,000$ times per second, so that the switch is acting as a large number of space-division switches and the power of time-division switching is thereby realized.

As the number of time-division transmission lines to be switched increases, it becomes more economical to employ multistage networks composed of high-speed crosspoints. These networks are usually nonblocking or nearly nonblocking. Because of their topology, TMS is known as the *space* or S stage of a time-division switching system.

The time-slot switching pattern, or permutation, is repeated for each frame until the call requiring the association is no longer required and is replaced by an association required by a new call.

17. Time-Slot Interchange (TSI)

The multiplexed channels to be interconnected are not necessarily in the same time slot. Therefore it is necessary for the switch to associate the two different time slots. This can be achieved by delaying the message sample in one time slot

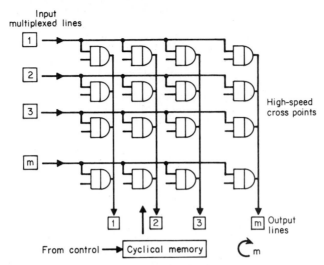

FIG. 12.9 Time multiplexed switching. Cyclical memory contains information for the operation of the high-speed gates resulting for each time slot in a permutation of the input multiplexed lines with respect to the output multiplexed lines.

so that it may be used in the desired time slot of the next frame. This technique is known as *time-slot interchange* (TSI).

The most popular method for achieving delay where the message signals are digital is to store the sample momentarily. Samples are read as received into a bulk-storage medium, such as an integrated-circuit memory, and in the next frame are read out in a rearranged order. As was the case for TMS, this rearrangement is repeated for each frame until the connection association is no longer required.

18. Time-Division Switching Networks

Since TSI involves the use of time delay, stages where this function is performed are known as *time* or T stages.

Time-division switching networks are made up of combinations of T and S stages. (See Fig. 12.11.) The order of the stages is a designer's choice. In general, it has been found that smaller systems may better employ S-T-S networks, where large systems use T-S-T.

Integrated-circuit chips are available that combine multiplexing and T stages so that, typically, eight incoming 24- or 32-channel lines may be connected to eight outgoing lines, thereby switching as many as 256 time slots on a single chip. While these multiplex stages are only 8 by 1, they are still capable of performing an S switching function.

In transmission lines, the digital signals travel only serially along the line. However, within a switch the bits representing the samples may be sent serially and/or in parallel. For example sending the eight-bit samples through the switch on a parallel basis means that a switch with 1.544 Mbs capability may serve 192 time slots rather than only 24.

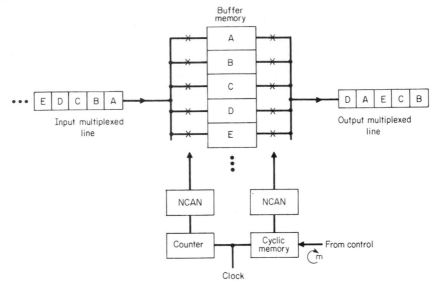

FIG. 12.10 Time slot interchange. Counter demultiplexes the input line, placing the signal samples into buffer memory locations. At the same time the cyclic memory takes the information out of the buffer memory locations, remultiplexing the signal samples in a different order.

Also, within a switch, additional bits may be added to the sample for such functions as parity checking or signaling. Buffering of lines to provide frame synchronism of all time slots entering the switch takes place at the digital line interface circuit between the transmission system and the switching system. This circuit also provides for the conversion from serial to parallel and the addition of additional bits.

The time-division digital network is actually two networks controlled by a single network control circuit (memory). The networks are two-sided with incoming transmission lines appearing on one side while outgoing lines appear on the other. Interconnecting the receiving end of one line with the transmitting end of another line is only half of the connection. To connect the other half, a reciprocal relationship is established within the same network so that the inverse connection may be established. (See Fig. 12.12.)

Since time-division networks employ many active elements, it is not unusual for these networks to include redundancy similar to that designed into the control portions of the system. This redundancy is in addition to that required for the reciprocal relationship.

Within the switching network the number of time slots may exceed the total number presented by the connected lines. This expansion, sometimes called *decorrelation*, is used to compensate for any blocking which may occur in matching time slots in the S stages of the network. The time slots used within the network as connections are made are not necessarily the same time slots used for messages in transmission media.

In summary, time-division switching networks use a combination of time slot interchange and time multiplex switching elements to associate time slots for transmitting and receiving between different digital transmission lines. Buffers are required between the lines and the switching elements to establish synchronism between lines before the signals enter the switching system.

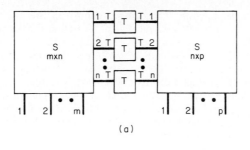

(a)

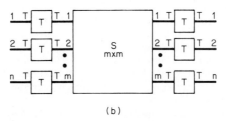

(b)

FIG. 12.11 Time division switching network architectures. (*a*) Space-time-space architecture: $n > m$ and $n > p$. "S" stages may be multistage networks. (*b*) Time-space-time architecture: $m > n$ to deload blocking multistage "S" stages.

SYSTEM CONTROL

There are two aspects to control: control of the networks and call information processing.

19. Network Control

Historically, the control of the switching center networks has been related to the types of switches or crosspoints employed in them and the type of signaling used

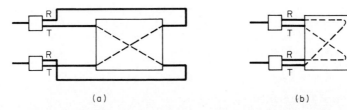

(a)   (b)

FIG. 12.12 Time division switching network reciprocity. (*a*) Two-sided network; (*b*) folded network. Two-sided and one-sided (folded) networks are used with two connections, one in each direction, to carry a complete telephone connection. This is the equivalent of two separate networks. Furthermore, duplicate networks may be required for service redundancy.

from the station to the switch. The remote control of the switching-center networks is basic to the principle of locating the switch at a central point or node.

The principal forms of network control in the past have been used for the control of gross-motion electromechanical switches. For most applications these switches were set *progressively*. This means that the switches were organized in stages and connected in sequence. In many cases the switches were controlled *directly* by the calling device, usually a dial, at the calling station.

If the switches were of a type not compatible with the usually decimal calling device, the control signals would be recorded in a *register* that in turn would provide the nondecimal code to control the switches. This type of network control is known as *indirect progressive*.

While stages of crosspoint matrices can be controlled progressively, some form of *common control* for a complete multistage switching center network has become the more accepted form of network control. One type of common control is known as *end marking*. Here, opposing electrical polarities are placed on the active input and selected outputs of a two-sided switching-center network. An idle set of links between the crosspoints of each stage is automatically selected to complete the required connection. If more than one suitable set of links form an idle path, a contention circuit selects one path.

The form of common control utilized more generally is known as *link access*. Here a network control access network is used to connect the possible link paths and crosspoint controls to a common *network control* (NC) circuit. This circuit makes a match of links and selects and establishes a suitable path across the SCN.

Associated with each crosspoint is some form of memory. It may be inherent as in a PNPN crosspoint. It may be a locking contact as part of a relay-like set of contacts, as in a crossbar switch. Or it may be a *hard* magnet that holds the crosspoints operated, as in sealed-reed contact matrices. In electronic switching systems the crosspoint memory is more frequently dependent upon a common bulk memory, such as an integrated circuit *random-access memory* (RAM). This memory may be separate for network control or as part of the general RAM requirements of the system. In either case this application of the RAM is known as the *network map*.

20. Call Information Processing (CIP)

Call information processing is the most complex of the functions performed in a switching system. For this reason it has been the greatest challenge to system designers. There are many variations of CIP controls. These variations are due to the many different principles and techniques that have been conceived and the architectures resulting from their application. It is in the controls where most of the logic and memory are deployed for implementing the services and features that are required of the switching system.

21. Control Techniques

Electromechanical switching systems have been defined by the type of techniques used in the switching center networks, such as crossbar. Electronic switching systems have been defined on the basis of the control techniques and architecture employed. There are three control techniques that have been widely used in the control of switching systems. These are

Wired logic: The call processing decisions and actions are determined by configurations of logic and memory circuits. Such systems make extensive use of logic gates in circuits that are especially designed for the needs of the system. Most electromechanical and some small electronic switching systems use these techniques.

Action translator: The sequences of possible system actions are stored in memory while the call processing is performed in logic circuits specific to the system. Macroinstructions are stored in memory and executed by the logic. "Establish a network connection" might be an example of a macroinstruction for which the logic circuits have been designed. Microprocessors may be used to implement the logic.

Stored program control (SPC): System decisions and actions are taken by a general-purpose logic circuit, known as a *processor,* by reference to a sequence of instructions stored in memory. The degree of SPC is dependent upon the relative autonomy of all of the control circuits of the system.

There is a spectrum of system controls from wired logic to SPC. No one need be used exclusively in a given system. In some systems wired logic or action translation is used for one portion of the system actions, while SPC is used for others.

22. System Control Architectures

The concentration of all call processing in a single system location is known as *central control.* The control portion of the system is where most of the services and features are implemented. As system requirements have grown, so too has the time required for a system to make all of the decisions necessary for a single call. The use of electronic logic and memory has greatly increased the capabilities of a single active central control. However, logic speeds have not grown commensurately with the growth of system requirements, and for many applications a single active central control is insufficient.

The following system architectures are employed to increase system capacity. At the same time, some also add intangible potential advantages by permitting the system to accommodate future changes in the system requirements.

Multiprocessing: Employing more than two active central controls that may divide the offered load in either of two ways. (See Fig. 12.13.) One is *traffic division* where calls are distributed evenly between (for two) or among the several active processors. The other is *functional division* where the offered load is divided by service or feature characteristics, such as originating and terminating calls or call processing and administrative features.

Hierarchical: Here one or more processors associated with call signal processing operate through an intracontrol access network. (See Fig. 12.14.) One or more of these processors may serve an entire office. They are usually used to assimilate the incoming signals, such as the called number.

Distributed control: The call signaling and hierarchical controls may be closely associated so that each serves the same input lines, ports, or trunks.

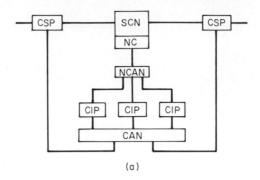

(a)

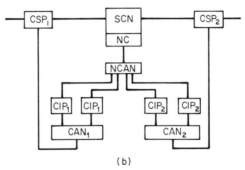

(b)

FIG. 12.13 Nonhierarchical multiprocessing central control architectures. (*a*) Traffic division. The call information processors (CIP) are equally capable of serving all offered calls. The control access network (CAN) allocates calls to the processors based upon anticipated traffic demand; (*b*) Functional division. The service is provided by different groups of call information processors, e.g., CIP₁ and CIP₂. The offered load may be from sources that differ functionally, and the groups of processors are accessed through different CANs.

When so configured, the combination is currently known as a *module*. These modules may connect with the central control directly through a CAN or indirectly by using the *switching center network* (SCN) as an ICAN. Systems need not employ a central control. With indirect access, they must provide for the direct progressive control of the switching center network. Distributed control modules may be located remotely from the main switching center network. Signaling between modules and the central control or, in the case of fully distributed control, between modules may take place over a regular message path or a separate signaling path used as an ICAN similar to common channel signaling (see below).

Distributed switching: In some systems one or more functions of the switch-

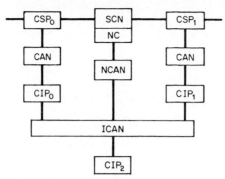

FIG. 12.14 Hierarchical control architecture. Call information processor CIP_2 is the central control. CIP_0 and CIP_1 are controls that preprocess call information, thereby forming a hierarchy of controls. These preprocessors divide the traffic but may also differ in function, e.g., lines and trunks. The intracontrol access network (ICAN) provides paths for moving call information between the controls.

ing center network may be placed in the distributed modules. (See Fig. 12.15.) When this occurs, the module is said to perform both the control and switching functions. The switching may be confined to providing for signal access such as for dial tone and signaling receivers, or it may be used to provide the SCN function between its own terminations. This intramodule connection ca-

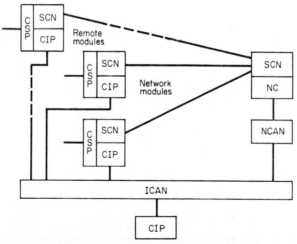

FIG. 12.15 Distributed switching and control architecture. Modules containing the switching, control, and signaling functions are interconnected by the switching center network (SCN). In an hierarchical arrangement a central control (CIP_2) is accessed directly through an intracontrol access network (ICAN), as shown, or indirectly through the central SCN.

pability is important in modern remote switching applications where a high community of interest places a high traffic requirement upon trunks to the host office or in the case when all trunks are out of service.

23. Software

Stored program control added a new dimension to the design of the control portion of switching systems. Although highly dependent upon the use of a general-purpose logic and memory processor complex, the software dimension has grown to cover many other aspects of system design that have been computerized. This dimension now includes the following items.

1. Specification of system service and feature requirements
2. Specification of program instruction set
3. Choice of an operating system and executive or scheduling algorithm
4. Choice of high-level language and where to employ machine language design
5. Design of development utilities
6. Definition and interface standardization of macroinstructions
7. Programming and, where necessary, coding
8. Simulation of program operation
9. Call processing capacity
10. Design and management of data bases
11. Specification of office parameters
12. Program and office data assembly
13. Program revision and distribution
14. Updating data bases
15. Adding features and services

Programs for typical central office switching systems may vary in size depending upon the system architecture. The software is generally divided into three major parts. The first is the executive software required for running the program. While this is usually small in size, it runs constantly, consuming a small portion of real time.

The second portion of the software is that devoted to call processing. While this may consume no more than 25 to 35 percent of the total instructions, it accounts for between 65 and 80 percent of the real time.

The final portion of software is devoted to keeping the system operating when troubles develop in either the hardware or software. These programs can be quite large, in some cases accounting for as much as 90 percent of the total. They consume very little real time. Together with hardware signals or flags they detect incorrect system functioning, switching defective units from service and providing automatic trouble locating. When defective units are repaired they are restored to service and, where necessary, the system is reinitialized.

In the same manner a well-designed system automatically checks software actions to ensure that the memory or other actions do not present the software with inaccurate data.

Some of this system operating software is devoted to the implementation of features that are not call processing. These include *administrative, maintenance, and operations* (AMO) features that not only help keep a system functioning, but also provide for local and remote human interfaces with the system to provide important data for service measurement, continuity, and engineering.

With the advent of microprocessors, AMO software is being provided as firmware directly in some hardware subsystems rather than in the main memory of the system.

Many of the logic and data base functions devoted to AMO are now being remoted into separate general-purpose computers known as "operation support systems."

SIGNALING AND INTERFACES

24. Introduction

Signaling is the sinew that binds switching systems with each other and their stations or terminals. It is the technology that enables the implementation of telecommunication networks.

For electromechanical switching systems, signaling was very much determined by the characteristics of the switching devices. More recently, interface and protocol standards have been agreed upon by designers and users throughout the world. These international standards have been translated into regional and national standards to meet their more specific needs.

Included in the subject of signaling are the tones and recorded announcements used to indicate call progress. The tones are also standardized, particularly since in recent years they have been required to be *machine-readable.*

There are two basic applications of signaling. One is for signaling to and from stations or terminals and the first switching entity. The other is signaling between offices, known as *interoffice* or *internodal* signaling.

25. Station Signaling and Interfaces

There are two types of electrical circuits used for signaling. The first circuits used in telephony were direct current. It was and still is used to indicate *receiver off-hook.* This *on-hook/off-hook* use of direct current is known as *supervision.* Off-hook is used to indicate not only the *request for service* but also to trip ringing when the call is answered. A prolonged on-hook is used to indicate the end of a call, or *disconnect.*

The dial calling device was invented to produce on and closure impulses, or *pulses,* of dc in the lines between the station and the central office. *Dial pulsing* sends serial openings and closures of the line at a rate of 10 to 20 pulses per second per digit with a minimum closure period of 0.75 s between digits. Currently a more generic term for pulsing is *addressing.*

Analog Line-Space Division Switch Interface.   What is described above is known as *loop* supervision and *dial pulsing.* It is the most common interface between the telephone used to convert dc steadily supplied from the central office to a varying electrical current that is the *analog* of the voice pressure on the telephone transmitter.

Ringing is used for alerting a called telephone. These signals are also *high potential,* employing, in North America, 20 Hz alternating current at 90 V rms.

To prevent the simultaneous seizure of a line used as a trunk to and from a *private switching exchange* (PBX), *ground start* supervisory signaling is used. (See Fig. 12.16.) The loop is protected from high-voltage lightning strikes and foreign potentials that may accidentally become crossed with telephone wires. In some applications of station signaling, a ground return circuit is used for such functions as ground start, party-line ringing and identification, and coin control. Care is then taken to prevent longitudinal currents from proving false line signals.

In locations where loops are exposed to the many vagaries of outside plant exposure, a *line lockout* feature is provided to enable the central office switch to relieve itself of a fortuitous continuous request for service.

For public switching systems, is it common to provide test access at the interface so that from a test desk it may be determined if a trouble is located within or outside of the central office.

To speed station pulsing, alternating current tone signals are used. These signals use two tones, one from a low group of four frequencies in the range from 679 to 941 Hz and one of four from a high group from 1209 to 1633 Hz in what is known as *dual tone multifrequency* (DTMF) signaling. Push buttons or key calling devices produce a maximum of 16 low-high frequency combinations or digits. The central office receivers are part of the *call signal processing* (CSP) and are capable of receiving the digits at a nominal rate of 10 digits per second.

Time-Division Digital Switch Interfaces for Analog Lines.   The interfaces between the high-current (20- to 200-mA) analog loops and a time-division digital switch used for a local central office are unusually complex. This interface is known as BORSCHT, which stands for the functions required of it: *b*attery feed, *o*vervoltage control, *r*inging, *s*upervision, digital *c*oding (and decoding), *h*ybrid

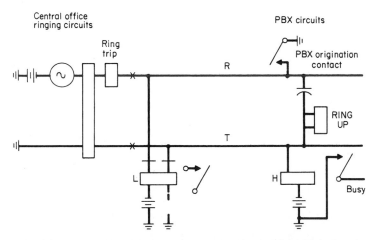

FIG. 12.16   Ground start supervisory signaling. The ground normaling appearing on the right winding of the line relay *L* in the central office is removed. The *L* relay operates over the *R* wire when a two-way trunk is taken into use at the PBX end. This allows the *T* wire to be used when the trunk is taken into use at the central office end by operating the *H* relay.

(to separate the two directions of transmission through the digital switch), and testing. In some cases to reduce the cost of the complexity and to improve service should the complex circuit fail, a line concentrator is introduced between the lines and the BORSCHT circuits. BORSCHT circuits are implemented in integrated-circuit chips. Much effort has been expended to reduce the number and cost of these chips. Unlike the analog-to-space-division interface, these chips require considerable power for each line even when the line is not active on a call. Automatic power reduction is introduced into these circuits to minimize central office power consumption.

The hybrid portion of this interface, unlike hybrids used in transmission systems, must operate with a wide range of line impedances. Generally the transmission requirement for local digital switches is zero transmission loss. This means that the line hybrid balance must provide an accurate image of the line impedance, automatically.

Time-Division Digital Switch Interfaces for Digital Lines.   An attractive reason for employing local time-division digital switching is to anticipate future use of station or terminal loops for digital rather than analog transmission and signaling. This will enable digital signals generated from computers, alarms, and other telemetry applications to be sent with digitized voice signals opening the application of telecommunication networks to data as well as voice. An international standard for digital lines has been proposed and is currently (1986) being field testing through the world. This standard is known as the *digital subscriber line* (DLS). There are two versions, each specifying B or speech channels and D or data channels. The *basic* standard is for individual lines employing 2B + D, where B is 64 kbs and D is a total of 16 kbs. The *primary* standard is for 23 (31 in some countries) B channels and one D channel. This is used to serve as many as 23 PBX trunks. The 2B + D signaling uses the two-wire two-way lines to avoid rewiring customer premises when digital loops are introduced. Means must be provided to send digital signals over the two wires in both directions. Several methods are being tested. They include *time-compression* signaling where bits are sent alternately in each direction, known as *ping-ponging* and *echo canceling* where signals in each direction are separated by hybrid methods.

26. Interoffice Signaling

Metallic Facilities.   Metallic facilities are used when central offices are close to one another. The same types of dc signaling used on loops, loop supervision, and dial pulses, are used. Many other varieties of dc signaling, such as high-low, wet-dry, simplex, duplex, composite, etc., have been used in the past. Many are obsolete.

The on- and off-hook signals sent over trunks are used for additional functions compared with those encountered in station signaling. These include the equivalent of dial tone known as *start dial* to indicate that pulsing may be sent forward. Also an off-hook signal is sent from the called office back over the connection to the billing information recording point as an *answer* signal.

Since trunk channels are common arteries for traffic, they may be reused very shortly after they are released from the previous call. To ensure that all signaling is completed for the previous connection, a timed guard period is included in the network control.

Also, longer trunk channels, particularly in North America, are two-way. Provision must be made to mitigate the effects of simultaneous seizure from both ends. This condition is known as *glare*.

For longer distances, tone signaling is used. For supervision, a *single* but different *frequency* (SF), usually one in the voice band such as 2150 Hz, is used in each direction to indicate on-hook. *Multifrequency* (MF) pulsing is used for addressing. It is different from DTMF in that it uses two frequencies out of six between 700 and 1700 Hz, also giving 16 combinations. Included in these combinations are KP to open the receiver and ST to indicate that all digits of the address have been sent. See Table 12.1.

Multifrequency signaling is also used to send calling line numbers over trunks between offices where needed by intermediate offices for billing or other purposes. Outside North America there are other forms of MF pulsing. These use terms such as *compelled* and *confirmed* to send MF pulses in the reverse direction over the trunk to ensure the correct reception of the digits as well as to indicate traffic conditions ahead in the network.

Analog Multiplex Interface.   The tone signals may be used over analog multiple-transmission facilities. To apply the SF signaling equipment in a generic manner, a basic interface has been established.

The CSP functions of a line or trunk in a switching system are allocated to a line circuit or a trunk circuit. The interface between the trunk (or line) circuit and an SF signaling circuit has been standardized. From the trunk circuit, battery potential is applied to a wire designated M to indicate off-hook or the closed period of a dial pulse. An E lead receives battery signals from the SF circuit to indicate off-hook from the distant end of the transmission channel. This interface has become known as *E&M lead signaling*. There are many varieties of the E&M lead interface. Sometimes *duplex* (DX) signaling is used over short metallic circuits to connect trunk circuits with signaling circuits.

Digital Multiplex Interface.   When digital facilities are used, the supervisory signals are sent over the channels by using one of the eight bits normally used in

TABLE 12.1 Pulsing Frequencies

DTMF (station)	Digits		MF (trunk)
941/1336	0		1300/1500
697/1209	1		700/900
697/1336	2		700/900
697/1477	3		900/1100
770/1209	4		700/1300
770/1336	5		900/1300
770/1477	6		1100/1300
852/1209	7		700/1500
852/1336	8		900/1500
852/1477	9		1100/1500
941/1209	*	KP	1100/1700
941/1477	#	ST	1500/1700
697/1633	A	ST′	900/1700
770/1633	B	ST″	1300/1700
852/1633	C	ST‴	700/1700
941/1633	D		

digitizing the voice. Instead of 64 kbs for the voice resolution, there are only seven bits for 56 kbs. In some digital formats, this bit is used for signaling only once in every six frames, or once per 0.75 ms. Between the digital transmission channel and a trunk circuit there is an E&M lead interface. This method of supervisory signaling over digital facilities is known as *in-(time) slot* signaling.

In some countries, separate multiplexed digital channels are used for the signals represented by the other channels. In particular, with 2.048-Mbs digital transmission lines, 32 eight-bit channels are multiplexed. Only 30 are used as message channels. The digital supervisory signals from the message channels are coded into one of the two remaining digital channels. This means that the message channels have a clear 64 kbs capability as originally coded. This technique is known as *out-of-slot* signaling.

If dial pulsing is used, the in-slot on- and off-hook signals represent the pulses. If MF pulsing is used, the signals are digitized along with the voice signals. If MF signals are sent from time-division digital switching systems, their digital representations may be taken directly from a read-only memory. They are converted to analog signals when they are sent over analog facilities.

27. Common Channel Signaling (CCS)

The signals described in the preceding sections are transmitted over the same paths as the message. Where there is a plurality of paths and trunks between the same points, a separate path or data link may be used to interchange signals between the two points about all of the trunks. The data link is known as a *common (signaling) channel*. The *signaling* messages include in coded form the identity of the trunks and the supervisory and address information.

The common channel may use an analog or digital facility. When analog facilities are used, the coded data is sent using modems, at typically 4800 b/s. An international standard for the signaling message has been agreed upon for analog common channels. It is known as *CCITT signaling system no. 6*. See Table 12.2.

When digital facilities are used for the data link, then digital signals generated within the switching system use digital channels. The data rates may then be 64 or 56 kbs. No modem is needed. The international standards for this form of common channel signaling is known as *CCITT signaling system no. 7*.

Common channel signaling has the advantages of being much faster, not dependent upon analog pulsing, not subject to simulation of signals from stations (usually fraudulent), and the ability to send digital information about the call in both directions.

Since with common channel signaling no signals are sent over the trunks used for messages, there is no check that these trunks or channels are operable. Therefore once the signaling messages are sent from one office to the other, a "path assurance" check is made that the desired connection has been established and is in working condition.

28. Signaling Networks

Signal Transfer Points.   While common channel signaling has many advantages, it is costly for small long-trunk groups. The data links are like the lines of a noncentral telecommunication system (see Sec. 5) requiring almost $n \times n$ data links, where n is the number of switching offices in a large network. Therefore

TABLE 12.2 CCITT Signaling System

Feature	CCITT signaling system number								
	3*	4	5	6	7	R1		R2	
						A†	B‡	A†	B‡
In-band signaling	Y	Y	Y	N	N	Y	N	Y	N
Out-band signaling	N	N	N	N	N	N	N	Y	N
Common channel signaling	N	N	N	Y	Y	N	N	N	N
Analog two-frequency signaling	N	Y	Y	N	N	N	N	N	N
Analog multifrequency signaling	N	N	Y	N	N	Y	Y	Y	Y
Digital	N	N	N	Y	Y	N	Y	N	Y
Suitable for operation over satellites	N	N	Y	Y	Y	Y	Y	N	N
Suitable for operation with TASI circuits	N	N	Y	Y	Y	N	N	N	N
Recommended for operation between SPC exchanges	N	N	N§	Y	Y	N§	N§	N§	N§

*Obsolete.
†Analog version.
‡Digital version.
§May be used on links connecting SPC and electromechanical exchanges.
Y = yes; N = no.

the signaling data links may be connected to a central signaling message switch. This switch is called a *signal transfer point* (STP). The STP may be basic to an additional function of a switching system.

The signaling messages are data packets. The central message switch is therefore a packet switch (see Chap. 14). The *signaling network* then consists of one or more data links from each office with common channel signaling capability. These data links are used for calls that are served by trunks to offices that are also equipped for common channel signaling. More than one data link may be required for redundancy to ensure reliability and if the signaling messages exceed the capacity of a single data link.

A signaling network usually has duplicated STPs that are not collocated. All STPs of the signaling network are completely interconnected with data links.

Signaling messages originate at the first office that has common channel signaling capability. The office first selects a trunk to serve the call and then passes the called number and the trunk number to its serving STP. If the office at the other end of the trunk is served by the same STP, it retransmits the signaling message directly to it. If the office is associated with another STP, it forwards the signaling message to that STP. Supervisory messages representing answer and called-party-disconnect signals are sent over the signaling network in the reverse direction.

Network Control Points.   Once the signaling network is established, it may be used for nonsignaling messages. Generally these messages pass through the STPs between only the intended offices. They are known as *direct* signaling messages. The protocols for these messages may be of the X.25 type rather than in the form of CCITT CCS standard messages.

The term *network control point* is used when the network of STPs includes data link terminations to points that are not switching offices. For example, some operator or maintenance systems may access the signaling network for direct messages.

29. Signaling Data Bases

With the expanded use of the signaling network has come the introduction of data bases into the STP. Among other things, these data bases are used to translate call addresses. For example, in North America, certain addresses such as 800 IN-WATS numbers may be referred to a data base to determine the routing of a call depending upon the calling address, the time of day, the day of the week, etc. The called address is changed in the process.

The data bases may be used to check other message data, such as credit card numbers. Eventually interactive messages may be used at "action points" between callers and data base information. Announcements at the action point prompt the caller to dial additional information. On the basis of this information, as referred to the data base, the call routing is determined. In many respects the flexibility that SPC brought to the control of switching systems has now been brought to an entire switched telecommunication network. Some call this the *SPC network*.

REFERENCES

1. Clos, C. "Study of Non-Blocking Switching Networks," *Bell System Tech. J.*, Vol. 32, pp. 406–424, March 1953.
2. McDonald, J. C., *Fundamentals of Digital Switching,* Plenum Publishing, New York, 1983.
3. Fink, D. G., and D. Christiansen, *Electronics Engineers' Handbook,* 2nd ed., Sec. 22, McGraw-Hill, New York, 1982.
4. Syski, R., *Introduction to Congestion Theory in Telephone Systems,* Oliver and Boyd, London, 1960.
5. Bellamy, J. C., *Digital Telephony,* John Wiley & Sons, New York, 1982.

CHAPTER 13

Message Switching and Handling Systems

Jack Branch
Chief Executive Officer, CW Incotel, Ltd.

CONTENTS

INTRODUCTION

1. Evolution of Message Switching

Manual Torn-Tape Systems. Less than 20 years ago, the main method of transferring messages electronically between two company locations was by utilizing a dedicated circuit between the locations with a teleprinter connected at each end (Fig. 13.1).

This configuration was adequate if messages were to be sent only between the two locations, but when messages had to be sent between many locations, the circuit and teleprinter costs became prohibitive (Fig. 13.2). This type of requirement led companies to designate a central location for the receipt and transmission of all messages between locations, as shown in Fig. 13.3. In this configuration all messages were received in New York on both hard copy and paper tape. Those messages destined for New York were delivered locally and those messages destined for one or more locations other than New York were tape-

FIG. 13.1 Dedicated circuit between two locations.

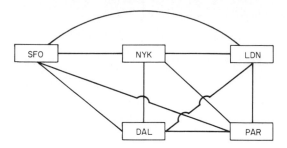

FIG. 13.2   Dedicated circuits between multiple locations.

relayed—they were torn off the paper tape and transmitted on the appropriate teleprinter. Hence the term *torn-tape* system. Therefore message communications became a centralized operation (see Fig. 13.4).

Computerized Systems.   The torn-tape method of operation was satisfactory for those companies with less than 10 locations, but when the number of locations was much higher, the central communications operation was continuously buried in a sea of paper and paper tape. When the retransmission of a message was required, baskets of paper tape had to be searched to avoid rekeying the message.

The international record carriers, which had approximately 100 point-to-point circuits with different countries just for cablegram traffic alone, had an even more overwhelming need to automate the message switching operation. The technology of the computer was applied to meet this requirement, as shown in Fig. 13.5. The computer allows the messages to be received, stored, processed (routed), and then transmitted to the addressed destination.

Initial Users of Message Switching Systems.   The initial purchasers of message switching systems were the communications carriers, for handling their telegram and cablegram traffic, and the airlines, for communicating with their reservations, operations, and administrative offices in locations throughout the world. The communications carriers also recognized that many of their customers required message-switching systems, but such systems were too costly at the time. The communications carriers introduced shared message-switching systems, also commonly referred to as *closed user group* (*CUG*) systems, which allowed many different customers to utilize a single system on a shared basis. These systems were operated and maintained by the communications carriers. The CUG systems were designed to prevent messages from inadvertently being sent between

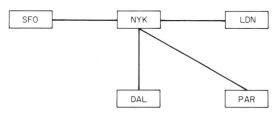

FIG. 13.3   Routing by the "torn tape" system.

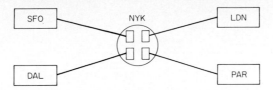

FIG. 13.4 Centralized operation of message communications.

different customer networks. There are a number of CUG systems still in use today.

STORE AND FORWARD

2. The Store-and-Forward Function

Store and forward is the primary function of a computer-based message-switching system. It allows messages to be transmitted to the system even though the destinations to which the messages are being sent are not available to receive them. Therefore, the sender of a message does not have to wait and continually retry sending a message when the destination terminal is occupied with receiving messages from other terminals. The message is received by the system where it is stored and then forwarded to the destination when it becomes available; see Fig. 13.6. In order to perform a store-and-forward message exchange, the message switch provides three basic functions.

Input Processing. Message detection, format validation, and message safe-store functions are performed by the message switch during the input processing cycle.

Routing and Queuing. Normally, upon detection of an end of message signal, the message switch will prepare to route a valid input message, on the basis of the address criteria found in the address field, to specific destinations. Once the routing is determined, the message will be queued to the destination designated by the address. Messages are queued by destination and by priority within a destination and are held in an in-transit storage file while awaiting delivery.

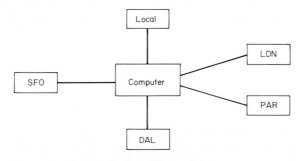

FIG. 13.5 The introduction of the computer.

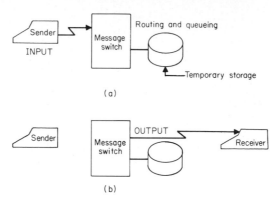

FIG. 13.6   The store-and-forward function. (*a*) sender deposits message with switching ''mechanism'' where it is placed in temporary storage; (*b*) switching mechanism delivers message to receiver, from storage.

Output Processing.   Message delivery is performed by the message switch during the output processing cycle. The output program monitors the activity on the output circuits. When an operational output circuit becomes available, the destination queue associated with that circuit is examined to determine if there are any messages queued to the destination. The queues are checked in order of priority from the highest to the lowest. Messages to be delivered are fed out by the message switch, on a first-in–first-out basis, by priority. As part of output processing, messages may undergo reformatting before transmission.

NETWORK INTERFACES

3.   The Interface Function

To provide a cost-effective message-transfer operation, computer-based message-switching systems support a multitude of different network interfaces. These interfaces permit the system to exchange messages with terminals or other systems connected to like or dissimilar networks. The system accomplishes this task by connecting with each network by means of a communications circuit and then emulating either a terminal or host computer operation (Fig. 13.7).

Some of the most common networks between which message switches provide the interface are

International telex

Telex I (domestic telex)

Telex II (TWX)

Direct distance dialing (PSTN)

International direct distance dialing (IDDD)

Public and private packet networks (X.25)

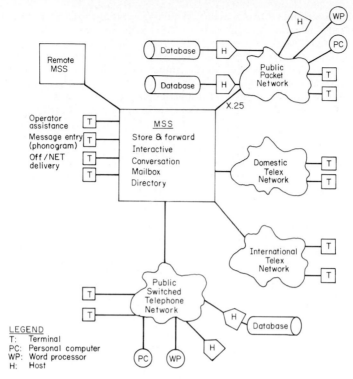

FIG. 13.7 The network interface function.

4. Interface Conversions

In order to provide a message exchange across these networks, the message switch must work with different terminals, code sets, speeds, and communication protocols. Thus, the computer-based message-switching system must perform all of the speed, code, and protocol conversions which are required.

Speed Conversion. As an example, a message input from a telex II terminal would be received by the message switch at 110 baud or 10 characters per second (CPS). If the message is to be sent to a terminal connected to the international telex network, the message switch must convert the input speed of 10 CPS to an output speed of 50 baud or 6.6 CPS when delivering the message, since the baud rates supported by each of the networks are not similar.

Code Conversion. In addition to differences in speed, the message switch must compensate for networks and terminals which use different data codes. ASCII, Baudot, and EBCDIC are examples of code sets commonly used for the sending and receiving of messages. The message switch utilizes code-conversion tables which enable it to accept input messages in one code set and output the message in a different code set. Thus, a message received in ASCII code can be converted and then delivered in Baudot or EBCDIC code as required by the receiving terminal.

Protocol Conversion.   Most message-switching systems support both asynchronous and synchronous transmission or communication. Asynchronous transmission is often referred to as *start/stop transmission* because each character is framed with a start bit in front and stop bits at the end. In asynchronous communication the system depends on recognition of the start and stop bits to differentiate between characters. Synchronous communication uses timing to differentiate between characters and thus eliminates the need for the start/stop bits.

For both asynchronous and synchronous communication there are a multitude of different protocols which are used to allow terminals and computers to talk with one another. Where a code set is used to identify information after a communication link has been established, it is the protocol which establishes the communication link and the manner in which the information will be conveyed. The protocol conversion provided by the message switch makes it possible for a terminal to send a message to a terminal or terminals on different networks with different protocols.

FORMAT HANDLING

5. Types of Standard Formats

All messages transmitted to and from a message switch must conform to a format structure. A format consists of a sequence of fields as defined by a data structure or format table. There are a number of standard formats utilized for message switching. These formats are designed so that the information required to automatically transfer a message between two points is easily identifiable by the message-switching system. Another requirement is that the format be easy for the users to understand. Message formats normally consist of header, text, and ender sections.

IATA Format.   One of the most simple and best-designed formats is that which was developed by the airlines for the interairline and intraairline transfer of messages. This International Air Transport Association (IATA) format is still very much in use today by the airlines. Many other industries have also adopted it for their use with either no modifications or only minor changes that serve their particular requirements. An example of a simple IATA-type formatted message is as follows:

Line 1	ZCZC	DEF456	ABC123
Line 2	QN	NYCRRTW	SFORRAA	PHLOWUA
Line 3		DALRRUA	111222G
Line 4	TEXT
Line 5	111223G
Line 6	NNNN

Lines 1 to 3 represent the header section.

Line 1 is the numbering line and consists of a start-of-message signal, output station identifier and sequence number, and input station identifier and sequence number.

Line 2 is the address line and consists of a message priority indicator and address codes. In IATA format, the address code is divided into three elements. The first three characters define the city, the second two characters define the organization's function in the city, and the last two characters define the airline. Therefore, the address code, NYCRRTW is the main reservations office (RR) of TWA (TW) in New York City (NYC). When airlines switch messages within their own network they normally omit the last two characters of the address which defines the airline itself. Normally up to four address lines of address codes are permitted in a message.

Line 3 is the originator line and consists of an end-of-address signal, the address code of the originator of the message, and an *input date time group (IDTG)* representing the time the message was entered into the system. The IDGT consists of the two-digit date, followed by the four-digit time based on a 24-hour clock, followed by the time zone.

Line 4 represents the text portion of the message. In modern message-switching systems there is usually no limit to the size of the message other than practical limits. Normally system parameters for message length are set between 3000 and 100,000 characters.

Line 5 represents the time the message was transmitted by the system.

Line 6 is the end-of-message signal.

ICAO Format. This format is similar to the IATA format. It is utilized by airports as a means of moving message traffic on the Aeronautical Fixed Telecommunication Network (AFTN). An example of the International Civil Aviation Organization (ICAO) normal format is

```
ZCZC     NRA062     111225
FF     LGATKLON
111222     LFPGAC
TEXT
NNNN
```

As can be seen from the example, the major differences between ICAO and IATA formats are the use of a six- or eight-character address and an originator line starting with an input date time group followed by the originator code. In ICAO format the address code consists of a four-character location code followed by a two-character organization code, and, optionally, by a two-character department code. Although most AFTN systems presently allow only one address line and eight addresses, this is changing as newer systems are installed.

F.31 Format. This format is used internationally between all countries for the handling of public telegram traffic. It also consists of a header, text, and ender section as shown in the following example.

Line 1: Numbering line	ZCZC SPA108 RCA015 TIG
Line 2: Pilot line	PSCB CO URNY 013 ABC12345
Line 3: Preamble line	New York 013 230900 RPX
Line 4: Address	ABC CO
Line 5:	3 Rizal Cebu City (Philippines)
Line 6: Text	Please send file 1053AB soonest
Line 7: Signature	ABCCO NYC
Line 8: Collation	Col 3 1053AB
Line 9: End of message	NNNN

The first three lines of the message contain the entire header section. The first line of the header section is called the *numbering line* and consists of the *start-of-message* (SOM) followed by the output channel designator and sequence number, the input channel designator and sequence number, and the *telegram identification group* (TIG). In the more modern systems the input channel designator takes the place of the TIG.

The pilot line is the second line in the header section. It contains the destination indicator, priority and tariff indicator, origin indicator, the number of chargeable words, and the customer account number. The third line in the header section is called the *preamble line*. It consists of the originating office, the number of actual words in the message, a date/time group, and any special service instructions. In the above example, the RPX in the special service instruction field indicates that a reply has been prepaid by the sender. The text section includes the information shown in lines 4 through 8. Included in this section is any optional service indication, such as LT for night letter, a street address or registered address, the message text, a signature, and any number collation requested by the sender.

The ender section in line 9 includes the end-of-message indicator (NNNN) followed by 10 line-feed characters for message separation purposes.

Other Variations. In addition to the formats described above, which are industry-specific and defined by accepted standards, many other formats are used by government, the military, and the general business community. Some of these formats are based on accepted standards while others are user-defined for a specific application. These formats can range from simple to complex. As a rule, formats contain a header field, a text field, and an ender. An example of a very simple format variation using only these basic fields is

        TO:
        FROM:
        TEXT
        END

6. Format Elements—Mandatory and Optional

Within a format structure, the fields are defined as either mandatory or optional. A mandatory field contains information which the message switch requires to perform message detection, validation, routing, queueing, and delivery functions.

An optional field may or may not be present in an input message. If an optional field is present, the message switch will take one of two specific actions:

• Validate the information in the field and perform additional functions.

• Deliver the information without validation.

The specific action taken by the message switch is based on the optional field function as delineated in the format table. Some optional fields contain information which affects message processing, whereas other optional fields do not.

Station Identification and Message Sequence Number. These are used by message-switching systems to provide for message accountability. Station identification usually consists of a unique three- or four-character code which identifies a particular sending or receiving location to the system. The station identification is often referred to as the *station ID*. In some message-switching systems, two separate unique codes are assigned to each location for station identifications. These codes are usually referred to as the *input station ID* (*ISID*) and the *output station ID* (*OSID*). The message sequence number may be either a three- or four-character numeric group. Message numbering is sequential starting with either 001 or 000 and ending with 999 or 9999. The message number may be reset to zero at midnight, or after the last number combination in the sequence has been assigned, for example, 999 or 9999. An input message will contain an *input sequence number* (*ISN*). A message delivered by the message switch will contain an *output sequence number* (*OSN*). The input station identification and input sequence number are usually inserted by the originator or automatically assigned by the originator's terminal. The following input message in IATA format is an example:

- (SOM) (ISIDISN)
- ZCZC ABC002

The output station identification and output sequence number are automatically assigned by the system for each output message. The input station identification and sequence number may also be retained by the system in the delivered message to provide an end-to-end message audit trail. The following output message in IATA format is an example:

- (SOM) (OSIDOSN) (ISIDISN) ZCZC DEF006 ABC002

If the system receives an out-of-order sequence number from an input station, it will either send a sequence alarm note to the originator station and switch the message or reject the message back to the originator. The actual action taken by the system is determined by the error processing routines.

Addresses.   An address code normally will consist of 2 to 16 alphanumeric characters. Each line of the address field can contain one or more address codes which are usually separated by a space. The address field in IATA format is normally sized at four lines with a maximum of eight address codes per line. There are normally eight characters per address code. Therefore, a single message usually does not have more than 32 address codes in the address field. An input message may contain a single address, multiple addresses, a group address, or a broadcast address.

1. *Single address:*   A single address in the address field of an input message results in a delivery to a single destination.
2. *Multiple address:*   If the address field in an input message contains more than one address, each address may result in a delivery to a single destination or to multiple destinations.
3. *Group address:*   A group address is a single address which points to multiple destinations in the routing table. The group address scheme provides the system user with the capability to input a single message which the system will deliver to multiple predefined destinations.
4. *Broadcast address:*   A broadcast address is a single address which also points to multiple destinations in the routing table. It is similar to a group address and is often referred to as a group address. In general, group addresses are used to send a message to a specific group of destinations or common interest groups. On the other hand, a broadcast address is used to send a message to all destinations on the system.

The system routes outbound messages to their destination according to the address codes appearing in the address field of the input messages and the associated destinations as indicated in a routing table.

Most message-switching systems will accept input messages containing multiple addresses, some of which can be group addresses. Thus, even though the system may set a limit of 32 addresses in the address field of an input message, the number of actual message deliveries resulting from one or more addresses could be substantially higher. For this reason, most systems will set a limit on the maximum number of destinations that can be associated with each group and/or broadcast address.

Priority Codes.   These are used in message-switching systems to enable the message originator to indicate the urgency for each input message. Priority indicators usually consist of two alpha characters which indicate a specific delivery priority, for example:

Priority code	Delivery
QX	Emergency
QU	Urgent
QK	Normal
QD	Deferred

The above priority codes are listed in a descending order of priority. The message switch delivers messages, which have been routed and queued for delivery on a first-in–first-out basis in priority order. Therefore, the oldest, highest-priority message queued for delivery to a destination is the first message the system will deliver to the destination. In IATA format the priority field is an optional field which appears at the beginning of the address line:

ZCZC ABC001
QX ADDRESS

The priority code is the first entry on the address line and is separated from the address code by a space. If the originator does not insert a priority code, the message is handled by the system as a normal-priority message.

Security Codes.   In many message-switching systems, particularly government and military systems, classified or secret information is exchanged in a message. These systems utilize security codes to protect against the unauthorized transmission or reception of classified data. Security codes are assigned for each station and the code assigned designates the security level for each station, for example:

Security codes	Security level
TS	Top secret
SS	Secret
CL	Classified

The security codes shown above are listed in descending order. A location assigned a security code of SS will be permitted to send or receive secret and/or classified messages but not top secret messages. On input the system will check the security code SS in the security field of the message against the security code table. If the table indicates the originator has a clearance of secret or top secret the system will accept the message. Prior to delivery, the system will check the security classification for each address/destination against the security code table. If a valid security level is not obtained for a destination, then the message will not be delivered and an appropriate alarm is issued by the system.

The use of passwords to prevent unauthorized access to the system is another form of security employed in message-switching systems. Each user is assigned a unique password identification which is stored in the system. When the user connects to the system, the system will require the user to send a password identification code which the system must validate (a log-on) prior to message transmission. An invalid password entry will cause the system to terminate the connection prior to message transmission.

Originator Codes.   An originator code is an address which identifies the origin of a message. In most message formats, the originator field follows the address field and is preceded by an end-of-address sequence. The originator field is an optional field in most formats, and if it is present, the system will check the field contents for syntax. Some systems will also validate the originator code. If it is received in the input message, the originator code normally will be included in the output message.

Input Date Time Group/Output Date Time Group.   A date time group usually consists of an eight-character group which represents the month/date/hour/minute. The system appends an *input date time group* (IDTG) to each received message and an *output date time group* (ODTG) to each transmitted message. By subtracting the IDTG from the ODTG, one can easily determine the length of time a message was in the system prior to delivery.

7. Format Delineators

Format delineators consist of either control characters or strings of characters which a message switch looks for to identify format fields in a message.

Some of the format delineators which would appear in a typical IATA message format are

Function	Delineator
Start of message	ZCZC (space)
Start of address	(CR)(LF) at the end of the number line
End of address	(CR)(LF) period
Start of text	(CR)(LF) at the end of the originator line
End of message	NNNN

Since all circuits connected to a message switch do not necessarily use the same format, the system uses format tables to convert delineators from one format to another when performing a message exchange between locations with different message formats.

The following example will illustrate a typical conversion between different input and output formats.

Function	Input	Output
Start of message	ZCZC(space)	(CR)(LF)(LTR)
Start of address	(CR)(LF)	(CR)(LF)(LTR)
End of address	(CR)(LF)period	(CR)(LF)(FIGS)M(LTR)
Start of text	(CR)(LF)	(CR)(LF)BT
End of message	NNNN	(FIGS)H(LTR)

8. Format Analysis

Inbound messages are analyzed and subjected to validation. If a format error is detected in a message, the message switch will send an error response message back to the originator or to the originator's intercept station along with the message itself. The error response message will identify the type of error found and serve as a notice to the originator to correct the error and resend the message.

Format Relaxation. The message switch provides some format relaxation to allow for common operator errors. For example, if a start of message is expected to be ZCZC and the system receives CZCZ or ZCCZ it will accept the sequence as a valid start of message. If an *end-of-line* (EOL) sequence is received as (LF)(CR)(LTR) instead of (CR)(LF)(LTR) it also will be accepted by the system as a valid EOL sequence. In the address field where the system is expecting each address to be separated by a single space character, it will also accept multiple space characters as a valid address separator. Basically, format relaxation can be described as a technique used in message switching to make the system more tolerant of common input format errors or operator-introduced mistakes, which the system can recognize, interpret, and handle.

Error Handling. In message switching there are two basic levels of error processing, (1) errors detected as a result of communication data handling and (2) errors detected in message formats. Communication data handling errors may occur during input processing or output processing, while format errors are detected during input processing. Associated with each error condition is an error-handling routine which generates an error note and/or alarm and also ensures accountability for each message in which an error condition has been detected. Most message-switching systems provide a series of options for error message handling. These options include rejecting the message back to the originating station or to a specific intercept station with an error notice appended. In the latter instance, some systems reject the message, clearing it out of the system, while other systems display the message and error note at the terminal but retain the message in the system until the necessary correction is entered by the terminal operator.

9. Special Format Handling

In addition to the format delineator conversion previously mentioned, a message switch can provide special format handling features such as wild addressing, address stripping, envelope generation, and crypto/transparency.

Wild Addressing. This is a technique employed to send a message to a dial terminal which is not registered in the system. The originator of the message enters a predefined wild address indicator in the address field of the message and then enters the dial digits and answerback of the destination terminal in the designated wild dial field. Upon recognition of the wild dial indicator, the system will look in the wild dial field for the routing information. On output, the system will initiate a call on the network associated with the wild dial indicator, establish a connection using the dialing information from the wild dial field, and request an answerback from the dialed terminal. If the received answerback matches the answerback in the wild dial field the system will deliver the message to the connected terminal.

ZCZC ABC001
RCAWILD*
.ABCCO
225620 Answerback**

where * is the wild dial address indicator and ** is the wild dial information field.

Address Stripping.   When an input message contains multiple addresses, it is sometimes necessary that the system delete some of these addresses when delivering the message to a particular destination. If a channel is configured for address stripping, the system will strip out of the address field any address which points to a destination other than the destination to which the message is to be delivered. This feature is often used on a channel which connects two message switching systems to prevent the ping-pong routing of a message between systems.

Input address field (input message to switch A)	Stripped address field (output message to switch B)
ZCZC ABC002	ZCZC FAX002 ABC002
ACC ADF FAB FAX	FAB FAX
.ABCCO	.ABCCO
TEXT	TEXT
NNNN	NNNN

Envelope Generation.   This is used primarily for exchanging messages between two message-switching systems which use different formats. It is also used as an alternative to address stripping when the receiving system requires that all addresses in the original input message must be indicated in the message output to the receiving system.

Input address field (input message to system A)	(Output message to system B)
ZCZC ABC003	ZCZC FAX004 ABC003
ACC ADF FAB FAX	FAB FAX
.ABCCO	
TEXT	ACC ADF FAB FAX
NNNN	.ABCCO
	TEXT
	NNNN

Crypto/Transparency.   Message-switching systems are often used to switch coded or encrypted messages between destinations, in both military and commercial applications. To switch a message from origin to destination the system relies on the information contained in the header and address fields in the input message. This information is validated by the system and provides the routing information required for message delivery. Coded or encrypted text is usually preceded by a delimiter such as QQQQ. All data received between the QQQQ delimiter and the end-of-message signal are processed by the system as transparent data. Text error checking and text editing routines normally used to process nontransparent text are inhibited, to preserve the integrity of the coded text.

OTHER MESSAGE-SWITCHING FUNCTIONS

10. Delivery/Nondelivery Notification

Message switching systems generate delivery and nondelivery notifications back to the originator of a message, after a message has been delivered by the system or when it cannot be delivered by the system. A delivery notice is sometimes referred to as a delivery confirmation. The delivery notice contains an input message reference and the output station identification and sequence number under which the message was delivered. It also contains the output date time group and a terminal answerback if the delivery was made to a dial station. If a station was configured for delivery notices, the system would automatically generate a delivery notice (for every message input by that station) for each delivered message. For stations which are not configured with the delivery notice options, the message switch can recognize a special delivery notice request when it appears in the specified field, normally the priority field of an input message. This special priority code invokes the delivery notice option but only for messages in which the special priority code is present. Nondelivery notifications are automatically generated by the system for messages which cannot be delivered. The nondelivery notice, which contains a reason for nondelivery, is appended to the message which the system returns to the originator or the originator's intercept position for further operator handling.

11. Communications Network Error Handling

The message switch is designed to recognize communication network errors, as well as the formatting errors which were previously discussed. Parity errors, open circuit, and interrupted circuit are examples of common error conditions which can be detected by a message switch.

Parity Errors.   If a channel is configured for parity checking, the system checks each received character to make certain that the proper parity bit setting (odd, even, or none) is present. If a character is received with invalid parity, the system stores a ? in place of the invalid character. The message is processed normally and is delivered by the system with a "possible garble" note appended.

Open Circuit.   When the message switch detects an open condition on the incoming side of a communication channel, the system starts a time-out period. If a valid character is not received within this time-out period, the line is declared open and an appropriate alarm is generated. If an inbound message was in progress when the open line condition occurred, the system would force an end of message signal and then intercept the message and append an *open line* notification. If a message was in the process of transmission on the associated outbound side of the communication channel, the message would be terminated and the output line put on a hold condition. The terminated message would be requeued and it would be delivered with a *possible duplicate message* (PDM) note appended when the line is restored to normal operation.

Interrupted Circuit.   If, in the process of receiving a message, the line goes into an idle state for some predetermined time limit, the system issues an interrupted circuit alarm and forces an end-of-message signal to terminate the incomplete message. The message is normally rejected with an *interrupted message* note appended. The system can also be configured to forward the message to the destination, if the header has been validated with a *possible incomplete* note appended to the message. Without this error-handling procedure, a single interrupted message could disable an entire communication channel indefinitely.

12. Journaling

Message journaling is a procedure whereby a message switch writes all messages handled by the system to a journal file on disk and/or magnetic tape media. Journaling provides a system data base which is used for message retrieval and accountability purposes.

Short Term.   Messages which have been journaled and can be retrieved from on-line storage are referred to as being in *short-term* or *on-line storage.* The number of messages which can be held in short-term storage is dependent on message lengths and capacity of the storage media.

Long Term.   Messages which have been journaled and cannot be retrieved from on-line storage are referred to as being in *long-term* or *off-line storage.* Long-term storage is provided on removable disk or magnetic tape media for archival purposes.

Input Journal.   When required, the message switch can store each input message received by the system in an input journal file on disk or tape. The input message is safe-stored in its original form and can be retrieved when required.

Output Journal.   In output journaling, the system writes each delivered message to a journal file on storage media. The delivered message is journaled with an index containing the input and output message references, input and output date time groups, and other relevant information which may be required for statistical accounting purposes. The on-line output message journal is available for retrieval purposes.

13. Retrieval

When a message is received garbled at its destination because of communication line problems or when a retransmission of the message is required for other reasons, the message can be requested from the computer by the receiving point.

This request for a message rerun is referred to as a *retrieval* or *retransmission request message*. The request must contain sufficient message identification information for the system to locate the requested message in the journal file, and the retrieval request must be in proper format.

Some of the types of message retrieval requests which are supported in message switching systems include the following.

ISID/ISN. The input station identification and *input sequence number* (ISN) can be used in a retrieval request to identify a single specific message which is to be retrieved from the system.

OSID/OSN. A single message can also be retrieved by specifying the output station identification and *output sequence number* (OSN) in the retrieval request message.

Output Time Range. A retrieval request which specifies a time range will result in the system rerunning all of the messages which were previously delivered to that destination during the specified time period.

Keyword. In addition to retrieving messages using specific message identifiers, which are found in the message header field, a special form of retrieval, which provides the system with a keyword or string of characters, can be requested. The system would perform a text search of the journal file and rerun all messages found on the journal in which the keyword appeared in the time period specified.

Retransmission. A retransmission performs the same function as a retrieval. The basic difference between the two forms of message reruns are

1. A retrieved message is assigned a new output sequence number and is journaled. The rerun message can be sent to the original recipient or to an alternate destination if specified in the retrieval request.
2. A retransmitted message is delivered under the original output sequence number and is not journaled. A retransmission can only be sent to the original recipient destination.

Trace. A trace is a request made to the system to search both the in-transit file and the journal file to ascertain the status of all of the deliveries associated with a specified input message. Upon completion of its search, the system generates a trace disposition report message to the requester which indicates the current disposition for each message delivery. The disposition indicates message delivered or waiting to be delivered. If delivered, the output station identification, output sequence number, and output date time group are also provided in the trace disposition report message.

14. System Supervision

The control and supervision functions supported by message-switching systems fall into two main categories: those functions required for control and supervision of the system and those functions required for control and supervision of the communications network.

The *system control and supervision functions* relate to the computer system and its internal functioning. The *network control and supervision functions* relate to the external network which is connected to the system.

Commands. These are entered into the system from a supervisory position. System commands are entered from a system supervisory position, and network commands are entered from a network supervisory position. These are logical positions and therefore can be the same or different terminal positions. System commands are entered to request reports and control the system. Some typical system command functions include the opening and closing of storage devices; starting, stopping, and shutting down the system; and requesting system resources reports. Network commands are entered to request network reports and to control the flow of traffic through the system. Network commands are used to put a line on hold, take a line off hold, skip polling, and start polling and for the alternate routing of traffic from one location to another. Network-related reports such as status and statistical reports are also provided in response to network commands. All commands contain a command mnemonic.

Alarms. Two types of alarms are generated in message-switching systems: alarms which indicate system problems and alarms which indicate network problems. System alarms are automatically generated by the system to indicate disk failure, tape failure, queue warnings, resource level warnings, and other abnormal system conditions. Network alarms are automatically generated by the system to indicate network problems. Some typical network alarms include: open line, interrupted message, sequence number discrepancy, no response to poll/select invalid answerback, and data link failure.

Each network alarm message contains a date time group stamp, a station identification code, and a diagnostic code. This indicates the time the error was detected, the station affected, and a description of the error condition. Most alarms are limited in length to a single line for clarity. Alarm messages are normally journaled for record purposes.

Reports. System and network reports are issued automatically on a predetermined time basis or in response to a command requesting a report. System reports include information related to the system operation. A typical system report is the system resource utilization report. This report provides information regarding the status of peripheral devices, number of blocks used/available on the intransit file, the current journal status, output message queue status, *central processing unit* (CPU) utilization, number of buffers used and available, journal utilization status, and retrieval utilization status.

Network reports provide various status and statistical information which the system accumulates. This information is formatted and printed automatically on a periodic basis or in response to a command entered by the network supervisor. Network reports can be requested for the full system or for an individual station or channel.

Reports which include information pertaining to conditions and queues are called status reports. Condition reports provide information concerning the current status of channels/stations. Conditions such as open line, link down, interrupted message, output hold, and poll/select failure are reported in a condition report. A queue report lists the number of messages waiting in each queue for each channel or station. An indication is provided if a queue threshold is exceeded. The network statistical report is usually generated each hour and every 24 hours. It contains statistical information such as

1. The number of messages and characters received
2. The number of messages sent
3. The number of messages retrieved
4. The number of messages rejected during the report period

The report also contains dialed-circuit statistics such as

1. Total number of calls and connection times inbound and outbound
2. Total number of attempted calls
3. Total number of busy or incompleted calls

If a network report is requested for the full system, they are sent to a network reports position. However, network reports requested for a single station (by command) are generally sent to the requesting stations. All reports are journaled for record purposes.

Intercept. In many systems, a special intercept-and-repair position can be configured as an intercept position for any or all stations in a network. The position would receive all reject messages for stations so configured. If, as a result of error analysis, the system intercepts a message as being invalid, the message would be rejected to the intercept queue with an error note appended. The intercept operator would read the messages from the queue, correct the errors, and reenter the messages into the system for normal processing. The alternative to this special intercept handling function is to have the system reject an invalid message back to the originator for message correction and reentry. Most systems support both types of intercept handling, and each station can elect to be configured for whichever type is preferable.

15. Load Handling

The message switch constantly monitors system resources to determine the level of resource depletion. According to the depletion level, certain alarms are generated which require operator intervention. In other instances, system functions are automatically suspended until normal levels are restored. System throttling and drain are two functions which are employed in message switches to manage and control work scheduling and resource availability.

Throttling. System throttling occurs when the system buffer blocks or in-transit file blocks reach a prespecified resource depletion level. Upon recognition of this condition, the system automatically suspends low-priority work scheduling until the resource levels return to normal. If the resources are depleted to exceptionally low levels, the system will automatically put controlled circuits in a stop-new-input state, until the resources are restored to normal levels.

Drain. When a channel or station becomes inoperable and the system cannot deliver messages to it, a large message queue can build in the in-transit file, and resources can become depleted. In this instance the system would issue a channel queue warning alarm. The system supervisor can use the *drain channel/station* command to move the message queue temporarily from the in-transit file to the journal file, and in effect restore the in-transit resources to normal. When the inoperable channel or station is restored, a *no-drain channel station* command would be entered to restore the messages to the in-transit file for normal delivery. A drain command is normally used when alternate routing is not possible.

16. Restart and Recovery

Restart and recovery is a function of the on-line support software. It is primarily responsible for an orderly shutdown and restoral of the system to the state that existed before the scheduled or unscheduled system shutdown.

Message-switching systems generally support two types of system recovery, automatic restart and recovery. In an *automatic restart,* the system restores itself to the state that existed just prior to the failure by performing a reload of static system tables and executable code from disk. The dynamic table information is reloaded from a checkpoint file on disk. During normal operation the system periodically writes the contents of selected memory areas to the checkpoint file on disk. The checkpoint data include information relating to disk queue control, station sequence numbers, peripheral devices, file control information, and internal message identification information.

If an automatic restart attempt fails because there has been excessive corruption of data, the system operator can initiate a more elaborate recovery procedure. This procedure involves scanning the in-transit file disk queues and rebuilding all corrupted queues. The dynamic tables are also reconstructed by the recovery program. When the automatic restart or recovery procedure is complete the system performs the following action:

1. Any messages which were in the process of being transmitted by the system when the failure occurred are retransmitted with a possible duplicate message note appended.
2. The sequence number of the last received message from a station (prior to failure) is sent to the station with a request for retransmission of all messages after that number. This action protects against the loss of any input message which may not have been received by the system because of the failure.

17. Special Features/Application Interfacing

The primary function of a message switch is to switch messages from origin to destination. As by-products of the switching function, many special features and application interfaces can be provided for a message switch.

The following describes some of the features and how they can be utilized.

Personal Address Line. This is where a string of data such as "To: Bill Jones, Director Sales" is preprogrammed into a data base which is pointed to by a stored address code, such as "NYC12," in the routing code database file. Each time the system receives an input message addressed to NYC12, it would fetch the personal address line "To: Bill Jones, Director Sales" from the data base and insert this string of data in the output message when it is queued for delivery. This feature relieves the operator from the task of having to type a personal address line on each message input to the system. It also reduces the input connect time and the associated transmission cost.

Work in Progress.   It is extremely important for many industries to have the message switch provide control for the entry, movement, approval, and release of data in the system. The message switch utilizes available switching features (such as password protection, implied routing, and message queue linking) which can be modified to provide a specific work-in-progress application.

A typical work-in-progress application which is often handled by a message switch is *order matching.* This is an application which is used by the brokerage community for matching each buy and sell order with a corresponding buy or sell confirmation. In this application, the message switch would accept a buy or sell order message from broker A which would be routed by the system and delivered to broker B for execution. A copy of the delivered order would also be retained in the system, in an in-progress queue. Once the order has been executed, broker B would send a confirmation notice to broker A confirming execution of the order. The system would deliver the confirmation message to broker A and match the confirmation with the corresponding order in the in-progress queue. Once the match has taken place the system would send a system-originated, transaction-completed message to a predesignated station, remove the entry from the in-progress queue, and journal the transaction-completed message to disk or tape for archival storage.

The contents of the in-progress queue can be monitored by a periodic system-generated report or displayed upon request.

Similar work-in-progress applications, such as electronic funds transfer for banking, reservations for airlines and travel, and vessel or car movements for transportation can be provided by the message switch.

Categórization.   This is a feature which can be extremely useful to a large organization which has many remote warehouses and where each warehouse stores different parts, rather than having a complete inventory of every part. The location of suppliers, parts requirements of different plants, and types of storage facilities are taken into consideration when developing a parts-distribution scheme. In the above scenario, categorization would work as follows:

1. The requisitioner would prepare a list of required parts in the form of a message which would be sent to the message switch.

2. The message switch would sort the requested parts numbers by warehouse locations. It would then automatically format and transmit system-generated messages to those warehouses, listing the part numbers required and where they are to be sent. This feature relieves the requisitioner from having to look up the storage location of each part and then prepare individual requests for each location where these parts are inventoried.

Billing.   One of the key resources of a message switch is its ability to produce data collection records as it processes messages. These records contain the information which the system uses to produce statistical data and generate bills. In a data-collection operation, the system would create a record during the input processing cycle for a message. Upon completion of the output processing cycle (message delivery) the record is updated with the delivery information. These completed records are maintained in the master billing file. Specific rate information is also stored in the billing data base file. A report generation program collates and reduces the data into a form suitable for producing the billing report. The report can be produced in hard copy form on a line printer or on a magnetic tape for entry into an external billing system for additional processing.

The form and content of the billing report is variable- and application-dependent.

The information to be collected, the charging methodology, the number of sorts required, the number of fields in the billing report, the number of billing records, and the frequency of the billing cycle are examples of variable- and application-dependent considerations which determine the complexity of a billing program.

Generally, the billing report produced by the message switch is used by a company to charge back to users of the system their share of the cost associated with the service provided by the message switch. The costs which are most often considered are depreciation, operations, communication facilities rental, and transmission usage expense.

The actual billing matrix is usually based on criteria such as

1. A per-message charge
2. A per-character charge
3. A combination of both
4. The input connect time
5. The output connect time
6. An on-net delivery rate
7. An off-net delivery rate

The billing report can be structured to produce an accounting by

1. Input station identifier
2. Originator code
3. Departmental billing code
4. Password identifier

By utilizing the prestored rate information from the rate table data base, the message switch can calculate and produce itemized billings sorted by date and by carrier network for each designated billing entity. The billing entity itself could be a station, an originator, a department, or an individual.

18. The Message Switch as a Network Node

Types of System Interfaces. An application well-suited for the message switch is one in which the message switch (MSS) is used as a regional hub in a multinode communications network. For example,

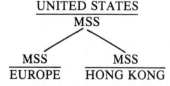

When functioning as a stand-alone message switch, the system is responsible for the end-to-end message exchange. However, when operating as a network node, the system is responsible for the end-to-end message exchange within its nodal region and also for interworking with other nodal switches which are responsible for the delivery of messages within their nodal regions. The message switch employs a variety of interfaces and controls which enable it to effectively work with other message switches in a multinodal environment. These interfaces and controls are used primarily to provide adequate safeguards against the loss of a message which transits multiple nodal switches in a network, before it is delivered to the final destination. Interface protocols used for communicating between nodal message switching systems must provide the following safeguards for a message exchange:

1. Guarantee that both systems are operational
2. Message exchange acknowledgment (ACK/NAK)

Some examples of controls or features which are provided to ensure message integrity and delivery assurance in a node-to-node message exchange include:

1. System-to-system message numbering
2. Delivery confirmation/nondelivery notification

System-to-System Translation.   Very often nodal message-switching systems on the same network support different message formats. When this occurs, a system-to-system translation technique is employed to provide the format conversion to support a message exchange. A format conversion routine is used which translates the format of a message received on a specific input channel to the format required on the specified output channel.

ELECTRONIC MAIL

19. Types of Electronic Mail

Electronic mail is a catch-all term for many different kinds of message exchange technology ranging from facsimile to communicating word processors, personal computers, telex systems, and computer-based messaging systems.

The computer-based message system is essentially a message switch that also incorporates word processing capabilities and data base management for storage and retrieval of information. Generally, these systems provide a user-friendly interface and utilize the word processing capability to assist the user with message composition (preparation of documents, files, or forms) and data base management for storing and retrieving information in electronic mailboxes.

20. Common User versus Dedicated

Communications carriers and time-sharing companies employ *computer-based messaging systems* (CBMS) to provide a common user or public electronic mail service. The services are offered to the business community and to the general public on a subscription basis.

In this type of offering, the service provider shares the resources of the CBMS over a large number of individual networks or subscriber groups. Many potential system purchasers initially subscribe to common user system services and use the service as a pilot program to evaluate their needs for a dedicated system.

Dedicated systems are purchased by companies with large user populations. The dedicated system approach to electronic mail offers the purchaser software which is customized to specific requirements, full control of system operations, and a fixed cost for internal electronic mail.

21. The User Interface

The user interface for electronic mail systems is designed to be user-friendly. Most systems provide basic and advanced user interfaces. The basic interface provides a menu mode of operation whereby the system displays a menu, the user selects an application, and the system prompts for the required inputs. This is an on-line operation designed for interaction with nonintelligent-type terminals. The basic interface is also used as an electronic training aid for novice users. The advanced interface is generally a common structured interface designed to work with intelligent terminals and the more experienced system users. Messages can be composed and edited in the terminal by the user, prior to establishing a session (connection and log-on) with the system. If message composition and editing are done off-line, the on-line connect time and corresponding cost are reduced and the system efficiency is increased in direct proportion to the reduced connect time.

22. Standards for Interconnection of Electronic Mail Systems

The installation of electronic mail programs on a number of separate computers provided by different vendors and linked by communications lines quickly raised the question of computer communication compatibility. The layered approach to computer communications, in which application software is independent of the physical mode of communications, was generally accepted as a necessary design principle. Some of the better known examples of such layered communications protocols include the International Standards Organization (ISO) open systems interconnection model; the IBM System Network Architecture (SNA), and the International Consultative Committee of Telegraphy and Telephony (CCITT) X.25 and X.400 recommendations. The CCITT X.400 recommendation on message handling has gained international acceptance as the standard for interconnection of electronic mail systems. Most major computer manufacturers and service providers have now publicly announced their intentions to conform with the CCITT X.400 recommendations on message handling.

EQUIPMENT CONFIGURATIONS

23. Typical Configurations

CPU, memory, peripheral storage devices, and communication multiplexers are the major equipment components used in message-switching systems. An equipment configuration designed around a single communication processor is referred to as a *simplex configuration* (Fig. 13.8). A system which includes two communication processors in the equipment configuration is generally referred to as a *redundant* or *duplex configuration* (Fig. 13.9).

Most message-switching systems are implemented in a redundant equipment configuration with all of the major components duplicated. In effect, the configuration consists of a primary system, a back-up system, and switching capability to provide either system with access to the communication network and peripheral devices. The duplex equipment configuration provides the highest level of redundancy and therefore results in the best possible (availability/reliability) equipment performance ratios. Some key operational advantages which can be

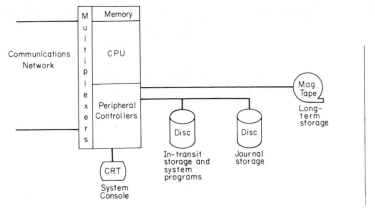

FIG. 13.8 Simplex communication processor configuration.

utilized in a duplex configuration but which are not available in a simplex configuration include the following:

1. The sharing of peripheral devices between systems.
2. Switchover from the prime system to the back-up system in the event of a processor failure.
3. Communications line switching to move a single line, a group of lines, or all of the communication network between systems.

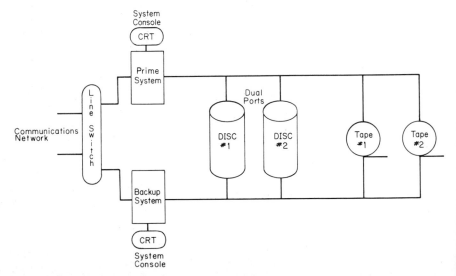

FIG. 13.9 Duplex communication processor configuration. Disc #1: system programs, in-transit and short-term journal storage; disc #2: mirror image of disc #1; tape #1: long-term journal storage; tape #2: standby spare for tape #1.

THE FUTURE OF MESSAGE-HANDLING SYSTEMS

24. Computer-Based Electronic Messaging

Since its introduction in 1980, computer-based electronic messaging has shown a respectable growth rate. Electronic messaging software is provided as a value-added product for large computer systems, personal computers, and intelligent terminals. Electronic messaging is also available as a service provided by communications carriers and time-sharing companies for their subscribers.

The systems and services which evolved have proven to be user-friendly, faster, more efficient, and less costly to use than the traditional communication services they are targeted to replace. Although the growth of corporate electronic mail (e-mail) systems and carrier e-mail services has been spectacular, the growth has been significantly slower than its potential rate.

25. Technical Developments

In order for the growth of electronic messaging to reach its full potential, two primary technical hurdles, compatibility and mixed media service integration, must be addressed and overcome by the participating vendors. The first major step toward developing a global communications standard for message-handling systems was undertaken by the CCITT. The committee defined a set of communications standards (X.400 series) that will provide universal electronic mail service and permit the automation of message handling across dissimilar computer-based systems. The standard has been endorsed by the Electronic Mail Association, major system suppliers, and communication carriers worldwide. Guidelines including algorithms to permit the handling and conversion of mixed-media communication—data, voice, and image—have been recommended and will be improved upon. The next item on the CCITT agenda is to develop a formalized directory scheme for internetwork or intersystem communications. In the near future, message-handling systems will function as automated information-handling systems. At the department or organizational level, the systems will provide users with a package of generic office tools which each group of users would then be able to customize for its own purposes.

Integrated mixed-media communications will be fully supported to provide visual aids in such forms as text, forms, image, and voice communications. Gateway software for off-net communications will be X.400 compatible. The systems will provide global messaging services, access to a global directory, and mixed-media messaging and conversion facilities.

CHAPTER 14
PACKET SWITCHING

Thomas H. Scholl
Senior Vice President, Hughes Network Systems

CONTENTS

INTRODUCTION

1. Applications of Packet Switching

Packet-switching technology was developed in the late 1960s and the 1970s to provide an efficient means for exchanging digital data between computers and, later, between computers and computer terminals. It can also be used for switching analog signals, such as voice, after conversion into a digital format.

2. Packet and Circuit Switching

The basic configurations of packet and circuit switches are shown in Fig. 14.1. A circuit switch conducts analog or digital signals transparently by connecting a single input circuit to an output circuit. A packet switch divides digital input signals into data segments called packets. Packets from multiple input circuits can be switched and multiplexed into one or more output circuits.

Complex networks have been developed around each of these basic configu-

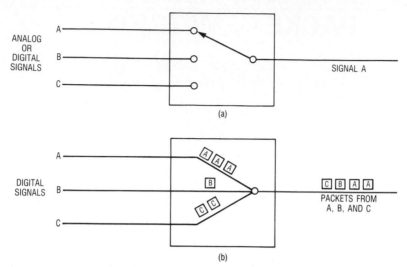

FIG. 14.1 Basic configurations of packet and circuit switches. (*a*) circuit switch; (*b*) packet switch.

rations. Circuit-switching technology is described in Chap. 12. This chapter describes packet switching and includes a comparison of the two.

3. Forms of Packet-Switching Networks

Packet switching is a versatile technology which can be used with a wide variety of network forms. It can be used for *local-area networks* (LANs), *metropolitan-area networks* (MANs), and *wide-area networks* (WANs). With the selection of appropriate protocols, packet switching can be used with diverse transmission media such as twisted pair or coaxial cable, fiber-optic cable, radio links, and satellite links.

Although packet switching has not been used extensively to date for voice networks, it is feasible for many voice applications.[1] The *integrated services digital network* (ISDN) includes a special form of packet switching to provide voice call signaling functions.

Finally, packet switching is adaptable to different network topologies, such as star, interconnected mesh, ring, and bus (Fig. 14.2).

4. Examples of Packet-Switching Networks

Since the development of the well-known ARPANET network begun in 1969,[2] many different types of networks based on packet switching have emerged. The Telenet and Tymnet networks in the United States, the Datapac network in Canada, and the Transpac network in France are examples of the more than 30 international *public* packet-switched networks.

The number of nonpublic, or *private*, packet-switched networks has grown

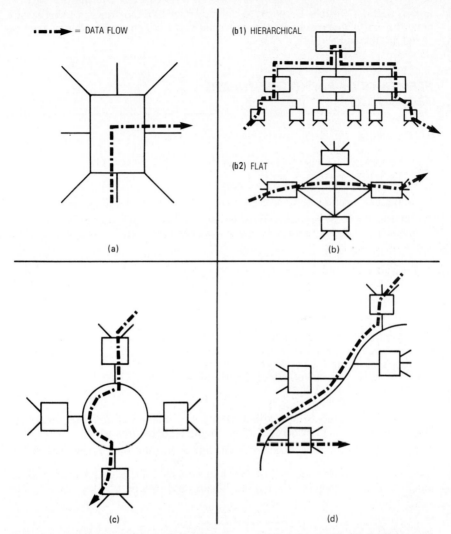

FIG. 14.2 Common network topologies. (*a*) star or hub-based network, (*b*) interconnected mesh network, (*c*) ring network, (*d*) bus network.

rapidly in the 1980s. Examples of private data networks are those belonging to major domestic and international corporations which utilize packet switching to interconnect geographically distributed data processing resources. *Hybrid* networks attempt to combine the benefits of both public and private networks while providing a common network appearance to users.[3,4] The Defense Data Network (DDN), which includes—and is the successor to—ARPANET, uses packet switching for U. S. military communications.[5]

Although the terms *packet switching* and X.25 are frequently used synonymously, IBM's System Network Architecture (SNA) and Xerox's Ethernet are nevertheless further examples of popular networks which are based upon packet-switching techniques.

ELEMENTS OF PACKET-SWITCHING TECHNOLOGY

5. Concepts

Packet-switching technology consists of three essential concepts:

1. Creating packets from a digital stream.
2. Statistically multiplexing packets on a common transmission channel.
3. Relating the basic services offered to users of packet-switched networks to the internal functions provided by packet-switched networks, including packet relay and routing.

   Each of these is described below.

6. Creating Packets

Computer data is already in a form that can be packetized, but analog information sources, such as audio and video, must first be digitized to create a digital data stream as shown in Fig. 14.3. The packetizing algorithm in the box of Fig. 14.3 creates a packet whenever any of the following is true:

1. A maximum of N bits (bytes) has been received.
2. A particular sequence of bits (bytes) or a combination of bit (byte) sequences is detected through inspection of the incoming data stream.
3. Some bits (bytes) have been received, and a time duration T has expired.

   Since packets are "pieces" of the data stream, each packet includes something to indicate where the packet begins and where it ends, making the

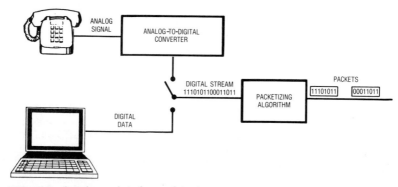

FIG. 14.3 Creating packets from a data stream.

"depacketizing" algorithm very simple: if a packet has been received, extract the source data contained within it.

7. Introduction to Multiplexing

Figure 14.4 illustrates three pairs of computer terminals or personal computers, which communicate using three *full-duplex* (i.e., simultaneously two-way), point-to-point transmission links. In this example, terminals 1, 2, and 3 are located in Dallas, and terminals 4, 5, and 6 are located in New York City.

The number of transmission links used to interconnect Dallas and New York can be reduced from 3 to 1 by adding a multiplexer in each location. The primary motivation to provide multiplexing is to reduce overall cost. This can be realized in the example so long as the cost of two multiplexers plus one transmission link is less than the cost of the three original transmission links. Figure 14.5 illustrates the basic operation of two *frequency-* or *time-division multiplexers,* while Fig. 14.6 shows how two packetizers can be configured to accomplish *statistical multiplexing.*

Frequency- and Time-Division Multiplexing. In frequency-division multiplexing (FDM), the link between the multiplexers is divided into different frequency bands. Each band is allocated for transmitting data between a pair of terminals—one terminal at each end of the link. A time-division multiplexer (TDM) divides the link into time slots, rather than frequency bands.

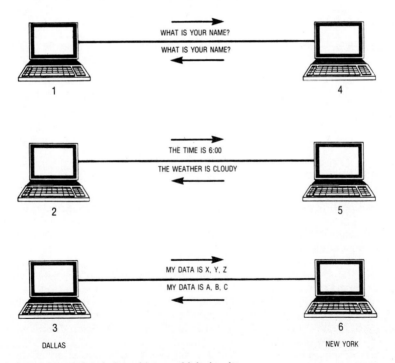

FIG. 14.4 Transmission without multiplexing data.

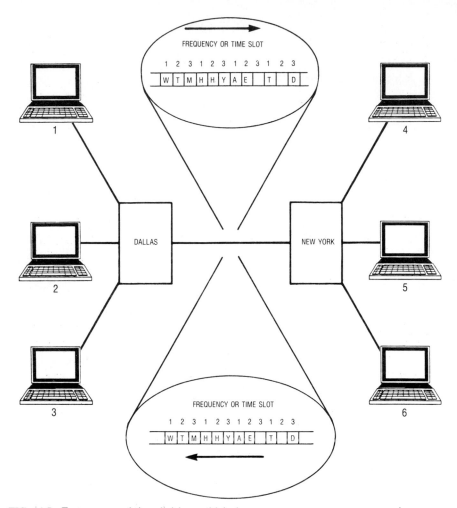

FIG. 14.5   Frequency- and time-division multiplexing.

The allocation of frequencies or time slots to pairs of data sources is fixed. As shown in Fig. 14.5, three frequency bands or three time slots in each direction are required, and data from terminal 1 are transmitted to destination terminal 4 in either frequency band F1 or time slot T1.

The communication line from each terminal is sampled at a rate commensurate with the speed of the terminal and replicated in the appropriate frequency band or time slot on the composite link. If each terminal transmits at the rate of 2400 bits per second (b/s), for example, the capacity of each frequency band or time slot is also 2400 b/s, and the capacity of the composite transmission link must be at least 3 times 2400 b/s for a total of 7200 b/s, in each direction.

Statistical Multiplexing.   Figure 14.6 illustrates the basic operation of statistical multiplexing. Each of the three terminal lines connected to a multiplexer is monitored, and a packet is created according to the packetizing algorithm described

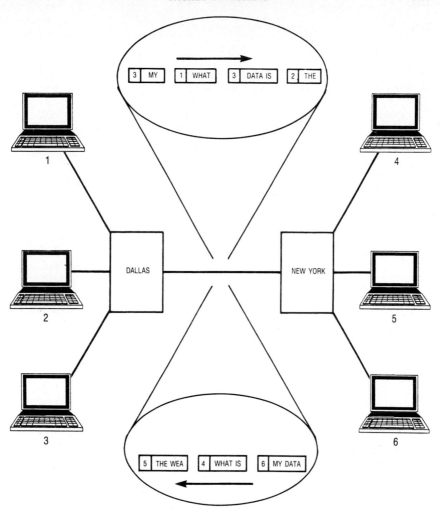

FIG. 14.6   Statistical multiplexing.

earlier. In order to identify which terminal generated the data contained in a packet, the originating multiplexer stores the number of the terminal in each packet. In this example, the terminal number represents what is called the *source address* of the packet.

Data in packets containing a source address of 1 is depacketized by the New York multiplexer and output to terminal 4. Data in packets with a source address of 2 are output to terminal 5, and so on.

8.   Advantages of Statistical Multiplexing

Table 14.1 summarizes the advantages of statistical multiplexing compared to

TABLE 14.1 Advantages of Statistical Multiplexing

Advantage	Reason	Condition
Dynamically assigned bandwidth	Higher utilization of lower-capacity transmission links	Data are bursty
Flow control	Prevents loss of data under load. Allows mismatch of source and destination capacity.	Flexibility of configuration is desirable
Error control	Ability to detect and correct errors	Transmission errors cannot be tolerated
Code and protocol conversion	Multiplexer contains intelligence which can be used for auxiliary functions.	Required processing power is available
Packets provide flexibility	Packet length can be varied to suit characteristics of the application or the transmission channel. Packets can be adapted to add new features and protocols.	Flexibility in application or transmission channel is required, and processing power and memory requirements can be met

frequency- and time-division multiplexing, and each major advantage is discussed in the sections following.

Dynamically Assigned Bandwidth.   In the multiplexer examples described above, data are generated whenever a human types some text on one of the terminals. However, if nothing is typed, there is no text to transmit. In other words, data consist of occasional bursts of characters or text, rather than a continuous stream of characters. In the case of an FDM or TDM, this means that there are times when frequency band F1 or time slot T1, for example, is idle, consequently wasting the bandwidth previously allocated.

In the statistical multiplexing example, bandwidth is assigned dynamically by using the transmission link only when there is a packet to send. Therefore, if the multiplexed data are bursty,* the overall capacity of the composite link may be proportionately reduced. For example, the capacity of the transmission link might be reduced by 50 percent from 7200 b/s to 3600 b/s. The use of lower speed links, usually resulting in lower cost, is a major advantage of statistical multiplexing when compared to fixed-assigned bandwidth techniques, such as frequency- or time-division multiplexing.

Flow Control.   As data are received by a statistical multiplexer, they are formed into a packet (packetized) and buffered in the memory of the multiplexer until the packet can be output. Since the statistical multiplexers configured in the example allow the aggregate input rate (3 × 2400 = 7200 b/s) to exceed the composite out-

---

*That is, the average data rate is significantly less than the maximum possible data rate.

put rate (50% × 7200 = 3600 b/s), there can be situations where input data are lost because no more memory is available to buffer packets.

Flow control is a technique used to ensure that input data are not accepted at a rate that is faster than it can be processed (buffered) and output. Three interfaces typically subject to flow control (Fig. 14.6) are

1. *Terminal to multiplexer:*   The multiplexer interprets a signal from the terminal to start or stop output to the terminal.
2. *Multiplexer to terminal:*   The multiplexer generates a signal to the terminal to allow or inhibit input.
3. *Multiplexer to multiplexer:*   One multiplexer signals the other multiplexer to start or stop the exchange of data across the link connecting them.

By exercising flow control at the expense of throughput (i.e., characters sent and received per unit time), the capacity of the link connecting the multiplexers in Figure 14.6 could be further reduced were there greater economic benefits in doing so.

Error Control.   Transmission links are typically susceptible to transmission errors. Suppose the *bit error rate* (BER) of a link is 10^{-4}, then on the average 1 bit out of every 10,000 bits transmitted will be corrupted. Since FDMs and TDMs transmit data transparently, data may be either lost or altered each time a bit error occurs. To solve this problem, additional information—called *error control*—can be added to a packet to detect and compensate for bit errors.

The general idea behind error detection and recovery is as follows. The source multiplexer retransmits any packet which it believes may not have been received correctly at the destination, and the destination multiplexer discards any packet which either contains an error or has already been received.* The capability to verify whether or not a packet has been received by the destination correctly is accomplished by sending a packet containing an *acknowledgment* back to the source.

Intelligence and Flexibility.   Although data processing and special-purpose minicomputers were used originally, the processing required by statistical multiplexing has been performed by *microprocessors* since the mid-1970s.[6] As discussed above, the direct benefit of this processing is increased utilization of lower capacity (and potentially lower cost) transmission links when compared to other forms of multiplexing. As microprocessors have become more and more powerful, the intelligence provided by them has resulted in indirect benefits. One indirect benefit, called *data conversion,* is the ability to connect dissimilar devices together. The degree of dissimilarity allowed depends in part upon how much processing power is required to convert from the format of the source to the format of the destination. Three common forms of data conversion are

1. *Speed Conversion:*   The ability to allow a low-speed device to communicate with a higher-speed device (and vice versa) by buffering and flow-controlling data.
2. *Code Conversion:*   The ability to convert character data from one format into another format.†

---

*See the discussion of the X.25 data link level below.
†Three common character formats are ASCII, EBCDIC, and Baudot.

3. *Protocol Conversion:* A *protocol* is a set of rules and formats governing communications among communicating entities. Protocol conversion allows devices using different sets of rules and formats to communicate.

Packets are highly flexible units of information. Although a packet must have a maximum acceptable length, packets can expand or contract in size, providing a convenient vehicle for adapting to different kinds of data streams, transmission links, media access methods, and protocols.

9. Disadvantages of Statistical Multiplexing

Table 14.2 summarizes the disadvantages of statistical multiplexing when compared to frequency- and time-division multiplexing.

Packetizing Delay.   A packet cannot be sent until it has been created and it cannot be depacketized until it has been fully received. This delays the transmission of data to its destination. All multiplexers introduce delay, but in some network applications, the delay due to packetizing may be unacceptably long, and statistical multiplexing cannot be used as the technology for these networks.

Queuing Delay.   Suppose, referring to Fig. 14.6, that data arrive from each of the three terminals all at the same time so that three corresponding packets are created. While one packet is being transmitted, the other two packets have to wait. This additional source of delay is called *queuing delay*.

Queuing delay is one form of delay that makes packetized voice difficult to handle. Human speech, like computer data, also exhibits a bursty characteristic due to pauses between syllables and words, and the fact that human communication typically alternates between listening and talking. Modern digital techniques, therefore, can be used to convert what appears to be a continuous speech stream into a bursty data stream. However, in order for speech to be intelligible (or at least pleasant to listen to), the time which passes between bursts at the source must be accurately

TABLE 14.2   Disadvantages of Statistical Multiplexing

Disadvantages	Reason	Condition
Data are delayed	Time it takes to packetize and depacketize. Time a packet waits before being transmitted.	Timing of original source data must be faithfully reproduced
Packet overhead is added	Less transmission capacity available for conducting useful data	The ratio of packet overhead to packet size is high
Intelligence required	Needed to create and handle packets	The speed of the source data is too fast or costly to process.
Many performance parameters are probabilistic, rather than deterministic	Performance difficult to predict	Performance models are complex

reproduced at the destination; with statistical multiplexing, this is difficult to guarantee.*

Packet Overhead. Packetizing delay can be reduced by creating smaller packets, but this aggravates the problem of *packet overhead*. Each packet must include essential overhead data (packet boundary, source address, error control, etc.), and as the packet is shortened the ratio of the overhead data to the useful data increases. This effectively reduces the capacity of the transmission link available for sending useful source data.

The ratio of packet overhead to packet size can often be decreased if a single packet is constructed to contain data from multiple data sources, so long as those data sources can be uniquely identified. This is the idea behind a popular form of statistical multiplexing called *statistical time-division multiplexing* (STDM).[7]

It is important to note, however, that some forms of overhead data need not be transmitted in every packet and that packets can be constructed to contain overhead data only in order to remove redundant overhead which might appear in a stream of data packets.

Probabilistic Performance. Statistical multiplexing requires processing and buffering (intelligence). But how much processing power and how many buffers are needed in order to guarantee the required performance? Since high line utilization is desired, how much transmission capacity is actually required in order to achieve an acceptable queuing delay?

Answering such questions is more difficult with statistical multiplexing rather than time- or frequency-division multiplexing because performance predictions are probabilistic instead of deterministic.[8] On the bright side, performance monitoring can be built into a statistical multiplexer because of its inherent intelligence, so that alarms can be generated when operating under peak load or overload conditions.

10. Introduction to Packet-Switched Networks

In the multiplexer examples illustrated in Figs. 14.5 and 14.6, connections are fixed, and therefore the address of the destination for each packet is implied. In a switched network, connections can be established dynamically and a *destination address* must be provided explicitly.

The statistical multiplexers shown in Fig. 14.6 can be transformed into packet switches by:

• Providing an interface to terminal users in order to request new connections or clear existing connections.

• Providing an interface between packet switches to make or break connections, in accordance with user requests.

• Modifying the contents of packets as required by the above interfaces, including the addition of a destination address.

• Providing the ability for each switch to inspect the destination address and relay the packet to the terminal indicated.

---

*Error control is also less significant if packetized voice, rather than computer data, is multiplexed. Depending upon the encoding scheme used, digitized voice can withstand a limited number of transmission-induced bit errors and still remain intelligible.

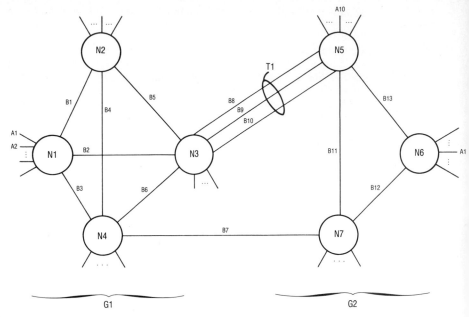

FIG. 14.7   A mesh packet switching network.

Figure 14.7 illustrates a packet-switching network, based upon a mesh topology, containing the following components:

- Two switching *groups* or areas, G1 and G2.
- Seven switching *nodes,* N1 through N7.
- Thirteen *backbone links,* B1 through B13, which are used to interconnect the switching nodes.
- One *transmission group,* T1, consisting of multiple backbone links.
- Numerous *access links,* A1 through A*n,* which are used to access the network by network users.

The capabilities of packet-switched networks will be presented below in light of two simplified contexts: *user services* refer to the basic capabilities available to *users* attached to access links, while *internal functions* refer to the capabilities provided between switching nodes across backbone links or within a switching node itself.

Packetizing and depacketizing are performed "outside" the network (Fig. 14.7). Packets associated with users enter the network directly via access links, and each access link can be statistically multiplexed with packets belonging to many different users.

As packets arrive from an access link attached to a node, they are inspected and switched to the node attached to the destination access link. The source and destination nodes involved in this packet exchange are called *edge nodes* because they represent the edge of the network from the viewpoint of a network user.

Intervening nodes which participate in the packet exchange between edge nodes are called *transit nodes*.

11. Services Offered by Packet-Switched Networks

Packet-switching networks today can provide two basic types of user services: *connection-oriented service* and *connectionless service*.[9] A particular packet-switching network may provide only one of these or both. If a network is capable of supporting both types of service, the decision as to which service should be used is a function of the user's application and the capabilities available in the equipment and software employed by the user.

Connection-Oriented Service. A *connection* is an association of two or more communicating entities for the purpose of controlling the exchange of information. The connection-oriented service provided by packet-switched networks is similar in nature to the *call* in circuit-switched networks. However, precisely what a connection is, how it is identified, and how it is established and released depend heavily upon the specific collection of protocols utilized. Generally, the connection-oriented service has the following characteristics:

1. A connection must first be established before users can exchange information.

2. Establishment of a connection implies that the parties exchanging information have agreed to the "terms and conditions" which may have accompanied the connection.

3. Multiple connections established for or by the same user are distinguished by a unique *connection identifier* contained in each packet, assigned at the time the connection is established.

4. Depending upon the protocol, connection identifiers may serve as an "abbreviated" form of addressing in data packets, thereby reducing the packet overhead otherwise needed to contain "full" user addresses.

5. Establishing a connection is a method to achieve a form of synchronization among cooperating users, providing a convenient vehicle to:

   a. Reserve transmission or processing resources to fulfill the connection in advance of needing them.

   b. Sequentially number each packet so that out-of-sequence, missing, or duplicated packets can be detected and optionally recovered.

   c. Provide an acknowledgment capability based upon packet sequence numbers.

   d. Selectively implement flow control techniques based upon individual connections.

The connection-oriented service may appear in a wide variety of forms. Most often it is associated with the capability to provide acknowledged, sequenced packet communications for applications requiring a fairly reliable exchange of a packet stream—that is, situations in which packets are expected to be received during a connection in the same order they were sent.*

---

*A connection-oriented service may not be inherently reliable. Different forms of connection-oriented protocols may offer differing degrees of packet transmission reliability. For example, a connection-oriented service may be desirable for handling packetized voice, but not recover packets which have been lost—assuming the probability of losing a packet is acceptably low.

Two common forms of the connection-oriented service* are the *switched virtual circuit* and the *permanent virtual circuit.* A switched virtual circuit, sometimes known as a *virtual call,* passes through the three familiar communications processing states of *connection establishment, data transfer,* and *connection release.* The desire to initiate or release a connection is indicated through the use of *supervisory* packets sent from one user to another.

The *connection request packet* may contain an array of parameters regarding the connection which can be negotiated among users and the network, before establishing the connection. Subsequent to the connection negotiation, vital resources such as processing or transmission capacity can be reserved for exchanging packets during the data transfer phase. The connection can be cleared by either the source or destination user or by the network itself, by generating a *connection release packet.*

Establishing and releasing a connection requires considerable supervisory overhead if users wish to exchange only a small amount of data. To compensate for this problem, a variation of the switched virtual circuit allows user data to be contained in connection request or release packets, resulting in fewer packet exchanges between source and destination user than would otherwise be required.†

As opposed to a switched virtual circuit, a permanent virtual circuit is neither established nor cleared by the user, but rather is maintained on behalf of the user by the network. So long as a connection has not failed, it is perceived by the user to be in the data transfer state.

A variation of the permanent virtual circuit is the *semipermanent virtual circuit,* in which the circuit is available for data transfer only at predefined or scheduled times.

Connectionless Service.   As its name implies, the connectionless service provides for the transfer of data among users without first establishing a connection. Characteristics of this service are

1. Since no connection is established, no dynamic connection agreement is exchanged, and no connection identifier is assigned. Consequently, users cannot negotiate the terms by which data are to be exchanged.

2. Everything necessary to accomplish the transmission of user data is contained in a single packet.

3. The connectionless service imposes no requirement that packets exchanged among users be received by destination users in the same order they were sent by the source.

Since the connectionless service does not incur the overhead required in establishing a connection, connectionless protocols tend to be less complex than connection-oriented protocols and are conducive to environments which are characterized by broadcast or multipoint communications, high-speed communications, transmission links that provide a low BER, or applications that rely on transactions which fit in a single packet.

Interestingly, limited synchronization can be achieved using a connectionless service to accomplish acknowledgments and packet numbering, but this does not imply that numbered packets have to be delivered from source to destination in sequential order.

---

*The following examples are based loosely on the X.25 protocol.
†See the X.25 fast select facility described later.

Two popular forms of connectionless service are the *datagram* and the *acknowledged datagram*. A datagram contains addressing and control information, as well as user data. Control information may consist of quality or class-of-service parameters to be used by the network in determining how to transport the datagram to its destination.

In the case of an acknowledged datagram, the network may provide supervisory information as to whether the datagram was successfully transmitted to the destination user (positive acknowledgment) or not (negative acknowledgment). Typically, a datagram can be acknowledged so long as it contains a unique *data unit or reference identifier*. Negative acknowledgments are also sometimes accomplished by copying a portion of the original datagram into the data field of the acknowledging datagram.

The absence of a dynamic agreement between users of a connectionless service does not mean that no agreement is in effect. Such agreements may exist among users a priori and are, therefore, unknown within the scope of such a service. Furthermore, if a network inherently offers a high probability that packets will not be lost or that packets will be received in the order they were sent, these benefits might be assumed by connectionless users of that network, in spite of the fact that such benefits are not explicitly requested.

12. The Internal Operation of Packet-Switched Networks

The concepts that characterize network user services can also be applied recursively to internal network functions. Thus, for example, each node shown in Fig. 14.7 can be considered to be a "user" of "services" offered by one or more successor nodes (comprising a "network"), or a "network service provider" for predecessor "user nodes." Packet transmission between nodes can be described as connection-oriented or connectionless.

However, even though concepts may be applied in a similar fashion to either context, the actual internal operation of a network may differ considerably from the services such a network offers. The definition of a virtual circuit service as described above, therefore, does not include the requirement that all packets composing the virtual circuit follow the same route through the network. In fact, packets belonging to a connection-oriented service may be transported through a network using a form of connectionless packet transmission. These points are often a source of confusion, even to those familiar with packet switching, because the terminology employed is frequently dependent upon a presupposed frame of reference.*

The Interface between User Services and Internal Functions.   User services and internal functions can best be distinguished by examining the interface that resides between them, required in order to convert—or map from—the capabilities of one context into the capabilities of the other.

Some examples of the features contained in this interface are:

• *Address Translation:* Addresses contained in user packets may have been constructed for the convenience of the user, or to protect the security of the network. If user addresses cannot be utilized directly for the purposes of rout-

---

*Specific protocols provide the best context for the precise definition of terms and are therefore the subject of the next major heading in this chapter.

ing (see below), they must first be translated into a format which can be utilized by the network internally.*

• *Packet Segmentation and Reassembly:* The maximum size of packets exchanged between nodes on backbone links may be optimized to take into account the characteristics of those links (e.g., capacity, BER, and delay). If the maximum packet size of user packets entering the network is larger than can be accommodated internally, the interface must provide the ability to segment user packets into smaller internal packets, and to reassemble internal packets into user packets.

• *Transmission Mode Conversion:* As mentioned earlier, it may be necessary or desirable to convert from a connection-oriented mode of transmission on the user side of the interface to a connectionless mode of transmission on the other side, or vice versa.

Other issues which affect the complexity of this interface are the ability to propagate flow control across the interface in either direction, the assigned scope of packet retransmission and acknowledgment, and the need to resynchronize transmission on both sides of the interface resulting from a transmission failure. The complexity of the interface is further compounded by the fact that the features it provides are typically not entirely resident in only an edge node, but may be distributed throughout the network.

One of the most important internal functions is *routing,* which consists of identifying the path a packet should take through the network (route selection), and then relaying the packet according to the chosen route.

Route Selection. Route selection usually consists of considering many different parameters, including the destination address contained in the incoming packet. As shown in Table 14.3, there are several possible categories of parameters which can be taken into account when a route is selected:

• *Network User Parameters:* Parameters are included as supervisory overhead in packets that invoke a user service.

• *Network Administration Parameters:* These are parameters which may be used as "defaults" (if not specified by a network user), or which may "override" or complement network user parameters that have been specified.

• *Network Configuration Parameters:* These are parameters that specify what is contained in the network, including all network components and features.

• *Network Utilization and Traffic Parameters:* These parameters are based upon measuring the utilization of transmission and processing capacity, and monitoring traffic flows in the network.

If the route selection process dynamically adapts to changes in parameters such as network traffic load or network component failures, it is called *adaptive routing.* If the route selection process does not adapt to such changes, it is called *deterministic routing.* An example of adaptive routing is the motorist who takes a detour because a radio announcer has just broadcast the fact that the direct route is currently congested with traffic. An example of deterministic routing is the motorist who decides how to drive from Washington, D. C., to New York by using a roadmap. A routing strategy may be "mostly deterministic" or "somewhat

---

*Address translation is described more fully in the discussion of ANET below.

TABLE 14.3   Inputs to Selecting a Route

Parameters specified by a network user or by the network administration

Destination address
Source address
Type of service
Explicit path
Cost (e.g., billing code)
Priority
Security
Quality of service, in terms of delay, throughput, and failure probability
Class of service, including a network-defined combination of any of the above
Maximum size of access packets

Parameters based upon the network configuration

Network topology
Network interconnectivity—may be static or dynamic
Type and characteristics of all network components, including transmission links and switching nodes
Number of hops (links) from source to destination
Maximum size of internal network packets

Parameters based upon network utilization and traffic statistics

Utilization of transmission links
Number of connections on a link
Delay incurred on a link
Number of packets waiting for transmission on a link
Observed BER of a link
Quantitative traffic flows per unit time
Processor and memory utilization

adaptive,'' depending upon the kind of information that feeds back into the selection process, and how frequently this feedback takes place.

Selecting routes may be a time-consuming task, depending upon how big the network is and how many of the parameters described above are taken into account. These factors influence the decision as to which network component should be responsible for route selection, yielding several possibilities:

- The network user who generated the packet.

- The first switch that receives the first packet associated with an invocation of service (i.e., an edge node).

- Each switch along the route, which contributes by selecting the next leg of the route.

- One or more specifically designated *routing servers* or *network managers,** including, perhaps, a human operator of the network.

---

*The concepts of network servers and network management are discussed in conjunction with the description of ANET below.

In some networks, routes may be preselected. In other networks, the route selection process may be divided into smaller tasks, which can be performed by different network components in concert. *Centralized routing* means that the route selection process is centralized within a single network component. *Distributed routing* means the route selection process is distributed over more than one network component. If the route selection is done one leg at a time (from the viewpoint of a switch), it is called *incremental* or *step-by-step routing*. If the route selection is done all at once by the source (e.g., a network user or an edge node), it is called *source path routing*. *Fixed routing* means that routes are preselected and cannot be changed (except perhaps under unusual circumstances such as a backbone link failure, resulting in the selection of a predefined *alternate route*). *Dynamic routing* is a term which has been used ambiguously to mean either that the routing strategy is adaptive, or that one of several routes may be dynamically selected based upon deterministic parameters (for example, there are multiple roads which can be used to drive to New York).

Packet Relay and Transmission.   When packets arrive at a switch to be relayed from link to link, the switch must inspect each packet to determine how to relay it. In most switches today, the inspection process is performed by software rather than hardware (in order to provide sufficient flexibility in terms of packet format), and it is, therefore, a major contributor to the overall *switching delay* which each packet incurs.

The information in the packet to be inspected might be any one of the following:

- *Destination Address:* The destination address (usually in combination with other parameters in the packet such as quality or class of service) is converted into a path to be used for transmitting the packet on its way to its destination. This conversion is usually accomplished by using a table or directory-lookup algorithm, and consequently has been called *directory routing.** The contents of the directory may be static, periodically updated, or calculated as needed at the time a packet is to be relayed.

- *Explicit Path or Explicit Route:* The packet contains a field that designates the path or route to be taken. In other words, the switch must step through and decode this field in order to relay the packet. The path or route specified could consist of either a low-level or high-level representation of the route to be taken. For example, if components in the network can be addressed in a hierarchical manner, only components at the top of the hierarchy might be represented. Thus, in Fig. 14.7, the explicit route designation A1-(G1-G2)-A10 is a high-level specification, while the designation A1-(N1-N4-N3-N5)-A10 is a low-level specification of the route.

- *Connection Record Identifier:* An internal connection identifier contained in the packet is used to find a previously created *connection record,* stored in the receiving node. The connection record, in turn, contains an indication of the path to be taken out of the switch.

Connectionless packet transmission between nodes can be accomplished using a packet relay process based upon either converting a destination address or decoding an explicit path.

---

*The description above assumes the destination address is a *physical address*. If the destination address is a *logical address,* it must be translated first into a physical address. The mapping of a logical address into a physical address is called *address translation*.

Similarly, either of these methods can be used for connection-oriented packet transmission to relay the initial connection request packet. As each node receives the request to establish a connection, a connection record is created and a locally significant connection identifier is assigned accordingly. Subsequently, the connection identifier is used to relay packets during the data transfer phase of the connection.

Internal connectionless or connection-oriented packet transmission can be implemented in a wide variety of ways, depending upon how these two modes of transmission are integrated with the many different route selection and packet relay strategies described above. Table 14.4 compares some of the features and issues of each transmission mode, when used internally.

Since establishing an *internal connection* is not identical to establishing a *user connection,* it is important to describe the circumstances which could cause an internal connection to be established. Some of the possibilities are

- Internal connections could be established and released in synchrony with the user connection.

- A single user connection might result in employing multiple internal connections (*splitting*) in order to meet the throughput requirements of the user connection. Furthermore, if the traffic load associated with a user connection is

TABLE 14.4 Comparison of Internal Packet Transmission Modes

Based on connectionless packet transmission	Based on connection-oriented packet transmission
If coupled with a form of dynamic routing, high backbone link utilization can be achieved because of the effect of load balancing on a per-packet basis	If coupled with dynamic routing, load balancing is achieved on a per-connection basis
Highly resistant to path failures, if coupled with adaptive routing	Connection path is susceptible to path failures. (Reconnect feature needed to recover from failures; see below.)
If coupled with dynamic routing, packets received at the destination node must be resequenced before transmission to a sequenced, connection-oriented user service	Packets are received by the destination during the data transfer phase in the same order they are sent by the source
If coupled with adaptive routing, effective flow control policies may be difficult to initiate, implement, or manage	Per connection flow control techniques can be used, aiding in the identification of offending users
If interfaced with a connection-oriented user service, it may be difficult to ensure that all packets can be delivered through the network with the same quality or class of service required by the user	Transmission and processing resources can be reserved at the time the connection is established, and all packets follow the same path selected to meet the service requirements
All packets are relayed in a common manner, usually having a common packet format	Two styles of packet relay—one to establish the connection, and another to transfer data once the connection is established
Time to inspect a packet in order to relay it is increased	Once the connection is established, packet relay can be based upon inspecting a single connection identifier

asymmetrical, one internal connection could be used for traffic going in one direction and another internal connection could be used for traffic going in the opposite direction.

- Internal connections could be preestablished based on a priori knowledge of certain traffic flows. Thus, such a connection could be used to transport connectionless user traffic, or to *multiplex* multiple user connections.

- Internal connections could be established by recognizing specific characteristics of user packets. For example, a connection might be established to transport a single, large, connectionless user packet which has been segmented into multiple internal packets, or such a scheme could be used to transport user "messages," where a message typically consists of more than one user packet.

- If it is determined that no traffic has been flowing through an existing internal connection for some period of time, the connection could be released, consequently releasing resources reserved for the connection. Upon receipt of subsequent user traffic, a new internal connection could be initiated.

- Existing internal connections could be reviewed periodically. If a route is found that is better than the one currently in use by a connection, that connection could be released, and a new connection—based on the new route—could then be established.

- Since an internal connection is subject to path failures, a new connection could be established upon the detection of such failures. This feature, called *reconnect,* allows user connections to be maintained in spite of an internal connection path failure.

Because of the many possibilities, there are correspondingly numerous alternatives to consider in designing an effective interface to user services. For example, interfacing a connection-oriented user service to more than one internal connection is similar to the problem of interfacing a connection-oriented user service to an internal connectionless mode of transmission (see Table 14.4). That is, the interface must provide the ability to detect duplicate packets or resequence packets received out of order, before handing them over to the user.

Congestion Avoidance and Control.   A motorist may select a highway for commuting into town because it has multiple traffic lanes or it is well-paved for high-speed driving. If other motorists have made the same decision, the selected route may become congested with traffic. Similarly, a poor routing strategy used in a packet switching network may lead to *network congestion.*

Since packet-switched networks strive to achieve high transmission line utilization, it is difficult to avoid all forms of congestion while also maintaining an acceptable throughput. Congestion is a relative rather than an absolute term, and is usually defined according to the implementation of internal network functions. For example, congestion could mean that the quality of service achieved for a particular user is less than required, or that a specific node has no more buffers available for holding packets. Mechanisms used to avoid or control network congestion are

- *Clear Existing User Connections:*   This, of course, is highly undesirable, but may be a viable alternative if network traffic has been categorized into different *priorities.*

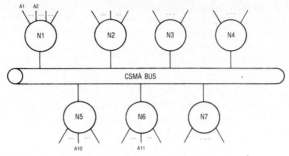

FIG. 14.8   A local area network based on CSMA.

- *Discard Data:*   If a particular switching node is congested, packets can be discarded—so long as this strategy does not violate the network integrity expected by the network user.

- *Block User Requests for Services:*   User requests for services can be rejected outright with an indication of network congestion.

- *Flow-Control Data:*   Data generated by existing connections can be flow-controlled or throttled up to the point that no new data are accepted by the network.

- *Reroute Existing Traffic:*   The detected congestion may be localized in only part of the network. Traffic can be moved from congested routes to uncongested routes. (This approach must be tempered to avoid simply moving congestion from one place in the network to another.)

- *Use Adaptive or Dynamic Routing:*   An attempt is made to "spread" traffic across the network to prevent transmission or processing bottlenecks. This concept is called *load balancing*.

- *Reserve Resources:*   Transmission or processing resources can be reserved in advance of needing them through the use of connection-oriented internal packet transmission.

13. Multiple-Access Broadcast Channels

The backbone links shown in Fig. 14.7 could be replaced using a multiple-access broadcast channel shared among all switching nodes. *Multiple access* means that multiple nodes can share a common channel in the transmit direction. *Broadcast* means that nodes share a common channel in the receive direction. The receive and transmit channels may be either the same physical channel (as is the case for a token ring local-area network), or different channels (as in a TDM/TDMA* satellite-based network).

Figure 14.8 (see above) illustrates a local-area network based on *carrier-sense multiple access* (CSMA)[10] that has the same number of nodes and access links shown earlier in Fig. 14.7. Each node is permitted to transmit packets on the common bus whenever the bus is not already in use.

---

*Time-division multiplexing/time-division multiple access.

Routing in this kind of network is simplified when compared to the mesh network discussed above. Since all nodes share the same backbone link (the CSMA bus), each node need only determine whether a received packet should be relayed to an access link or broadcast on the common backbone link. Packets are received by each node attached to the CSMA bus in one of two ways:

- *Packet Boundary Detection:* The node receives all packets transmitted on the link on the basis of its ability to identify packet boundaries. If this method is used, the contents of each packet must be inspected to determine whether to relay or discard the packet.

- *Destination Address Filtering:* The node receives only the packets that contain the unique address of that node, an address indicating a group of nodes (of which the receiving node is a member), or an address indicating all nodes.

14. Circuit Switching versus Packet Switching

One way to compare circuit and packet switching is to compare a specific circuit switch with a specific packet switch. While this approach is simplest to understand, it may be misleading in its conclusions because it is easy to confuse the embodiment of a switching technique with the essential attributes of that technique. For example, today's telephone network is frequently cited as the paradigm for circuit switching, but this network was originally designed for carrying voice traffic, not data traffic. Therefore comparing the telephone network to a public data packet-switched network is an apples versus oranges type of comparison.

Another way to make a comparison is to select a specific application and attempt to understand which switching technique handles that application the best. In order to present this type of comparison, the network application, the chosen transmission channel, and the proposed switch equipment and accompanying software must be carefully analyzed. There are many variables to consider, and not the least of these are the costs of procuring, operating, and maintaining the resulting solution—all of which, unfortunately, are beyond the scope of this chapter.

Circuit- and packet-switching techniques can be compared and differentiated, however, on the basis of three general characteristics: data transparency, bandwidth assignment, and flexibility. Tables 14.5, 14.6, and 14.7 expand upon each of these characteristics.

Data transparency can only be defined on a relative scale rather than absolutely because tolerance to data distortion is a function of the communicating entities (e.g., computer devices or human beings). Common forms of distortion are errors and delay.

Circuit and packet switching employ different methods for assigning bandwidth. Packets can be viewed as self-descriptive units of data, capable of instructing a switch to adapt to an optimum use of bandwidth for many different applications.

Flexibility may be packet switching's strongest suit, but the telecommunications industry is still learning how best to quantify this capability at the time a specific network technology is selected. For example, often a network is initially

TABLE 14.5   Circuit versus Packet Switching—Transparency of data

Circuit switching

Circuit switches may be analog or digital.

Average traffic rate is measured in terms of bits per second for the duration of a connection.

Susceptible to data distortion due to transmission or switching errors. (However, at the expense of using additional circuit capacity, *forward error correction* can be performed.)

Nontransparent features—such as speed, code, or protocol conversion, and extensive buffering or queuing—must be added to the network externally, if required.

Packet switching

Analog data must be converted into a digital format before it can be switched.

Average traffic rate measured as the product of bits per packet times packets per second for the service duration. (However, because of limitations on the lower bound of the time it takes to switch a packet, a packet switch which can switch 14×1000-byte packets/second may not be capable of handling 1400×10-byte packets/second, even though both quantities appear to be equal.)

Susceptible to data distortion due to packetizing, queuing, and switching delays.

---

justified on the basis of its suitability for a single application, but is later evaluated in terms of its adaptability to multiple applications.*

PACKET-SWITCHING PROTOCOLS

A protocol is a set of rules and formats governing communication between two or more communicating entities. Due to the many different forms of packet switching, many possible packet switching protocols can be imagined. However, defining and developing new protocols is a difficult and labor-intensive task, and the demand for standard protocols has therefore evolved much like the demand for standard programming languages.

15. Protocol Standards

The development of the *open systems interconnection* (OSI) reference model was begun in 1978 and approved as a standard in 1983. Although not a protocol itself,

---

*The characteristics of a packet-switching system are similar to those of a *message-switching system* when one message is equal to a single packet—for example, a datagram.

Message-switching systems, however, are usually associated with a variety of *store-and-forward* techniques. A complete message is received and stored by a switching node (typically on an external magnetic medium, such as a disk, rather than in the internal memory of the node) before it is forwarded to the next node or user of the system. Messages, therefore, may be very large and can remain in transit (stored) for long periods of time.

Instead of being viewed as a technology which competes with packet switching, message switching functions are increasingly being designed as value-added applications dependent on an underlying packet-switched transport. Messages are dissassembled into one or more packets, and those packets are subsequently recombined into messages.

TABLE 14.6 Circuit versus Packet Switching—Method for Assigning Bandwidth

Circuit switching

Only one call per circuit.

New calls are blocked when there are no more circuits available.

Once a call is established, it is not subject to congestion, but congestion can occur in an out-of-band signaling network during call supervision.

Because of the transparency requirement, circuit switching is connection-oriented and the data path between source and destination must be completely established before data can be exchanged.

Call setup and clearing times are typically limited by the time it takes to physically establish or release the corresponding circuit. Circuit and switch capacity reserved while establishing or clearing a connection are wasted since no useful data are exchanged during these times.

Out-of-band signaling must be used if a high degree of signaling flexibility is required, but there must exist an additional means to synchronize the signaling circuit with the data circuit.

Based on fixed-assigned multiplexing techniques such as frequency- or time-domain multiplexing—see Table 2.

Packet switching

Connection-oriented or connectionless packet transmission.

Multiple connections and packets per circuit.

Can distribute data from one connection over multiple physical circuits.

User connections (like packets) can be queued rather than blocked.

Allows a single source device to generate multiple application connections on a single circuit. (With the advent of very low cost personal computers, it is usually not feasible to provide a physical circuit for each personal computer application or "application window.")

Must strike a balance between acceptable network congestion and required data transmission throughput.

Based on dynamically assigned bandwidth techniques which characterize statistical multiplexing—see Table 1.

the reference model provides an important framework for defining protocols. An *open system* is a system that employs *OSI protocols* conforming to the model, to enable communication with other open systems.

The OSI reference model clarifies the notion that communication between systems consists of many functions which can be grouped into *layers*. The definition of an N- layer ($1 \le N \le 7$) can then be set forth based upon the "services" expected from the layer below it (the $N - 1$ layer), and the services required to be provided by the layer above it (the $N + 1$ layer). Specific protocols are defined within the constraints of a layer, and although the layering of functions in the OSI reference model is static, individual protocols can be "stacked" or threaded together in a wide variety of ways.

Many excellent articles have been written describing the reference model,[11,12] so it will not be reviewed further here, except for Fig. 14.9, which illustrates the organization of the seven functional layers.

TABLE 14.7 Circuit versus Packet Switching—Flexibility

Circuit switching

Not inherently failure resistant. Providing resistance to failures implies adding more equipment.

New call capacity is added to an existing circuit switching network by either (1) adding more physical circuit capacity or (2) creating more circuits with existing equipment by decreasing the capacity per circuit. In the first case, the result is additional equipment and cost, and in the second case the network user pays a performance penalty all the time, instead of being penalized only under peak loading.

Network management and control functions require supervisory circuits to be added on to the network.

Packet switching

Inherently failure resistant.

Network owner may choose to allow degraded network performance in times of peak network loading in order to avoid acquiring additional network equipment at an added cost.

Easier to offer dramatically different applications without changing fundamental network architecture because of the ability to tune performance parameters or add features such as protocol conversion.

Network management and control functions are more easily built into packet-switched networks rather than circuit-switched networks (see the description of ANET).

Many new protocols and products are evolving which are based on underlying packet switching technology. Therefore, in order to use these products, networks based upon packet switching will be preferred.

Standards Organizations. There are a myriad of organizations presently establishing protocol standards. Fortunately, many of these organizations have agreed

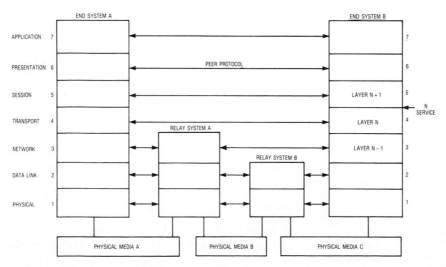

FIG. 14.9 OSI reference model for open systems.

to cooperate in "standardizing on standards." For example, the OSI reference model was first approved by the International Organization for Standardization (ISO) in a standard called ISO 7498, and later it was approved as Recommendation X.200 by the International Telegraph and Telephone Consultative Committee (CCITT) in 1984.

ISO and CCITT are international standards organizations. In 1918, the American National Standards Institute (ANSI) was founded in the United States.[13] ANSI sponsors standards committees and governs the approval process required in order to convert a committee proposal into an American National Standard (ANS). Figure 14.10 illustrates the general relationship between ANSI and other standards organizations.*

In addition to international and national standards, there are also *vendor standards*. These are standards promoted by a particular communications equipment or software supplier. Such de facto standards include IBM's System Network Architecture and Digital Equipment Corporation's DECNET, for example.

By-Products of Protocol Standardization.   Many advantageous by-products have been created as the result of the protocol standardization process:

- The vocabulary of protocols is also becoming standardized, which in turn simplifies the task of defining or understanding protocols.

- Efforts are underway to develop *protocol certification* procedures. Such procedures are useful in determining whether a particular protocol implementation meets the corresponding protocol standard.[15]

- It is difficult to define a protocol in a completely unambiguous and error-free manner while allowing different implementation approaches. Therefore, *formal description techniques* and *specification design languages* are being investigated.[16,17] (See also CCITT Recommendation X.250 and the Z.100 series.)

- Protocols need to be evaluated in light of the performance they offer when used for different applications. Recognizing this, standards are being developed to specify performance measurement parameters.[18,19]

16.   The CCITT X-Series Recommendations

Study Group VII of the CCITT is responsible for developing recommendations for data communications networks, and these are known as the X-series recommendations. A total of 72 recommendations (numbered X.1 through X.430) were approved by the CCITT Plenary Assembly and published in the so-called 1984 CCITT Red Books.

Historically, the X-series recommendations have been intended for use by public networks. A *public data network* is a network for use by the general public and may be owned or operated by a public or governmental entity in a particular country. Because of the deregulation of such networks in the United States and elsewhere, it has become economically feasible to develop, own, and operate *private data networks*. The distinction between public and private data networks has, therefore, become blurred and the CCITT is gradually addressing the applicability of its recommendations to private, as well as public, data networks (see the discussion of X.121 below).

---

*For an excellent reference concerning standards organizations and background on the OSI reference model see Ref. 14.

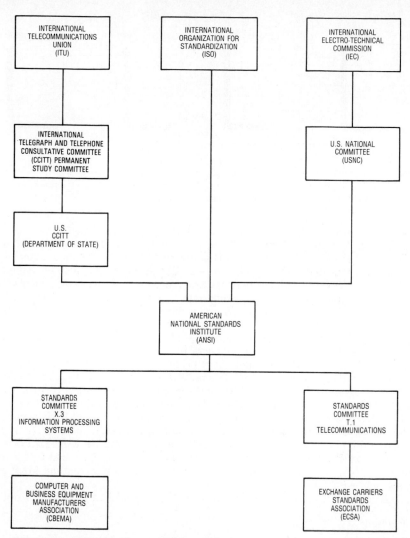

FIG. 14.10   ANSI and other standards organizations.

Recommendation X.25.   Recommendation X.25 was first approved by the CCITT in 1976 and was revised in 1980 and 1984. This recommendation describes the "interface between data terminal equipment (DTE) and data circuit-terminating equipment (DCE) for terminals operating in the packet mode on public data networks." (The word *terminal* is used to mean any communicating device, including a computer.)

As shown in Fig. 14.11, Recommendation X.25 consists of three levels: the *physical level,* the *data link level,* and the *packet level.* These three levels fit into the first three layers of the OSI reference model, shown in Fig. 14.9.

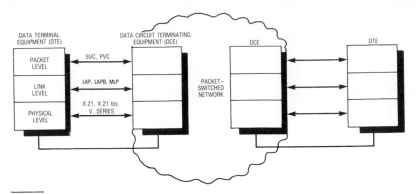

FIG. 14.11 The three levels of Recommendation X.25.

Physical Level. The physical level provides for the transparent transfer of bit streams between the DTE and DCE, using a physical medium. Rather than defining the physical level within Recommendation X.25, the recommendation cites two other applicable recommendations—X.21 and X.21 bis. Recommendation X.21 is for a digital interface, first approved in 1972, and has not gained wide acceptance in the United States at this time. Recommendation X.21 bis, on the other hand, specifies the use of V-series modems (i.e., voice-grade modems which are based on the CCITT V-series recommendations). V.24, which is compatible with the popular U.S. EIA* RS-232C standard, and V.35 for transmission speeds of 48 kb/s and above, are prevalent examples.

Data Link Level. The data link level protocol actually consists of three different connection-oriented *procedures*—the *link access procedure* (LAP), the *link access procedure balanced* (LAPB), and the *multilink procedure* (MLP). LAP and LAPB are also referred to as *single-link procedures* (SLPs). An SLP is a procedure that operates over a single physical link between DTE and DCE, while an MLP is a procedure that allows operation on multiple physical links (e.g., a transmission group as shown in Fig. 14.7.) LAP was first defined in 1976, LAPB was added in 1980, and X.25 MLP has been introduced in 1984. LAP and LAPB are mutually exclusive. LAPB is the preferred implementation today, and MLP coexists only with LAPB.

The data link level of Recommendation X.25 provides the following functions:

- Framing
- Transparency
- Synchronization
- Identification
- Initialization and termination
- Flow control
- Error control and recovery

The data link level is a full-duplex, point-to-point, bit-oriented protocol. The fundamental unit of information exchanged across a link is the *frame,* which is

---

*The Electronic Industries Association is a U.S. trade organization also involved in the development of standards for data communications.

illustrated in Fig. 14.12. The boundary of a frame is indicated with an opening and closing *flag,* and a single flag may signify the closing of one frame, as well as beginning the next.

Because of the transparency requirement, it must be possible to transmit any sequence of bits provided by the packet level occupying the information field of the frame. Since the eight-bit flag sequence serves to identify a frame, it follows that such a sequence cannot appear inside the frame. This apparent contradiction is solved through a technique called *zero-bit insertion* or *bit stuffing.* Since the flag sequence contains a 0 followed by six 1s and a 0, the transmitter inserts a 0 after five contiguous 1s appearing anywhere in the frame between two flag sequences. Similarly, the receiver discards any 0 bit that is discovered immediately following five contiguous 1 bits.

Even though the address field is large enough to support 256 addresses, it is defined to support only two (plus two additional addresses if MLP is used). Interestingly, the address does not identify the DTE or DCE. Rather, the address field is used to indicate whether the frame contains a command or response.

Specific commands and responses are found in the control field. The control field also indicates one of three types of frames—*information, supervisory,* and *unnumbered.* Information frames carry packet level data and are numbered, providing a vehicle to implement flow control and to ensure that an out-of-sequence frame can be detected. Supervisory frames are used to indicate whether a receiver is busy or not or to reject an information frame which is out of sequence. Unnumbered frames are used to establish or disconnect the link or to reject an invalid frame.

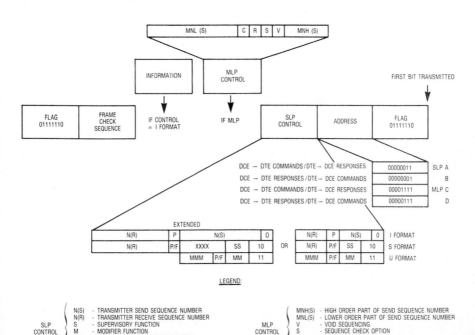

FIG. 14.12 Data link level frame format.

Flow control is accomplished by permitting the receiver to command the transmitter to stop sending frames (the *receiver-not-ready* or RNR command) or to command the transmitter to continue (by issuing the *receiver-ready* or RR command). In addition, a "windowing" scheme is used that causes the transmitter to stop sending frames as soon as K consecutive frames have been sent, but not yet acknowledged. (That is, K is the number of frames "in the window" which have been transmitted but not yet acknowledged.)

Besides rejecting invalid frames or frames containing an invalid sequence number, error control is based on an *automatic repeat request* (ARQ) technique called *go back N*. In this case, N represents the number of frames that must be retransmitted, beginning with the frame whose sequence number is one frame greater than that of the last correctly received frame. The *frame check sequence* (FCS) at the end of the frame is used to detect that a frame contains an error. If a frame is found to contain an error, it is discarded (i.e., no negative acknowledgment is sent to the transmitter), and it is therefore up to the transmitter to discover that the faulty frame must be retransmitted.

Packet Level.   The X.25 packet level is generally considered to be a user-network *access protocol,* rather than either an intranetwork *backbone protocol* or an internetwork *gateway protocol.* (However, with the advent of private packet-switched networks, X.25 is being used today in a wide variety of situations, encompassing all of these protocol environments.) In 1980, the packet level provided two services—a *datagram* service and the *virtual circuit* service. However, the datagram service was subsequently dropped in 1984.*

There are two types of virtual circuits—*switched virtual circuits* (SVCs), and *permanent virtual circuits* (PVCs). SVCs are also called *virtual calls.* As described earlier, virtual calls transition through the three call processing states of call establishment, data transfer, and call clearing. PVCs, on the other hand, do not require a call establishment or clearing state since they are permanently assigned.

In order to allow multiplexing of multiple virtual circuits on a single logical link, virtual circuits are identified with a locally significant connection identifier,† called the *logical channel identifier.* The logical channel identifier is composed of a *logical channel group number* (which is less than or equal to 15), and a *logical channel number* (which is less than or equal to 255).

Many different packet formats are defined, but every packet must contain at least three fields of data (also occupying three bytes or octets): the *general format identifier* (GFI), the *logical channel identifier,* and a *packet-type identifier.* Table 14.8 lists all the types of packets presently defined (as of 1984).

Each virtual circuit is treated independently, and therefore many of the protocol features introduced by the data link level of X.25 must be duplicated in the packet level. For example, exercising flow control at the data link level affects all virtual circuits, but at the packet level each virtual circuit can be independently flow-controlled.

Once a virtual call is established, it can be distinguished by its logical channel identifier alone. Prior to call establishment, a call-request packet contains the address of the called DTE. (The call-request packet may or may not contain the

---

*Ironically, the network layer of OSI is currently undergoing review to consider providing both a connection-oriented mode of operation (i.e., virtual circuit service) and a connectionless mode of operation (i.e., datagram service). It remains to be seen what effect, if any, this will have on X.25 in 1988.[20,21]

†That is, significant to the local DTE-DCE interface.

TABLE 14.8 X.25 Level 3 Packet Types

FROM DCE TO DTE	FROM DTE TO DCE
Call setup and clearing*	
Incoming call	Call request
Call connected	Call accepted
Clear indication	Clear request
DCE clear confirmation	DTE clear confirmation
Data and interrupt	
DCE data	DTE data
DCE interrupt	DTE interrupt
DCE interrupt confirmation	DTE interrupt confirmation
Flow control and reset	
DCE RR	DTE RR
DCE RNR	DTE RNR
	DTE REJ
Reset indication	Reset request
DCE reset confirmation	DTE reset confirmation
Restart	
Restart indication	Restart request
DCE restart confirmation	DTE restart confirmation
Diagnostic	
Diagnostic	
Registration	
	Registration request
Registration confirmation	

*Virtual Calls Only

address of the calling DTE.) There are many ways in which DTE addresses can be defined and formatted, and this is reflected in the fact that DTE addresses are preceded by a length field in the call-request packet. In order to make it possible to interconnect multiple packet-switching public data networks, addresses may conform to CCITT Recommendation X.121 (see below).

At this point in the discussion on CCITT protocols, one may pause to reflect on how many different combinations of features have already been described. However, over 30 additional packet level facilities are presently (1984) defined in Recommendation X.25, and these are listed in Table 14.9.

Although each of these facilities cannot be explained here, the *fast select* facility is particularly important. It is this facility which allows call-request and clear-indication packets to contain up to 128 bytes of user data, thus accomplishing an exchange of data similar to the previously extant X.25 datagram service.

TABLE 14.9 Optional User Facilities (Packet Level)

On-line facility registration
Extended packet sequence numbering
D bit modification
Packet retransmission
Incoming calls barred
Outgoing calls barred
One-way logical channel outgoing
One-way logical channel incoming
Nonstandard default packet sizes
Nonstandard default window sizes
Default throughput classes assignment
Flow control parameter negotiation
Throughput class negotiation
Closed user group related facilities:
Closed user group
Closed user group with outgoing access
Closed user group with incoming access
Incoming calls barred within a closed user group
    Outgoing calls barred within a closed user group
    Closed user group selection
Closed user group with outgoing access selection
Bilateral closed user group related facilities:
Bilateral closed user group
Bilateral closed user group with outgoing access
Bilateral closed user group selection
Fast select
Fast select acceptance
Reverse charging
Reverse charging acceptance
Local charging prevention
Network user identification
Charging information
RPOA (recognized private operating agency) selection
  Hunt group
Call redirection
Called line address modified notification
Transit delay selection and indication
CCITT-specified DTE facilities to support the OSI network service:
Calling address extension facility
Called address extension facility
Quality of service negotiation:
Minimum throughput class facility
End-to-end transit delay facility
Expedited data negotiation facility

Recommendation X.32. The X.25 interface described above normally assumes the existence of a dedicated physical circuit. Circuit-switched access (e.g., dial-up) had not been precluded by 1980 X.25, but such access was not well-defined. Although Recommendation T.71 defines a so-called half-duplex LAP-X, X.25 LAPB assumes full-duplex operation.

X.32 was approved by the CCITT in 1984 to provide both dial-in (DTE to DCE) and dial-out (DCE to DTE) access to X.25 users. Since a switched-access circuit may be utilized for multiple network users, X.32 addresses the problem of how to identify network users. Several possibilities have been included:

- No identification required.
- Reliance on the use of the DTE address field, which already exists as defined in X.25.
- Identification provided by the network.
- Revision of the data link level to include an *identification exchange* frame (called the XID frame).
- Use of the 1984 CCITT packet level *registration* facility. (See Table 14.9.)
- Use of the 1984 *network user identification* (NUI) facility. (See Table 14.9.)

Although X.32 has been approved in its present state, the CCITT recognizes that it is incomplete. Therefore, interim reports will be published to update X.32 until it is submitted for approval again in 1988.

The Triple-X PAD Recommendations. Figure 14.13, shows the relationship between X.3, X.28, and X.29, which make up what has been called the triple-X-series recommendations. These recommendations allow non-X.25 devices to communicate with an X.25 DCE through the operation of a *packet assembly/disassembly* (PAD) facility. A PAD interfaces asynchronous, start-stop terminals on one side to an X.25 DCE on the other side. In other words, it contains the functions of packetizing, statistical multiplexing, and (limited) packet switching as defined earlier in this chapter. It could also be called a protocol converter, but it is important to note that the term PAD is not generic as defined by the CCITT, but rather implies a device that follows the appropriate recommendations. (The CCITT has not defined facilities to allow converting, for example, from IBM's Bisync or SDLC to X.25.)

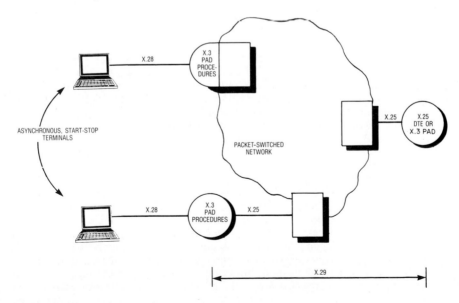

FIG. 14.13 The triple-X protocols in a network.

Recommendation X.3. Recommendation X.3 defines over 20 parameters used by a PAD to control the terminal. If there is more than one terminal connected to a PAD, each terminal has its own set of these parameters.

Two of the parameters are the *selection of data forwarding signals* and *selection of idle timer delay,* which are used in concert (along with the maximum number of user data characters in a packet) to determine when a packet will be created from the data stream (see Fig. 14.3).

Another important parameter is the *PAD recall character*. Once a terminal is in the data transfer mode, it must be possible to "escape" to a mode in which the PAD can be commanded by the terminal (e.g., to end a call); otherwise, all characters would be packetized and ignored by the PAD. The PAD recall character is an in-band signaling technique used to accomplish this mode change.

Recommendation X.28. This recommendation contains the procedures to be used for a terminal and PAD to communicate. It defines PAD *command signals* issued by the terminal to the PAD and PAD *service signals* issued by the PAD to the terminal. Examples of commands are the establishment and clearing of a virtual call, the selection of X.3 parameters, or requesting the status of a circuit. Service signals provide status information to the terminal or acknowledge PAD command signals. The character set specified by the CCITT to be used for communicating between the terminal and the PAD is called the *International Alphabet Number 5* and is defined in Recommendations V.3 and X.4. This alphabet is equivalent to the ASCII character code set.

Recommendation X.29. Recommendation X.29 specifies procedures to exchange control information between one PAD and another PAD, or between a PAD and a packet-mode DTE, as shown in Fig. 14.13. Use of X.29 is indicated by filling in the call-user-data field of an X.25 call-request packet with a special code for protocol identification. In addition, PAD control messages are subsequently provided in X.25 data packets by setting a bit called the *qualified data bit* (Q bit) to 1. Data intended for the terminal is indicated by setting the Q bit to 0.

Recommendations X.75 and X.121. Recommendation X.75 was created to allow two packet-switched public data networks to communicate. The interface at each network is referred to as the *signaling terminal* or STE.

X.75 is very similar to the three levels of protocols already described for X.25, and in 1984 the CCITT decreased the number of differences between X.25 and X.75. For example, in 1980 only the SVC was defined for X.75, but in 1984 the PVC capability was added. Similarly, the *multilink procedure* (MLP), which is now contained in the 1984 X.25, was first defined only for X.75 in 1980, since the desirability of interconnecting networks with multiple transmission links was easily recognized.

As in the case of the worldwide telephone network, an adequate addressing or numbering plan must be established for public data networks in order to recognize the location of the called party. Such a plan is put forth in Recommendation X.121. This Recommendation defines a subscriber address which is a maximum of 14 decimal digits in length. Countries, so-called world zones, escape codes to indicate telex or telephone network interworking, and nonzoned services such as maritime mobile services are indicated in the first three digits, collectively known as the *data country code* (DCC). The subsequent 11 digits may be formatted in a variety of ways, but the format used cannot be determined from the number itself. The fourth digit may be used, for example, to signify a particular network in a given country. Then the first four digits are called a *data network identification code* (DNIC). Since the X.121 addressing scheme was originally intended to be used only by public data networks, there was no allowance for private data network addresses. This problem has been tackled in the 1984 recommendation, which provides for a *private network identification code* (PNIC), containing up to six digits following the DNIC.

17. ISDN Signaling Protocols

The integrated services digital network is a networking concept providing for the integration of voice, video, and data services using digital transmission media and combining both circuit and packet switching techniques. Recommendations concerning ISDN are coordinated by Study Group XVIII and are published in the *I-series recommendations*.

The I-series recommendations define two types of user-network digital channels—those used for carrying user data, and those used to carry signaling information associated with data channels. That is, all information having to do with circuit-switched signaling or circuit management and administration is carried in an *out-of-band* signaling channel, rather than in the data channel (*in-band*) itself. However, the use of X.25 can be accommodated in either type of channel. Data channels are

- *B Channel:* 64 kbs
- *H0 Channel:* 384 kbs
- *H1 Channel:* Either 1536 kbs (H11) or 1920 kbs (H12)

And associated signaling channels are

- *D Channel:* Either 16 kbs or 64 kbs
- *E Channel:* 64 kbs

Channels can be combined in different configurations called *interface structures*. Typical examples are one D channel, one B channel and one D channel together (B + D), two B channels and one (16 kbs) D channel together (2B + D), or 23 B channels and one (64 kbs) D channel together (23B + D). In some cases—for example with H channels—signaling information may be provided either in the same interface structure containing data channels, or in a different interface structure.

The CCITT has defined two *signaling protocols:* the ISDN user-network protocol and signaling system no. 7. Although these protocols provide a form of data communications, they are not intended to be general-purpose packet switching protocols in the sense of X.25, but rather are targeted specifically at providing out-of-band supervisory information concerning the data channels defined above. ISDN signaling protocols, therefore, address issues which are not presently considered in the OSI reference model (X.200) and, although these protocols are based on the principles of the reference model, their relationship to it has raised many, as yet unanswered, questions.

The ISDN User-Network Protocol. Different layers of this protocol are not contained in a single CCITT recommendation. Layer 1 is described in Recommendations I.430 and I.431, which define an ISDN-specific interface. An alternate V-series interface (providing an evolutionary means to connect to an ISDN) is also defined in Recommendation I.463/V.110.* Layer 2 is described in I.440/Q.920 and I.441/Q.921, and layer 3 is described in I.450/Q.930 and I.451/Q.930.

Layer 2 is called the *link access procedure on the D channel* (LAPD). Both the frame structure and the terminology of LAPD are similar to that of X.25 level 2. However, there are also many interesting differences between LAPD and X.25 LAPB.

LAPD, for example, does not support the multilink procedure supported by X.25/X.75. However, LAPD can be operated in either a point-to-point or a point-to-multipoint configuration. To support multipoint operation, multiple LAPD connections may coexist on the same physical circuit. This is equivalent to providing multiple SLPs which share a common link—and, in fact, provisions have even been made to allow LAPB and LAPD to coexist on the same physical link. In addition, LAPD provides a packet broadcast capability so that a single data link frame can be received by multiple data link entities attached to the link.

A data link connection is established as soon as the user side of the protocol has been assigned a so-called *terminal endpoint identifier* (TEI). After the connection establishment, three different modes of operation are possible:

1. *Unacknowledged Operation:* Layer 3 information is transmitted in *unnumbered information* (UI) frames.

2. *Acknowledged—single-frame operation:* Layer 3 information is sent in either of two newly defined frames: *sequenced information 0* (SI0) or *sequenced information 1* (SI1). Only one unacknowledged frame may be outstanding at a time.

3. *Acknowledged—multiple-frame operation:* This mode of operation is the same as X.25 LAPB, providing both modulo 8 and modulo 128 information-frame sequencing.

Layer 3 of the ISDN user-network protocol is for use on both D and E channels.† Layer 3 procedures are quite different from X.25 level 3, since the intent is to provide for the control of circuit-switched connections, user-to-user signaling connections, and possible packet-switched connections. Thirty different call-control messages are defined for circuit-mode connections, which are generally exchanged in a sequence reflecting the progress or state of the call being established on the data channel.

Signaling System No. 7.   The description of signaling system no. 7 (SS7) is contained in the Q.700-series recommendations. SS7 is intended for use in *common channel signaling* (CCS) applications—that is, call-control and circuit-management applications similar to those conveyed over a D channel, but typically provided between large digital (telephone) exchanges, or between PBXs and a local exchange, using 64 kbs digital circuits.

Although SS7 is an integral part of the ISDN concept, its acceptance does not necessarily depend upon the future success of ISDN. Rather, SS7 should also be viewed as the successor to previous CCS protocols—for example, CCITT no. 6

---

*Since ISDN consists of combined multiple interfaces and protocols, some of which are contained in other CCITT recommendations provided by different study groups, many ISDN recommendations are identified with two numbers—an I-series number and the number assigned in a different series.

†E channels, however, use signaling system no. 7 for lower layers.

recommended in 1972*—and as the vehicle necessary to provide new telephone and data services. (In the United States, credit-card calling and advanced 800-number calling are examples of such services.)

SS7 recommendations not only describe the details of an SS7 protocol, but also define a *signaling network structure,* including the identification of physical network components and (unlike X.25) specific network functions such as routing, packet relay (called *message discrimination and distribution*), and even network and traffic management. Since SS7 is to be used in interconnecting international, as well as national, networks, the real-time performance of the SS7 network is critical to the usefulness of the protocols implemented. SS7 recommendations therefore specify a complete packet-switched network consisting of *signaling points* (SP)—i.e., access nodes—and *signaling transfer points* (STP)—i.e., transit nodes.

As presently defined, the SS7 protocol consists of the following parts:

- *Message Transfer Part (MTP):*   Recommendations Q.701–Q.709
- *PABX Applications:*   Recommendation Q.710
- *Signaling Connection Control Part (SCCP):*   Recommendations Q.711–Q.714
- *Telephone User Part (TUP):*   Recommendations Q.721–Q.725
- *Data User Part (DUP):*   Recommendation Q.741/X.61
- *ISDN User Part (ISDN UP):*   Recommendations Q.761–Q.766
- *Operations and Maintenance Application Part (OMAP):*   Recommendation Q.795

In each part, performance objectives are also set forth. In addition, Recommendation Q.791 itemizes measurements which can be taken in monitoring the performance of the MTP.

Although X.200 was obviously used in preparing the recommendations concerning LAPD, the description of SS7 more closely follows the terminology of previous CCITT signaling protocols. For example, layer diagrams are represented in a horizontal graphic, rather than vertically, and layer 2 frames, called *signal units,* contain no address field.

Nevertheless, it is the message transfer part which contains the functions associated with OSI layers 1, 2, and a portion of layer 3. Higher-level layer 3 functions are provided in the *signaling connection control part,* containing procedures for connectionless as well as connection-oriented modes of transmission. MTP and SCCP together are, therefore, called the *network service part* of SS7.

The various user parts and OMAP make up level 4 of SS7 and are oriented to specific control procedures, as their names imply. It is not clear at this time how SS7's level 4 maps into higher-level OSI layers.

THE ARCHITECTURE OF A PACKET-SWITCHING SYSTEM

Up to now, packet-switching network capabilities have been presented from the viewpoint of providing data transmission functions only. However, the design and

---

*Note that SS7 is not assigned a corresponding I-series number.

consequent operation of a real-world packet switching network is significantly affected by requirements to manage and control it as well. This section extends the concepts developed earlier by examining a model network, ANET, which is based on actual hardware and software.

18.   The ANET Network

The ANET network is illustrated in Fig. 14.14, and consists of the following *network components:*

- *Ports:*   The physical inputs and outputs characteristic of network equipment.
- *Links:*   The physical transmission circuits used to interconnect network equipment ports. The circuits may be either satellite or terrestrial, and either dedicated or dynamic.
- *DTEs:*   X.25 DTEs are users of ports. A port can have more than one DTE, and one DTE can share more than one port.
- *Network Users:*   Network users in turn are users of DTEs. A network user may be equivalent to a DTE or associated with one or more DTEs.
- *PADs:*   CCITT packet assemblers/disassemblers may be either stand-alone or integrated into a node as shown in Fig. 14.13.

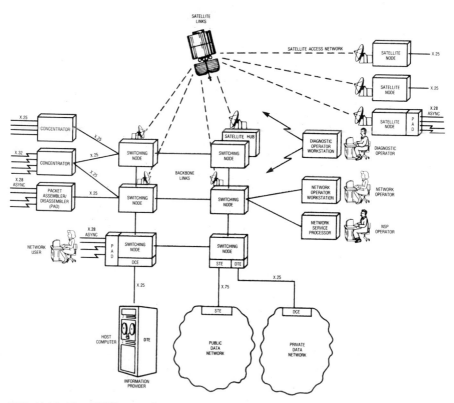

FIG. 14.14   The ANET network.

- *Concentrators:* Devices which concentrate multiple X.25 links into fewer X.25 links.*
- *Switching Nodes:* Devices which have the ability to relay packets to a terrestrial or satellite backbone link.
- *Satellite Nodes:* Devices which are capable of relaying packets via a *satellite access network.*
- *Satellite Hub:* The equipment that resides at the center of the star-based satellite network.
- *Network Service Processors (NSP):* Computers which provide services to the network and network operators.
- *Auxiliary Service Processors (ASP):* Data base processors which are adjuncts to switching nodes, capable of performing a subset of network services.
- *Public Data Network Gateways:* X.75 ports in the network, capable of being linked to a public data network.
- *Private Data Network Gateways:* X.25 ports in the network which can be configured to be linked to either a private data network or a public data network.
- *Network Operator Workstations (NOW):* Workstations which provide the human interface for network operators.
- *Diagnostic Operator Workstations (DOW):* Workstations which provide the human interface for diagnostic operators.
- *Network Operators:* Operators who can manage, control, or monitor the network via NOWs.
- *Diagnostic Operators:* Operators who can perform special diagnostic functions via DOWs.
- *Software Packages:* Includes both the executable software programs and configuration data which are loaded into the network.

ANET includes both terrestrial and satellite circuits in a single system. Satellite circuits, however, are used for either of two purposes: (1) to provide high-speed, dedicated backbone links interconnecting switching nodes or (2) to provide multiple-access channels, configurable into low-cost *subnetworks,* composed of *very small aperture terminals* (VSAT) or satellite nodes.[22]

ANET combines circuit and packet switching by accommodating dynamic (*dial-up*) as well as dedicated (*leased*) circuits, and both types of circuits may be either satellite-based or terrestrial.

Since satellite circuits impose a transmission delay of approximately 270 ms, packet switching protocols are adapted to account for this delay by achieving the proper balance of packet size, error control, and flow control.[23] In addition, the routing algorithm recognizes that delay-sensitive traffic can not be routed over satellite backbone links.

Routing in ANET uses connection-oriented protocols internally, as described earlier in this chapter, and routes are calculated from many different parameters, including class of service, monitored delay and congestion, and transmission link type. The reconnect feature mentioned previously provides the capability to reroute around failed equipment without dropping user calls.

Certain devices, such as PADs, can be configured in a stand-alone or inte-

---

*Such devices have also been called *packet multiplexers.*

grated arrangement; providing the flexibility required to make the best possible economic tradeoffs.

The gateway to public data networks is based on Recommendation X.75. In order to interconnect ANET with a private data network (or a public data network not supporting X.75), an X.25 gateway is provided. The gateway is configurable to operate in either a DTE or DCE mode, and correlates packet addressing, formatting, and optional facilities between the two networks.

Network equipment consists of *packet switching equipment* to provide data transmission functions and *network service equipment* necessary to provide complementary network server and management functions.

19. Packet-Switching Equipment

Packet-switching equipment (PSE) is based on a common hardware and software architecture using microprocessors. The underlying hardware technology is designed either for high performance (i.e., high throughput and high availability) or for low cost (i.e., equipment with fewer ports and features). This design approach allows PSE to be constructed in a modular, building-block fashion while also permitting a network designer to select an equipment configuration with an optimum price/performance ratio. Thus, for example, X.25 PADs or concentrators may be constructed out of either technology.

The switching node shown in Fig. 14.14 is typically constructed from high-performance technology components and consists of three building blocks—*intelligent modules, I/O modules,* and *buses.* Since low-cost technology components used in the satellite node, the PAD, or the concentrator generally consist of collapsed or repackaged versions of these blocks, they will not be described further, except by implication.

Intelligent Modules. Modules are called intelligent if they contain the microprocessors, memory, and associated circuitry necessary to be programmed with software. There are three types of intelligent modules: the *line interface module* (LIM), the *peripheral interface module* (PIM), and the *cluster interface module* (CIM). Features of these modules are shown in Fig. 14.15.

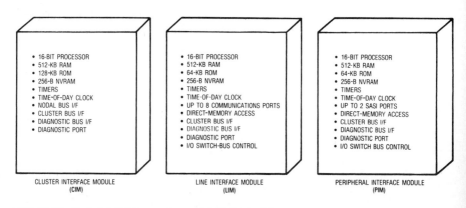

FIG. 14.15   Packet-switching–equipment-intelligent modules.

The LIM contains up to eight ports which may be connected to different kinds of communications links.

The PIM contains one or two ports and may be connected to different kinds of peripherals, such as disks or magnetic tape units.

The CIM is used to interconnect LIMs and PIMs into a configuration called a *cluster.* The CIM also provides an interface used to communicate with other clusters.

I/O Modules.   I/O modules are used to convert port signals to the physical interface required by either communications links or peripherals. Three examples of I/O modules are the RS-232C module (eight ports), the RS-449 module (four ports), and the V.35 module (two ports). The peripheral I/O module is used to attach disks or tape units to a PIM. Finally, a LAN I/O module may be attached to a CIM to provide access to a local-area network based on CSMA or to another switching node.

Buses.   A switching node contains up to four different electronic buses: the *cluster bus,* the *nodal bus,* the *diagnostic bus,* and the *I/O switch bus.*

Up to eight intelligent modules can communicate with one another using the cluster bus. Figure 14.16 illustrates a cluster containing one CIM and three LIMs with different I/O modules.

The redundant nodal bus is used to interconnect multiple clusters to form a switching node. Figure 14.17 shows a node with eight clusters. Unlike the cluster bus, the nodal bus allows clusters to communicate using packets rather than sharable memory. In other words, a switching node can be viewed as a LAN subnetwork of independent clusters.*

As shown in the diagram, each intelligent module supports a single diagnostic port, normally used in software development as a debugging or diagnostic aid. Once a switching node is fully configured for deployment, however, all diagnostic ports can be daisy-chained together via the diagnostic bus. The diagnostic bus therefore allows access to any intelligent module through a single diagnostic port.

Figure 14.17 illustrates a cluster supporting 24 communications links. If that cluster fails, the costly transmission links connected to it are rendered unusable. This situation is typically overcome by inserting a manual patch panel or a *port matrix switch* between the links and the PSE. When a failure is detected, an operator causes the links to be switched to alternate or standby equipment. The built-in I/O switch bus is provided in order to avoid the cost of the inserted equipment. This bus allows one or more active clusters to be configured with a standby cluster. A failure in an active cluster is detected by the standby cluster, and the physical transmission links are automatically switched to it for continued operation.

20. Network Service Equipment

Network service equipment (NSE) provides added capabilities to access large data bases in real time, to perform non-real-time tasks, or to perform centralized high-speed processing, as required. In ANET, there are two types of NSE, called *network service processors* or (NSPs) and *auxiliary service processors* (ASPs).†

---

*In fact, clusters communicate using protocols based upon IEEE 802 standards. IEEE 802 is an ANSI-recognized standards committee developing LAN protocol standards.

†ASPs are normally integrated into switching nodes, so they are not shown in Fig. 14.14.

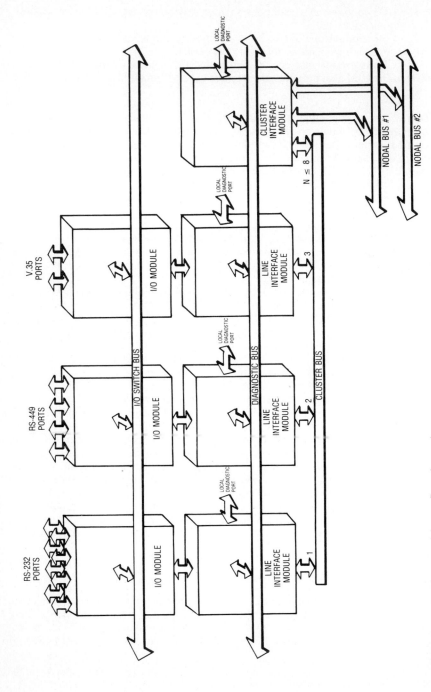

FIG. 14.16 Intelligent modules configured as a cluster.

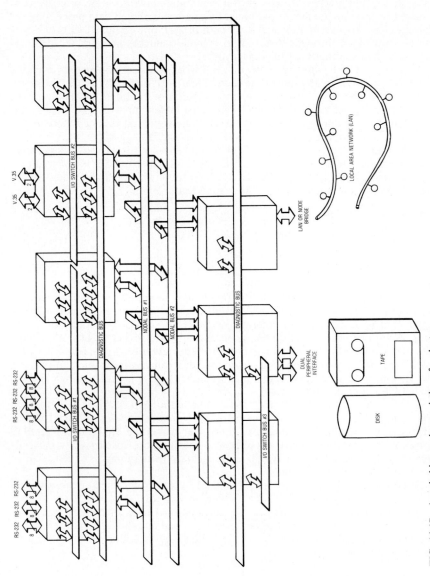

FIG. 14.17 A switching node consisting of a cluster.

An ASP is a cluster consisting of a CIM and one or more PIMs, which in turn provides disk or tape storage for data base functions. NSPs, on the other hand, consist of mainframes or minicomputers and associated peripherals, such as tape units, disks, and printers. Just as packet switching networks were originally envisioned to provide the means to efficiently share distributed data processing resources (e.g., host computers and their data bases), one or more NSPs or ASPs can be distributed throughout ANET to provide valuable network server functions in a load-sharing, fault-tolerant manner.

In some packet-switched networks, network service equipment provides server functions such as route calculation. That is, periodically, traffic flows are monitored by the NSE, and route directory tables are created for each node in the network in a centralized fashion. Directory tables must then be loaded into each node for use during call setup. In ANET, however, all routing functions are adaptive and distributed, so that no NSE assistance is necessary.

ASPs and NSPs are used in ANET for two other purposes: (1) network management and control functions (described below) and (2) extended call setup features.

Extended call setup features are optional features which may be requested by a network user during call setup. Three such features are

- Address translation
- Network user verification
- Network access restrictions

Generally, these features require access to extensive data bases available in the NSE.

In many packet switching networks, the calling or called DTE addresses contained in the X.25 call request packet are *physical addresses*. A physical address is one that is based upon the internal hardware or software structure of a network. Such addresses are required in order to route packets because they indicate the location of the network component in the network—i.e., a particular port associated with a node. However, physical addressing is inflexible and may be cumbersome, since addresses are not assigned with any particular significance to the network user or administrator.

A more flexible scheme of addressing is called *logical addressing*.* A variation on logical addressing is an optional *abbreviated addressing facility*,† where DTEs can be identified by an easy-to-remember name like MAIL or MYHOST. If logical or abbreviated addresses appear in a call request packet, they must ultimately be translated into a physical address, and this capability is called *address translation* or *address resolution*.

Network user verification is accomplished by checking whether the network user is allowed to access ANET via a given access port.

Network access restrictions provide an additional level of network security. Depending on the call request packet, various combinations of requested facilities, the calling party's address, and the called party's address are analyzed to decide whether a call is valid.

21. Network Management and Control Functions

As packet-switched networks become larger and larger or more and more geographically distributed, the need for *network management and control* (NMC)

---

*Note that X.25 DTE addresses may represent either physical or logical addresses—a decision made by the network designer or administration—and still generally conform to Recommendation X.121.

†This facility is presently "for further study" in the 1984 X.25 Recommendation.

functions increases dramatically.*

Packet switching, as opposed to circuit switching, provides the opportunity to fully integrate NMC functions in the network equipment by utilizing the switching functions inherently provided. In ANET, this opportunity is realized by associating a so-called *NMC-DTE* with every intelligent module in the network. The NMC-DTE is a software program which can initiate or receive X.25 calls like any other network DTE. Because of the distributed nature of these DTEs, the collection of them together is called the *NMC supervisory network*. The NMC supervisory network (which includes transport and higher-layer protocols—see Fig. 14.9) may be thought of as an overlay network internal to ANET, and is the source of many of the network management and control functions which ANET exhibits.

Network management and control functions can be categorized as follows:

- Human interfaces
- Equipment and software configuration management
- Network data collection and analysis
- Equipment and administrative control functions
- Test, measurement, and diagnostic functions

Each of these functions is described below.

Human Interfaces. In many respects, the network management and control features offered in a network are only as good as the human interface provided to access those features. In the case of ANET, human interfaces are based on easy-to-use programmable workstations. Two types of workstations are provided, depending upon the function to be performed—the *network operator's workstation* (NOW) and the *diagnostic operator's workstation* (DOW).

The NOW is not a traditional console connected to a host computer, but rather many NOWs may be located throughout the network by connecting to any available X.25 port in ANET. The NOW is menu-driven, and typical menus are shown in Fig. 14.18. An operator skips from menu to menu using one of four methods—pressing a function key, entering the abbreviated name of a menu, using a "help" menu tree, or accessing an operator procedure consisting of a "canned" sequence of menus.

An operator who uses a NOW is called a *network operator,* and many different types of network operators are possible. For example, some operators may be responsible for entering the network configuration, others may be responsible for monitoring network alarms, and still others may be responsible for extracting user billing data. Since the type of operator is associated with that operator's job responsibility, defining types of operators is the prerogative of the network operators themselves. This is accomplished in ANET by restricting access to specified menus and function keys based on up to 16 different operator-definable operator types.

The *diagnostic operator's* workstation is based upon a low-cost, portable personal computer, and allows a diagnostic operator to access any node in ANET via the diagnostic port and diagnostic bus which were described earlier. The DOW serves two purposes

---

*In some texts, the term *network management and control* is used to mean management or control of transmission facilities only. The term NMC is used here in a broad sense to mean all of the functionsrequired in order to operate or maintain a packet-switched network.

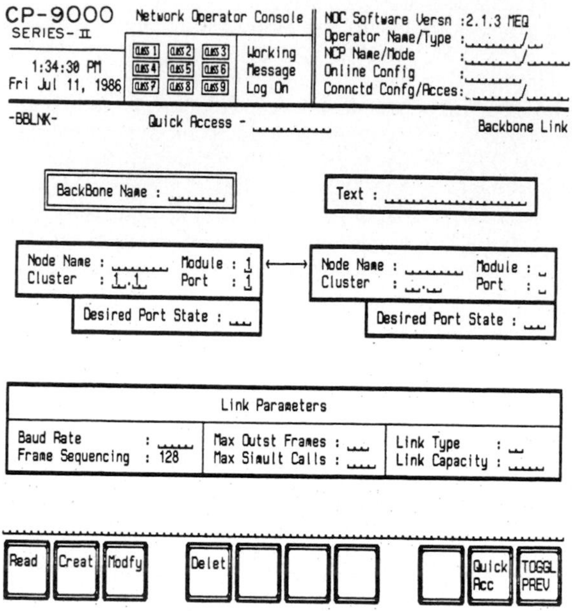

CP-9000 SERIES-II Network Operator Console NOC Software Versn :2.1.3 MEG
Operator Name/Type :
1:34:30 PM NCP Name/Node :
Fri Jul 11, 1986 Online Config :
Connctd Confg/Acces:

CLASS 1 CLASS 2 CLASS 3 Working
CLASS 4 CLASS 5 CLASS 6 Message
CLASS 7 CLASS 8 CLASS 9 Log On

-BBLNK- Quick Access - Backbone Link

BackBone Name : Text :

Node Name : Module : 1
Cluster : 1.1 Port : 1
Desired Port State :

Node Name : Module :
Cluster : Port :
Desired Port State :

Link Parameters

Baud Rate : Max Outst Frames : Link Type :
Frame Sequencing : 128 Max Simult Calls : Link Capacity :

Read Creat Modfy Delet Quick Acc TOGGL PREV

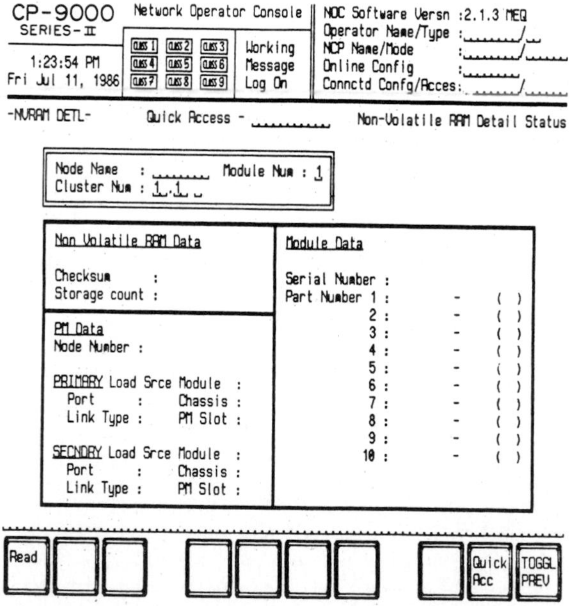

CP-9000 SERIES-II Network Operator Console NOC Software Versn :2.1.3 MEG
Operator Name/Type :
1:23:54 PM NCP Name/Node :
Fri Jul 11, 1986 Online Config :
Connctd Confg/Acces:

CLASS 1 CLASS 2 CLASS 3 Working
CLASS 4 CLASS 5 CLASS 6 Message
CLASS 7 CLASS 8 CLASS 9 Log On

-NVRAM DETL- Quick Access - Non-Volatile RAM Detail Status

Node Name : Module Num : 1
Cluster Num : 1.1

Non Volatile RAM Data Module Data

Checksum : Serial Number :
Storage count : Part Number 1 : - ( )
 2 : - ( )
PM Data 3 : - ( )
Node Number : 4 : - ( )
 5 : - ( )
PRIMARY Load Srce Module : 6 : - ( )
Port : Chassis : 7 : - ( )
Link Type : PM Slot : 8 : - ( )
 9 : - ( )
SECNDRY Load Srce Module : 10 : - ( )
Port : Chassis :
Link Type : PM Slot :

Read Quick Acc TOGGL PREV

FIG. 14.18 Menus on the network operator's workstation.

1. It may be carried into the field as an aid to install or locally troubleshoot packet-switching equipment.

2. It may be operated from a central location, providing remote diagnostic functions.

Under normal circumstances, a network operator can access any switching node by simply calling the NMC-DTE via the NOW. However, should a node become unreachable because of transmission link failures, it can be accessed via the DOW using a terrestrial dial-up circuit connected via an autoanswer modem to its diagnostic port.

Equipment and Software Configuration Management. The entire configuration of ANET, including all equipment and software, is stored in a *relational data base management system* resident on network service processors.

For intelligent modules, configuration data consist of both executable software packages and required configuration tables. Once a complete configuration has been entered into the NSP via the NOW or loaded from a tape, it can be *down-line–loaded* into all of the intelligent modules in the network.

Although there can only be one on-line or master configuration, an operator may create many different off-line configurations. An off-line configuration can be changed to an on-line configuration by entering a single command at the NOW, and in that case, the original on-line configuration is saved as the *fallback* configuration. Furthermore, the current on-line configuration can be modified dynamically at any time, and changes are automatically down-line–loaded into the intelligent modules affected by the changes.

Configuration data, which is specific to a particular node, can also be down-line–loaded to auxiliary service processors. An ASP can then act as a *down-line load server* for that node. In this manner, the distribution of configuration data is not dependent on a single NSE component, nor are transmission links overburdened by carrying internal network traffic instead of network user data.

Network Data Collection and Analysis. Because of the distributed intelligence in ANET, a great deal of data can be collected, stored, and later analyzed. Such data can be categorized into:

- Status data
- Event data
- Statistical data
- Billing data
- Measurement data

Status data consists of information about the present condition of all of the network components listed earlier. Status data is collected automatically by the NSE, but can also be requested by entering a NOW command. Status data is used to determine which transmission links have failed, which are operating normally, or which are experiencing a poor BER, for example.

When something happens in ANET of concern to a network operator, an *event* is generated by the network component that detects it. An event is a message that is sent from anywhere in the network to one or more NSPs. Events are categorized by *type, severity,* and *class.* The type of event is identified by a unique number. The severity of an event is assigned by NOW operators to indicate the significance of that event to different operator types. Some events may be considered to be major or minor alarms, while others may be noteworthy for record-keeping purposes only.

Events that require an action to be taken must be explicity "cleared" or acknowledged by operators. Classes of events are also assigned by operators, and usually indicate the general cause of the event—e.g., switching node, transmission equipment, etc. As events are received by an NSP they are stored on disk and printed on an event printer. Using the relational data base, operators may also view specific events using operator-defined query commands. Finally, operators can annotate events or create events themselves for use in communicating with other operators.

Statistical data are data that are accumulated over a dynamically configurable time period. When the time period ends, the data are sent to an NSP for storage and a new collection period is begun. Representative statistical data consist of the number of calls which have been completed, the number of X.25 level 2 transmission errors which have occurred, and the total number of packets sent on a link per unit of time.

Billing data include the information required to track or bill network users. For each network call—including NMC-DTE supervisory network calls—a record is sent to the NSP which contains the number of packets sent or received, the duration of the call, the identities of the calling and called parties, the reason the call was cleared, the route which was used during the call, and so on.

From time to time, it is important to conduct special tests to measure specific characteristics of ANET. Such data, called measurement data, are similar to normal statistical data, except that they are provided only as the result of executing special network tests.

All of the data described above can be dumped to a magnetic tape for later analysis. A typical analysis program is executed in non-real-time to provide histograms of network activity over a particular analysis time interval, such as a 24-hour period, a week, a month, etc.

Equipment and Administrative Control Functions. Components in ANET may be in one of several states:

- The component is unknown or *undefined*
- The component is *out of service*
- The component is currently undergoing *maintenance*
- The component is presently *in service*

Control functions provide the ability for a NOW operator to command a network component to change from one state to another. For example, a specific port may be removed from service by putting it into the out-of-service state. In addition, intelligent modules can be restarted, forcing them to automatically request a new down-line load software image from the NSE server.

Test, Maintenance, and Diagnostic Functions. These remaining functions consist of the actions taken by a network or diagnostic operator under unusual circumstances or during times new equipment or software is being installed or tested.

Examples of these functions are the ability to *up-line–dump* the contents of memory belonging to any intelligent module, the ability to force a physical or virtual circuit loop-back, or to create a software "patch" to fix a problem temporarily. Because ANET uses source path routing, network operators can also force test calls to use a specific route or physical data path* to perform data transmission diagnostics.

22. Ten Steps to Set up an X.25 Call through ANET

An examination of the major steps to successfully establish an X.25 virtual call through ANET provides a convenient vehicle to summarize the packet switching concepts described throughout this chapter:

1. A PAD or X.25 DTE generates a call request packet which contains a locally assigned logical channel number, the called DTE X.121 address, the required ANET class of service, and optional X.25 facilities as desired. The call request packet (see Table 14.8) is then transmitted to the appropriate edge switching node. (To get to an edge node, the packet may have been sent directly via a leased or dial-up telephone line, via a concentrator, or via one of the VSATs comprising the satellite access network.)

2. Upon receiving the call request packet, the source edge node inspects, analyzes, and validates the packet. If the called address is a physical address, it is converted directly into an *internal network address* (INA), and call setup proceeds to step 4. If it is a logical address or if the abbreviated addressing facility has been indicated, the NMC-DTE in the source cluster (resident in the CIM) sends an address translation request to one of its assigned NSE servers using the preestablished supervisory network. (Should no response be received from the first NSE server, the address translation request is sent again to another NSE server in the network, if any exists.)

3. The address translation request is received by the responsible NSE, analyzed, and validated (possibly according to the access restrictions described earlier). The logical DTE or abbreviated address is then resolved to an INA, and an address translation response packet is returned to the source NMC-DTE in the source cluster.

4. The source cluster inspects its route configuration table (continuously updated to contain the latest status on network interconnectivity and utilization), and a route is dynamically calculated on the basis of the INA and other information contained in the original X.25 call request packet. A *backbone connection request packet* is then constructed, containing the selected (explicit) route and the X.25 call request packet, and sent to the destination edge node.

5. As each transit node receives the backbone connection request, a backbone connection identifier is assigned, appropriate resources (connection control records and buffers) are allocated, and the packet is relayed according to the explicit route provided by the source node.

6. The destination cluster receiving the backbone connection request further analyzes and validates the packet, and if the packet is valid, the X.25 call request packet is extracted and relayed to the X.25 access link associated with the called DTE.

7. The resulting incoming call packet is received by the DTE, and, if it is acceptable, a DTE-DCE call-accepted packet (see Table 14.8) is returned to the destination edge node.

---

*This feature is called *dictated routing*.

8. When the destination edge node receives the call-accepted packet it forms a *backbone-connection-accepted packet,* enveloping the packet from the DTE. The backbone-connection-accepted packet follows the same route originally taken by the backbone connection request packet, in the reverse direction.

9. When the backbone-connection-accepted packet is received by the source cluster, the backbone connection is established. The X.25 call-connected packet is extracted and sent to the source DTE completing the X.25 user call establishment, and allowing data transfer to proceed.

10. At the end of the call (and periodically), an X.25 *call record* containing the billing and statistical data described above is sent via the supervisory network from the NMC-DTE in the original source cluster to one or more NSE servers for storage.

FUTURE TRENDS IN PACKET-SWITCHING TECHNOLOGY

23. General Trends

Packet-switching technology is relatively new, and developments are taking place which will significantly increase its usage for applications previously dominated by circuit switching. These developments include:

• Continued decline in hardware processing costs
• Increasing use of all-digital transmission facilities. (Eliminating both the need to convert from analog to digital formats at the switch and the need for analog transmission through the switch.)
• Availability of higher-speed, lower-cost, and lower-error-rate transmission media based on fiber optics, satellites, and T-carrier facilities
• Advances in digital compression and *digital signal processing* (DSP) techniques.
• Innovations and refinements of protocols
• Migration of standardized software functions into hardware VLSI* components.

In response to these developments, new switching architectures are being designed.[24] Hybrid circuit-packet-switched networks, for example, can be developed in which voice signals are circuit-switched and data signals are packet-switched.

24. Fast Packet Switching

A new switching technique, called *fast packet switching,* has been proposed to supplant circuit switching by creating a packet-switched data transport capable of carrying voice, video, and data simultaneously.[1] Although based on the concepts described in this chapter, fast packet switches are further characterized by:

---

*Very Large Scale Integration

- The ability to handle throughputs in excess of 1 Gbs, resulting in switching speeds of kilo- or mega-packets per second[26]
- Architectures which minimize packet, queuing, and switching delay
- Simplified packet switching protocols, when desired, which take advantage of the increasing availability of high-speed, low-error-rate transmission media
- Protocols especially adapted to handle delay-sensitive signals such as voice
- Hardware-assisted packet inspection and relay

ACKNOWLEDGMENTS

Thanks are due to Tony Heatwole, Bill Witowsky, and Stan Kay for their careful reading of the manuscript and subsequent suggestions; to Tom Hsu, Mike Skeith, Ashok Mehta, Ranjan Pant, and Pradeep Kaul for many useful discussions; and to John Stodola, Rick Gardner, Sylvia Simpson, Dan Wendling, and Susan Hitz for their help in preparing the manuscript.

TRADEMARKS

- SNA is a registered trademark of International Business Machines, Incorporated.
- DECNET is a registered trademark of Digital Equipment Corporation.
- Ethernet is a registered trademark of the XEROX Corporation.

REFERENCES

25. Cited References

1. Turner, J. S., "Design of an Integrated Services Packet Network," *Proceedings of the Ninth Data Communications Symposium*, **15**(4):124–134, September 1985.
2. Roberts, L. G., "The Evolution of Packet Switching," *Proceedings of the IEEE*, **66**(11), November, 1978.
3. Sun, M., P. Kaul, and S. Revkin, "The Middle Ground between Public and Private Networks," *Data Communications*, **14**(8):115–119, July 1985.
4. Castiel, D., "Hybrid Networks Hold Great Promise if Designed with the Proper Tools," *Data Communications*, **14**(8):121–126, July 1985.
5. Herman, J. G., and J. M. McQuillan, "How to Expand and Modernize a Global Network," *Data Communications*, **14**(8):171–190, December, 1985.
6. Heatwole, A., and P. Kaul, "The CP9000—A Multiprocessor for Data Communications," *Proceedings of the National Electronics Conference*, pp. 413–417, October 1979.
7. Scholl, T. H., "The New Breed—Switching Muxes," *Data Communications*, **10**(6):99–105, June 1981.

8. Kleinrock, L., "Principles and Lessons in Packet Communications," *Proceedings of the IEEE*, **66**(11), November 1978.

9. Chapin, A. L., "Connections and Connectionless Data Transmission," *Proceedings of the IEEE*, **71**(12):1365–1371, December 1983.

10. Metcalfe, R., and D. Boggs, "Ethernet: Distributed Packet Switching for Local Computer Networks," *Communications of the ACM*, **19**(7):395–404, July 1976.

11. Day, J. D., "The OSI Reference Model," *Proceedings of the IEEE*, **71**(12):1334–1340, December 1983.

12. de Jardins, R., and J. S. Foley, "Open Systems Interconnection: A Review and Status Report," *Journal of Telecommunication Networks*, **3**(3):194–209, Fall 1984.

13. Cerni, D. M., *Standards in Process: Foundations and Profiles of ISDN and OSI Studies*, U.S. Department of Commerce (NTIA/ITS), 1984.

14. Folts, H. C., et al., *Open Systems Handbook*, Omnicom Information Service, 1985.

15. Linn, R. J., and J. S. Nightingale, "Testing OSI Protocols at the National Bureau of Standards," *Proceedings of the IEEE*, **71**(12):1431–1434, December 1983.

16. Vissers, C. A., R. L. Tenney, and G. V. Bochman, "Formal Description Techniques," *Proceedings of the IEEE*, **71**(12):1356–1364, December 1983.

17. Dickson, G. J., and P. E. Chazal, "Status of CCITT Description Techniques and Application to Protocol Specification," *Proceedings of the IEEE*, **71**(12):1346–1355, December 1983.

18. Seitz, N., D. Wortendyke, and K. Spies, "User-Oriented Performance Measurements on Public Data Networks," *Proceedings of International Conference on Communications*, pp. 365–371, 1985.

19. Seitz, N. B., and D. S. Grubb, NTIA Report 83-125, *American National Standard X3.102 User Reference Manual*, U.S. Department of Commerce, National Telecommunications and Information Administration, 1983.

20. Piscitello, D. M., and A. L. Chapin, "An International Internetwork Protocol Standard," *Journal of Telecommunication Networks*, **3**(3):210–221, Fall 1984.

21. Hemrick, C., "The Internal Organization of the OSI Network Layer: Concepts, Applications, and Issues," *Journal of Telecommunication Networks*, **3**(3):222–232, Fall 1984.

22. Lyon, D. L., "Personal Computer Communications Via Ku-Band Small Earth Stations," *IEEE Journal on Selected Areas in Communications*, **SAC-3**(3):440–448, May 1985.

23. Deaton, G., and D. Franse, "Performance Analysis of Computer Networks That Access Satellite Links," *Proceedings of ICCC-80*, Atlanta, October 1980.

24. Kulzer, J. T., and W. A. Montgomery, "Statistical Switching Architectures for Future Services," *Proceedings of ISS '84*, May 1984.

26. General References

Abramson, N., "Packet Switching with Satellites," *Proceedings of National Computer Conference*, pp. 695–716, 1973.

Bartoli, P. D., et al., "X-Series Recommendations for Public Data Networks (1981 to 1984)," *Journal of Telecommunication Networks*, **3**(3):159–193, Fall 1984.

Bernstein, S., and J. Herman, "NV: A Network Monitoring, Control, and Management System," *Proceedings of International Conference on Communication*, pp. 478–483, 1983.

Brodd, W., and R. Donnan, "Data Link Control Improvements for Satellite Transmission," *Conference Proceedings, Satellite and Computer Communications International Symposium*, Versailles, France, pp. 201–213, 1983.

The International Telegraph and Telephone Consultative Committee (CCITT), VIIIth Plenary Assembly, 8–19 October 1984, Red Book, Volume VI.7, *Specifications of Signaling System No. 7*, 1985.

The International Telegraph and Telephone Consultative Committee (CCITT), VIIIth Plenary Assembly, 8–19 October 1984, Red Book, Volume VI.8, *Specifications of Signaling System No. 7*, 1985.

The International Telegraph and Telephone Consultative Committee (CCITT), VIIIth Plenary Assembly, 8–19 October 1984, Red Book, Volume VIII.2, *Data Communication Networks: Services and Facilities*, 1985.

The International Telegraph and Telephone Consultative Committee (CCITT), VIIIth Plenary Assembly, 8–19 October 1984, Red Book, Volume VIII.3, *Data Communication Networks: Interfaces*, 1985.

The International Telegraph and Telephone Consultative Committee (CCITT), VIIIth Plenary Assembly, 8–19 October 1984, Red Book, Volume VIII.4, *Data Communication Networks: Transmission, Signaling and Switching, Network Aspects, Maintenance and Administrative Arrangements*, 1985.

The International Telegraph and Telephone Consultative Committee (CCITT), VIIIth Plenary Assembly, 8–19 October 1984, Red Book, Volume VIII.5, *Data Communication Networks: Open Systems Interconnection (OSI), System Description Techniques*, 1985.

Chou, W., A. Bragg, and A. Nisson, "The Need for Adaptive Routing in the Chaotic and Unbalanced Traffic Environment," *IEEE Transactions on Communications*, **COM-29**(4):481–490, April 1981.

Coates, K., and K. Mackey, "The Evolution of Network Management Services in the Bell Labs Network: Throes and Aftermath," *Proceedings of IEEE Computer Society International Conference*, pp. 220–230, Fall 1982.

Decima, M., et al. (eds.), Special Issue on Integrated Digital Network: Technology and Implementations—II, *IEEE Journal on Selected Areas in Communications*, **SAC-1**(6), November 1986.

Ebert, I. G., "The Evolution of Integrated Access towards the ISDN," *IEEE Communications Magazine*, **22**(4):6–11, April 1984.

Falk, G., et al., "Integration of Voice and Data in the Wideband Packet Satellite Network," *IEEE Journal on Selected Areas in Communications*, **SAC-1**(6):1076–1083, December 1983.

Gerla, M., ct al., "A Cut Saturation Algorithm for Topological Design of Packet Switched Communication Networks," *Proceedings of the National Telecommunications Conference*, pp. 1074–1085, December 1974.

Gerla, M., "Deterministic and Adaptive Routing Policies for Packet-Switched Computer Networks," *Proceedings of the Third IEEE Data Communications Symposium*, pp. 23–28, November 1973.

Gerla, M., and L. Kleinrock, "Flow Control: A Comparative Study," *IEE Transactions on Communications*, **COM-28**(4), April 1980.

Gerla, M., and R. Pazoo-Rangel, "Bandwidth Allocation and Routing in ISDNs," *IEEE Communications Magazine*, **22**(2):16–26, February 1984.

Gitman, I., and H. Frank, "Economic Analysis of Integrated Voice and Data Networks: A Case Study," *Proceedings of the IEEE*, **66**(11), November 1978.

Kelly, W. J., and C. Omidyar, "Analysis of Flow and Congestion Control in Packet-Switched Networks," *IEEE Transactions on Communications*, February 1983.

Kermani, P. and L. Kleinrock, "Virtual Cut-Through: A New Computer Switching Technique," *Computer Networks;* pp. 267–286, 1979.

Kleinrock, L., and F. Kamoun, "Hierarchical Routing for Large Networks," *Computer Networks*, pp. 155–174, January 1977.

Kronz, R., S. Lee, and M. Sun, "Practical Design Tools for Large Packet-Switched Networks," *Proceedings, Infocom*, pp. 591–599, 1983.

McQuillan, J., "Interactions Between Routing and Congestion in Computer Networks," *Proceedings, Flow Control in Computer Networks,* pp. 63–75, 1979.

Mier, E., "Packet Switching and X.25—Where to From Here?," *Data Communications,* pp. 121–138, October 1983.

Montgomery, A., "Techniques for Packet Voice Synchronization," *IEE Journal on Selected Areas in Communications,* SAC-1(6), December 1983.

Pouzin, L., "Methods, Tools, and Observations on Flow Control in Packet-Switched Data Networks," *IEEE Transactions on Communications,* **COM-29**(4):413–425, April 1981.

Roberts, L. G., "Dynamic Allocation of Satellite Capacity Through Packet Reservations," *AFIPS Conference Proceedings,* p. 42, 1973.

Roberts, L. G., "Data by the Packet," *IEEE Spectrum,* pp. 46–51, February 1974.

Rosner, R. D., *Packet Switching, Tomorrow's Communications Today,* Lifetime Learning Publications, 1982.

Taylor, D., "An Annotated Bibliography of Congestion Control in Packet-Switched Communications Networks, Royal Signals and Radar Establishment," Report No. 8011, November 1981.

Tobagi, F., R. Binder, and B. Leiner, "Packet Radio and Satellite Networks," *IEEE Communications Magazine,* **22**(11):24–40, November 1984.

Weir, D. F., J. B. Holmblad, and A. C. Rothberg, "An X.75 Based Network Architecture," *Proceedings of the ICCC,* 1980.

P · A · R · T · 2

ELECTRONIC COMMUNICATIONS SYSTEMS

CHAPTER 15

OPERATING, PERFORMANCE, AND INTERFACE STANDARDS FOR VOICE AND DATA CIRCUITS

Michael C. Goldstein
General Electric Information Services Company

CONTENTS

INTRODUCTION

1. Overview

All telecommunications systems consist of transmission facilities which are connected either directly to end user equipment or interconnected through some form of switching. This chapter describes the operating and performance standards which are commonly utilized for the design, specification, and operation of such systems. The intent is to provide telecommunications managers a framework within which to plan a network that will integrate facilities obtained from a variety of sources and to successfully construct and manage the resulting network.

To set the stage for a discussion of factors necessary to specify transmission and switching facilities, there is a brief description of the alternative sources for network components and services. This provides a framework for selecting the components of the overall system and is relevant to the problem of network management. It is followed by a brief review of the four principal transmission media which are utilized—cable, microwave, satellite, and fiber optics.

The next part of the chapter defines the factors which should be included in the specification of operational requirements of a communications system. After defining the operating characteristics which must be specified, the chapter discusses performance standards. These are principally parameters which apply to individual voice and data circuits, but system-level parameters are also considered. While the relevancy of most parameters is independent of the transmission medium, the special considerations that apply to each are discussed.

With deregulation of the telecommunications industry during the 1970s, a vast selection of transmission and switching components from which to construct private systems became available. It soon became apparent, however, that the biggest difficulty with implementing and operating such systems was interfacing between the components and, where required, with the commercial telephone system. The final sections of this chapter discuss interface requirements for various transmission facilities and the overall problem of managing circuits in a private network. Included are specifications to ensure compatibility with other network components, considerations for the implementation of transmission facilities, and methods to assure performance criteria are being met.

NETWORK, SWITCHING, AND TRANSMISSION ALTERNATIVES

2. Private Networks

In order to transmit voice and data for a large organization, the manager is faced with a basic decision between using existing public networks and service offerings or building a private network. In the latter case there are many switching and transmission components to choose from, including both public service offerings and equipment or systems which may be leased or purchased. Thus, even a private network may consist entirely of components which are themselves services that are shared with other users but for which administrative control is exercised by individual users. This section will review the networking alternatives and transmission media which are available in order to set the stage for a description of the parameters used to specify and monitor network quality.

3. Switching Alternatives

Some kind of switching capability is central to all private networks. In the simplest example this could be a set of data multiplexing devices in which all paths are predefined. Such a network would be used to support a centralized data center where terminals are relatively low-speed and are concentrated in dispersed geographic locations. More generally, however, there will be switches which route traffic from an incoming to an outgoing trunk according to the particular session or packet requirement. Private voice switching capability has been available since the 1960s from *local telephone companies* (telcos) which market such services as *private branch exchanges* (PBXs) on site or centrex on telco premises and from AT&T which offered tandem switching arrangements such as CCSA and later EPSCS for interconnecting PBXs. During the 1970s, numerous equipment manufacturers began offering on-site data switches and PBXs of all sizes (10 to 10,000 lines) with a variety of features including both local and tandem switching and the integration of voice and data on the same facilities.

Descriptions and reference data for circuit, message, and packet switching systems are contained in Chaps. 12, 13, and 14.

4. Transmission Alternatives

Transmission facilities are generally leased from common carriers, which can typically provide this service at lower cost since they can combine many users onto high-capacity carrier systems. In some cases a user cannot specify which facilities will be used, but only the performance desired; in others, the specific facilities can be requested, and the cost reflects the medium being used. Finally, in some instances it is cost-effective for large users to build their own transmission systems. A technical and economic comparison of available transmission media is contained in Chap. 9.

Cables.[1,2] There are two kinds of conductive transmission lines—wire pairs and coaxial cable. A coaxial cable consists of an inner conductor surrounded by a grounded outer conductor which is held in a concentric configuration by a dielectric. Wire pairs are an economical method of handling a single voice channel, but, when multiple channels are required, crosstalk becomes a problem when the conductor separation approaches half the wavelength of the operating frequency. Thus for carrier systems, which operate at microwave frequencies to obtain sufficient bandwidth, wire pairs are not used, and coaxial cables, which can operate at up to 18 GHz, become the desirable choice.

Coaxial cables are susceptible to deformation leading to changing impedance and signal degradation, and to temperature changes. More significantly, however, with the increased use of microwave transmission, coaxial systems have become limited to video and wideband digital circuits where microwave is impractical, e.g., where microwave frequencies are limited, in fixed facilities such as building wiring, and over short distances for applications such as *community antenna television* (CATV).

Microwave Systems. Microwave systems (see Chap. 4) transmit in the microwave radio band a signal which has been modulated by multiplexed combinations of voice and wideband source signals. Early microwave systems employed *frequency modulation* (FM), *frequency-division multiplexing* (FDM), and other analog transmission methods. The growing use of T carrier, a system which carries digital information in a standard format, has not affected existing microwave plant, as these signals are converted to analog before transmission. T_1 systems carry data at 1.544 Mbs, divided into 24 channels of 64 kbs, the bit rate used to transmit digitized voice using *pulse-code modulation* (PCM). An individual user, however, now has available T_1 service offerings which allow division of this bandwidth as required into voice, medium- and high-speed data, and even video channels (see Chap. 11). This flexibility and the growing availability of equipment to accomplish these functions make such bulk bandwidth attractive to a large user. In fact, in some cases the use of T_2 at 6.3 Mbs or even T_3 at 44.7 Mbs (672 voice channels) may be warranted.

Communications Satellites. The successful placing of a satellite in geostationary orbit in the mid-1960s provided the capability of transmitting signals to a large number of points nearly halfway around the globe at a cost which is independent of the distance between them (see Chap. 5). During the next 20 years the capacity of a single satellite grew from only 240 voice circuits to over 30,000 while the cost of a channel dropped nearly two orders of magnitude.[3] The attractiveness of satellite technology for long-distance transmission results from its cost insensitivity to distance, its broadcast nature, the relatively large bandwidths available, and the low error rate for digital transmission since the signals are traversing a path which is nearly atmosphere-free.

The economy of point-to-multipoint service by satellites results in their heavy use for program distribution for CATV and commercial television stations, and this still accounts for a substantial amount of their usage. For the corporate manager, satellites offer an opportunity for private video services, but the high bandwidth requirements plus the cost and effort of supplying the television studio have limited the use of private video systems. However, *bandwidth compression techniques,* such as slow scan and freeze frame, are continuing to reduce the cost of the transmission portion of such systems[4] (see Chap. 18). The relatively large bandwidths available from satellites result from the use of microwave carrier frequencies, typically 4/6 GHz for older satellites and 12/14 GHz for the more recent ones. This allows transponders to handle signals with bandwidths of 40 or even 80 MHz, and today's satellites contain many transponders. Error rates in favorable weather conditions have been as good as 10^{-11} on the transponder link, leading to error rates on individual data channels of 10^{-6} or even 10^{-7}.

Satellite capacity is available on a voice circuit basis or on a bandwidth basis, with earth stations located either on the user's premises or on the carrier's premises, with the capacity being shared with other users as is typical of any private-line arrangement. Large bandwidth requirements are generally met with on-site earth stations, which give the user more responsibility for placing data on the satellite and controlling the overall environment. Individual channels are principally employed for voice use and terminate on voice equipment such as PBXs or tandem switches.

Fiber Optics. Optical transmission systems consist of a light source which is pulsed on and off in response to the digital signal being transmitted, an optical fiber which carries the signal between repeating points, and a sensor to convert the pulses back to electrical signals (see Chap. 8). Since frequencies of lightwaves are being employed, very large bandwidths (compared to radio or cable systems) are possible. Typical systems carry several hundred megabits per second of information in a single fiber, and a number of fibers are placed in a cable.

OPERATIONAL REQUIREMENTS

5. Definition of Operational Requirements

The first step in any system design is defining and specifying the requirements the system is meant to meet. The operational requirements which must be fully specified include the voice and data characteristics the system must handle, the geographic area to be covered, environmental conditions to be encountered, and flexibility to adapt to changing requirements. Further, in order to keep the system operating in the prescribed manner, requirements for network reliability and maintenance must be defined.

6. Voice and Data Characteristics[5]

Ideally, a communications system should produce an exact replica of the originating signal at the destination.* Thus a voice user would perceive the other
party to be talking normally in the same room while data would be transmitted
completely error-free. Unfortunately the process of encoding these signals onto a
carrier and transmitting them through space or other media introduces distortions
into the original signal. For voice traffic these are interpreted as noise and more
generally as "poor quality," while for data they result in errors in the bits being
received. To define acceptable quality, many parameters are specified in CCITT
Recommendations and in the United States in Bell System specifications. The
latter are generally internal documents, but some of these are published in tariffs,
giving users a convenient method of describing the quality of private lines desired. With the deregulation of the telecommunications industry in the United
States, however, there has yet to be developed and published operating standards
that all carriers agree to meet.

Bandwidth and Bit Rate.   The fundamental parameter associated with a transmission path is its bandwidth. Bandwidth is defined as the frequency band within
which performance characteristics are within (or above) defined limits. More specifically, for voice use it is the range of frequencies where the signal does not fall
more than 3 dB below a reference level (normally that at 1004 Hz). To reproduce
all the sounds discernible to the human ear, frequencies from 20 to 20,000 Hz are
needed, and this is the goal of good-quality audio reproduction systems. For
voice-grade channels, however, CCITT Recommendation G.132 recommends a
standard of 300 to 3400 Hz. In the United States, the basic AT&T voice-grade
private line has a 3 dB frequency range of 300 to 3000 Hz. Wideband analog channels are also available with a nominal bandwidth of 48 kHz (used by carriers as a
"group" of 12 voice channels). These are of declining interest as their principal
private use has been for high-speed data (specifically 19.2, 40.8, and 50 kbs)
which can now be transmitted over digital channels.

Digital channels are not specified by a bandwidth but by a capacity in bits per
second. Thus a channel referred to as narrowband does not have a specified
bandwidth at all but is one which can carry data at up to 300 b/s.

Low-speed data channels are typically specified to be 110, 300, and 1200 b/s,
medium-speed are 2.4, 4.8, 7.2, 9.6, and 14.4 kbs, and high-speed offerings include 19.2, 32, 56, 64, and, more generally, 8×2^n kbs. Even when ordering
voice-grade channels, a user can specify minimum requirements for transmitting
data (e.g., 4.8 kbs) but this capability is dependent in part on the user's data set.
In the evolution to digital networks, 64 kbs PCM has been accepted as a voice
standard. Other techniques have been developed to transmit voice at 32 or even
16 kbs. It is likely that a 32-kbs *variable-slope delta modulation* (VSDM) method
will become the next standard, but such bandwidth-saving measures should be
tested to assure they will not degrade overall voice quality, especially when such
circuits carry data.

Attenuation Distortion.   Having specified a bandwidth requirement and the signal
level at some point in the band (1004 Hz is typically used for a voice channel),
there is need to assure that the level does not vary significantly in the rest of the

---

*An exception occurs in some data communication services where the signal is deliberately altered,
e.g., by protocol conversion.

band. This variation from the reference point is referred to as *attenuation distortion*, and common carriers will provide limits for different service offerings. As an example, an AT&T private line is specified to have a variation of -2 to $+8$ dB from 500 to 2500 Hz and of -3 to $+12$ dB from 300 to 3000 Hz. A slightly different description is stated in CCITT Recommendation G.132, which recommends that relative to 800 Hz the variation be less than 9 dB between 400 and 3000 Hz.

Envelope Delay Distortion.   A characteristic which has little impact on voice but which can severely degrade data performance is *envelope delay distortion* (EDD), or *phase shift*. EDD occurs because signals take a finite amount of time to pass through bandpass filters, and this time varies with frequency. It is defined as the difference in time delay between any frequency and a reference frequency (typically 1800 Hz). Such differences in delay will produce intersymbol interference in many data signals and limit the modulation rate which can be used. It is a phenomenon to which the human ear is not sensitive. As with attenuation delay, common carriers will provide specifications for EDD.

Signal Level.   Ideally the signal level at the receiver would be identical to that of the transmitter on a communications link. Since the transmission system will contain points which amplify or regenerate the signal in its entirety, the level is not only dependent on the media being traversed but also on the output levels of the various repeaters. While it is clear that too low a signal level will result in conversations which cannot be heard, too high a signal level can cause equally severe problems. Specifically, high levels will overload amplifiers, leading to nonlinear behavior which will increase intermodulation products and crosstalk between adjacent channels.

Noise.   In addition to attenuation distortion, most other degradations of signal quality are referred to as *noise* since the listener hears sounds other than those that were transmitted. *Thermal noise* results from random electron motion and is present in any electronic device. *Impulse noise* is characterized as noncontinuous, irregular, high-amplitude pulses of short duration. *Intermodulation noise* comes from the presence of intermodulation products which result from passing signals at different frequencies through nonlinear filters. *Crosstalk* is an unwanted coupling of signals and exhibits itself as information from one channel appearing at a low level on another channel. It is a result of electrical coupling between media, poor filters, or nonlinearity of performance. The effect of crosstalk in voice systems is unwanted side conversations, while a signal carrying binary data will incur a higher bit error rate.

   The single measure which is used most often to describe the effects of noise in a communications system is the *signal-to-noise ratio*. This is the amount (in decibels) by which the signal level exceeds that of the noise.

Echo.   Long-distance transmissions are nearly always done in a full-duplex manner using four-wire paths, while local loops are principally half-duplex as the result of their two-wire configuration. A conversion must thus take place at both ends, and imperfect impedance matches cause a portion of the received signal to be reflected back to the transmitting end. This is referred to as *echo*.

Switch Performance.   A related topic which will not be covered here is parameters for switch performance. These are important since distortion introduced by switches is as significant as that caused by poor transmission facilities.

7. Network Reliability and Availability

Public voice and data networks in the United States have a reputation for high reliability, and, as a result, users expect telecommunications facilities to give high-quality service which is virtually uninterrupted. Since private networks are utilized to reduce costs or provide added capabilities or features, care must be taken not to achieve these goals at the expense of reliability or availability.

Reliability is a measure of unexpected failure of network components. The most common measure is the *mean time between failures* (MTBF) which is usually expressed in hours or days. *Availability* is defined as the percent of time a system or component is available for use, or the percent of time it will perform at some minimum level. It includes the effect of both technical and operational failures, including system overload.

8. Communications Security

With the increasing sensitivity of the information that corporations communicate, there is a growing concern with protecting it from outside listeners. Placing information on cables or fiber, which are broadband and underground, in itself offers a high level of security against eavesdropping. By comparison microwave and especially satellite can be easily intercepted, although the cost of retrieving the original signal can be prohibitive. To assure information is not being received by unintended sources, users are turning to encoding of information before transmission. The most common method is use of the *data encryption standard* (DES)[8] algorithm published by the National Bureau of Standards. See Chap. 19 for further discussion.

PERFORMANCE STANDARDS

9. Performance Standards

This chapter subdivision describes performance standards for the parameters defined in the previous subdivision, Operational Requirements, which will ensure good transmission quality. It also describes commonly used tests which are employed to measure their adherence to these standards. Throughout, power measurements are given in decibels (dB) where dB $= 10 \log P_2/P_1$, or as dBm for $P_1 = 1$ mW, dBrn for $P_1 = 1$ pW (-90 dBm), or finally dBrnC when P_1 refers to a C-message* signal.

10. Bandwidth-Related Parameters

The bandwidth-related parameters are capacity, attenuation distortion, and envelope delay distortion. The capacity of an analog channel, referred to as its bandwidth, is specified in Hz, and allowable attenuation and envelope delay are specified over given frequency ranges. To meet the varied requirements for voice and low- to medium-speed digital transmission, AT&T and other carriers offer voice

---

*A C-message signal of 1 pW introduces less than 5 dB attenuation in the range of 600 to 3000 Hz.

circuits which have variable degrees of conditioning to meet different quality requirements. Typical bandwidth parameters for the basic and conditioned C-message circuits, referred to as C1 through C5, are shown in Table 15.1. These standards are for baseband signals; modulation and multiplexing methods employed for their transmission are described in Chap. 11. In practice, C2 is generally chosen for medium-speed point-to-point transmission while C5 is selected if the circuit is part of a switched network or involves an international connection.

The capacity of digital transmission is normally specified by bit rate. While the basic channel transmits 64 kbs, signaling considerations reduce the available capacity to 56 kbs for data. For lower bandwidth, additional bits are needed for signaling, leaving only 48 kbs which can be subdivided into five 9.6-, ten 4.8-, or twenty 2.4-kbs channels.[9] For these circuits, attenuation and envelope delay are not relevant, but error performance is (see Sec. 14).

11.   Circuit Loss and Noise Parameters

The acceptable ranges for circuit loss at 1004 Hz and various noise parameters for a voice-grade circuit are shown in Table 15.2. Where limits on noise or nonlinear distortion are critical, D conditioning, D1 and D2, gives lower limits on both C *notched noise* and *intermodulation* (IM) *distortion*. Specifically, the level of C notched noise must be more than 28 dB below the received test tone (versus 24 dB for the basic channel), and the level of second- and third-order IM distortion must be down by more than 35 and 40 dB respectively (versus 27 and 32 dB for the basic channel).

TABLE 15.1 Bandwidth Parameter Limits of Conditioned C-Message Channels

Channel conditioning	Attenuation distortion (frequency response) relative to 1004 Hz		Envelope delay distortion	
	Frequency range, Hz	Variation, dB	Frequency range, Hz	Variation, μs
Basic	500–2500	−2 to +8	800–2600	1750
	300–3000	−3 to +12		
C1	1000–2400	−1 to +3	1000–2400	1000
	300–2700	−2 to +6	800–2600	1750
	300–3000	−3 to +12		
C2	500–2800	−1 to +3	1000–2600	500
	300–3000	−2 to +6	600–2600	1500
			500–2800	3000
C3	500–2800	−0.5 to +1.5	1000–2600	110
	300–3000	−0.8 to +2	600–2600	300
			500–2800	500
C4	500–3000	−2 to +3	1000–2600	300
	300–3200	−2 to +6	800–2800	500
			600–3000	1500
			500–3000	3000
C5	500–2800	−0.5 to +1.5	1000–2600	100
	300–3000	−1 to +3	600–2600	300
			500–2800	600

Source: Bell System Technical Reference Publication 41004, Data Communication Using Voiceband Private Line Channels, AT&T, New York, October 1973.

TABLE 15.2 Noise Specifications for Transmission Channels

Parameter	Acceptable Ranges		
1004-Hz loss	No more than ± 4 dB long term and no more than ± 3 dB short term		
C-message noise	Facility miles:		Max. noise, dBrnC:
	0–50		28
	51–100		31
	101–400		34
	401–1000		38
	1001–1500		40
	1501–2500		42
	2501–4000		44
C-notched noise	At least 24 dB below received 1004-Hz test tone		
Impulse noise	Threshold with repeat to received 1004-Hz test tone, dB:		Max. counts allowed in 15 minutes:
	6		15
	2		9
	− 2		5
Phase hits	≤ 8 in 15 min, ≥ 20°		
Gain hits	≤ 8 in 15 min, ≥ 3 dB		
Dropouts	≤ 2 in 15 min, ≥ 12 dB		
IM distortion	Second-order 27 dB, third-order 32 dB		
Phase jitter	Facility length, mi:		Max. degrees
		20–300:	4–20 Hz:
	0–250	2	5
	251–500	4	5
	501–1000	6	5
	1001–2000	8	5
	2001–4000	10	5

Source: Bell System Technical References 41008, 41009, AT&T, New York.

A description of the noise parameters in Table 15.2 follows.

C-Message Noise. Since modems are more affected by noise in the middle of the voice band, weighting filters are utilized to isolate this area. The most common weighting filter is the C-message filter, which introduces less than 5 dB attenuation in the range of 600 to 3000 Hz. In Europe, a slightly different weighting function, the psophometric, is used.

The noise measurement is made by inserting the filter ahead of the decibel meter and measuring the level with no signal applied. This is referred to as idle channel noise and is measured in dBrn, where 0 dBrn = 90 dBm. Maximum allowable levels are a function of the length of the line (see Table 15.2).

C-Notched Noise. While C-message noise is a measure of noise with no signal applied, a more significant measure is to apply a signal and then to measure noise with the signal removed at the distant end.[10] Such noise results from the way in which signals are digitized and processed before modulating the carrier.

These tests are accomplished by applying a 1004-Hz tone at one end of the circuit and removing it at the measuring end with a narrowband notch filter. Thus, to measure C-message notched noise, a test tone is inserted in the channel and the signal-to-noise ratio is measured with the tone notched, or filtered, out.

Impulse Noise.   Impulse noise is characterized by large peaks in the total noise waveform and is measured by counting the number of peaks exceeding a predetermined threshold over an interval of time. Available instruments measure over three different thresholds simultaneously. The actual measurement is made by inserting a 1004-Hz test tone at one end of the circuit, then removing it with a C notched filter at the other end and using impulse-counting instruments to count the number of occurrences of signals exceeding the three levels.

Phase and Gain Hits.   These are moderate changes in the phase or amplitude of a signal which last at least 4 ms and return to acceptable levels within 200 ms. A voice-grade circuit should register no more than eight phase hits ($\geq 20°$) or 8 gain hits (≥ 3 dB) in any 15-minute period. Gain hits of greater than 12 dB are referred to as dropouts, as they are perceived as a loss of signal by a listener. There should be no more than two in any 15-minute period. Gain hits and dropouts are most noticeable in voice use, since they result in the listener not being able to hear to some degree. Phase hits, by contrast, cause bit errors in digital signals.

Intermodulation Noise[6].   When two signals at different frequencies F_1 and F_2 pass through nonlinear filters, *intermodulation products* are produced. Second-order products are at frequencies $F_1 \pm F_2$, third-order at $F_1 \pm 2F_2$ and $2F_1 \pm F_2$, and so on. This results in unwanted signals at the output. To test for these signals, two equal tones are applied at the circuit input, usually at a frequency ratio of 3:5, such as 1500 and 2500 Hz. A spectrum analyzer is then used at the distant end to measure the power of the intermodulation products. The difference in decibels of unwanted tones to the received signal is the signal-to-distortion ratio. For voice-grade channels, second-order products should be less than 27 dB below the signal, third-order less than 32 dB. Higher-order products are generally ignored.

Phase Jitter.   Phase jitter occurs when successive cycles of a constant tone arrive at irregular time intervals. The most common causes are frequency components of 20-Hz ringing current or 60-Hz commercial power. The phase jitter caused by these sources is measured in the 20 to 300 Hz band. Another cause of phase jitter is phase instabilities in FDM oscillators which take place below 20 Hz. To measure this low-frequency jitter, measurements are taken in the range of 4 to 20 Hz. Acceptable measures are given in maximum degrees and are a function of circuit length in different frequency bands (see Table 15.2).

12. Echo Return

Echo is a reflection, or "return," of the signal resulting from an imbalance at the two- to four-wire hybrid at the distant end.[7] The degree of annoyance caused by echo is determined by both its loudness and delay. If the delay in receiving this signal is sufficient, it will be perceived as a separate signal. This is considered to be the case for delays in excess of 50 ms, which occur for terrestrial transmission at distances in excess of approximately 1800 mi. For satellite circuits with delays of approximately 250 ms, the effect is most severe. Echo suppressors, automatic "push-to-talk" devices, are satisfactory for terrestrial circuits. They are marginally satisfactory for satellite circuits, and newer devices known as *echo cancelers* must be used to ensure sufficient attenuation of the return signal.

The magnitude of the echo return is measured by the *return loss* which is defined as the ratio of the incident to the reflected signal. *Balanced return loss,* described in CCITT Recommendation G.122, also referred to as *echo return loss,* is the loss over the range 500 to 2500 Hz. For local circuits this should be more than 6 dB ± 2.5 dB. For long-haul terrestrial circuits, a satisfactory value is considered to be 11 dB.

13. Switching Equipment

Criteria for signal degradation in switching equipment were originally developed for electromechanical systems. Unfortunately, with hybrid systems many of these standards are too loose and do not apply at all for digital signals. The published standard for these criteria is CCITT Recommendation Q.45, and a more stringent set is given in an ITT Labs publication.[11] Table 15.3 shows some of the parameter ranges allowed by these organizations. All items have the same definition as for circuit degradation, with *loss dispersion* meaning the variation in loss from calls with the highest loss to those with the lowest.

14. Digital Error Rates

The fundamental measure of the quality of digital transmission is the ratio of bits received in error to the total number of bits sent and is referred to as the *bit error*

TABLE 15.3 Specifications for Telecommunication Switches

Item	CCITT Q.45	ITT Labs
Loss	0.5 dB	0.5 dB
Loss, dispersion	< 0.2 dB	< 0.2 dB
Attenuation		
300–400 Hz	− 0.2/ + 0.5	− 0.1/ + 0.2
400–2400	− 0.2/ + 0.3	− 0.1/ + 0.2
2400–3400	− 0.2/ + 0.5	− 0.1/ + 0.3
Impulse noise	5 in 5 min, > − 35 dBm0	5 in 5 min, 12 dB above noise
Crosstalk		
Go and return path	60 dB	65 dB
Any two paths	70 dB	80 dB
Impedance variation		
300–600 Hz	15 dB	
600–3400	20 dB	
300 Hz	18 dB	
500–2500	20 dB	
3000	18 dB	
3400	15 dB	

rate (BER). These errors can be caused by virtually any of the circuit impairment problems described in this section.

The method used to measure the BER on a line is to insert a pseudo-random bit sequence of length $2^n - 1$ and measure the resulting errors at the distant end. Pseudo-random sequences include various combinations of 1s and 0s of length n. Specific sequences are contained in CCITT recommendations. The most common lengths are 511 for $n = 9$ and 2047 for $n = 11$. The bit error rate for a voice-grade channel should be no worse than 1 in 10^5 (10^{-5} BER). Digital channels should provide quality no worse than 10^{-6} BER.

Transmission impairments often occur from large bursts of noise or interference such as *radio-frequency interference* (RFI). When this happens there will be a large number of bit errors on consecutive bits. When a protocol is being utilized which has block error recognition (and this includes all synchronous protocols such as the HDLC (High-level Data Link Control) of the X.25 standard and IBM's binary synchronous and SDLC (Synchronous Data Link Control) procedures), the result will be a single block in error. In these cases, it is important not how many bits were in error, only that the block needs to be retransmitted. To reflect this measure of transmission quality it is useful to measure *block error rate* (BLER). For BLER, the error test is performed by dividing the input bit stream into blocks and measuring the ratio or error-free blocks to total blocks sent.[12] A related measure is *error-free seconds*. This is the number of seconds in a given period of time that all bits are transmitted without errors. DDS service specifies an average of greater than 99.5 percent error-free seconds at 56 kbs and better at 9.6, 4.8, and 2.4 kbs.[9]

15. Availability and Continuity

Availability.   Communication circuits can be expected to be available in excess of 99 percent of the time, with typical figures ranging from 99.5 to 99.8 percent. (DDS, or Digital Data Service, specifies an average channel availability greater than 99.96 percent, i.e., annual downtime less than 0.04 percent.) In all cases, the availability figures should include outages both for scheduled maintenance and for unforeseen problems.

For some applications where real-time transmission is particularly vital, e.g., the distribution of program material for the major television networks, an even higher degree of reliability, up to 99.99 percent, is desired. In order to achieve an availability which approaches unity, e.g., 99.98 percent, there must be an order-of-magnitude improvement in equipment reliability, conservatism of system design, and system redundancy. Incorporating such concepts can increase the cost of the circuit by as much as 2 to 1 or more.

Continuity.   The most basic problem that can affect a transmission path is loss of continuity between endpoints. For a facility which is directly connected to user equipment at both ends, this will be exhibited as an inability to establish a connection or the termination of an established connection. Even under these conditions, however, it is not immediately clear whether the transmission path is open or whether there is a problem with a data set or the user's terminal equipment. One measure of intermittent failure that can be made on data circuits is a carrier loss count, which is the number of times the carrier is lost for a given time period. When the circuit is part of a switched network, alternative paths are generally available and the unavailability of a single circuit is not readily noticeable. Therefore switching equipment must notify operations of such occurrences.

Where the user has access to the end of a data circuit, a remote loop-back test can be performed.[10] This is accomplished by applying a test tone, usually 2713 Hz, to command the remote data set to assume loop-back mode. Then a 1004-Hz tone is applied, and the continuity can be determined at the originating point. This arrangement can also be used to measure the loss in the circuit by applying the test tone at 0 dBm and measuring the level of the received signal. A standard voice channel is designed for a loss of 16 dB ± 1 dB with a long-term variation of no more than ± 4dB. Testing voice circuits clearly requires different techniques, since they are not connected to equipment which generally performs loop-back functions. One method is to have two test sets which operate in a master-slave mode connected to either end of the circuit with the master unit initiating tests and the remote unit responding to commands. A more sophisticated method is to have a control center establish a dial connection to test sets at either end of a circuit and initiate tests by commands over the dial circuit. This arrangement is not dependent on the quality of the circuit under test, and such tests can be performed on many circuits during a single off-peak period with all results available at a central point.

16. Performance Monitoring

Specifying the operating, performance, and interface requirements desired from voice and data circuits is only the first step in attaining and maintaining satisfactory transmission performance. The system must be designed and constructed (see Chap. 19). The management of traffic on the system and implementation of system modifications to accommodate changing traffic patterns is an ongoing task (see Chap. 16). Last but not least, the performance of the system must be monitored regularly (and in some cases continuously) to ensure that the desired standards are being maintained.

A number of procedures for measuring specific aspects of transmission and switching performance were described in the preceding sections. This section contains additional suggestions for ongoing performance monitoring.

Methods used to monitor circuit performance will differ greatly, depending on how many circuits the user is monitoring, their importance, and whether they are used for voice or data. A single voice circuit, such as a tieline between two corporate locations, can be evaluated simply by using it and observing (or waiting until users complain about) its quality. Where there are multiple circuits between two locations, however, a bad circuit often will go unnoticed because either users will simply place another call to obtain a different circuit or the switch will not select it.

For this arrangement, it is usually worth the expense of test instruments to perform some of the tests described in previous sections. Note that the circuit cannot be in normal operation while performing these tests. The most common procedure, which has minimal operational impact, is to test a group of circuits during off hours by taking circuits out of service one at a time. Many users have written programs for personal computers which perform this function automatically and compile the results of the tests on disk. These programs are run at night and the results are processed by other programs the next day to determine which circuits do not meet the prescribed parameters.[13] The most sophisticated method of performing such tests is with the switches themselves. Where equipped, the switches can be commanded to perform tests on circuits when they are unused and store the results for later analysis. At a minimum, however, switches should be capable of reporting when they cannot place a call over a circuit, since this represents an immediate problem which requires attention.

Error measurements on data circuits, and on voice circuits used for data, often can be performed during normal operation, since most are connected to intelligent devices at both ends. When the circuit is connected to two programmable devices, e.g., data switches or computers, they can be programmed to count bit or block errors on normal data or special test data. Whatever method is used, these devices should periodically produce reports on circuit performance or at least store the information where it can be accessed when desired. More powerful measurement systems, often referred to as network management systems, are growing in popularity as the difficulties of keeping track of the condition of a large number of circuits become evident. These systems, operated from a central site, monitor the condition of the circuits in a network, sounding alarms to indicate out-of-service conditions, and collect and analyze performance data in order to indicate deteriorating conditions before an outage occurs.

NETWORK AND INTERFACE REQUIREMENTS

In many instances, a telecommunications circuit can be operating within specified performance limits and still render a level of service which is unacceptable to the end users. This can result from the cumulative degradation of a number of system components operated in tandem. It can also result from an improper interface between the circuit and user equipment or other system components. The previous chapter subdivision described the methods of specifying individual circuit parameters to ensure that overall performance will be within acceptable limits. This subdivision addresses the requirements of network and system interfaces, especially with end-user equipment and local loops.

17. Networks

When a network consists only of point-to-point circuits, the performance which is achieved on each circuit is that which is perceived by the end user. But when circuits are connected in tandem through some kind of switching facilities, the cumulative effects of degradation from each segment and from the switching equipment will be observed. With a monopoly network supplier who provides all facilities, as in most countries, end-to-end performance can be specified; for a private network, however, the user takes responsibility for integration of the different components.

Achieving acceptable performance from a switched network requires more stringent specification of the parameters in each component such that their cumulative effect is within set limits. As an example, consider the connection through a switched voice network in which a telephone connected to a PBX makes a call to a distant telephone either on or off the network. The following losses will be encountered:

1. Telephone to PBX: < 1 dB
2. PBX to network switch: 0 dB if four-wire to four-wire, 2.5–5 dB if two-wire to four-wire
3. Node to node: normal circuit loss, e.g., 16 dB

4. Node to public telephone system: 3.5 to 5 dB

Since the user has little control over losses encountered in the public telephone system, the losses on the other parts of this system must be tightly controlled both in terms of initial specification and ongoing performance. In order to provide acceptable voice quality, the particular parameters which must be met are:

1. Total loss should not exceed 20 dB, with 30 dB the maximum

2. Frequency response should be 300 to 3000 Hz

3. Signal-to-noise ratio should be 20 to 30 dB

The integration of satellites and terrestrial facilities raises some unique concerns. The first is the compatibility of multiplex arrangements on the terrestrial links. Then there are potential synchronization problems when interfacing with digital links. Finally there are the problems associated with propagation delay which require modifications to signaling, different error control schemes in data networks, and voice echo. When earth stations are located on the user's premises, proper echo control on the channel is sufficient to eliminate this degradation. However, if the channel connects to terrestrial channels or the public telephone system at one end, the echo performance of that section plus the mismatch between them become important. Thus, while echo is generally not a problem, even on satellite circuits, such interconnections can lead to poor performance which will be difficult to control unless each component strictly adheres to its design criteria.

In order to lower costs on transmission systems which contain more than one circuit, schemes have been developed which allow more voice connections than circuits. This is accomplished by switching the user off a circuit when there is no conversation, then switching to an available one when conversation resumes. Both analog and digital methods have been developed and are referred to as *time-assigned speech interpolation* (TASI) and *digital speech interpolation* (DSI). Degradations from this equipment occur when it cuts off the beginning of a phrase of speech because it does not activate quickly enough. This is referred to as clipping. (Similar behavior is also evidenced by faulty echo suppressors which do not properly switch the active side of the four-wire connection through the hybrid.) The user should be aware of the potential for clipping to ensure the supplier of such a service is providing adequate quality.

A final area of concern at the network level is reliability, since failures of individual components have varying effects on overall performance. Additional measures which are used to evaluate quality include *mean time between major failures* (MTBMF), *mean time between nodal failures* (MTBNF), and *mean time between nodal subsystem failures* (MTBNSF).

18.   End-User Equipment Compatibility

Dedicated circuits will terminate on either voice or data equipment. Voice equipment can range from a single telephone handset up through key systems and various size PBXs. When PBXs are utilized, care must be taken to ensure compatibility not only with that switch but also the actual devices which will be connected to the circuit on demand. *Data communications equipment* (DCE) will typically be data sets, modems, data service units (in the case of digital circuits), and data switches, although these usually require a data set to interface to communication lines. Unlike many PBXs, data switches generally isolate the user equipment from the circuit, and a proper interface to the switch is sufficient to ensure that the circuit can be properly utilized.

The connecting of *customer-provided equipment* (CPE) to facilities provided by common carriers has been permitted in the United States since 1968 (the Carterfone decision). This regulatory policy preceded the availability of a competitive transmission industry by several years. As a result, equipment design must provide for ready connection to standard voice-grade and digital circuits provided by common carriers, and the requirement for compatibility is usually given to the equipment supplier rather than the circuit provider. In addition to being able to operate over circuits meeting the parameters specified in Performance Standards above, the following items should be considered:

1. Terminating impedance to avoid power loss from improper match to the circuit

2. Signal power to compensate for loss on the circuit but not overload it and cause nonlinear distortions or excessive echo

3. Capability to generate, process, and pass required signaling information when connected to a central office

4. Any other item required by the equipment manufacturer to assure proper operation over the selected transmission path.

19. Local Loops

Long-distance circuits which terminate at the carrier's premises will require a loop either to the user's location or to a local telco central office. To reach the user's location, the connection will most often be made by the local telephone company under published tariffs. In this case, the circuit supplier is held accountable for end-to-end performance and must design the long-distance portion to interface properly with the local loop. Performance of the local portion is not the user's responsibility and generally cannot even be measured. Initial implementation as well as ongoing maintenance is the responsibility of the primary circuit supplier.

When the connection is to a central office, the same considerations hold, and in this case the user has even less "visibility" of the circuit to perform testing. In particular, if digits are passed to the central office in order to establish connection to a testing point, there is no way to differentiate between the performance of the leased circuit and its associated local loop and that of the local telephone system needed to complete the connection. In this case, only cooperative testing between the circuit supplier and local telco can isolate problems.

The alternative to using telco-provided local loops is to use the so-called *bypass* suppliers who are building facilities such as private microwave systems, coaxial cable, and fiber optics. In this case the user will have to contract for specific performance parameters to ensure that end-to-end performance objectives will be met. Since one end of the facility will likely be on a carrier's premises, however, cooperative efforts will be required between the two suppliers to ensure proper installation, ongoing maintenance, and problem resolution.

A variant of the bypass strategy is for the user to build a private system. This might be the best arrangement when a large number of channels are needed from a carrier location to the user. In this case the contract with the provider would specify relevant parameters, but it will ultimately be the user's responsibility to ensure proper operation of this segment of the transmission path.

REFERENCES

20. Cited References

1. Pooch, Green, and Moos, *Telecommunications and Networking,* Little, Brown, Boston, 1983.
2. Flood, J. E. (ed.), *Telecommunications Networks,* Peter Peregrinus, New York, 1975.
3. Campanella, S. J., and J. V. Harrington, "Satellite Communication Networks," *Proc. IEEE,* **72**(11):1506, November 1984.
4. Sabri, S., and Prasada, B., "Video Conferencing Systems," *Proc. IEEE,* **73**(4):677, April 1985.
5. Freeman, R. L., *Telecommunications Transmission Handbook,* Chap. 1, John Wiley & Sons, New York, 1975.
6. Freeman, R. L., *Telecommunications Transmission Handbook,* sec. 4.7, John Wiley & Sons, New York, 1975.
7. Freeman, R. L., *Telecommunications Transmission Handbook,* Chap. 2, John Wiley & Sons, New York, 1975.
8. FIPS Publication 46, Data Encryption Standard, National Bureau of Standards, Washington, D.C., 15 January 1977.
9. Bell System Technical Reference Publication 41450, AT&T, New York, November 1981.
10. Fredrick, R. L., "Measurement of Voice Frequency Transmission Impairments," *Telecommunications,* **19**(3):85, March 1985.
11. *Telecommunications Planning Document,* ITT Laboratories, Spain, 1973.
12. Lenk, J. D., *Handbook of Digital Communications,* Chap. 6, Prentice Hall, Englewood Cliffs, N.J., 1984.
13. Scott, S., "GE tests own CCSA net links to assure high quality service," *Communications News,* May 1982.

21. Further Reading

Telecommunications Transmission Engineering, Vol. 1, *Principles,* 2d ed., AT&T, New York, 1977.

CHAPTER 16

PRIVATE COMMUNICATIONS SYSTEMS PLANNING AND DESIGN

H. Charles Baker

President, Telecommunications Engineering, Inc.
Professor of Electrical Engineering, Southern Methodist University

CONTENTS

SCOPE OF CHAPTER

1. Introduction

A private communications system involves transmission, switching, and/or signaling components that can be used only by a specific group of users and are not available to the general public. Such a system might be as simple as a point-to-point voice-grade circuit linking two PBXs or as complex as a multinational corporate telephone network. It may include user-owned components, such as microwave towers, satellite earth stations, and in-house coaxial cable, or it may consist only of equipment and services that are leased from public carriers and other suppliers. Many large private telephone systems contain both owned and leased components, and most such systems interconnect with the public telephone network at one or more points.

Probably the most basic form of private communications system is the *private branch exchange* (PBX).* Figure 16.1 illustrates the underlying concept of a PBX, which is to provide on-premises and off-premises communications while reducing the total number of trunks (telephone company "business lines") between the PBX and the *central office†* (CO).

RATIONALE FOR PRIVATE COMMUNICATIONS SYSTEMS

While there are potentially many reasons for establishing a private system, most reasons generally may be grouped into three categories: *economics, special requirements,* and *availability.*

2. Economics

The term *economies of scale* is particularly applicable to communications systems. Economies of scale means that the more of something one has, the less it costs incrementally to add more. Economies of scale is a characteristic of telecommunications partly because of the remarkably high capacities now available in modern transmission and switching systems and partly because of a phenomenon known as *trunking efficiency* to be covered later in the traffic engineering portion of this chapter. Because of its inherent concentration capability, the PBX of Fig. 16.1 has the potential to save money in terms of CO trunking requirements *if not all of the telephone stations connected to the PBX need a trunk at the same time.* Because of economies of scale, large communications users usually benefit by lumping their communications requirements together along high-density routes or into high-capacity nodes. The public networks are intended to provide basic communications services to a large number of geographical locations. Their economies of scale derive not from heavy demand of a few users but from light demand of many users.

A corporate user having PBXs in more than one city may find it economical to interconnect them via a leased voice-grade circuit. Figure 16.2 illustrates this concept. The amount of demand required to justify such a private system economically is not necessarily high. For example, on the basis of the comparative cost of long-distance calls, two hours' daily volume of communications traffic between two domestic locations is probably adequate to justify a private circuit between those locations. However, usage patterns could greatly influence the practicality of this justification: If the two hours' equivalent daily volume came from many users, all in the same five-minute period, a single private circuit could not handle the traffic load without requiring queuing of callers. Such a situation raises the twin questions of urgency and of cost of delays.

Whenever a call is attempted at a moment that the system has insufficient capacity available to handle it, the call is said to be *blocked.* Keeping blockage at an acceptable level and providing an acceptable disposition of blocked calls are the

---

*Sometimes the term *PABX* is used to denote an automatic PBX, but the distinction is moot because manually operated PBXs are rare.

†The *central office,* also known as the *end office* or *wire center,* is the location of the first switch encountered by a call entering the public network.

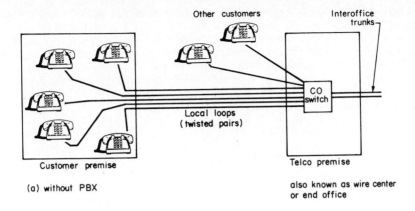

(a) without PBX

also known as wire center
or end office

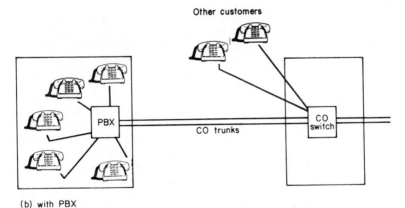

(b) with PBX

FIG. 16.1 The concentration capability of a PBX allows a reduction of the number of lines to the central office.

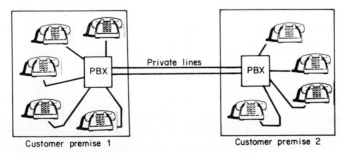

FIG. 16.2 A simple private network.

primary issues of traffic engineering, which is discussed in Secs. 11 to 20 of this chapter.

As already indicated, PBXs take advantage of economies of scale by reducing the number of circuits required between a customer's premises and the telephone company central office. Each such circuit represents a significant monthly cost. Since only a small percentage of telephones on the premises actually require a line at any given time, the PBX provides concentration of the active telephones into the available lines.

Key systems also provide the same type of concentration function, but there is a substantial difference between PBXs and key systems: a key system lets the user select a specific business line; a PBX does not. Also, PBXs generally provide better intercom flexibility. Figure 16.3 shows a typical key system.

3. Special Requirements

There are many situations in which the public networks do not serve the *required locations*. Most early private networks were established for this reason. Certain types of industries, notably oil, mining, and rail, have significant communications requirements outside the populated areas that the public carriers tend to serve. Half a century before the word *interconnect* came into popular use, companies of this type had established their own communications networks that were fully interconnected with appropriate public networks.

The basic PBX provides switching, intercom, and BORSCHT functions (*b*attery feed to stations, *o*vervoltage protection, *r*inging voltage, *s*ignaling, *c*oding, *h*ybrid, and *t*esting). In addition, modern computer-controlled PBXs offer a large variety of *customized calling features and administrative aids* such as abbreviated dialing or "speed-call," message services, network access authorization codes, automatic least-cost network route selection, and premises directory printing.

The rapid growth of *data communications requirements,* both on- and off-premises, has been the stimulus for numerous private systems. Most older analog PBXs require the use of expensive modems for handling data, while most digital PBXs generally digitize speech and then handle voice and data with equal ease. Unfortunately, excessive costs, inadequate capacity and/or the lack of widely accepted and flexible standards for establishing and operating data communications sessions have frustrated many users' attempts to fully integrate their data communications requirements with their voice systems.

For many users, the solution has been the creation of special private systems for handling data communications requirements separately from voice. These range from leased digital circuits to complete private packet networks.

For the *local site,* specialized data PBXs and numerous shared-medium (usually coaxial cable) schemes are available, the details of which are beyond the scope of this chapter. Stallings[1] has documented many of these systems.

4. Availability

Today's public telephone network has achieved a remarkable record of availability. There are times, however, when service on the public network is unavailable, either because of *outage* or because of *blockage* due to extraordinarily heavy usage. A private system may offer a degree of additional availability. Some users

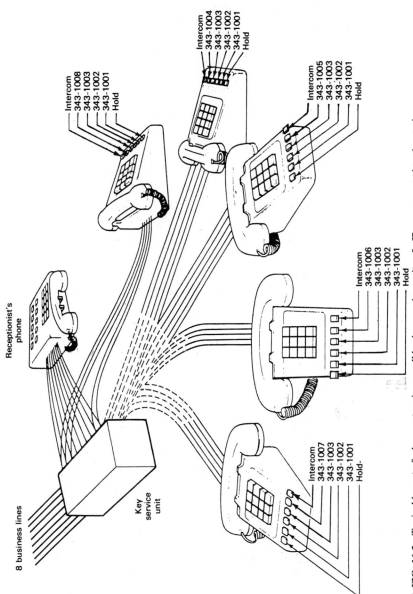

FIG. 16.3 Typical layout of a key system in a small business or in a suite of offices connected to the station side of a PBX.

lease a private transmission facility and use the public network for backup. Others lease two diversely routed facilities. Still others build their own primary facility and use either leased or public facility backup.

Two precautions regarding private facilities should be noted where availability is the goal:

1. *Diverse routing* of the local lines between the user premises and telephone company central office is as important as diverse routing elsewhere. The most secure arrangements use separate premises-access points, with the two trunk groups routed to different central offices; this involves extra installation and monthly charges.

2. The planner must carefully read and understand the public carrier's written *policies* regarding availability, diverse routing, and restoration of the respective public or leased services, particularly during extraordinary conditions. A special disaster contingency service known as *line-load control* is available from the public carrier in most areas and should also be taken into account.

ELEMENTS OF A PRIVATE COMMUNICATIONS SYSTEM

A private network generally consists of the same basic building blocks as the public network, namely, transmission, switching, and signaling. The basic principles for both private and public networks are the same. Since switching can be of more than one type, a special review of switching types follows.

5. Switching

There are three basic types of communications switching, any of which can be used in a private system.

Level 1 or *physical level switching* is equivalent to the circuit switching discussed in Chap. 12. *Even if there is no electrical connection between the input and output ports of the switch, if all signals entering an input channel port are transferred by the switch to a specific output channel port for the duration of that connection, the result is called* physical level switching.

Level 2 or *link level switching* is done on multipoint digital circuits. This technique is sometimes called *broadcast networking.* Most practical applications of link level switching are in data communications, although some satellite-based digital voice systems use it. With link address headers, blocks of information bits are addressed to a specific receiving terminal on the multipoint circuit. While all receiving terminals sense the bits, only the specifically addressed terminal passes the message through to the intended recipient.

Compared with physical level switching, link level switching is much simpler, but it has the disadvantage that only one message can be sent through the switching system at a time. For this reason, most link switching systems use buffers and burst the information bits through the channel at a high rate. Figure 16.4 shows a satellite-based telephone system using link level switching. Bus-type and ring-type data networks, including IEEE 802 systems,[2] also use link level switching.

Level 3 or *network level switching* is used in all store-and-forward systems. The two main forms of network level switching are message switching, discussed

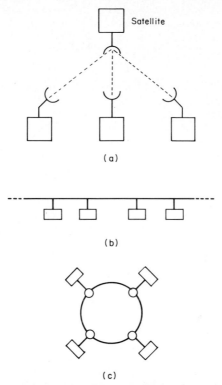

FIG. 16.4 Link level switching (also known as *Broadcast Networking*). (*a*) satellite network, (*b*) bus local network, (*c*) ring local network.

in Chap. 13, and packet switching, discussed in Chap. 14. Level 3 switching has numerous advantages in terms of efficiency and flexibility but presently is well-developed only for data communications as opposed to voice.

Message and packet communications carriers offer both public and private services in their respective fields. In addition, many users have established their own private message and packet networks in both local and wide-area situations.

Whether the switches are analog or digital, all significant terrestrial telephone system switching today is level 1, and this is the type of switching discussed in the remainder of this chapter. It should be noted, however, that the signaling information used to control the level 1 switches in modern public networks is normally sent on a separate level 3 (packet) network. Such an arrangement is sometimes called *common-channel interoffice signaling* (CCIS).

6. Tandem Networks

The basic form of private telephone network is the tandem network based on PBXs and transmission facilities, with some appropriate form of signaling, as shown in Fig. 16.5. Early versions of tandem networks used a *progression con-*

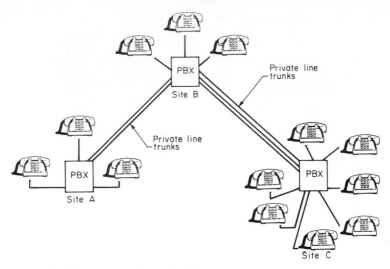

FIG. 16.5   A three-point tandem network.

trol. A caller at site A wishing to communicate with site C would first dial a code that would provide dial tone from site B; then another code would provide dial tone from site C; and finally, dialing the extension number for the destination telephone would provide for completion.

Progression control has a number of disadvantages, the most obvious being nonuniform dialing; i.e., to call a specific destination requires different dialing sequences from different sites. Modern computer-controlled PBXs have the capability of *common-control* tandem networking. As discussed in Chap. 12, this feature allows uniform dialing of a given destination without regard to place of origin of the call. Common control also makes it possible to provide a number of other desirable system features leading to more efficient use of the network and to prevention of the types of abuse discussed in Sec. 16.9.

7. Off-Network Access Strategies

There are very few situations in which 100 percent of the calls placed within a private voice network will be to points on the network. Private voice networks almost always provide for access to the public network; otherwise, each user would be required to have two telephone instruments—one for calls to telephones on the private network and one for calls to other telephones. Calls placed from within a private network to a telephone on a public network are called *off-net* calls. Several strategies are available; whichever one is best depends on calling patterns and applicable tariffs.

The basic form of off-net access line is the *CO trunk,* a local loop from the central office that terminates at a trunk port of a PBX or switch instead of a telephone instrument. Such an arrangement is sometimes called a *local off-net access line* (LONAL). Figure 16.6 illustrates how both local and long-distance calls may be placed using LONALs.

Another access strategy makes use of the *foreign exchange* (FX) circuit. An FX circuit is like a local loop that terminates in a geographical area outside the

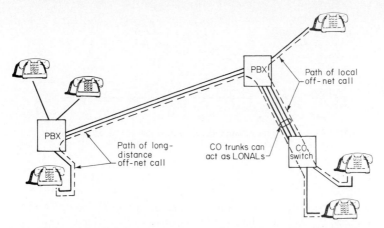

FIG. 16.6 Both local and long-distance off-net calls can be completed via LONALs.

normal area of its central office. Figure 16.7 shows an FX whose *open end* is in Dallas and whose termination end is Kansas City. The user is billed monthly for the local service in Dallas, plus mileage and other charges associated with extending that service to Kansas City. Although the telephone instrument is located in Kansas City, toll calls placed from this telephone are billed as if it were in Dallas.

Figure 16.8 illustrates an FX terminating on a PBX or switch. Such an arrangement is sometimes called an *off-net access line* (ONAL).

Long-distance toll services that are billed on a bulk rather than a per-call basis are generically called *wide-area telecommunications service* (WATS). WATS lines can be terminated on individual telephone instruments, PBXs, and network switches, and can be used as a form of ONAL.

Figure 16.9 shows a long-distance call placed from Dallas to Sacramento on a private network having switches in Dallas and San Francisco. In Fig. 16.9a, the call incurs a long-distance or WATS toll charge from Dallas to Sacramento; this routing is sometimes called *head-in–head-out*. In Fig. 16.9b, the call incurs a long-distance or WATS toll charge from San Francisco to Sacramento; this routing is sometimes called *head-in–tail-out*. The latter scheme obviously results in lower toll charges but may require greater transmission capacity between Dallas and San Francisco.

The determination as to which routing is more economical requires a detailed analysis based on actual *tariffs*. The answer may also depend on the time of day, since both the toll tariffs and the transmission facility loading are time dependent. Private networks often contain software that provides for automatic selection of

Kansas City   FX private line   CO

Dallas

FIG. 16.7 Foreign exchange (FX) circuits extend local telephone service to a customer outside the local exchange (CO) area.

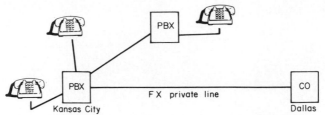

FIG. 16.8 Foreign exchange (FX) circuits can act as *off-network access lines* (ONALs).

the best route between various combinations of city pairs for different times of day. The term for this capability is *automatic route selection* (ARS).

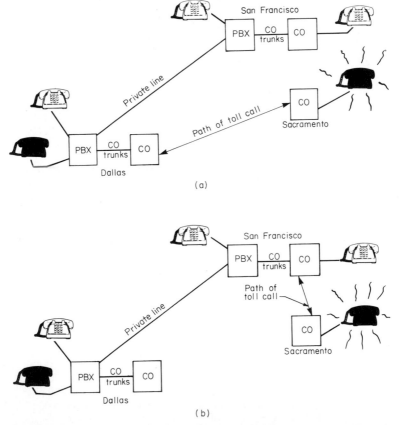

FIG. 16.9 Two diverse strategies for routing the same call. (*a*) head-in–head-out off net toll call, (*b*) head-in–tail-out off-net toll call.

8. On-Network Access

The counterpart of off-net calling is the call that originates in the public network and uses part of a private network in completing the call to a private network destination. The mechanism for *on-net* access can be via business line, FX, 800-WATS, etc.

Many companies that have private networks provide on-net access as a money-saving alternative for their traveling executives and for independent suppliers, dealers, etc. Many such networks are designed to allow on-net calls to go to off-net destinations, resulting in *on-to-off–net*. Figure 16.10 shows how on-net and on-to-off–net calling may result in savings when compared with a pure long-distance toll call.

9. Abuse Prevention

People who use corporate private networks regularly sometimes fall into the trap of thinking, "Since it's paid for, it's free." Not only is this attitude completely false in the case of many on-net and off-net calls, but it leads to *increased usage* and eventual *loss of any economic advantage* of having a private system. Good managers of such systems usually develop equitable ways to charge usage back to the corporate departments of the users.

Basic to the on-network access concept is the issue of *authorization*. Most managers of private networks would agree that simply having an unlisted telephone number for the access line at the gateway into the network would not be sufficiently prudent. Normally, once a user dials an unlisted number to get into the private network, special security equipment at the gateway requires the caller to dial a personal identification/authorization code before dialing the destination telephone number. An eight-digit authorization code is typical, with only a small percentage of possible numerical combinations being valid. Managers change both the access numbers (the unlisted telephone numbers) and authorization codes from time to time.

To illustrate another potential area of abuse, consider the following situation. A telephone in an employee lounge area of a corporate headquarters building is restricted by the PBX from being used to place toll calls on the public network. *However, for company business purposes, this phone can be used for calls to company offices in other cities on the company's private network.* A person

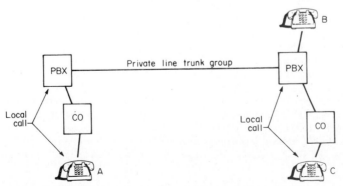

FIG. 16.10   A–B is an on-net call. A–C is an on-to-off–net call.

might try to dial to another PBX on the private network, obtain central office dial tone in that city, and then dial a toll call. If completed, the toll call might be billed to the corporate office in the other city, and the company might have no way to trace the source of the call. To prevent this kind of abuse, most modern PBXs are designed to provide *traveling class marks,* also called *traveling class-of-service codes;* i.e., when the call is passed through the private network to the other PBX, toll restriction information is passed in the signaling, resulting in the distant PBX refusing to complete the call.

Much abuse of private networks is unintentional but nevertheless costly. Consider again the case of the FX circuit arranged to allow calls into the Dallas area from Kansas City. If that circuit were inadvertently used to place a call to a Kansas City number, the path of the call would be from Kansas City to Dallas via the FX circuit, thence from Dallas back to Kansas City via the public toll network. The caller would be charged for the toll call even though the conversation might be with a person only next door! Furthermore, another user might have been deprived of the legitimate use of the FX line during that period.

On-to-off–net calls involving nonlocal segments at both ends often result in inadvertent misuse. Consider the traveling executive who dials an 800 number to get into the company network and then dials an off-net number that involves a WATS line or toll call. Seldom is this type of call cheaper to the executive's company than a straight toll call.

The thrust of these examples is to point out that an efficient private network should include a well-planned and painstakingly administered accounting system together with traveling class marks and ARS capabilities for calls of the various types to ensure that facilities are not being wasted or abused.

SYSTEM PLANNING

In the present context, *planning* refers to strategic planning that typically involves long-range (5 to 10 year) considerations. *Design* implies shorter-range (2 to 4 year) tactical planning. Strategic planning usually employs numbers only for general projections; tactical planning involves an effort to provide numerical detail concerning location, quantity, and cost.

10. Strategic Planning for Communications

The success of the modern corporation is often closely related to its ability to communicate among its employees and with its public. Furthermore, a private communications system can represent a significant commitment of financial and human resources. It is vital, therefore, that the creation and development of a communications system plan be approached with the same degree of formality and thoroughness as any other major strategic business plan.

The basic approach to business planning—from the initial statement of mission to the creation of the executive summary—is covered in numerous business and management textbooks.[3,4] *All the steps of such a generic planning cycle process can and should be applied in creating a communications system plan.* In addition, there are 10 special considerations that should be applied to planning in this special field.

1. Many companies whose main business is *not* telecommunications-related have found ways to gain a *competitive edge* and improve corporate finances by using private telecommunications systems productively. The following is a checklist of strategies for improving the corporate *bottom line* through telecommunications:

    a. Reduce costs while maintaining quality
    b. Improve response time
    c. Reduce excessive inventory
    d. Support customers
    e. Increase productivity
    f. Create new business opportunities

2. A *data base* must be created showing all current and probable future uses of telecommunications by the company. The result will be a matrix showing quantity, type, priority and timing of traffic between various locations. This step requires numerous "brainstorming" sessions and in-depth interviews with persons who are intimately familiar with all aspects of the company's operations. Thoroughness in this step will be rewarded.

3. Potential *new applications* of communications should not be ruled out too early in the planning stage. The following are examples of "brainstormed" applications that each brought big bottom-line results to the organization that developed them:

    a. An airline that made its on-line reservations system available to travel agents
    b. A hospital supply company that put on-line order terminals in health-care facilities
    c. A retailer that developed a point-of-sale network to improve customer services and credit authorization, transmit inventories, speed warehouse orders, etc.
    d. An auto maker that created a satellite-based private video network for training auto mechanics
    e. A university that made graduate-level courses available to employees in remote plant sites via closed-circuit TV with audio talkback

4. Identify in as much detail as possible the corporate and organizational *objectives* in establishing a private system as opposed to using the public system. Consider establishing a hybrid system, partly public and partly private, including the possibility of using a public backup in case of failure or inadequacy of the private system. Make sure that an all-private system is clearly justified before entirely excluding use of the public network.

5. Reliability and restoration of service are more crucial for communications than perhaps any other corporate system. These issues require especially detailed *contingency/backup planning,* with particular attention to the hidden costs and impact (such as lost business) in the event of outage. Any private network, particularly with user-owned subsystems and/or with components provided by multiple vendors, must provide for sophisticated testing and diagnostic procedures by qualified technicians.

6. Vendors can be very helpful in the creation of alternative system plans during the early stages of the planning process. Not only do they help to establish the bidder list, but they also generate ideas that can be collected and evaluated. However, vendor information is no substitute for solid technical expertise in formulating the system architecture. Whether or not this expertise exists in-house, consider obtaining an objective *independent evaluation* of the plan by a qualified consultant.

7. The proliferation of alternative suppliers of subsystems underscores the importance of *compatability,* particularly at the points of interface between subsystems provided by different vendors. One of the best ways to achieve compatibility is to ensure that each vendor adheres strictly to industry-wide interface standards. (See Chap. 15.) Such adherence applies to both hardware and software interfaces. If standards do not exist, then independent certification of interoperability may be possible. In *any* case, an iron-clad guarantee of interoperability should be included in contract terms. The onus should always be on the suppliers—not the customer—to make subsystems work together.

8. While the temptation in today's competitive environment is to pick and choose subsystems from among numerous suppliers, the complexity of operating a system increases dramatically with increasing numbers of suppliers involved. It is better to choose carefully a *small number of suppliers* and develop a strong working relationship with these.

9. *Information flow at the local level* must not be overlooked. While local-area private systems other than PBX-based systems are beyond the scope of this chapter, many companies have improved information flows within and between offices on the local level through telecommunications.

10. Because removal and relocation of communications system components can be costly, special consideration should be given to the *amortization period* of system financing in relation to anticipated long-range business expectations and to the potential for offices being relocated during this period.

SYSTEM DESIGN

Design is intimately related to planning. One can never be done without the other. In general, planning of a communications system answers the question of *what,* whereas design answers the *where* and *how many.* Obviously, neither can be done without the other; if done properly, the resultant system will be the product of an iterative process between planning and design.

Taken together or separately, system planning and design constitute an *art,* not a science. There is almost never a unique correct answer. For example, one plan might involve a number of terrestrial foreign exchange circuits; the design would attempt to determine how many such circuits will be needed, and to what locations. To install only FX circuits, however, will mean forcing calls to destinations outside the FX localities into the public toll network. Perhaps a combination of WATS and FX is justified.

Even if the *what* and *where* were definitely determined, answering the question of *how many* is itself an art. The issue here is to provide "enough" capacity of the *what* to the *where.* Unfortunately, to provide "enough" during peak communication requirements is to have expensive unused capacity during the rest of the time. So the answer to the question, "How much is enough?" is a subjective balance between (1) availability of sufficient capacity when needed and (2) cost. Spending more money to provide additional capacity increases the probability that a communication requirement can be handled when needed, but there is a law of diminishing returns: the actual percentage of utilization of that additional capacity will always be less than that of former capacity, leading to lower efficiency.

The class of techniques available for assisting the designer in resolving these issues is called *traffic engineering,* the subject of the next several sections.

11. Basic Principles of Traffic Engineering

Traffic engineering is the art of providing the "proper" whole-number amount of a user-shared resource that will yield an acceptably low probability of that resource not being available when users need it; i.e., an acceptable level of blockage.

Traffic engineering techniques are used in the design of private networks to determine the number of trunks required in a trunk group, the number of dial-in computer ports required in a computer center, the number of computer terminals required for a group of students to do their homework, the number of FX or WATS lines needed for off-net access, and numerous other communications-related concerns. This type of design activity is called *sizing* or *dimensioning* the resource. *For purposes of illustration in this chapter, the resource being sized is a trunk group between two PBXs or a trunk group from a PBX to a public network switch, but the techniques are much more general than just for that limited type of application.*

Traffic engineering is an art, not a science, because the "correct" answer is always a matter of judgment. Adding more trunks will increase the probability of finding a trunk available on the first attempt (i.e., not being blocked), but it will cost more. Thus, traffic engineering is really a matter of satisfying whoever is paying for the resources that the number of resources provided is the "proper" amount.

Figure 16.11 illustrates the basic concept of *blockage*. Some assumed amount of *traffic* is offered at the entrance to a trunk group; a certain amount is carried, while the remainder is blocked. An increase in offered traffic will likely increase the carried traffic, since an added call may arrive just as another call is being disconnected, but the addition will almost certainly increase the probability of blockage.

Closely related to the issue of an acceptable probability of blockage is the issue of an acceptable disposition of blocked users. Are they to be turned away, placed in a queue, or overflowed to another type of trunk?

The basic inputs to the traffic engineering problem are thus threefold:

1. Traffic pattern (number of would-be users, their arrival and holding times, and the number of trunks needed by each)
2. Disposition of blocked calls
3. Relative economics of the cost of providing additional trunks versus the cost (including possible lost productivity) that will be incurred by blockage

Although there are useful and accurate simulation and modeling methods, the most convenient traffic engineering methods are analytical. These analytical methods find their roots in probability theory, but their application is fairly me-

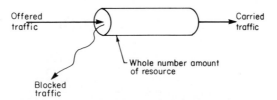

FIG. 16.11   Offered traffic − Blocked traffic = Carried traffic.

chanical and does not require any in-depth comprehension of the theory. It does, however, require a thorough understanding of the *assumptions* and *limitations* implied in the use of the various techniques.

Measuring the Traffic.   The first step is to measure the traffic. In most systems, the traffic level has certain peaks and valleys over a period of time. In the case of typical business voice traffic, there are often daily peaks in the morning and afternoon, as shown in Fig. 16.12. The 60-min period in which the heaviest traffic occurs is known as the *busy hour* and is generally the only period of interest in the design process. That is, if an acceptable probability of blockage can be achieved during the busy hour, then the probability of blockage will be less at all other times, assuming that the number of trunks is constant at all times.

There are numerous theories as to the proper time to measure the busy-hour traffic. Some would choose to sample traffic during the busiest hour on the busiest day of the year. Others would design for a more "typical" day. Obviously, the nature of the business will have significant impact on the choice: travel-related businesses, retailers, and accountants will have quite different annual peaks.

There are a variety of ways to measure traffic. Telephone toll records are dependable if no private alternative currently exists. Traffic usage recorders can be installed on older PBXs, while newer PBXs usually come equipped with a call accounting feature. Total traffic during the busy hour is the summation of the number of call-seconds that will be used by all callers if all calls are handled on the first call attempt (without blockage). If each call occupies the same number of trunks (e.g., one voice-grade trunk), then the traffic in call-seconds is simply (number of calls during busy hour) $\times H$, where H = the average holding time in seconds.

CAUTION: It must be noted that T is greater than the conversation time if part of the trunk group being sized is tied up during the signaling and ringing phase prior to call completion and for calls that do not complete (busy or no answer). Traffic data based on toll charges do not include this time overhead, even though a trunk may have been tied up by the call being set up.

The mathematical unit of traffic is the *erlang*. One erlang is defined as the amount of traffic that can be carried by one trunk. For example, one voice-grade trunk can

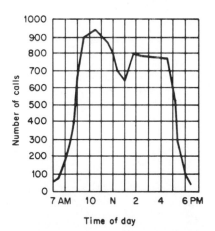

FIG. 16.12   Variation of business communications traffic during a typical business day.

carry one call continuously, or 3600 call-seconds during the busy hour. If 100 callers placed 3-minute calls during the busy hour, the traffic generated would be

$$\frac{(100 \text{ callers}) \times (180 \text{ seconds})}{3600 \text{ seconds/hour}} = 5 \text{ erlangs}$$

Neglecting call setup time, these 5 erlangs might be handled by five trunks *if the calls could be coordinated in such a way that whenever one call terminated, making a trunk available, another call would be placed at that same instant*. Unfortunately, calls do not arrive that way. Callers attempt their calls independently, and their holding times are also independent.

It should be noted that some communications system designers formerly used the CCS as a unit of traffic measurement instead of the erlang. One CCS is equivalent to 100 call-seconds during the busy hour. (The first C is the Roman numeral for 100.) Since one erlang equals 3600 call-seconds during the busy-hour, then

$$1 \text{ erlang} = 36 \text{ CCS per hour}$$

The concept of the CCS as a measure of telecommunications traffic was in common usage before the subject of traffic engineering was well defined in mathematical theory. When traffic is measured in CCS, it is sometimes expressed as number of *unit calls,* where

$$1 \text{ unit call} = 1 \text{ CCS}$$

Modern traffic engineering is based on the mathematics of probability which uses the erlang. When traffic is measured in erlangs, it is sometimes expressed as number of *traffic units,* where

$$1 \text{ traffic unit} = 1 \text{ erlang}$$

Traffic Distribution. It has already been pointed out that distribution of traffic can be as important as quantity of traffic as far as blockage is concerned. Most traffic engineering is based on two special conditions governing distribution:

- *Poisson distribution:* of call attempts during the busy hour.
- *Exponential distribution:* of holding times.

The mathematical theory is well-developed[5] for cases in which these two conditions are satisfied.

Poisson distribution implies three assumptions about call attempts:

- That attempts are independent of each other—that no single event will trigger a rush of call attempts
- That the attempts are randomly distributed throughout the busy hour—that the probability of occurrence of any one call is the same for all times during the busy hour
- That the total number of calls is so large that individual calls do not make a significant difference; i.e., a single call does not make a noticeable lump on the traffic chart.

Holding Time. *Exponential distribution of holding times* implies that whenever the number of calls is large, if the number of calls of length t is plotted versus t, the resulting curve obeys the formula ae^{-bt}, where a and b are constants. Figure 16.13 shows such a mathematical curve and a curve of actual typical[6] call holding times.

For small values of t, the "actual" curve usually peaks above the exponential, then dips somewhat below it as t increases. The peak represents false starts such as wrong numbers and intended party not in. The subsequent dip reflects the fact that, if the intended party is in, then some minimum amount of time is necessary to conduct a meaningful conversation.

Research involving numerous actual situations has shown that the exponential model is sufficiently correct for typical voice conversations. Other types of communications do not necessarily obey the exponential holding-time distribution. Many data systems, packet networks, and computer timesharing systems have holding-time distributions that are significantly different from exponential. The methods, curves, and examples shown in the sections that follow are accurate only for the exponential distribution.

Disposition of Blocked Calls.   A call is said to be *blocked* if there is no available trunk *in the trunk group being sized* at the moment the call is first attempted. Every blocked call will be disposed of in some way. The method of disposition is important to the subject of traffic engineering.

Among the theoretical models of disposition, two *extremes* are possible:

- *Blocked calls cleared* (BCC) means that blocked calls immediately disappear from the trunk group being sized, never to appear again during the busy hour.

- *Blocked calls delayed* (BCD) means that blocked calls are placed in queue indefinitely until the moment a trunk becomes available. Such a model is sometimes called an *infinite-queue* model, but since only the busy hour is being considered, a trunk may become available after the traffic level begins to subside, if not sooner.

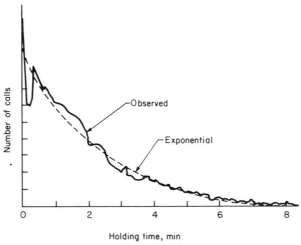

Holding time, min

FIG. 16.13 Typical distribution of holding times for a large number of calls.

Blocked Calls Cleared Model. The BCC model finds many practical applications in *overflow systems*—systems that automatically overflow a blocked call to a different type of trunk besides the one being sized. Examples of systems for which the BCC model applies include:

- A PBX that attempts to place a network call to a certain city on a private-line group to that city and overflows blocked calls to the public network.
- A PBX that attempts to place all long-distance calls on a group of WATS lines and overflows blocked calls to the public network.

Clearly, these examples fit the BCC model perfectly because the overflowed calls immediately "disappear" from the group being sized.

The BCC model has been treated[7] analytically in probability theory, resulting in a solution known as the *Erlang B* formula:

$$P_B = \frac{T^N/N!}{\sum_{i=0}^{N} (T^i/i!)} \tag{16.1}$$

where P_B = probability of blockage (between 0.0 and 1.0) during the busy hour
$\quad\ \ T$ = first-attempt traffic in erlangs distributed over the busy hour
$\quad\ \ N$ = number of trunks in the system being considered

It should be noted that the underlying assumptions about the traffic (namely, *Poisson* distribution of call attempts and *exponential* distribution of holding times) must be true for this formula to yield accurate results.

Blocked Calls Delayed Model. The BCD model also finds many practical applications in modern telephone systems that provide for automatic queuing of blocked calls. In the "infinite queue" extreme in which no maximum waiting time cap is imposed, probability theory[7] has yielded the *Erlang C* formula:

$$P_B = \frac{(T^N/N!)[N/(N-T)]}{\left[\sum_{i=0}^{N-1} (T^i/i!)\right] + [(T^N/N!)[N/(N-T)]]} \tag{16.2}$$

As in the previous case, the Erlang C formula applies to Poisson distribution of attempts and exponential holding times.

Other Models. Between the BCC and BCD extremes are two other important classes of blocked call disposition:

- User retrials, in which users force themselves through the system by redialing at various intervals. The analytical solution to this case is called *extended Erlang B* (EEB).
- Limited queuing of blocked calls, with a specified maximum waiting period after which the unsuccessful caller is automatically overflowed to an alternative facility such as the public network. The analytical solution is called *equivalent queued extended Erlang B* (EQEEB).

12. Probability Curves

Figures 16.14 and 16.15 show the Erlang B results of Eq. 16.1, the BCC

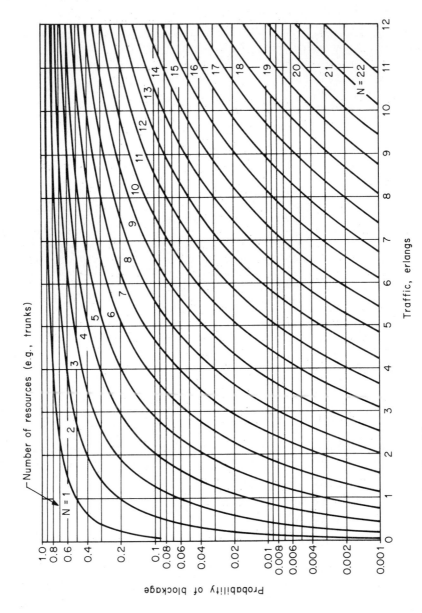

FIG. 16.14 BCC (Erlang B) model.

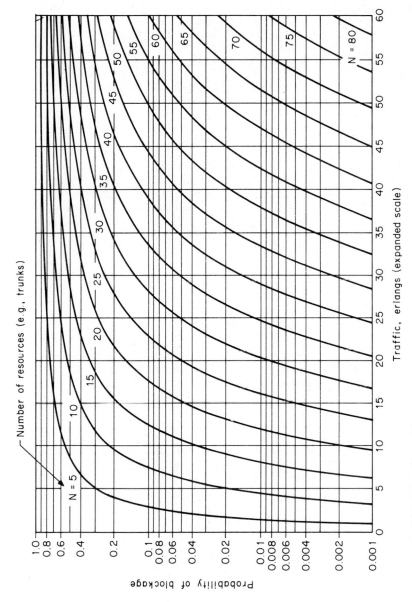

FIG. 16.15 BCC (Erlang B) model.

model. These curves show the probability of blockage P_B versus traffic level T in erlangs for various numbers of resources, N. Figures 16.16 and 16.17 similarly show Erlang C information for Eq. 16.2, the BCD model.

CAUTION: *Accurate results may be expected only if Poisson distribution of call attempts and exponential holding times apply.*

Jewett et al.[8,9] have investigated and tabulated a large number of useful charts showing various probabilities and relative level of loading on facilities under a variety of blocked call disposition models. Figure 16.18 indicates which set of tables from Jewett et al.[9] should be used in a given situation. Other useful information is given in these references concerning the effect of different call arrival and holding patterns.

13. Grade of Service

The probability of blockage on *first attempt during the busy hour* is sometimes referred to as the *grade of service*. It is expressed as Px, where x is some fraction of 1; e.g., $P.05$ means 5 percent probability of blockage during the busy hour. *Ironically, a high grade of service means a high probability of blockage.*

14. Trunking Efficiency

The percentage of resource capacity actually used in a resource system being sized is called the *trunking efficiency* η, also called *group efficiency*.

Since 1 erlang by definition is the amount of traffic one resource can carry,

$$\eta = \frac{\text{carried traffic in erlangs}}{\text{number of resources}} \times 100\% \qquad (16.3)$$

Note that the numerator of Eq. 16.3 is *carried* traffic, not offered traffic. High trunking efficiency means that the resources are being well-utilized but that the probability of blockage (and grade of service) will be high.

15. Tandem Resources

Figure 16.19 illustrates the situation in which a call is jeopardized by potential blockage at more than one point along its route.

The total probability of blockage is

$$P_{SC} = \frac{B_1 + B_2 + B_3}{A_1} \qquad (16.4)$$

If the levels of blockage are *small,* this total is approximately

$$P_B \approx \frac{B_1}{A_1} + \frac{B_2}{A_2} + \frac{B_3}{A_3} = \sum_i P_{B_i} \qquad (16.5)$$

In applying these simple relationships, it should be noted that the various blockage probabilities must be for the same busy hour. Traffic entering at different nodes in a large network involving multiple time zones and calling patterns can create inaccuracy in applying Eq. 16.4 and 16.5.

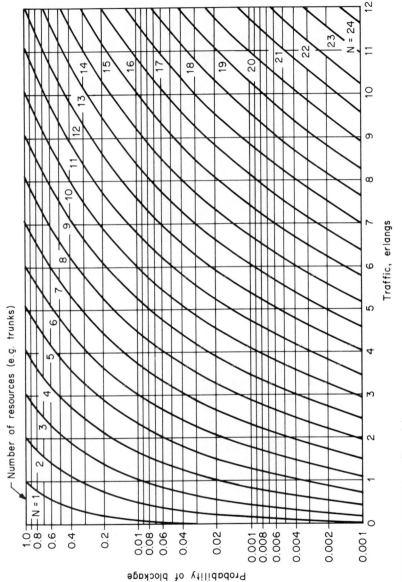

FIG. 16.16 BCD (Erlang C) model.

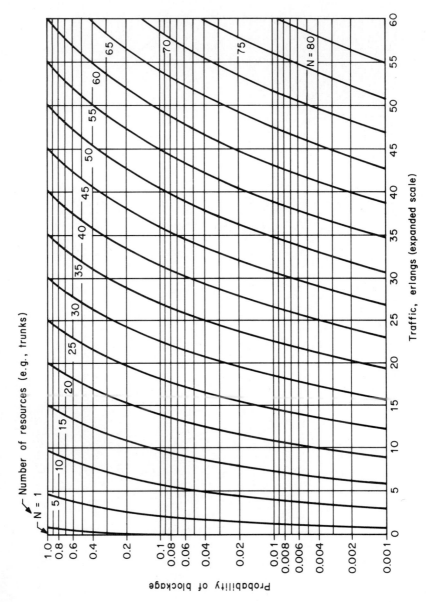

FIG. 16.17 BCD (Erlang C) model.

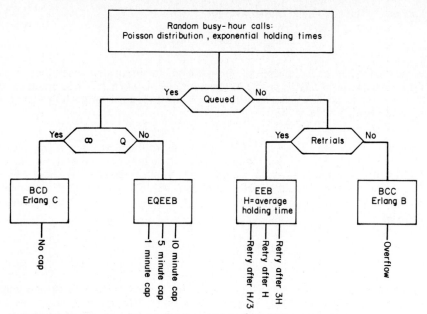

FIG. 16.18 Decision tree for determining which traffic table or curves to use.

16. BCC Example

Problem. In an effort to reduce long-distance toll costs between two of its offices, a certain company wishes to install voice-grade private lines between the PBXs at those sites. The history of toll bills indicates an average of 170 calls averaging 3.5 min each between the two sites during the period from 10:20 to 11:20 a.m. each business day, with lesser traffic levels for all other 1-h periods. Because of the time urgency of the company's business, call attempts for which there is no private line available are to be overflowed immediately to the public *direct distance dialing* (DDD) network. How many private lines will be required to limit DDD to 10 percent of the busy-hour calls?

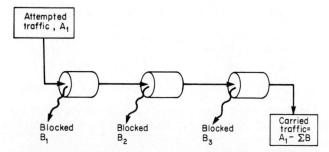

FIG. 16.19 Tandem blockage. B_1, B_2, and B_3 are the probabilities of blockage at the respective stages.

Solution.   The busy-hour traffic level is

$$\frac{170 \text{ calls} \times 3.5 \text{ min}}{60 \text{ min/h}} = 9.92 \text{ erlangs}$$

Since the blocked calls "disappear" from the private line system when they overflow to DDD, this is a BCC problem. From Fig. 16.14, when $T = 9.92$ erlangs and $P = 0.1$, the value of N is between 12 and 13; call it 13, since only whole numbers of trunks are possible.

Analysis.   Since the value of N was rounded up to the next integer, the actual probability of blockage during the busy hour will be lower than 0.1. Returning to Fig. 16.14, the blockage for 9.92 erlangs and 13 trunks is found to be about 0.082. The grade of service would be referred to as *P*.082. The carried traffic (through the 13 trunks) during the busy hour is $(1 - 0.082) \times 9.92 = 9.11$ erlangs, giving a trunking efficiency

$$\eta = \frac{9.11}{13} \times 100\% = 70\%$$

Cost Comparison.   Since only 70 percent of the capacity of the 13 private lines was actually used, and that during the busy hour, the question arises as to whether any money was really saved in comparison with DDD. To answer that question, a knowledge of comparative costs and the off-peak traffic pattern is required. For the purpose of illustration, assume a DDD cost of 50 cents per minute or fraction thereof and $1000 per month per private line, counting all termination and other charges. Further assume that the company averages 21 working days per month, not counting holidays, and that there are a total of 460 non-busy-hour calls spread equally over the 7 nonbusy hours (probably an oversimplification, but it illustrates the technique).

The traffic during each nonbusy hour is

$$\frac{460 \times 3.5}{7 \times 60} = 3.83 \text{ erlangs}$$

For 13 trunks, Fig. 16.14 gives a blocking probability of less than 0.001; i.e., overflow during nonpeak hours is negligible. The total cost of all calls without the private lines would be

$$(170 + 460) \text{ calls/day} \times \$2.00/\text{call} \times 21 \text{ days/month} = \$26,460/\text{month}$$

With the 13 private lines, the cost would be the $13,000 for the private lines, plus the DDD cost of overflowed calls. Since the overflow amounted to 0.082 of the 170 calls or about 14 calls per day, the DDD cost would be

$$14 \text{ calls/day} \times \$2.00/\text{call} \times 21 \text{ days/month} = \$588/\text{month}$$

Thus, the total estimated cost with the 13 private lines is $13,588/month, compared with $26,460 without the private lines. A healthy saving, but is it optimal?

Optimization Example.   The optimization question is generally one of "educated trial and error." It should be obvious that, since the amount of overflow cost in

TABLE 16.1 Results of BCC Example

Characteristic	Number of private lines							
	13	12	11	10	9	8	7	6
Busy-hour blockage	0.082	0.117	0.16	0.21	0.27	0.33	0.41	0.48
Calls/day over-flowed in busy hour	14.0	20.0	27.2	35.7	45.9	56.1	69.7	81.6
Off-peak blockage	Small	Small	0.0015	0.0041	0.011	0.025	0.054	0.105
Calls/day over-flowed off-peak			0.7	2.0	5.0	11.5	25.0	48.3
Total calls over-flowed daily	14	20	28	38	51	68	95	130
DDD cost at $2/call, 21 days	$588	840	1176	1596	2142	2856	3990	5460
DDD cost incre-ment		$252	336	420	546	714	1134	1470

the above calculation is $588 per month, which is less than the cost of adding a single private line, the optimum number is not more than 13.

The question is, at what point does the incremental cost of DDD amount to more than the saving created by removing a private line? Table 16.1, based on Fig. 16.14, shows what happens in this particular example for different numbers of private lines.

Removing trunk number 8 from the group would increase monthly DDD cost by more than $1000. Therefore, from the information given, it appears that eight is the right number of trunks. The total cost of trunks and DDD tolls will be $10,856 per month. At eight private lines, the trunking efficiency is

$$\eta = \frac{(1 - 0.33) \times 9.92}{8} \times 100\% = 83\%$$

The grade of service is $P.33$.

Before leaving this example, it should be noted that two factors were ignored:

- The time a private line is tied up before completing a call. This factor can inflate the apparent usage by up to a half-minute per call.

- The possibility of increased trunking requirements between the PBXs and COs as DDD overflow is allowed to rise.

Both of these neglected items, if taken into account, would tend to make the optimal number of trunks larger. If the two offices in this problem are located in different time zones, then the busy hours for the CO trunks may be different from the busy hour for the private line group, thus negating the second item.

17. BCD Example

Suppose now that instead of blocked calls overflowing to DDD, they are placed in a queue until a private line becomes available. While it lowers telephone costs, such a system can result in long delays, disgruntled personnel, and lost productivity on unusually heavy days. Many managers of such systems provide for either an executive override or a maximum time in queue or both.

Queued systems are a classic example of the subjectivity aspect of traffic engineering. That is, how much queue delay is appropriate to save the premium cost of DDD?

The curves given in this chapter do not reveal how long a queuing delay to expect. Martin[10] has tabulated these delays.

From Fig. 16.16, the blockage levels for 9.92 busy-hour erlangs for different numbers of trunks are

Number of trunks	13	12	11	10	9
Probability of blockage	0.27	0.43	0.66	0.97	1.0

Such high blockage levels would probably be unacceptable in a corporate environment without some override feature for high-priority calls. In a situation in which no DDD alternative existed, the BCD model would be more practical. Likewise, if the DDD cost per minute were higher, it would make the DDD overflow alternative less attractive and BCD more practical.

OTHER CONSIDERATIONS

The technical complexity and myriad alternatives for private communications systems planning and design preclude more than a basic description in one chapter. Future digital technologies and services may change much of the terminology, but the concepts emphasized here apply equally to analog and current digital systems, including integrated services digital networks. In the paragraphs that follow are several special considerations that may apply to current and future systems.

18.   Blockage in Switching

Electromechanical switch contacts, whether or not they are controlled by computer, place a premium on the number of *simultaneous connections* that can be made in a given switch. Switching systems of the past were sized on the number of erlangs of busy-hour traffic in much the same way as trunk groups are sized. Total tandem blockage included both trunk and switch blockage according to Eq. 16.4 and 16.5.

Modern digital integrated circuit technology has significantly reduced the actual cost of the switching matrix. The cost of today's switches is much more related to software and other protocol-sensitive factors. In most cases, the incremental switching cost to reduce or eliminate blockage is much smaller than the incremental cost of trunks to yield similar results.

While no longer a major concern, potential blockage even in modern switches is an important consideration. The "rule of thumb" of the last generation of voice network designers was to allow from about 0.15 to about 0.20 erlangs of busy-hour traffic per telephone station, depending on the type of business. When no hard data on usage were available, the analyst usually would just assume a level of usage.

For several reasons, the era of data communications has invalidated all traditional rules of thumb. *First,* holding times for switched data connections tend to be longer, especially for interactive and timesharing applications. *Second,* holding times for data tend to be of relatively constant length for a given system, negating the exponential holding time assumptions used in classical traffic engineering. As data traffic becomes an increasingly significant portion of total traffic flow, the exponential holding time assumption becomes less valid. *Third,* integrated voice/data systems tend to combine multiple simultaneous uses (by time-division multiplexing) into one switch port. *Fourth,* the push for faster data speeds has caused some users of analog transmission facilities to dial repeatedly each morning until they attain an adequate link and then leave the connection "up" all day!

As mentioned in Sec. 9, Abuse Prevention, good management of corporate communications systems includes finding an equitable way to charge users of these systems. As mentioned in Sec. 10, Strategic Planning, system suppliers and consultants can be helpful in suggesting ways to obtain and evaluate valid traffic data.

19. Higher-Level Switching

Section 5, Switching, discussed both *level 2* (link level) and *level 3* (network level) switching. All discussion on design has so far been directly applicable only to *level 1* (physical or circuit level) switching.

Since level 2 switching is done in a broadcast environment in which all points on the link "hear" all messages but only the intended user terminal "receives" them, there can be only one message in the link at a time. In order to yield sufficient capacity most link-type switching systems burst the messages through the transmission medium at speeds much faster than the terminals can send or receive them and use digital buffers to make up the difference between burst rate and terminal rate. Many types of local-area data networks use this bursting principle, including *carrier-sense multiple-access with collision detection* (CSMA/CD) systems and collision-avoidance systems such as token bus and token ring.

The capacity, transit delay, and other crucial factors in level 2 switching are critically related to amount of usage, distances, and other variables in ways that are completely different from analogous relationships in level 1 switching. Planners are cautioned that a thorough understanding of the technical aspects and limitations of such systems is vital. Stallings[1,11] has documented many of these phenomena. As with any prospective system, it is useful to require a would-be supplier to *demonstrate* the prospective application running satisfactorily on a working system prior to any contractual arrangement.

Level 3, or *packet,* switching uses store-and-forward techniques in which the actual route of a message (sometimes called a *packet*) through a network is dependent on a header at the beginning of that message. While there can be numerous economic advantages to packet switching in many data applications, as explained in Chap. 13, the technology of the 1980s makes it impractical for certain real-time applications. For example, variations in routing and buffer delays have so far prevented widespread application of packet switching of digitized speech. The promise of faster packet processors may make packet voice practical in the future.

20. Simulation Techniques

The analytical design approach presented in this chapter required certain mathematical assumptions about the traffic, the disposition of blocked calls, time zones, etc. While this analytical approach can be a very useful tool for the initial design phase, inaccuracies and exceptions in any of the assumptions can invalidate capacity predictions.

Besides the analytical approach, there is another entire class of design techniques involving *simulation*. In simulation, all of the significant factors in a real system are represented inside a computer.

A certain memory location is designated as the *master clock* and is advanced at intervals representing times (e.g., seconds) throughout the business day. Other memory locations are designated as *individual resources* to be sized (e.g., trunks). As the master clock advances, calls are generated and destroyed according to actual historical records or some other technique. While a call exists, flags are placed in the resources in combinations representing the tandem topology of the actual network. Whenever a call is generated for which no unflagged end-to-end path is available, that blocked call is disposed of appropriately (e.g., placed in a queue or in an overflow counter). To simulate a real first-in–first-out queuing system, calls already in queue are given priority over calls that arrive at a later time on the master clock. As the master clock continues to step through the business day, significant data are recorded, such as maximum length of queue, maximum time in queue, resource utilization, and calls overflowed.

Any system for which call data can be gathered can be simulated. The problem is that the computer programming of the actual simulation must be customized for each network topology. Graybeal and Pooch[12] have summarized the state of the art of simulation technology.

21. Practicality of Private Communications Systems

Total traffic volume alone cannot justify a private communications system. A company might spend millions of dollars per month on long-distance calls, but only when there is sufficient *geographic concentration* in and/or between specific areas can a private system be justified economically.

Even given sufficient concentration of communications requirements, a private system does not necessarily mean that every application must be included in a private network. Planners must avoid the tendency to include every conceivable application in order to justify a network. Beware of including in a network those heavy applications that would best be served by a high-capacity point-to-point circuit.

Also beware of the tendency to assume that all information flow must necessarily be handled by electronic communications. There are numerous low-priority bulk data transfers that are best handled by shipment of a truckload of magnetic tape rather than by hundreds of megabits per second. Even high-priority bulk data is sometimes best handled by the planeload rather than by electronic means. The planner must maintain an *unbiased* broad view of alternatives.

A CHECKLIST-SUMMARY OF CLASSICAL NETWORK DESIGN PROCEDURE

Designing a good nontrivial system requires long experience and other factors too numerous to cover in one chapter. However, we present the following as a general summary:

1. Adhere to solid planning-cycle procedures (Sec. 10).
2. Create a traffic matrix, including quantity, type, and priority of traffic (Secs. 10 and 11).
3. Determine a philosophy of off-net and on-net access (Secs. 7, 8, and 9).
4. Determine the appropriate disposition of blocked calls (Sec. 11).
5. Eliminate those communications applications that should not be networked (Sec. 21).
6. Combine clusters of nearby sites into optimum single-node regional networks. This is usually done by exhaustively testing all sites in a cluster as a potential switching node for that region.
7. Design a backbone network to tie the regional nodes together.
8. Recheck to see whether certain pairs of distant sites having a high level of common traffic need to be regrouped into common clusters.
9. Recheck the design regarding alternative routes, number of intermediate switching sites allowed for a given call, time zone effect on busy hour, etc.
10. Remove from the network sites for which traffic is too light to justify inclusion.

REFERENCES

22. Cited References

1. Stallings, W., *Local Area Networks,* 2nd ed., Macmillan, New York, 1987.
2. *ANSI/IEEE 802 Standards,* distributed by Wiley-Interscience, New York, 1984 and later.
3. Steiner, G., J. Miner, and E. Gray, *Management Policy and Strategy: Text, Readings, and Cases,* 3rd ed., Macmillan, New York, 1986.
4. Glueck, W., *Business Policy and Strategic Management,* 3d ed., McGraw-Hill, New York, 1980.
5. Beckmann, P., *Introduction to Elementary Queuing Theory and Telephone Traffic,* The Golem Press, Boulder, Colo., 1968.
6. *Switching Systems,* AT&T, New York, 1961.
7. Briley, B., *Introduction to Telephone Switching,* Addison-Wesley, Reading, Mass., 1983.
8. Jewett, J., et al., *Designing Optimal Voice Networks for Business, Government, and Telephone Companies,* Telephony Publishing Corp., Chicago, 1980.
9. Jewett, J., et al., *Traffic Engineering Tables: The Complete Practical Encyclopedia,* Telephony Publishing Corp., Chicago, 1980.
10. Martin, J., *Systems Analysis for Data Transmission,* Prentice-Hall, Englewood Cliffs, N.J., 1972.
11. Stallings, W., *Data and Computer Communications,* Macmillan, 2nd ed., New York, 1988.
12. Graybeal, W., and U. Pooch, *Simulation: Principles and Methods,* Little, Brown, New York, 1980.

23. General References

Datapro Reports, Datapro Research Corp., Delran, N.J. (There are several series of subscription reports on a variety of subjects relating to private communications systems planning and design. Contact Datapro for a current list of series.)

Freeman, R., *Telecommunication System Engineering,* John Wiley & Sons, New York, 1980.

Gurrie, M. and P. O'Connor, *Voice/Data Telecommunications Systems,* Prentice-Hall, Englewood Cliffs, N.J., 1986.

Martin, J., *Telecommunications and the Computer,* 2d ed., Prentice-Hall, Englewood Cliffs, N.J., 1976.

CHAPTER 17
TELEVISION COMMUNICATION SYSTEMS DESIGN

Edward D. Horowitz
Senior Vice President, Technology and Operations
Home Box Office, Inc.

Virgil D. Conanan
Senior Systems Engineer, Networking Engineering
Home Box Office, Inc.

CONTENTS

INTRODUCTION

1. Scope of Chapter

This chapter describes the essential elements that make up the television communication medium, particularly the systems for the origination and transmission of television signals. In the interest of brevity, much of the information has been condensed into capsular form. The reader may obtain additional information from the extensive list of references at the end of the chapter.

TELEVISION SIGNAL FORMATS AND SPECIFICATIONS

2. Luminance and Chrominance Signals

In monochrome television, an image is represented in terms of its luminance (brightness) level alone. The luminance voltage generated at the studio is proportional to the instantaneous brightness on each point of the scanned screen.

In color television, the qualities of color are conveyed simultaneously with the luminance information. This results in compatibility; i.e., a monochrome receiver is able to display the brightness attribute of a color signal as though it were receiving monochrome while a color receiver can reproduce either a monochrome or color picture.

The basic function of the color camera is to separate the red (R), green (G), and blue (B) components of the optical image and generate the corresponding electrical signals. These RGB signals are processed (encoded) to form the luminance and chrominance signals.

The luminance signal, E_y, is produced by adding specific portions of the RGB signals:

$$E_y = 0.30E_r + 0.59E_g + 0.11E_b \tag{17.1}$$

The chrominance signal includes two basic components, the I and Q signals, which are transmitted simultaneously with the luminance signal:

$$I = 0.60E_r - 0.28E_g - 0.32E_b \tag{17.2}$$

$$Q = 0.21E_r - 0.52E_g + 0.31E_b \tag{17.3}$$

These three simultaneous equations are solved continuously in a receiver matrix circuit to recover the original RGB signals.

The I and Q signals are band-limited before transmission; this takes advantage of the eye's relative insensitivity to fine detail in color and particularly its insensitivity to detail in the hues represented by the Q signal. This makes it possible to band-limit the I signal to 1.6 MHz and the Q signal to 0.6 MHz.

The I and Q signals modulate the color subcarrier in phase quadrature with double-balanced modulators which suppress the subcarrier as well as the baseband I and Q signals. Thus, the combined output contains only sidebands. The composite waveform E_m can be expressed as

$$E_m = E_y + [E_I \cos (\omega t + 33) + E_Q \sin \omega t + 33)] \tag{17.4}$$

The process for encoding the luminance and chrominance signals is shown in Fig. 17.1. Note that the Q subcarrier is delayed from the I subcarrier by 90°, and both carriers are delayed by 57°. To maintain this relationship during transmission and reproduction, a reference signal called the *color burst* is added to the video signal.

3. Timing Signals

Standard synchronization signals are employed to ensure that the television camera and receiver have identical scanning processes. They are called the *vertical*

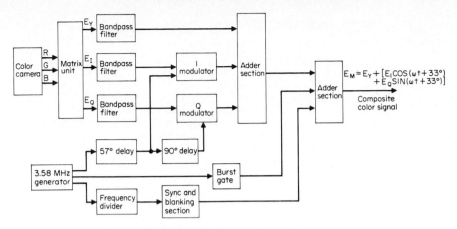

FIG. 17.1 The generation of luminance and chrominance signals.

and *horizontal blanking/synchronizing* pulses and are added to the picture signal at the studio.

The NTSC standard specifications for scanning and synchronization signals are shown in Fig. 17.2.

4. The NTSC Standard Composite Color Video Waveform

The composite video signal contains all the information required to reproduce the original scene faithfully. It includes the vertical and horizontal synchronization and blanking pulses, the luminance and chrominance information, the color burst, and other signals utilizing the vertical blanking interval. Figure 17.3 shows the standard National Television System Committee (NTSC) composite video waveform.

This signal must be transmitted with minimum degradation. A number of parameters have been defined to evaluate the performance of transmission systems on an objective basis. The performance standards for key parameters as established by the Electronic Industries Association (EIA)[13] for television relay systems (microwave and satellites) are summarized in Table 17.1. Standards for other services and methods for measurement, are described in Ref. 1 to 14.

1. *Insertion gain* is defined as the total peak-to-peak amplitude change between the input and output of a video facility. Excessive gain can cause overloading which results in such defects as clipping and sync compression. Insufficient gain results in a decrease in picture brightness and/or a degraded signal-to-noise ratio. It can also make the system susceptible to interference.

2. *Line-time waveform distortion* is determined by the low-frequency performance of the transmission system—below a few hundred Hertz. It affects the shape and tilt of the white bar test signal. Excessive line-time distortion results in horizontal streaking and smearing in the picture.

3. *Short-time waveform distortion* is determined by the high-frequency performance of the transmission system. It affects the horizontal resolution and the sharpness of the reproduced picture.

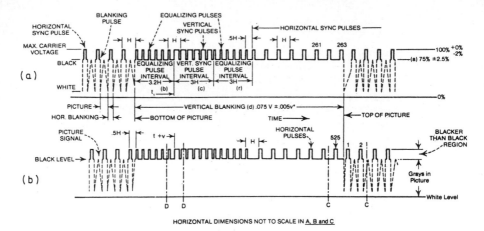

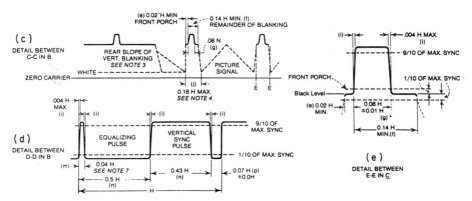

FIG. 17.2   The NTSC standard scanning and synchronization signals.[8]

4. *Chrominance-luminance gain inequality* occurs when the chrominance component of the signal is attenuated or becomes higher relative to the luminance component. The colors of the reproduced picture become more pale or more vivid as compared with the original picture.

5. *Chrominance-luminance delay inequality* is caused by a delay or advance of the chrominance component with respect to luminance. It results in misregistration of the color in relation to the actual picture.

6. *Differential gain* is a measure of the change in amplitude of the color subcarrier as the luminance is varied from blanking to white level. Its effect on the picture is a change in color saturation or vividness with brightness.

7. *Differential phase* is a measure of the change in phase of the color subcarrier as the luminance is varied from blanking to white level. Its effect is a change in hue with brightness.

8. *Gain-frequency distortion* is the change in gain of the system over the video-frequency band. Depending on the frequencies affected, poor fre-

quency response generally results in loss of picture sharpness and various color impairments.

9. *Chrominance nonlinear gain distortion* is a measure of the variation in chrominance gain with the luminance amplitude at various luminance levels and *average picture levels* (APLs).

10. *Chrominance nonlinear phase distortion* is a measure of the variation in the phase of the color subcarrier at various luminance levels and APLs. This distortion affects color hues.

11. *Chrominance-luminance intermodulation distortion* is the change in ampli-

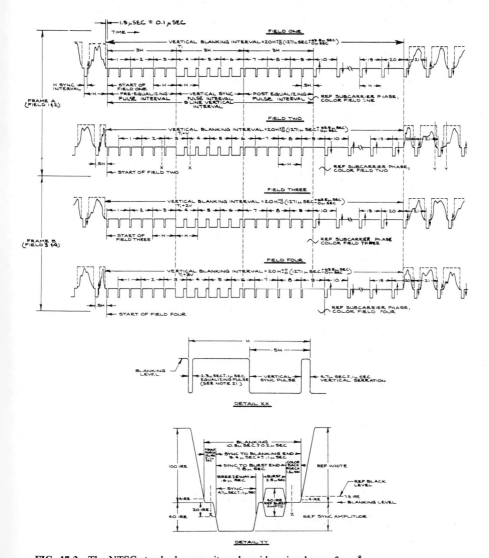

FIG. 17.3 The NTSC standard composite color video signal waveform.[8]

tude of the luminance signal upon superimposition of a chrominance signal at a specified amplitude and APL. It results from rectification of the subcarrier in highly saturated portions of the picture by nonlinear circuit elements. It causes portions of the picture to be lighter or darker than normal.

12. *Field-time waveform distortion* indicates the amplitude distortion at the field frequency of 60 Hz. It results in shading from top to bottom of the picture.

13. *Signal-to-random-noise ratio (S/N)* is the ratio, luminance weighted, measured in decibels, of the peak-to-peak luminance signal amplitude to the rms amplitude of the noise. The visual effect of excessive random noise is pronounced graininess or "snow" in the reproduced picture. *S/N* is measured

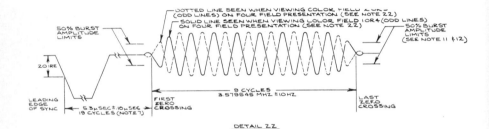

DETAIL ZZ

NOTES:

1. SPECIFICATIONS APPLY TO STUDIO FACILITIES. NETWORK AND TRANSMITTER CHARACTERISTICS ARE NOT INCLUDED.
2. ALL TOLERANCES AND LIMITS SHOWN IN THIS DRAWING PERMISSABLE ONLY FOR LONG TIME VARIATIONS.
3. BURST FREQUENCY SHALL BE 3.579545 MHz ±10 Hz.
4. HORIZONTAL SCANNING FREQUENCY SHALL BE 2/455 TIMES THE BURST FREQUENCY.
5. VERTICAL SCANNING FREQUENCY SHALL BE 2/525 TIMES THE HORIZONTAL SCANNING FREQUENCY.
6. START OF COLOR FIELDS ONE AND THREE IS DEFINED BY A WHOLE LINE BETWEEN THE FIRST EQUALIZING PULSE AND THE PRECEDING H SYNC PULSE. START OF COLOR FIELDS TWO AND FOUR DEFINED BY A HALF LINE BETWEEN THE FIRST EQUALIZING PULSE AND THE PRECEDING H PULSE. COLOR FIELD ONE IS THAT FIELD WITH POSITIVE GOING ZERO CROSSINGS OF REFERENCE SUBCARRIER NOMINALLY COINCIDENT WITH THE 50% AMPLITUDE POINT OF THE LEADING EDGES OF EVEN NUMBERED HORIZONTAL SYNC PULSES.
7. THE ZERO-CROSSINGS OF REFERENCE SUBCARRIER SHALL BE NOMINALLY COINCIDENT WITH THE 50% POINT OF THE LEADING EDGES OF ALL HORIZONTAL SYNC PULSES. FOR THOSE CASES WHERE THE RELATIONSHIP BETWEEN SYNC AND SUBCARRIER IS CRITICAL FOR PROGRAM INTEGRATION. THE TOLERANCE ON THIS COINCIDENCE IS ±45° OF REFERENCE SUBCARRIER.
8. ALL RISE TIMES AND FALL TIMES UNLESS OTHERWISE SPECIFIED ARE TO BE 0.140 μSEC ±0.02 μSEC MEASURED FROM TEN TO NINETY PER CENT AMPLITUDE POINTS. ALL PULSE WIDTHS EXCEPT BLANKING ARE MEASURED AT FIFTY PER CENT AMPLITUDE POINT.
9. OVERSHOOT ON ALL PULSES DURING SYNC AND BLANKING (VERTICAL AND HORIZONTAL) SHALL NOT EXCEED TWO IRE UNITS. ANY OTHER EXTRANEOUS SIGNALS DURING BLANKING INTERVALS SHALL NOT EXCEED TWO IRE UNITS, MEASURED OVER A BANDWIDTH OF 6 MHz.
10. BURST ENVELOPE RISE TIME IS 0.30 μSEC MEASURED BETWEEN THE TEN AND NINETY PER CENT AMPLITUDE POINTS. IT SHALL HAVE THE GENERAL SLOPE SHOWN.
11. THE START OF BURST IS DEFINED BY THE ZERO-CROSSING (POSITIVE OR NEGATIVE SLOPE) THAT PRECEEDS THE FIRST HALF CYCLE OF SUBCARRIER THAT IS 50% OR GREATER OF THE BURST AMPLITUDE.
12. THE END OF BURST IS DEFINED BY THE ZERO-CROSSING (POSITIVE OR NEGATIVE SLOPE) THAT FOLLOWS THE LAST HALF CYCLE OF SUBCARRIER THAT IS 50% OR GREATER OF THE BURST AMPLITUDE.
13. MONOCHROME SIGNALS SHALL BE IN ACCORDANCE WITH THIS DRAWING EXCEPT THAT BURST IS OMITTED, AND FIELDS THREE AND FOUR ARE IDENTICAL TO FIELDS ONE AND TWO RESPECTIVELY.
14. REFERENCE SUBCARRIER IS A CONTINUOUS SIGNAL WHICH HAS THE SAME INSTANTANEOUS PHASE AS BURST.

15. PROGRAM OPERATING LEVEL WHITE IS 100 IRE, +0, -2 IRE.
16. PROGRAM OPERATING LEVEL BLACK IS 7.5 IRE, ±2.5 IRE.
17. PROGRAM OPERATING LEVEL SYNC IS 40 IRE, ±2 IRE.
18. PROGRAM OPERATING LEVEL BURST IS 40 IRE, ±2 IRE.
19. BURST PEDESTAL NOT TO EXCEED ±2 IRE.
20. BREEZEWAY, BURST, COLOR BACK PORCH, AND SYNC TO BURST END ARE NOMINAL IN DETAIL BETWEEN YY. SEE DETAIL BETWEEN ZZ FOR TOLERANCES.
21. RATIO OF AREA OF VERTICAL EQUALIZING PULSE TO SYNC PULSE SHALL BE WITHIN 45 TO 50 PER CENT.
22. THERE WILL BE A 180 DEGREE REVERSAL OF PHASE WHEN VIEWING EVEN LINES ON A FOUR FIELD PRESENTATION. A FOUR FIELD PRESENTATION MEANS A DISPLAY DEVICE WHICH IS TRIGGERED BY FOUR FIELD (15 Hz) INFORMATION.

THIS DRAWING CORRESPONDS TO PROPOSED RS 170A VIDEO STANDARD. IT SHOULD BE RESTRICTED TO INTERNAL USE ONLY.

COLOR TIMING DATA:
1° = .776 ns
1ns = 1.289°
FOR CABLE WITH 66% PROPAGATION FACTOR,
1° = 6.035" = .503'
1ns = 7.778" .648'

Fig. 17.3 *(Continued)*

with specified band-limiting and weighting networks. Band limiting eliminates irrelevant out-of-band energy; weighting simulates the lower visibility of high-frequency, fine-grained noise.

The signal-to-noise ratio is a critical parameter, since the design of a television transmission or distribution system usually requires a tradeoff between cost and signal-to-noise performance. Commercial video distribution systems normally operate with a signal-to-noise ratio around 54 dB.

5. PAL Baseband Video Waveform

The *phase-alternating line* (PAL) system is another color television transmission system which is in wide use. It was developed to reduce the visible effect of transmission distortions.

The eye is very sensitive to the small changes in hue which result from phase distortion and quadrature crosstalk in the NTSC system. Phase distortion causes a change in hue; quadrature crosstalk, which results from attenuation of the color subcarrier's upper sideband, causes increased visibility of "cross-color" and colored edges in the picture. The PAL system reduces the visible effect of these deficiencies by reversing the phase of the $(E_r - E_y)$ signal by 180° on alternate lines. This results in complementary hue distortions in adjacent lines. Within certain limits, the viewer's eye "averages" the color errors, and the effects of transmission distortions are less annoying.

To enable PAL receivers to synchronize with the phase alternation process at the camera, the color burst phase is alternated ± 45°.

As with NTSC, the luminance signal is formed by combining the red, green, and blue components in accordance with Eq. 17.1. Different PAL systems vary slightly in the bandwidth of the luminance signal.

TABLE 17.1 RS-250-B End-to-End Video Performance Objectives

Parameters	Objectives
1. Insertion gain	± 0.5 dB hourly
	± 0.3 dB over 1 second
2. Line-time waveform distortion	2 IRE peak-to-peak
3. Short-time waveform distortion	7 IRE peak-to-peak
4. Chrominance-luminance gain inequality	± 7 IRE
5. Chrominance-luminance delay inequality	± 60 ns
6. Differential gain	10%
7. Differential phase	3 degrees
8. Gain/frequency distortion	0 dB, dc–3.0 kHz
	± 0.7 dB at 3 MHz
	± 0.3 dB, 3.3–3.9 MHz
	± 0.9 dB at 4.2 MHz
9. Chrominance nonlinear gain distortion	5%
10. Chrominance nonlinear phase distortion	5 degrees
11. Chrominance-luminance intermodulation distortion	4%
12. Field-time waveform distortion	3 IRE peak-to-peak
13. Signal-to-noise ratio	54 dB

The chrominance signal is expressed by

$$C_{PAL} = (U/2.03) \sin \omega t \pm (V/1.14) \cos \omega t \qquad (17.5)$$

where $U = E_b - E_y$, $V = E_r - E_y$, and $\pm$ implies phase reversal. The correct switching sense of the V signal is determined by the alternating burst phase.

U and V occupy equal bandwidths of about 1.3 MHz.

The subcarrier frequency for 625-line, 50-field television systems is given by

$$F_{sc} = (1134/4)f_h + (1/2)f_v \qquad (17.6)$$

The last term (25 Hz) is introduced to reduce the visibility of vertical dot patterns.

To ensure the correct phasing of the first color burst following the vertical interval, a pulse called the "meander gate" is used to shift the burst reinsertion by one line at the vertical field rate. The meander gate timing diagram is shown in Fig. 17.4.

6. Multiplexed Analog Components

Multiplexed analog components (MAC) is a baseband format which was developed to reduce or eliminate the inherent distortions and signal degradation resulting from the transmission of chroma as a subcarrier in the NTSC and PAL formats.

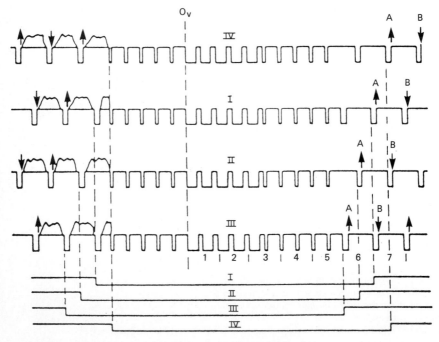

FIG. 17.4   The PAL "meander" blanking gate timing diagram.[1]

These include cross-luminance, cross-color, differential phase, differential gain, chrominance-luminance intermodulation distortion, and (for FM transmission) high chroma noise. These effects become pronounced in unstable and nonlinear transmission paths.

In the various MAC formats, the luminance and two chrominance signals are time-compressed in line-store devices and transmitted sequentially, line-by-line. This eliminates crosstalk between these signals and allows the chrominance to be transmitted as a baseband signal rather than as a subcarrier. The characteristics of the components of the MAC format are as follows:

1. *The MAC luminance component:* The MAC luminance component is time-compressed in a ratio of 3:2. This reduces the line transmission time from approximately 54 μs to 35 μs, and the luminance signal is transmitted at the end of each line (see Fig. 17.5). The bandwidth requirement is increased by the same ratio.

2. *The MAC chrominance component:* The chrominance signal components, I and Q or U and V, are time-compressed in a ratio of 3:1. This reduces the line transmission time to 17.5 μs, and this signal is transmitted at the beginning of each line. Only one chroma component is transmitted on a line; each alternates with the other on a line-by-line basis. This balances the vertical and horizontal chroma resolution. The chroma bandwidth is increased to about 2.1 MHz.

3. *MAC synchronization:* MAC eliminates vertical and horizontal sync pulses, and the sync information is transmitted in the form of digital words. The elimination of redundant synchronization signals results in substantial savings in bandwidth.

4. *MAC digital audio:* The bandwidth saved by eliminating the analog sync signal can be used for in-band digital audio. Up to four or eight channels of digital audio can be time-multiplexed with the MAC video. This not only results in the audio quality associated with digital transmission but also puts the signal in a form suitable for encryption.

The rest of the bandwidth can be used for additional digital messages mixed in the bit stream. Examples are decoder authorization, a data encryption algorithm for digital audio and video scrambling, teletext, and program-related control messages such as tiering, blackout, and pay-per-view.

A number of formats have been developed which employ the MAC principle. Two of the most common are B-MAC and C-MAC.

B-MAC inserts baseband sound and data at 1.8 Mbs during the 9-μs horizontal blanking period. A representation of the B-MAC signal in both frequency and time domains is shown in Fig. 17.5. Note that just over 6 MHz of bandwidth is occupied.

The C-MAC signal is shown in Fig. 17.6. The main difference between B-MAC and C-MAC is in the data component; C-MAC multiplexes the data at rf with a data rate in excess of 20 Mbs. Up to eight digital channels can be accommodated. The baseband signal bandwidth exceeds 10 MHz.

7. High Definition Television (HDTV)

It is expected that high definition television (HDTV) will develop in three phases. The first phase, which is underway today, will be its use as a tool for the production and editing of television programs and motion pictures. HDTV is especially convenient in post-production editing because of electronic special effects.

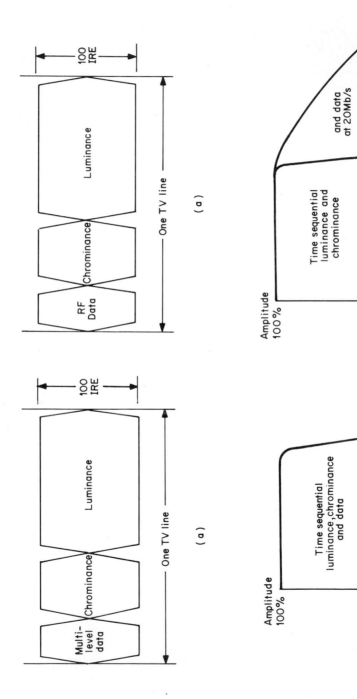

FIG. 17.5 The B-MAC signal in the time and frequency domains. (*a*) B-MAC format, (*b*) B-MAC baseband frequencies.

FIG. 17.6 The C-MAC signal in the time and frequency domains. (*a*) C-MAC format, (*b*) C-MAC baseband frequencies.

Further, its quality is sufficiently high to produce multigeneration copies without excessive degradation. After editing, the HDTV signal can be converted to a standard transmission format or rerecorded on film.

The second phase, which is expected to begin in the early 1990s, will be the use of HDTV for the display of pictures in commercial closed-circuit applications.

Finally, and probably many years in the future, HDTV may be used to deliver programs directly to the home with signal qualities and attributes similar to those found in theatrical films.

Because of its wider aspect ratio (16:9 for HDTV, 4:3 for NTSC) and greater image detail (see Fig. 17.12), HDTV requires more scanning lines and bandwidth. This will necessitate a major improvement in the performance of every element in the system—the origination equipment, such as cameras and tape recorders; the transmission medium, and the display devices. A number of production and display devices are already commercially available.

A comparison of the subjective picture quality of HDTV with that of motion picture film is shown in Fig. 17.7.

The lack of an accepted worldwide standard for HDTV contributed to its slow growth. In 1985, the U.S. Advanced Television Systems Committee (ATSC) endorsed the 1125-line/60 Hz system as a production (origination) standard.

The development of a standard transmission format is proceeding very slowly because of the complex technological and political issues involved. Some design criteria are the following:

1. The format shall be chosen so that the attributes of the reproduced picture constitute a signal that the public perceives as a significant improvement over the existing NTSC system;

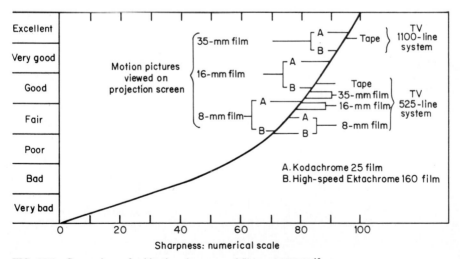

FIG. 17.7 Comparison of subjective sharpness of film and HDTV.[15]

2. Compatibility with existing NTSC television receivers shall be maintained;

3. If frequency spectrum reallocation is required, then an orderly transition shall be implemented to ensure that the general public is served;

4. The existing transmission equipment used for over-the-air, CATV, and satellite systems shall require minimum modification;

5. If additional receiving equipment in the home is required, then such equipment shall be made inexpensive and shall permit reception of other video services; and,

6. The selected transmission format shall allow a robust technique for securing the video service from unauthorized users without compromising signal quality.

8 . Other Baseband Signals

Vertical Interval Test Signals.   Vertical interval test signals (VITS) facilitate the in-service performance of video facilities. They were developed by the Network Transmission Committee (NTC), a joint committee of television network broadcasters and the Bell System, and are described in a document, NTC Report No. 7.

The Committee's recommendations resulted in the widespread use in the United States of two waveforms inserted in the video signal prior to transmission. Figure 17.8 shows the composite test signal, inserted on line 17, field 1. Figure 17.9 shows the combination test signal, inserted on line 17, field 2. Both figures list the transmission parameters which are monitored.

Other common test signals used worldwide are described in the references.

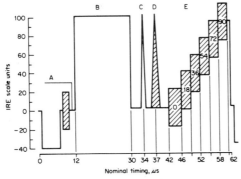

Composite VITS
(measurable parameters)
- Insertion gain
- Line-time distortion
- Short-time distortion
- Chrominance-luminance gain inequality
- Chrominance-luminance delay inequality
- Differential gain
- Differential phase
- Luminance nonlinear distortion
- Chrominance-luminance intermodulation
- Signal-to-random noise ratio

A = Blanking level
B = Luminance bar (reference white level)
C = 2T sine-squared pulse
D = Modulated 12.5T sine-squared pulse
E = Modulated 5-riser staircase

FIG. 17.8   The composite VITS test signal.

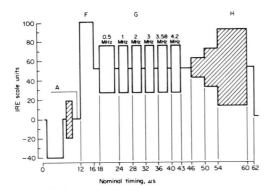

**Combination VITS
(measurable parameters)**

- Insertion gain
- Line-time distortion
- Short-time distortion
- Gain/frequency distortion
- Chrominance nonlinear gain distortion
- Chrominance nonlinear phase distortion
- Chrominance-luminance intermodulation
- Signal-to-random noise ratio

F = Reference bar
G = Multiburst
H = Superimposed three-level chrominance signal

FIG. 17.9 The combination VITS test signal.

Vertical Interval Reference Signal. The vertical interval reference (VIR) signal, Fig. 17.10, is used to maintain the original color quality of the video signal. It is inserted in line 19 of both fields of the composite color signal at the studio or transmitting facilities prior to transmission. Some color receivers use the VIR signal for automatic adjustment of hue and saturation.

Source Identification Signals. Source identification (SID) signals, Fig. 17.11, were developed by U.S. television networks to indicate the originating network

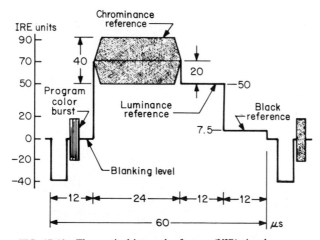

FIG. 17.10 The vertical interval reference (VIR) signal.

Source identification signal format
48 bits on line 20, field 1, frames 02 and 03

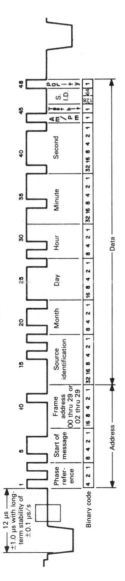

Typical waveform shown is for line 20, field 1, frames 02 and 03
and is coded starting at bit 8: frame 02, source identification 05,
6th month, 30th day, 11th hour, 55th minute, 30th second, PM, eastern
daylight time. Bit 48 indicates odd parity during this transmission.

Bits 1 through 12 and 48 appear on all frames. Bits 12 through 47
may contain different information on different frames as
assigned below.

Frame	Assignment
00	Not assigned: See note 10
01	Not assigned: See note 10
02	SID calendar and time
03	SID calendar and time
04 & 05	Uplink source name (call ltrs, licence no. etc.)
06 & 07	Destination name (call ltrs, co. name, etc.)
08 & 09	First half, contact info (tel. no. etc.)
10 & 11	Second half, contact info (tel. no. etc.)
12 & 13	First half, source location "t, zip, etc.)
14 & 15	Second half, source location (long. zip, etc.)
16-29	Next assigned

Source
Identification	Assignment
00-15	ABC
16-31	CBS
32-47	NBC
48-63	PBS
64-79	ABC
80-95	CBS
96-111	NBC
112-127	PBS
128-143	Not assigned
144-159	" "
160-175	" "
176-191	" "
192-207	" "
208-223	" "
224-239	" "
240-254	Not assigned
255	Non net source

Notes

1. Send on line 20, field 1, only.
2. Rise time 250 ns ± 50 ns. Overshoot, undershoot and spurious signals
   less than ± 2 IRE units.
3. Bit intervals 1 μs ± 0.1 μs, H.A.D. with cumulative error over 48 bits per line not to
   exceed 0.5 μs.
4. Data is in binary code.
5. "1" = 50 IRE units −0 + 10 IRE units.
6. "0" = 0 IRE units −0 + 10 IRE units.
7. AM = 0, PM = 1
8. Bit 45 is available for forcing a change line in decoder
   memory for system test.
9. Video not to scale.
10. Frame addresses 00 and 01 are omitted in certain
    seconds per drop frame time code. Deleting these
    frame numbers adjusts time frame count so that
    frame count time will track real-time seconds.
11. When present, frame address 00 identifies the
    first even frame number. When frame address 00
    is omitted, 02 is the first even frame address.
12. The data in adjacent frames 02 and 03 are identical.
    Data changes may occur following frame 03 and before
    the next frame 02.
13. Bit 48 will be a parity bit and be equal to a one when the
    sum of bits 1 thru 47 are even for that transmission.
    (odd parity)

FIG. 17.11   The source identification signal.

as well as the time of transmission. A total of 48 bits of digital information is encoded in line 20, field 1, frames 02 and 03. The information contains reference bits, the network initials, the current month, day, hour, minute and second, a.m. or p.m., and the time zone.

The first 12 bits are sent in each occurrence of the SID line; the rest is transmitted at the SID line only on frames 02 and 03 each second.

Closed-Caption Signals. Closed captioning is a system for providing text with television for the deaf or hearing-impaired. Alphanumeric characters are displayed on the screen using special closed-caption decoders. The text is transmitted digitally in 51.628 µs in line 21, field 1. The standard transmission format is shown in Fig. 17.12.

The seven 0.503-MHz cycles are frequency- and phase-locked to the caption data clock. They are followed by two data bits at zero logic level, then a logical one start bit. A frame code consisting of the last two cycles of the clock run-in and the three logical bits is used to indicate the start of data. The actual data consist of two bytes coded in a *non-return-to-zero* (NRZ) ASCII format with an eighth bit added for odd parity.

Videotex. *Videotex* is a generic term selected by the International Telephone and Telegraph Consultative Committee (CCITT) to describe the transmission of digitally coded text and graphics via television. There are two subsets of videotex: *teletext* and *videotext*.

Teletext is one-way videotex. It transmits text and simple graphics for display on the television screen. It is often termed *noninteractive* because the transmitter sends information in a constant cycle rather than on demand.

The subscriber retrieves information from a group of pages or "magazine" which is transmitted repeatedly. The user first accesses a list of the magazine contents or "menu" by means of a keypad, then selects a category and page number from the menu. The desired page is extracted and stored by the receiver and teletext decoder. It is then displayed on the television screen.

Videotext is a true *interactive* system that allows a business or home viewer to search and randomly select information from a central data base. The text and/or

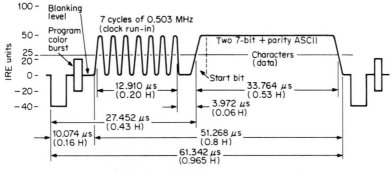

FIG. 17.12 The closed caption data signal format.[8]

graphics are then displayed in real time. The communication channels can be telephone lines, cable, or broadcast transmitters, or a combination of these media.

Videotex signals are digitally encoded and transmitted in data packets on one or more TV scanning lines. Lines 16 to 21 have been used without objectionable interference to present-day television receivers.

For optimum transmission efficiency, the data pulse spectrum is shaped to a 60 percent roll-off of raised cosine, truncated at 4.2 MHz. The optimum transmission rate is 5.727272 MHz.

9. Audio Signal Specifications

Audio is an indispensable portion of a television signal, and the television system designer must be familiar with audio principles and practices.

The most important audio measurement parameters and EIA recommendations for their limits in a television relay system are summarized in Table 17.2. Other standards are described in Refs. 1 to 14.

1. *Impedance:* The *input* or *load impedance* is that presented by the input terminals of an audio transmission facility. *Output impedance* is the internal impedance at the output terminals of the facility.

2. *Return loss:* Return loss is a parameter that permits the measurement of impedance errors. When impedance mismatches occur, reflections result and cause incorrect amplitude versus frequency characteristics. Audio input/output impedance is usually measured with a return loss bridge. Alternatively, return loss may be calculated as follows:

$$\text{Return loss} = 20 \log \left[(Z_o - Z)/(Z_o + Z) \right] \quad \text{dB} \qquad (17.7)$$

where Z_o = specified standard impedance and Z = measured impedance.

3. *Insertion gain:* Insertion gain is measured directly between the input and output terminals of an audio system. The standard test tone levels for measuring insertion gain range from 0 to + 18 dBm (reference 1 mW at the standard terminating impedance) at 1 kHz. The nominal program audio transmission level is set at least 10 dB and preferably 18 dB below the test tone level.

4. *Amplitude versus frequency response:* The amplitude versus frequency response is measured with respect to the response at 1 kHz using a test tone of constant amplitude (nominally at −10 dBm).

5. *Total harmonic distortion:* Harmonics result from nonlinearities in the system. Total harmonic distortion is measured with a standard test tone (typically 1 kHz) injected at a level of approximately 0 dBm. A distortion analyzer filters out the test tone, analyzes the harmonic content in a predefined passband, and expresses the result in percent total harmonic distortion.

6. *Signal-to-noise ratio:* The unweighted signal-to-noise ratio is the ratio of the rms signal voltage reference level to the rms noise voltage level at the system output terminals. In this measurment, the program audio input is removed and the input terminals terminated with the input load impedance (150 or 600 Ω). The signal now measured at the output is unwanted noise.

$$S/N = 20 \log \left[V_{\text{reference}}/V_{\text{measured}} \right] \quad \text{dB} \qquad (17.8)$$

TABLE 17.2   RS-250-B End-to-End Audio Performance Objectives

Parameters	Objectives
1. Input-output impedance	150 or 600 Ω, balanced to ground
2. Return loss	30 dB
3. Insertion gain	± 0.5 dB
4. Amplitude-frequency response	− 1 dB at 50 Hz
	0 dB from 100 Hz–7.5 kHz
	− 1.5 dB at 15.0 kHz
5. Total harmonic distortion	1%
6. Unweighted signal-to-noise ratio	56 dB
7. Channel separation (crosstalk)	Not set
8. Phase shift	Not set

7. *Channel separation:*   Channel separation or audio crosstalk is a stereo parameter. The input of the channel under test is terminated with its proper impedance, and its output voltage measured as the other channel is fully modulated with a 1-kHz test tone. The crosstalk into the left channel from the right is defined as

$$\text{Right-to-left crosstalk} = 20 \log \left[V_{\text{left}} / V_{\text{right}} \right] \qquad (17.9)$$

The crosstalk of the left channel into the right is measured with the right terminal terminated and calculated with the reciprocal equation.

8. *Phase shift:*   The phase relationship between the left and right channels of a stereo signal must be maintained throughout the transmission system to prevent sound cancellations in the reproduced sound. Extreme phase differences result in a lack of realism or loss of spatial separation.

10.   Video Baseband Frequency Spectrum

The frequency domain representation of the baseband video signal is necessary for planning the spectrum occupancy of the transmission channel.

Figure 17.13 shows the complete frequency spectrum of a standard NTSC color video signal. The chrominance information carried by the color subcarrier (dashed lines) is transmitted simultaneously with the luminance signal (solid lines) in a channel no wider than that used for monochrome transmission.

The energy produced by scanning an image is concentrated at whole multiples of the frame and line frequencies with levels decreasing at the higher harmonics. Figure 17.13 shows that nearly half the monochrome video spectrum is unoccupied. The location of these gaps is utilized for transmitting the *I* and *Q* sidebands of the chrominance signal.

In order to fill these gaps with the chrominance sidebands and thus achieve maximum compatibility between monochrome and color transmission systems, it is necessary to make the color subcarrier frequency an exact odd multiple of one-half the line frequency. This requires a change in the line rate from 15,750 Hz in monochrome systems to 15,734.264 Hz for color. This corresponds to a subcarrier frequency of 3.579545 MHz.

The field rate is reduced from 60 Hz to 59.94 Hz, but the effect of this reduction is negligible in modern TV sets.

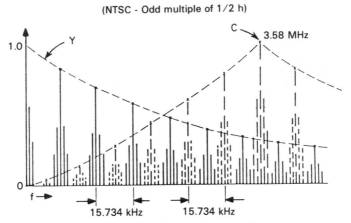

FIG. 17.13   The video baseband frequency spectrum.[1]

TELEVISION SIGNAL TRANSMISSION MODES

11.   Broadcast and CATV Transmission Modes

The technical standards for television broadcasting in the United States are defined by the Federal Communications Commission (FCC).

Each broadcast station is assigned a specific channel in the VHF band (54 to 72, 74 to 88, 174 to 216 MHz) or the UHF band (470 to 806 MHz). The FCC coordinates channel assignments so that cochannel VHF stations are 170 to 220 mi apart (170 mi in the densely populated northeast) and cochannel UHF stations are 155 to 205 mi apart. Adjacent channel separations are 60 mi for VHF and 55 mi for UHF.

The FCC also establishes transmission standards. These are summarized in Table 17.3.

The mode of transmission in cable television is different from broadcast. Since it is a "closed" and stable medium, the amplitudes of adjacent channel signals are kept approximately equal. This makes it possible to use all the broadcast channel frequencies within the CATV system without concern for adjacent channel interference. Additionally, the use of frequency converters at subscribers' locations makes it possible to utilize nonstandard channels which are either *incrementally related carriers* (IRCs) or *harmonically related carriers* (HRCs). In both cases, a 6-MHz separation between channels is always maintained. As a result, CATV systems can distribute 50 or more television channels.

Table 17.4 shows the broadcast and CATV channel assignments and designations. Modern TV sets and CATV converters are capable of tuning to these frequencies and display the corresponding designations.

12.   Digital Transmission

Transmission of television signals in the digital mode requires the conversion of *time-varying* (analog) signals representing images and sound into a series of *dis-*

TABLE 17.3   FCC Broadcast Transmission Standards

Parameter	Specification
Channel width	6 MHz
Picture carrier location	1.25 MHz $\pm$ 1 kHz above the lower channel boundary
Aural carrier center frequency	4.5 MHz $\pm$ 1 kHz above the picture carrier
Horizontal scanning frequency F_h	15,734.264 $\pm$ 0.044 Hz
Vertical scanning frequency	59.94 Hz
Polarity of transmission	Negative—a decrease in light intensity causes an increase in radiated power
Power ratio between picture and sound carriers	5:1–10:1 (7 dB to 10 dB)
Polarization of radiated wave	Horizontal, right-hand circular, or elliptical
Type of modulation:	
Picture	Vestigial sideband amplitude modulation of picture carrier by frequency-multiplexed luminance Y and chrominance signals I and Q
Sound (mono)	Baseband 50–15 kHz with 75 μs preemphasis, $\pm$ 25 kHz FM deviation
Sound (stereo)	Broadcast Television System Committee (BTSC)-encoded baseband frequency modulating the aural carrier up to $\pm$ 100 kHz
Main channel	Stereophonic sum (L + R) 50–15 kHz, with 75 μs preemphasis, $\pm$ 25 kHz deviation
Pilot subcarrier	15,734 Hz $\pm$ 2 Hz (F_h)
Stereo subchannel	$2F_h$ AM/suppressed-carrier modulated by (L − R) signal, 50–15 kHz, dBx-encoded for noise reduction
Second audio program (SAP)	$5F_h$, FM signal modulated by separate programming, 50–10 kHz, at 10-kHz deviation
Nonprogram-related channel (pro channel)	$6.5F_h$ $\pm$ 500 Hz, if both SAP and pro channel are transmitted; can be anywhere between 47 to 120 kHz $\pm$ 500 Hz if only stereo subcarrier is transmitted

crete (digital) pulses or bits. Once digitized, digital processing techniques are employed to produce the optimum format for the selected transmission medium.

Various *analog-to-digital* (A/D) and *digital-to-analog* (D/A) conversion and quantization techniques for both video and audio signals are described in Chaps. 10 and 11.

Either primary color component (RGB) or composite (NTSC or PAL) signals can be encoded in digital form. Each has advantages and disadvantages.

The transmission of component-encoded signals has the advantages described

Table 17.4 CATV Channel and Frequency Designations

Historical Channel Designation	EIA 1S-6 Channel Designation	Standard Video Frequency, MHz	IRC Video Frequency, MHz	HRC Video Frequency, MHz
2	02 or 2	55.25	55.25	54.00
3	03 or 3	61.25	61.25	60.00
4	04 or 4	67.25	67.25	66.00
4 + or A-8	01 or 1	N/A	73.25	72.00
5 or A-7	05 or 5	77.25	79.25	78.00
6 or A-6	06 or 6	83.25	85.25	84.00
A-5	95	91.25	91.25	90.00
A-4	96	97.25	97.25	96.00
A-3	97	103.25	103.25	102.00
A-2	96	109.25	109.25	108.00
A-1	95	115.25	115.25	114.00
A	14	121.25	121.25	120.00
B	15	127.25	127.25	126.00
C	16	133.25	133.25	132.00
D	17	139.25	139.25	138.00
E	18	145.25	145.25	144.00
F	19	151.25	151.25	150.00
G	20	157.25	157.25	156.00
H	21	163.25	163.25	162.00
I	22	169.25	169.25	168.00
7	07 or 7	175.25	175.25	174.00
8	08 or 8	181.25	181.25	180.00
9	09 or 9	187.25	187.25	186.00
10	10	193.25	193.25	192.00
11	11	199.25	199.25	198.00
12	12	205.25	205.25	204.00
13	13	211.25	211.25	210.00
J	23	217.25	217.25	216.00
K	24	223.25	223.25	222.00
L	25	229.25	229.25	228.00
M	26	235.25	235.25	234.00
N	27	241.25	241.25	240.00
O	28	247.25	247.25	246.00
P	29	253.25	253.25	252.00
Q	30	259.25	259.25	258.00
R	31	265.25	265.25	264.00
S	32	271.25	271.25	270.00
T	33	277.25	277.25	276.00
U	34	283.25	283.25	282.00
V	35	289.25	289.25	288.00
W	36	295.25	295.25	294.00
AA	37	301.25	301.25	300.00
BB	38	307.25	307.25	306.00
CC	39	313.25	313.25	312.00
DD	40	319.25	319.25	318.00
EE	41	325.25	325.25	324.00
FF	42	331.25	331.25	330.00
GG	43	337.25	337.25	336.00
HH	44	343.25	343.25	342.00

Table 17.4 CATV Channel and Frequency Designations (*Continued*)

Historical Channel Designation	EIA 1S-6 Channel Designation	Standard Video Frequency, MHz	IRC Video Frequency, MHz	HRC Video Frequency, MHz
II	45	349.25	349.25	348.00
JJ	46	355.25	355.25	354.00
KK	47	361.25	361.25	360.00
LL	48	367.25	367.25	366.00
MM	49	373.25	373.25	372.00
NN	50	379.25	379.25	378.00
OO	51	385.25	385.25	384.00
PP	52	391.25	391.25	390.00
QQ	53	397.25	397.25	396.00
RR	54	403.25	403.25	402.00
SS	55	409.25	409.25	408.00
TT	56	415.25	415.25	414.00
UU	57	421.25	421.25	420.00
VV	58	427.25	427.25	426.00
WW	59	433.25	433.25	432.00
XX	60	439.25	439.25	438.00
YY	61	445.25	445.25	444.00
ZZ	62	451.25	451.25	450.00
AAA	63	457.25	457.25	456.00
BBB	64	463.25	463.25	462.00
CCC	65	469.25	469.25	468.00
DDD	66	475.25	475.25	474.00
EEE	67	481.25	481.25	480.00
FFF	68	487.25	487.25	486.00
GGG	69	493.25	493.25	492.00
HHH	70	499.25	499.25	498.00
III	71	505.25	505.25	504.00
JJJ	72	511.25	511.25	510.00
KKK	73	517.25	517.25	516.00
LLL	74	523.25	523.25	522.00
MMM	75	529.25	529.25	528.00
NNN	76	535.25	535.25	534.00
OOO	77	541.25	541.25	540.00
PPP	78	547.25	547.25	546.00
QQQ	79	553.25	553.25	552.00
RRR	80	559.25	559.25	558.00
SSS	81	565.25	565.25	564.00
TTT	82	571.25	571.25	570.00
UUU	83	577.25	577.25	576.00
VVV	84	583.25	583.25	582.00
WWW	85	589.25	589.25	588.00
XXX	86	595.25	595.25	594.00
YYY	87	601.25	601.25	600.00
ZZZ	88	607.25	607.25	606.00
?	89	613.25	613.25	612.00
?	90	619.25	619.25	618.00
?	91	625.25	625.25	624.00
?	92	631.25	631.25	630.00
?	93	637.25	637.25	636.00
?	94	643.25	643.25	642.00

earlier for transmission of analog component signals (MAC). Further, there is greater flexibility in choosing an appropriate technique for bit rate or bandwidth reduction. For example, in RGB encoding it is not necessary to quantize all three signals with the same precision because at constant luminance the threshold signal-to-noise ratios are 36 dB for blue, 41 dB for red, and 43 dB for green. This implies that the bit rate can be reduced by using less quantizing bits for blue than those for red and green. Another advantage in component coding is that it eliminates the inherent distortions produced by chrominance-luminance interlacing such as cross-color (color patterns produced in fine detail), cross-luminance (dot structures on sharp edges of the picture), differential phase, differential gain, and chrominance-luminance intermodulation distortion. In addition, component coding makes it easier to exchange programs between countries having different color standards because each system uses essentially the same baseband components. Component editing has been adopted as a studio standard (CCIR Recommendation 601).

In instances where individual components are not accessible, such as signals derived from recordings or broadcasts, the composite signal can be directly coded. Since the composite video signal consists of the luminance signal Y and two chrominance signals I, Q or U, V, which are frequency multiplexed in the same spectrum originally assigned to the luminance signal, maintaining the signal in its composite form avoids problems of maintaining color balance in digital coders.

The sampling rate found most suited for coding composite video is an integral multiple of the subcarrier frequency. This avoids the intermodulation which may occur because of the simultaneous presence of the sampling and subcarrier frequencies in the A/D conversion process.

Whichever coding technique is employed, digital transmission requires much more bandwidth than analog.

A bit rate of 216 Mbs has been recommended (CCIR Recommendation 601) for component encoding. This rate has generally limited the use of this technique to studio applications.

Various techniques have been developed to reduce the bit rate for the transmission of composite video in the digital mode. Target bit rates of 44 Mbs (United States), 35 Mbs (Europe), and 33 Mbs (Japan) have been established for satellite transmission of standard broadcast signals.

Bit rates of 1.44 Mbs and lower have been used in teleconferencing applications (see Chap. 18), where compromises in picture definition and the portrayal of motion are acceptable.

The extremely large bandwidths which are available on fiber-optic systems make them particularly attractive for digital transmission.

The principal advantages of digital television are

1. It is relatively immune to noise and distortions
2. It is particularly adaptable to encrypted transmission
3. It can utilize large scale integrated circuits (LSIs) for signal processing and memory storage
4. It is compatible with integrated services digital network (ISDN) systems that combine video, voice, and data in one channel.

The disadvantages are

1. It requires extremely large bandwidth
2. It needs complex A/D and D/A converters
3. It is costly in point-to-multipoint applications

In the immediate future, the main applications of digital television will be in studio production and postproduction and in teleconferencing.

13. Fiber-Optic Transmission Modes

Fiber optics can be used for the transmission of either analog or digital television signals. Likewise, the transmission mode can be either analog or digital. The advantages and disadvantages of these modes are described in Chap. 9, in the references, and in Sec. 14.

Digital signals can be transmitted directly on fiber-optic systems. Analog signals can be converted to digital form for transmission or they can be transmitted in the analog mode. For analog transmission, the signal frequency modulates an rf carrier which in turn intensity-modulates the laser light source. The use of frequency modulation is desirable if not necessary because it is resistant to the nonlinear distortions which are inherent in the modulation of the laser.

For transmission of multiple channels, a combination of multiplexing in the time domain—*time-division multiplexing* (TDM) for digital, *frequency-division multiplexing* (FDM) for analog—and in the wavelength domain (WDM) can be used. Figures 17.14 and 17.15 are block diagrams of digital and analog multiplexing systems.

The frequency-modulated-carrier frequencies are chosen so that strong intermodulation products do not interfere with other carriers or subcarriers.

14. Microwave and Satellite Transmission Modes

While digital transmission of television signals is occasionally employed in microwave and satellite systems, the use of analog frequency modulation is by far the most common. The advantages of FM for this application are

1. It minimizes the effect of nonlinearities in the transmission channel

2. It is immune to AM noise

3. Power-limited systems can take advantage of the wider bandwidth to increase the S/N

4. Various processing techniques can be employed to optimize video transmission, e.g., multiplexing, preemphasis, and threshold extension.

The choice of the optimum modulation index m, which is the ratio of the FM deviation Δf to the highest modulating frequency f_m, is critical.

Increasing the deviation increases the bandwidth: Bandwidth $= 2(m + 1)f_m$.

On the other hand, a higher deviation increases the FM improvement factor and the S/N (see Sec. 17.21).

The spectral distribution of an FM signal as a function of m is shown in Fig. 17.16.

The selection of the optimum deviation must be based on the number of channels to be transmitted, the type of baseband signal (e.g., component or composite), the signal quality requirement, the power received, and the available bandwidth.

A typical value of peak Δf for a C-band satellite transponder with a nominal bandwidth of 36 MHz is 10.75 MHz.

For satellite transmission, in addition to the deviation of the carrier by the signal, it is usually necessary to subject the main carrier to a low-frequency devia-

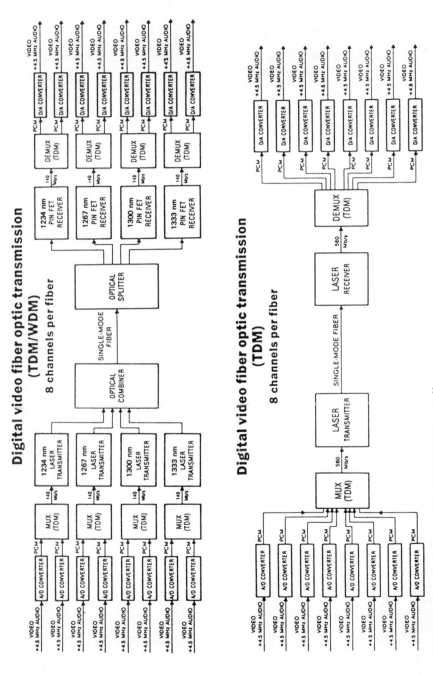

FIG. 17.14 Fiber-optic transmission in the digital mode. [16]

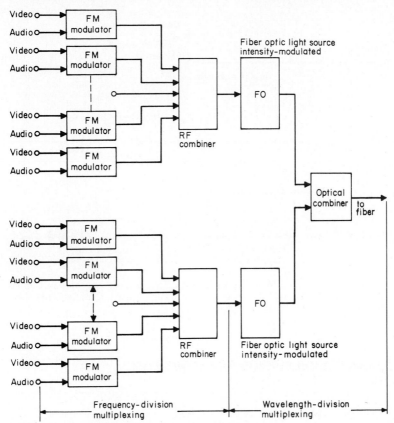

FIG. 17.15 Fiber-optic transmission in the analog mode.

tion. This spreads the high concentration of carrier and sideband energy over a larger range of the spectrum and permits higher satellite *effective radiated power* (ERP) without exceeding the FCC's limit on watts/meter^2/kHz downlink power density.

Preemphasis/deemphasis is employed in FM systems for the transmission of video to compensate for the increase in thermal noise with increasing baseband frequency. CCIR Recommendation 567 (based on Report 637) specifies a standard 75-μs preemphasis.

Threshold extension demodulation (TED) is a common technique used in FM receivers to reduce video impulse noise when the carrier-to-noise (*C/N*) ratio drops below the receiver's operating threshold. Above threshold, the receiver acts like a standard discriminator; when *C/N* drops below threshold, TED circuitry automatically switches to a narrow bandwidth.

A characteristic of FM is that the detected signal-to-noise ratio for the video signal is higher than its *C/N* ratio. This difference is the FM improvement factor. Both terrestrial microwave and satellite television take advantage of this improvement, as shown in Sec. 21.

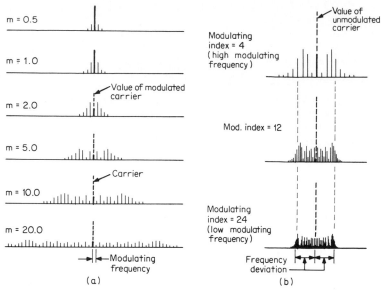

FIG. 17.16 Spectra of frequency modulated signals. (*a*) frequency spectra with increasing frequency deviation and constant modulating frequency, (*b*) frequency spectra with constant frequency deviation.

TERRESTRIAL TELEVISION TRANSMISSION MEDIA

15. Coaxial Cable

Coaxial cable finds widespread application for terrestrial television transmission because it is a "closed" medium, it is relatively wideband, it can support various modes of transmission, it is independent of terrain, and it has proven cost-effective for multichannel applications.

The diameter of the outer conductor of commonly used cables varies from 0.22 in (5.53 mm) to 1.63 in (41.3 mm).

The type of dielectric material that separates the inner and outer conductors influences the cable's propagation loss and frequency response. Polystrene, with its relatively low dielectric constant ($\epsilon = 2.25$), is widely used. Cables with diameters exceeding 0.5 in (1.27 mm) may be pressurized to prevent moisture penetration.

The characteristic impedance of coaxial cable in ohms is

$$Z_o = (138/\epsilon^{1/2}) \log (D/d) \qquad (17.10)$$

where ϵ = dielectric constant of separator
D = inner diameter of outer conductor
d = diameter of center conductor

Common impedances of commercially available cables are 50, 75, and 95 Ω. The losses in coaxial cable commonly used in CATV systems are shown in Table 17.5. They vary about ± 10 percent for every 10°F change in temperature. The losses also vary with frequency. To find the loss L_u at a frequency F_u:

$$L_u = L_k(F_u/F_k)^{1/2} \qquad (17.11)$$

where L_u = unknown loss to be determined for F_u and L_k = known loss at F_k.

The most extensive use of coaxial cable systems is in *cable* (originally *community antenna television*) *distribution systems* (CATV). Originally intended to distribute local broadcast signals, CATV systems now offer a variety of local broadcast, distant broadcast, pay TV, and other entertainment and commercial services.

16. Waveguides

Waveguides are transmission lines made of hollow tubes that transmit microwave signals. They are frequently used for the transmission of high power with low losses.

There are three basic types—rigid rectangular, rigid circular, and semiflexible elliptical. They are ordered from the manufacturer in exact lengths and specified for the desired frequency band.

Typical waveguide characteristics are shown in Fig. 17.17.

Rigid circular waveguides have the lowest loss. It is practical only in straight runs and requires complex networks for circular-to-rectangular transitions.

Rectangular and semiflexible elliptical waveguides have similar characteristics. They are widely used because they are easy to install in long continuous runs.

Transmission lines using coaxial cables or waveguides require careful attention to good installation practices. The number of bends, connections, and transitions should be kept to a minimum. Care should be exercised during installation to avoid slight deformations or misalignment; these can cause losses due to impedance mismatch or vapor penetration.

17. Terrestrial Microwave

Terrestrial microwave systems are used for short- and long-haul, high-capacity, terrestrial television transmission.

FM is generally used for repeatered long-haul applications because of its relative noise immunity and insensitivity to system nonlinearities. It requires more bandwidth, and this reduces the channel capacity.

Although AM is more susceptible to noise and distortions, it is used in short-haul (not exceeding two hops) multichannel transmission.

Short-haul point-to-point microwave systems are commonly used in *studio-to-transmitter links* (STL) at 12.95 to 13.2 GHz, in *cable television relay service* (CARS) at 12.7 to 12.95 GHz, and in common carrier services at 3.7 to 4.2 and 5.9 to 6.4 GHz.

The *amplitude-modulated link* (AML) is a specific application of the CARS band. Each CATV channel is upconverted to a microwave frequency and filtered

TABLE 17.5 CATV Coaxial Cable Propagation Losses

Cable size	Dielectric	Frequency, MHz							
		50	216	300	330	400	450	500	550
		Trunk and feeder cables							
0.412	Polystyrene	.62	1.35	1.63	1.71	1.88	2.00	2.10	2.21
0.412	1st-generation foam	.75	1.65	2.00	2.10	2.31	2.45	2.58	2.71
0.412	2d-generation foam	.71	1.52	1.81	1.90	2.09	2.22	2.34	2.45
0.412	Current generation foam	.62	1.35	1.63	1.72	1.92	2.05	2.10	2.21
0.500	Polystyrene	.50	1.10	1.32	1.38	1.52	1.62	1.70	1.79
0.500	1st-generation foam	.60	1.35	1.63	1.71	1.88	2.00	2.10	2.21
0.500	2d-generation foam	.56	1.24	1.49	1.56	1.72	1.82	1.92	2.02
0.500	Current-generation foam	.51	1.10	1.32	1.39	1.55	1.65	1.70	1.79
0.750	Polystyrene	.34	.75	.90	.94	1.04	1.10	1.16	1.22
0.750	1st-generation foam	.40	.96	1.15	1.21	1.33	1.41	1.48	1.56
0.750	2d-generation foam	.39	.87	1.05	1.10	1.21	1.29	1.36	1.42
0.750	Current-generation foam	.34	.75	.91	.96	1.07	1.15	1.22	1.23
0.625	Current-generation foam	.41	.90	1.08	1.14	1.27	1.35	1.44	1.46
0.875	Current-generation foam	.30	.66	.80	.85	.94	1.01	1.07	1.08
1.000	Current-generation foam	.27	.59	.72	.77	.85	.92	.98	1.03
		Drop cable*							
RG-59	Foam	1.78	3.59	4.27	4.47	4.98	5.30	5.61	5.89
RG-59	Solid	2.22	4.72	5.62	5.89	6.49	6.88	7.26	7.61
RG-6	Foam	1.43	2.90	3.46	3.62	4.04	4.31	4.57	4.71
RG-6	Solid	1.77	3.80	4.52	4.74	5.22	5.54	5.84	6.12
RG-11	Foam	.87	1.88	2.24	2.34	2.61	2.78	2.95	3.10
RG-11	Solid	1.20	2.61	3.12	3.27	3.60	3.82	4.03	4.22

*Values represent loss in decibels per 100 ft.

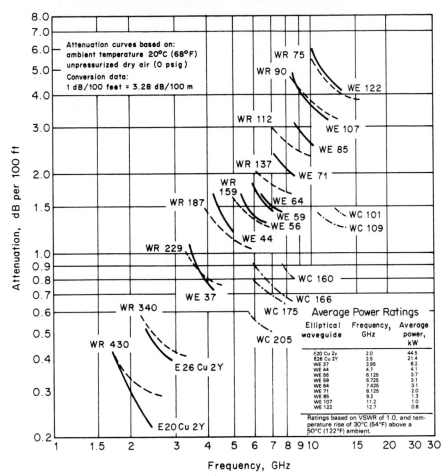

FIG. 17.17 Attenuation and power ratings of standard waveguides. (SOURCE: Cablewave Systems. Used with permission.)

to suppress the carrier and lower sidebands, passing only the upper sideband. This is called *suppressed-carrier single-sideband* (SCSSB) AM transmission. Up to 40 TV channels can be combined into a single waveguide output port. The rf signal can be transmitted to another headend or other receiving sites where the channels are demodulated and the original signals recovered.

A typical AML configuration and frequency plan is shown in Fig. 17.18.

Procedures for the design of point-to-point microwave systems are described in Chap. 4. These include the link calculations for determining the carrier-to-noise ratio. The conversion of C/N to S/N for video signals is similar for FM microwave and satellite systems and is described in Sec. 21.

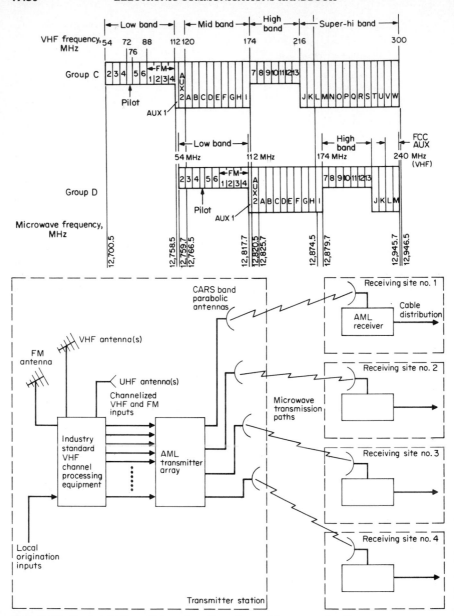

FIG. 17.18   Typical AML multichannel configuration and frequency plan.[17]

18.   Fiber Optics

There have been significant improvements in the cost and performance of fiber-optic transmission systems in recent years. As a result, they provide an alternative to microwave and coaxial cable for communication system designers.

The technology of fiber-optic systems is described in Chap. 8. The advantages and disadvantages of fiber as compared with microwave and satellites for voice and data circuits over 20 mi in length are described in Chap. 9.

This section compares fiber optics with coaxial cable for television transmission over short distances.

Advantages of Fiber Optics. Advantages are

1. *Immunity to electrical interference:* Since lightwaves are used as carriers, fiber-optic systems neither cause nor are susceptible to electrical interference.

2. *Security:* Fiber-optic systems are not susceptible to electromagnetic eavesdropping. Tapping a fiber-optic cable requires expensive and complex splicing equipment and special skills. Splicing of the cable invariably causes a disruption of service which alarms the operator.

3. *Capacity for digital transmission:* Fiber-optic systems have large bandwidths, and this eliminates one of the major problems in the use of digital signals for television transmission. Time-division and wavelength-division multiplexing can be used to carry several digital TV channels in a single fiber.

4. *Fewer repeaters:* Depending on the frequency and cable size, coaxial cable may require several line amplifiers or repeaters per mile. Analog fiber-optic systems which are commonly used for video transmission have a range or "reach" of several kilometers without amplification.

Disadvantages of Fiber Optics. Disadvantages are

1. *Requirement for special skills:* The techniques for installing and repairing fiber optic cables are more difficult. They require special instruments for aligning and fusing fiber ends, and small misalignments can accumulate and cause excessive losses. Unfamiliar test equipment such as an optical time-domain reflectometer is required.

2. *Cost:* At the present time, the cost of fiber-optic systems is higher than that of coaxial cable for most short-haul multichannel TV circuits, e.g., CATV trunk circuits. This disparity is expected to decrease significantly in the near future as the industry moves down the "learning curve" in the production of fiber cable, connectors, emitters and detectors, and electrooptical interfacing equipment.

Reach of Analog Fiber-Optic Systems. The *reach,* i.e., the distance that a signal can be transmitted without amplification, of typical intensity-modulated analog fiber-optic multichannel television systems is plotted in Fig. 17.19. This figure shows the range of S/N that can be expected in an average-grade, single-mode fiber, using standard CATV-FM modulators, deviated at 4 MHz. Note that the system reach is a function of the number of channels multiplexed in a single fiber.

Summary. The advantages of fiber optics over coaxial cable for short-haul video circuits are inherent and fundamental. The disadvantages will diminish with time. It can be expected, therefore, that fiber will increasingly replace coaxial cable in these applications.

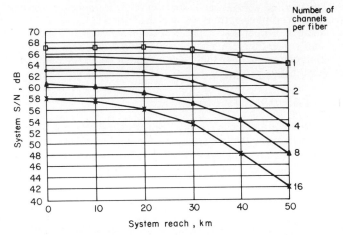

FIG. 17.19   The reach of multichannel FDM-FM analog video on fiber. (Catel Telecommunications)

SATELLITE TELEVISION TRANSMISSION SYSTEMS

19. Introduction

The technology of satellite systems with emphasis on their use for the transmission of voice and data traffic is described in Chap. 5. This section describes the characteristics of commercial satellites which are significant for the transmission of video signals.

The topology of satellite systems is particularly adaptable to point-to-multipoint services because receivers may be placed anywhere within the satellite beam coverage area or *footprint*. As a result, satellites are widely used for the distribution of television programs to broadcast stations and CATV systems. Also, satellites are beginning to be used for the distribution of programs directly to the home.

Satellites are equally adaptable to point-to-point services which require mobile transmitters since they may be located anywhere within the satellite beam. This capability has greatly enhanced a new form of journalism, *electronic news gathering* (ENG).

Finally, satellites have an important role in fixed point-to-point television services, particularly for international circuits.

20. C Band and K Band (FSS) for Satellite Television Transmission

C band (5925 to 6425 MHz up, 3700 to 4200 MHz down) and K band (1400 to 14,500 MHz up, 11,700 to 12,200 MHz down) are the two most commonly used bands for transmission of television frequencies allocated by the FCC for the *fixed satellite service* (FSS).

Two of the most important differences between C band and K band are the following:

1. The C-band FSS service shares frequencies with terrestrial microwave systems. This places constraints on the location of C-band earth stations, and it limits the permissible downlink power density.
2. K-band signals are subject to significant attenuation in heavy rainfall.

The advantages and disadvantages of C band and K band which result from these and other differences and from the relative newness of K-band technology are summarized in Table 17.6.

C band is used extensively for point-to-point relay service and for point-to-multipoint distribution of programs to CATV and *satellite master antenna television* (SMATV) systems, to broadcast network affiliates, and until recently to individual homes.

The use of K band for the transmission of television programs is more recent, but its characteristics make it particularly suited for point-to-multipoint distribution and mobile transmitters.

21. K-Band DBS

The World Administrative Radio Conference (WARC) and subsequently the FCC allocated K-band frequencies (17,300 to 17,800 MHz up, 12,200 to 12,700 MHz down) for transmitting directly to the home by *direct broadcast satellites* (DBS).

To reduce the size of receiving antennas at the home, the limits on downlink *effective isotropic radiated power* (EIRP) were increased to the range of 58 to 64 dBW, and the spacing between satellites was established at 9° (compared with 2° for K-band FSS).

Reliable and high-quality reception could be achieved from K-band DBS satellites operating at maximum permitted EIRP with receiving antennas having a diameter of 0.75 m or less. Antennas of this size are cheaper, easier to install, aesthetically more pleasing, and more tolerant of small mispointing than the larger antennas which are required for lower-power systems.

The technology for high-power DBS systems is available (although not completely proven), and its success will be determined by economic rather than technical factors. It may not be the optimum tradeoff between space and terrestrial costs. The satellites are heavier and more complex, and each can only accommodate three to six channels. It is not clear that the audience for this service will be large enough to make the high cost of designing and building the satellites and the even higher costs of programming services economically viable. Nevertheless, high-power DBS systems are still in their evolutionary stages and, given time, may prove to be commercially viable.

An alternative direct-to-home medium is FSS K band. FSS satellites have lower power (48 dBW EIRP), but they have proven to be reliable and cost-effective. Satisfactory reception can be achieved with antennas having a diameter of 1.2 m or less. It is a more mature and developed technology and can provide TV services at a fraction of the high-power DBS cost.

In forecasting the future of DBS technology, it must be recognized that the public has shown a surprising tolerance for large antennas and marginal picture quality if the program material is sufficiently attractive. It is estimated that there are more than 1 million C-band home receivers in the United States with anten-

TABLE 17.6   C-Band and K-Band Relative Merits for Satellite Television Transmission

C band		K band	
Advantages	Disadvantages	Advantages	Disadvantages
1. Hardware is less costly in the near term	1. Because satellite and terrestrial services share C band, it is now congested, making frequency coordination/licensing a requirement	1. Since band is only used for satellite communication, frequency coordination is not required	1. Affected by rain attenuation and depolarization
2. Less susceptible to rain outages	2. Requires larger antenna because of lower EIRP levels	2. Smaller antennas may be used because of higher gain at K band and higher satellite EIRP	2. Higher equipment cost in the short term
3. Established manufacturing infrastructure	3. Avoiding terrestrial interference makes site selection a difficult process	3. Easier site selection process because of smaller size of antenna and lack of interference	3. Narrow beamwidth of antennas may require more rigid mounts
4. Equipment more readily available	4. The use of artificial shielding that blocks interference can increase total system cost	4. Narrower antenna beamwidth is desirable in reduced orbital spacing	
5. Receiving technology is developed	5. Affected by Faraday rotation (See Chap.1)	5. Suitable for direct-to-home application	
6. Satellites have demonstrated reliable in-orbit operation	6. Satellite dispersal signal is required to prevent harmful interference to terrestrial stations, resulting in more stringent video receiver clamping specifications	6. Lower reception equipment cost in the long term	
		7. Flexibility in channelization plan	
		8. Not affected by Faraday rotation	
		9. No satellite dispersal signal	

nas 6 to 10 ft in diameter. They are used for the reception of the great variety of programs which are transmitted on C-band satellites, and neither their large size nor the degraded picture quality which is a consequence of the low power of C-band satellites has proved a deterrent to their sale.

22. Standards for Signal Quality

Numerous signal quality standards have been established as guidelines for system design and evaluation. Some, such as those specified by the FCC (Table 17.3), are mandatory. Others, such as those recommended by the Electronic Industries Association (EIA), National Cable Television Association (NCTA), and the Institute of Electrical and Electronics Engineers (IEEE) (Tables 17.1 and 17.2) are voluntary.

The establishment of a standard for a particular component of a television system, e.g., a satellite link, must take into account not only the cost-quality tradeoff but also performance of the other components of the system. On the one hand, no purpose is served in striving for a quality level which is an order of magnitude better than the poorest component of the system. On the other, the effect of cumulative degradation must be considered, and each component must be better than the overall system.

EIA Standard RS-250-B (Tables 17.1 and 17.2) provides a reasonable guideline for the satellite transmission of video and audio in most applications.

23. Calculation of Signal-to-Noise Ratio

The signal-to-noise ratio for a video signal transmitted over a satellite system is given by the equation:

$$S/N = C/N + f\,(\text{FM}) + f\,(\text{W}) \qquad \text{dB} \qquad (17.12)$$

where C/N = carrier-to-noise ratio, dB
$\quad f\,(\text{FM})$ = FM improvement factor (see below), dB
$\quad f\,(\text{W})$ = weighting and preemphasis factor (see below), dB

Calculation of C/N.   The calculation of C/N is known as *link analysis;* it is described in detail in Chap. 5. A simplified method involving some approximations but sufficiently accurate for most point-to-multipoint television services in the United States is given below.

$$C/N = \text{EIRP} + G - 20 \log f_{\text{GHz}}$$
$$- 10[\log (T_{\text{LNA}} + 50) + \log B_{\text{if}}] + 45 \ \text{dB} \quad (17.13)$$

where EIRP = effective isotropic radiated power of the satellite, dBW
$\quad G$ = gain of receiving antenna, dBi
$\quad T_{\text{LNA}}$ = low-noise amplifier noise temperature, K
$\quad B_{\text{if}}$ = receiver i-f bandwidth, Hz

Equations for calculating the antenna gain are given in Chap. 2. It is a basic antenna specification.

T_{LNA} is a basic specification of *low-noise amplifiers* (LNAs)

The approximations in the equations are in the two numerical constants.

The constant 50 is an average figure for the noise temperature T_a of the antenna system. At satellite frequencies under normal atmospheric conditions, the constant can range from 60 at an elevation angle of 5° to less than 30 at elevation angles in excess of 30° (see Chap. 2).

The constant 45 combines the Boltzmann constant (+ 229 dB) and an average figure for the frequency-independent term in the space-loss expression (− 184 dB). The latter figure can vary from −185.1 (very low elevation angle) to −183.6 (satellite at zenith).

The two terms in Eq. 17.13 which can be controlled by the system designer are the antenna gain G and the system noise temperature $T_{LNA} + T_a$. The ratio of these parameters G/T_s is often called the figure of merit of the system.

Calculation of the FM Improvement Factor. The FM improvement factor results from the use of frequency modulation and is given by the equation

$$f(\text{FM}) = 20 \log (\Delta f_v/B_v)$$
$$+ 10 \log (B_{if}/B_v) + 7 \qquad \text{dB} \qquad (17.14)$$

where Δf_v = peak deviation, MHz
B_v = highest baseband video frequency

This equation is only valid when the C/N exceeds the *threshold* for the receiver, typically 8 dB. Below the threshold, the S/N degrades rapidly; see Fig. 17.20.

Weighting and Preemphasis Factor. A standard noise-weighting factor is based on the reduced acuity of the eye for fine-grained noise at a viewing ratio of 4. The use of preemphasis increases the S/N of the higher-frequency components. With the standard weighting factor and standard preemphasis, the combined improvement in S/N is 12.8 dB (system M, United States and Canada).

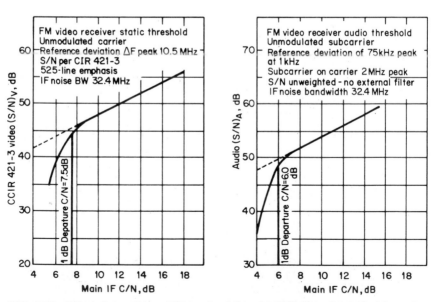

FIG. 17.20 Effect of operating an FM receiver below threshold (from *Scientific-Atlanta Satellite Communications Product Catalog,* 1985–1986, Scientific-Atlanta, Inc., Atlanta, Ga.)

24. Interference

C-band satellite systems are subject to cochannel interference from terrestrial microwave systems and adjacent satellites. Other sources of interference are sun outages and weather-induced interference such as the deterioration of cross-polarization by rain. Cochannel interference and sun outages are treated in Chap. 5 and precipitation depolarization in Chap. 1.

Cochannel interference is less serious at K-band, since there is no frequency sharing with terrestrial services and it is easier for receiving antennae to provide discrimination against radiation from satellites with 2° spacing.

Depending on the picture content and the tolerance of the viewer, the "just perceptible" *carrier-to-interference level (C/I)* ranges from 17 to 21 dB. The average *C/I* for "just objectionable" interference is 8 to 9 dB.

Interference from terrestrial microwave systems can be a difficult problem, particularly in urban areas. Despite this interference, earth stations can usually be operated successfully in these areas by taking advantage of the natural shielding offered by hills, buildings, trees, or depressions. Alternatively, artificial barriers can be erected. At C band, trees can reduce interference by 6 to 10 dB, earth mounds by 10 to 15 dB, and wire screens by up to 20 dB.

25. Availability Analysis

Availability is a statistical approximation of the total time a system will be operating above an arbitrary "outage" condition.

For C-band satellite systems, semiannual sun outages, human errors, and equipment failures are the major sources of outages. In well-designed and -operated systems, outage times are small.

For K-band systems operating above 10 GHz, precipitation-caused outages are the main source of reduced availability. To counteract this effect, the link's clear-sky *C/N* can be increased by a *rain margin* to minimize the amount of time that the instantaneous *C/N* falls below the user's arbitrary threshold.

The rain margin must overcome both the attenuation of the rainfall and the increase in background sky noise resulting from precipitation. The rain margin is

$$M_R = A + \Delta(G/T) \tag{17.15}$$

where A = path attenuation due to rain which is exceeded for the specified outage time and $\Delta(G/T)$ is the reduction in the receiving-station figure of merit as the result of noise from the rain cloud.

A number of models have been developed, partly theoretical and partly empirical, for the calculation of A and $\Delta G/T$. The CCIR model is described in Chap. 1. The equations below are based on a somewhat simpler model for the 12-GHz frequency range.

$$A = (H/\sin v) \, (\, 0.02R^{1.1} \,) \tag{17.16}$$

where H = latitude-dependent isotherm height, m
   v = elevation angle of antenna
   R = *average* rain rate over path for a given probability of exceedance, mm/h

H and R can be obtained from Fig. 17.21.

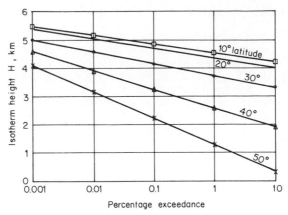

FIG. 17.21 Isotherm height as a function of latitude. (NASA)

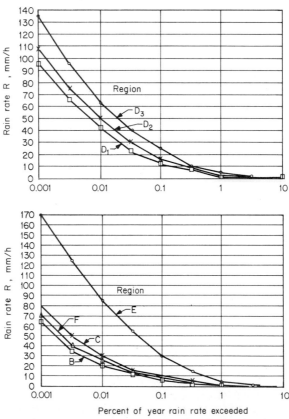

FIG. 17.22 Rain rates of various climatic regions. (See figure 23.) (NASA)

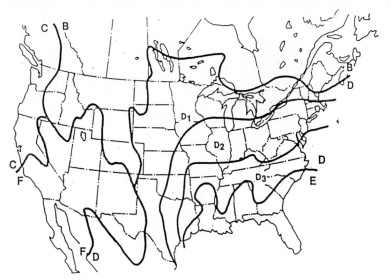

FIG. 17.23 Rain rate climatic regions.

The resulting decrease in the figure of merit can then be calculated:

$$\Delta G/T = 10 \log [1 + 290(A - 1)/AT_s] \qquad (17.17)$$

Availability analysis is performed iteratively with link analysis to determine an acceptable operating condition.

The prediction of rainfall rates (see previous page) and their conversion to signal attenuation is subject to considerable uncertainty. These equations, therefore, should be applied conservatively.

26. Receive-Only Earth Stations

A diagram of a typical receive-only earth station is shown in Fig. 17.24. The function of each component is as follows.

Antenna. The receiving antenna's function is to focus the electromagnetic signal into a feed point. On-axis gain, sidelobe rejection, elevation versus noise temperature, and the mechanical stability of the mounting structure characterize the quality of the antenna. If space is limited at the receiving site, multiple-beam offset-fed spherical antennas can be used. See Chap. 2 for a detailed description of satellite and microwave antennas.

Feed. The feed collects the focused energy from the antenna reflector and usually isolates the vertically and horizontally polarized radiation. Retrofits have been developed to permit reception from two or more adjacent satellites by a single antenna.

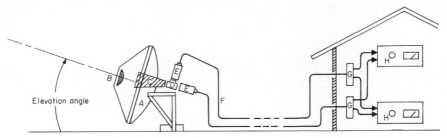

FIG. 17.24 Satellite receive-only earth station.

Low-Noise Amplifier.   The LNA amplifies electromagnetic energy while contributing a minimum amount of noise. Typical amplification is 45 to 60 dB. Its noise contribution is indicated by specifying its noise figure or equivalent noise temperature. As noted previously, the LNA is the predominant noise source in the receiving system, and a low noise figure is a basic characteristic of a high-quality unit.

Low-Noise Block Converter (LNBC or LNB).   If the distance from the antenna and receiver is large, the microwave frequencies can be downconverted to a lower intermediate-frequency band. Inexpensive cable rather than waveguides can then be used for the interfacility link. The most common downconverted frequency band is 950-1450 MHz.

Satellite Receiver.   Specifications for a typical satellite receiver are shown in Table 17.7. This lists the basic features which are desirable in a receiver. Other features to be considered are

- Skew control to compensate for the polarization twist (C band) caused by the earth's magnetic field.
- Antenna-positioning controller
- Digital readout
- Remote control
- Automatic *vertical-horizontal* (V/H) or *right-hand-circular/left-hand-circular* (RHC/LHC) polarization selection
- Signal-strength indicator
- Switchable predetection i-f filters
- Built-in TV modulator
- *Automatic-gain-control/manual-gain-control* (AGC-MGC) switch
- I-f output test port

TELEVISION SIGNAL SECURITY

27. Basic Concepts

The degree of protection against unauthorized reception provided to a television signal by scrambling or encryption can be classified as "soft" or "hard" security

TABLE 17.7 Typical CATV Satellite Video Receiver Specifications

---

Frequency: 3700 to 4200 MHz
Maximum level: -34 dBm
Channel selector
    Switch-selectable C- or K- band channelization plans
Impedance: 50 Ω, Type N or 75 Ω Type F
Return loss: > 20 dB
Noise figure: 15 dB maximum
Image rejection: > 60 dB
Local oscillator leakage: < -70 dBm
Intermediate-frequency signal:
    Frequency: 70 MHz
    Effective noise bandwidth: C-band 32.4 MHz nominal, K-band 24 MHz nominal; narrower bandwidth optional
    Impedance: 75 Ω, unbalanced
    Return loss at i-f monitor ports: ≥ 20 dB
    Dynamic operating range: 40 dB
Baseband:
    Deemphasis: 525-line (CCIR Rec. 405-1)
    Deviation range: 6 to 12 MHz peak at deemphasis cross-over frequency
Threshold extension demodulation: See Fig. 17.20.
Video:
    Video Level: 1 V p-p ± 3 dB adjustable
    Response (15 Hz–4.2 MHz): standard, ± 0.5 dB, with TED, ± 1.0 dB
    Impedance: 75 Ω, unbalanced
    Return loss: ≥ 26 dB
    Polarity: black-to-white, positive-going
    Clamping: 40-dB dispersal rejection
    Line-time waveform distortion: $< 1\%$ tilt
    Field-time waveform distortion: $< 1\%$ tilt
    Differential phase: $< \pm 1°$ 10 to 90% APL
    Differential gain: $< \pm 2.5\%$ 10 to 90% APL
Audio:
    Subcarrier frequency: 6.8 MHz standard, other frequencies available
    Frequency response: 30 Hz to 15 kHz ± 0.5 dB
    Deemphasis: 75 μs
    Output level: continuously variable, -10 to $+10$ dBm
    Impedance: 600 Ω, balanced
    Harmonic distortion: $\leq 1\%$
Operating temperature: 0 to 50°C (32 to 122°F)
Power requirements:
    100–130 V ac, 60 Hz, 100 W max.
    -19 to -32 V dc (with optional power module)
Mechanical:
    Standard 483-mm (19-in) EIA panel

---

depending on the difficulty in descrambling or decrypting the signal. A decision as to the degree of hardness which is optimum for a system is basic, and it requires a cost-performance tradeoff, heavily influenced by the value of the programming. The degree of hardness need not be the same for both video and audio, and the system designer has the option of employing, for example, soft scrambling for video and hard scrambling for audio.

Above all, the security system must not introduce a significant effect on the reliability of the entire transmission system.

In addition to protection of the signal against unauthorized reception, the communications plant must be physically secure. Means for providing physical security are described in Chap. 19.

28. Security Techniques

The basic methods which can be used to make the video portion of television signals secure are (1) varying the amplitude, (2) time-shifting components of the video waveform, (3) transmitting the signal in a nonstandard format, and (4) using various combinations of the above.

The various embodiments of these methods are summarized in Table 17.8.

TABLE 17.8 Security Techniques for Television

Video security technique	Application: Broadcast, CATV, MATV, Satellite	How it works	How signal is restored
1. Trapping (negative trapping)	C/M	At the pole, the picture carrier is trapped out, other channels may enter the home.	Trap is removed to allow desired signal to pass.
2. Jamming (positive trapping)	C/M	Jamming signal is inserted between picture and sound carrier of each secured channel.	Jamming signal is removed by installing traps for each desired channel.
3. Frequency conversion	B-(MDS)	Secured channels are placed beyond the tuning range of ordinary TV sets; P/S carriers may be inverted.	The desired block of channels is converted (heterodyned) into standard TV channels 2–6 or 7–12.
a. Block conversion	C/M		
b. Channel converters (indoors, on addressable)	B/C/M	Same as item 3.	Individual channels are converted into channel 3 or 4 by switchable converters.
c. Channel converters (outdoor addressable converters)	C/M	Converters are placed off subscriber premises and in-home keypad is given to subscriber.	Authorized channel is converted (one at a time) to channel 3 or 4 then sent to subscriber's home.
4. Baseband scrambling	B/C/M/S	Active video lines are inverted by line, field, or frame at fixed or pseudo-random rate.	The lines are reinverted at the receiving end according to the inverse of the scrambling algorithm.
a. Video inversion by line, field, or frame	B/S		

TABLE 17.8 Security Techniques for Television (*Continued*)

Video security technique	Application: Broadcast, CATV, MATV, Satellite	How it works	How signal is restored
b. Line rotation	B/S	Line is segmented, then shifted so that begin/end does not match the left/right scan.	Line-store memory detects the begin/end scan and restores it to the original left/right sequence.
c. Line permutation (one or more lines)	B/S	The standard line scanning sequence is altered in a fixed or time-varying fashion.	Field-store memory uses the permuting algorithm to restore the original scanning sequence.
d. Sync elimination	B/C/M/S	The horizontal and/or vertical synchronization signals are removed to render standard TV set unusable.	Synchronization signals are reconstructed to restore the original (e.g., NTSC) signal.
e. Sync alteration	B/C/M/S	The standard synchronization signal is either attenuated, altered, or replaced by other signals.	Standard sync is reconstructed using the inverse of the scrambling signal.
f. Time-multiplexed analog component video	B/C/M/S	The luminance, chrominance, and sync components are transmitted in time-sequence as separate signals.	Decoders reconstruct the signals back to NTSC for standard TV sets.
g. Line translation	B/C/M/S	Each horizontal blanking interval is time-shifted in pseudo-random manner, from zero to 2 × normal blanking.	Horizontal blanking interval is restored to normal by using the inverse of the time-shifting algorithm.
h. Time/frequency compression	C/M	Chrominance/luminance signals are both time/frequency compressed, then transmitted without redundant information. Sound format is retained.	Video components are time-expanded, reconstructed line by line, then sync is inserted to form a standard NTSC signal.

TABLE 17.8 Security Techniques for Television (*Continued*)

Video security technique	Application, Broadcast CATV MATV Satellite	How it works	How signal is restored
i. Digital	C/M/S	Video/audio waveforms are sampled, digitized, coded, and error-corrected; then transmitted in binary format.	The receiver demodulates, decodes, and converts the video/audio signals back into analog form.
5. Rf scrambling a. Gated sync suppression	C/M	At the CATV modulator, the horizontal sync is attenuated by 6–12 dB. The restoration signal is carried AM on the aural carrier or on a separate, out-of-band carrier.	The CATV converter/descrambler extracts the descrambling signal from either the aural carrier or out-of-band carrier and uses it to restore the sync level.
b. Sine wave sync suppression	C/M	Same as item 5a, except sine waves (15.75 or 31.5 kHz) are used to attenuate the sync signals as well as active video.	Same as item 5a.
6. Switched star topology a. Coaxial cable	C/M	The head end (node) is where signals are added or deleted; then transmitted via coaxial cables to subscribers.	Subscribers receive only authorized signals and may use cable-ready TV sets or standard TV consumer equipment.
b. Fiber optics	C/M	Fiber optics is used to connect subscribers to a common node (head end) where TV rf channels modulate fiber-optic transmitter.	Each authorized channel is converted from light waves into standard rf TV channels.

TABLE 17.8 Security Techniques for Television (*Continued*)

Video security technique	Application: Broadcast, CATV, MATV, Satellite	How it works	How signal is restored
7. Nonstandard transmission format			
a. AM time/ frequency compression	C/M	Channels are time/ frequency compressed, then combined with accompanying aural carriers. Two channels are transmitted over one 6-MHz channel. Restoration and control data signals are sent separately.	The restoration/ control signal is demodulated from a separate carrier, then used to expand and reconstruct two separate standard channels.
b. FM or AM analog components	B/C/M	Chrominance/luminance analog components are time-multiplexed. Digital audio control data modulates an rf carrier. The combined signals are transmitted FM or AM.	The receiver (AM or FM) demodulates the digital audio and control information, then produces analog audio, constructs RGB or standard signal with sync.
c. QPSK digital TV	C/S	RGB and audio signals are sampled, digitized, coded for error correction, then transmitted over cable or satellite.	The digital signals are demodulated, decoded/error corrected, then converted to their original, analog form for display by standard monitors/ receivers.

CITED REFERENCES

1. Pritchard, D. H. and J. J. Gibson, "Worldwide Color Television Standards—Similarities and Differences," *SMPTE Journal,* vol. 89, February 1980, pp. 111—120.

2. Weltersbach, W., "Systems and Realization Aspects of Multistandard TV Receivers," IEEE Transactions on Consumer Electronics, **CE-31**(4):617, November 1985.

3. "CCIR Characteristics of Systems for Monochrome and Colour Television—Recommendations And Reports," Recommendation 470-1 (1974–1978) of the XIVth Plenary Assembly of CCIR, Kyoto, Japan, 1978.

4. NTC Report No. 7, "Video Facility Testing, Technical Performance Objectives," prepared by the Network Transmission Committee of the Video Transmission Engineering Advisory Committee, June 1975, revised January 1976.
5. Strauss, T. M., "The Relationship Between the NCTA, EIA, and CCIR Definitions of Signal-to-Noise Ratio," *IEEE Transactions on Broadcasting,* **BC-20**(3):36, September 1974.
6. "Report of Panel 6, Levels of Picture Quality," Television Allocation Study Organization, NAB Library, January 1959.
7. EIA Standard RS-170A, Color Television Studio Picture Line Amplifier Output, Electronic Industries Association, Washington, D.C.
8. "Radio Broadcast Services," Federal Communications Commission, *Rules and Regulations,* vol. II, part 73.
9. IEEE Std. 511-1979, ANSI/IEEE Standard on Video Signal Transmission Measurement of Linear Waveform Distortion, December 1979.
10. Weaver, L. E., "Television Measurement Techniques," Peter Peregrinus, New York, 1971.
11. The Howard Sams Editorial Staff, *Color TV Training Manual,* Howard W. Sams, Indianapolis, 1980.
12. Tinnell, R. W., "Television Symptom Diagnosis, an Entry into Servicing," 2d ed., Howard W. Sams, Indianapolis, 1980.
13. EIA Standard RS-250-B, Electrical Performance Standards for Television Relay Facilities, Electronic Industries Association, Washington, D.C., September 1976.
14. EIA Standard RS-240, Electrical Performance Standards for Television Broadcast Transmitters, Electronic Industries Association, Washington, D.C.
15. Fink, D. G. and D. Christiansen, *Electronic Engineer's Handbook,* McGraw-Hill, New York, 1982.
16. Chiddix, J. "Fiber Optic Supertrunking," *Communications Engineering Design,* pp. 16–30, September 1985.
17. *AML Seminar Maintenance Manual,* Hughes Microwave Communications Products, Hughes Communications.
18. Kaul, R., R. Wallace, and C. Kinal, *A Propagation Effects Handbook for Satellite Systems Design,* chap. VI, NASA Communication Division, Washington, D. C. March 1980.

GENERAL REFERENCES

29. Television Signal Formats and Specifications

Asano, S., "Explaining Component Video," *TV Technology,* page 11, March 1985.
Berkhoff, E. J., U. E. Kraus, and J. G. Raven, "Applications of Picture Memories in Television Receivers," *IEEE Transactions on Consumer Electronics*, vol. CE-29, pp. 251–254, 256–258, August 1983.

Brainard, R. C., A. N. Netranali, and D. F. Pearson, "Predictive Coding of Composite NTSC Color Television Signals," *SMPTE Journal*, vol. 91, pp. 245–252, March 1982.

Bruch, W., "The PAL Colour TV Transmission System," *IEEE Transactions on Broadcast Radio and Television Receivers*, vol. BTR-12, pp. 87–100, May 1966.

CCIR Document T3/83, Proposal for a New Recommendation Worldwide HDTV Studio Standard, Revision 1, March 1985.

CCIR Document 11/398-E, Report of Interim Working Party, 11/6 High Definition Television (HDTV), October 1985.

EIA Standard RS-219, Audio Facilities for Radio Broadcasting Systems, Electronic Industries Association, Washington, D.C., April 1959.

EIA Standard RS-240, Electrical Performance Standards for Television Broadcast Transmitters, Electronic Industries Association, Washington, D.C., April 1961.

"HDTV Standard Closer," *TV Technology*, vol. 3, no. 12, December 1985.

Jackson, R. N., and M. J. J. C. Annegarn, "Compatible Systems for High-Quality Television," *SMPTE Journal*, 92(7):719, 1983.

Limb, J. O., C. B. Rubinstein, and J. E. Thomson, "Digital Coding of Color Video Signals—A Review," *IEEE Transactions on Communications*, vol. Com-25, pp. 1349–1384, November 1977.

Loughlin, B. D., "The PAL Color Television System," *IEEE Transactions on Broadcast Radio and Television Receivers*, vol. BTR-12, pp. 153–158, July 1966.

Lowry, J. D., "B-MAC: An Optimum Format for Satellite Television Transmission," 18th SMPTE Conference, Montreal, February 1984.

Lowry, J. D., "B-MAC Format for Satellite Transmission," *Communications Engineering Design*, pp. 14–20, January 1984.

Prezas, D. P., "Digital Television," pp. 213–214, *Television Technology Today*, T. S. Rzeszewski, ed. IEEE Press, New York, 1985.

Rhodes, C., "An Evolutionary Approach to High Definition Television," *Tomorrow's Television*, pp. 186–197, SMPTE, 1982.

Rhodes, C. W., "Interlaced Scanning Discussed," *TV Technology*, pp. 15–16, November 1985.

Rhodes, C. W., "MAC Offers New Hope for Video," *TV Technology*, p. 11, August 1985.

Rhodes, C. W., "Progressive Scan Has Potential," *TV Technology*, p. 13, December 1985.

Rhodes, C. W., "Time Compression Explained," *TV Technology*, September 1985.

Rhodes, C. W., "Widescreen Picture on Horizon," *TV Technology*, pp. 17–18, October 1985.

"Satellite Transmission of Multiplexed Analogue Component (MAC) Television Signals," CCIR Report AB/10-11, November 1983.

Stafford, R. H., *Digital Television-Bandwidth Reduction and Communication Aspects*, John Wiley & Sons, New York, 1980.

30. Television Signal Transmission Modes

Channin, D. J., "Optoelectronic Performance Issues In Fiber-Optic Communications," *RCA Review*, 46(2):240, RCA Labs, Princeton, N.J., June 1985.

Cunningham, J. E. *Cable Television*, Howard W. Sams, Indianapolis, 1980.

Kuecken, J. A., *Fiber Optics*, Tab Books, Blue Ridge Summit, Penn., 1980.

Reference Data for Radio Engineers, Howard W. Sams, Indianapolis, 1975.

31. Terrestrial Television Transmission Media

A Competitive Assessment of the U.S. Fiber Optics Industry, Publ. No. 0-473-816, U.S. Department of Commerce, Washington, D. C., 1984.

CATV, MATV, CCTV Coaxial Cable, p. 36, Alpha Wire Corporation, Elizabeth, N.J., 1980.

CCIR Report 215-5 "Broadcasting Satellite Service (Sound and Television)," vol. X and XI, part 2, Geneva, 1982.

"Component Considerations for Low-Cost Fiber Optic Links," *Professional Program Record,* No. 27, Electro 1982, Electronics Convention Inc., May 1982.

Engineering Considerations for Microwave Communications Systems, GTE Lenkurt Inc., San Carlos, Calif. 1975.

FCC *Rules and Regulations,* part 78, FCC, Washington, D. C.

"Frequency Sharing and Coordination between Systems in the Fixed-Satellite Service and Radio-Relay Systems," Recommendations and Reports of the CCIR, 1982, vols. IV and IX, part 2, XVth Plenary Assembly, Geneva, 1982.

Head, J., "Fiber Optics, A Systems Approach," *Professional Program Session Record,* no. 27, Electro 1982, Electronic Conventions, Inc., 1982.

IEEE Standard Dictionary of Electrical and Electronic Terms, 2d ed. IEEE, New York, 1978.

Kao, C. K., *Optical Fiber Systems: Technology, Design and Applications,* McGraw-Hill, New York, 1982.

"Lightwave Technology," special issue, *IEEE Communications Magazine,* vol. 23, no 5, May 1985.

Marsden, R. P. "Optical Fibers for Digital Studio Interconnections," *Broadcast Engineering,* pp. 300–308, March 1983.

OST Bulletin No. 60, Multichannel Television Sound Transmission and Audio Processing Requirement for the BTSC System, Federal Communications Commission, April 1984.

Prisco, J. J. and R. J. Hoss, "Fiber Optic Regional Area Networks," *IEEE Communications Magazine,* vol. 23, no. 11, November 1985.

RF Transmission Line Catalog and Handbook, Times Wire and Cable Co., Wallingford, Conn., 1972.

Stevenson, D., "An Approach to Low Cost Fiber Optic Components," *Professional Program Session Record,* no. 27, Electro 1982, Electronic Conventions, Inc. 1982.

Sylvania Pathmaker CATV Data, GTE Products Corp., El Paso, Texas.

Tyler, L. B., M. F. Davis, and W. A. Allen, "A Companding System for Multichannel TV Sound," *IEEE Transactions on Consumer Electronics,* CE-30:(4):633, November 1984.

Walker, M. J., "Cable System Upgrades," *Communications Technology,* pp. 19–26, September 1985.

32. Satellite Television Transmission Systems

Ananasso, F., "Coping with Rain above 11 Ghz," *Microwave Systems News,* pp. 58–72, March 1980.

"Application of International Satellite, Inc. for a North Atlantic Regional Satellite System," before the Federal Communications Commission, U.S.A, August 1983.

"Application of RCA American Communications, Inc., for a Ku-Band Satellite System," before the Federal Communications Commission, U.S.A., April 28, 1982.

Bhargava, V. K., *Digital Communication by Satellite,* p. 18, John Wiley & Sons, Toronto, 1981.

"Broadcasting-Satellite Service (Sound and Television); Technically Suitable Methods of Modulation," CCIR Report 632-2, Vols. X and XI, part 2, pp. 79–92, 1982.

CCIR report, nos. 561, 809.

Crane, R. K., "A Global Model for Rain Attenuation Prediction."

Engineering Considerations for Microwave Communications Systems, GTE Lenkurt, Inc., San Carlos, Calif., 1975.

"Handbook on Satellite Communications (Fixed-Satellite Service)," pp. 98–101, 123–143, CCIR, Geneva, 1985.

Highsmith, W. R., "Very Small Aperture Terminals," *Telecommunication Magazine,* pp. 57–64, December 1985.

Oguchi, T., "Electromagnetic Wave Propagation and Scattering in Rain and Other Hydrometeors," *Proceedings of IEEE,* vol. 71, no. 9, September 1983.

Olsen, R., D. Rogers, and D. Hodge, "The aR^b Relation in the Calculation of Rain Attenuation," *IEEE Transactions on Antennas and Propagation,* vol. AP-26, March 1978.

"Results of Video Interference Testing," presented to the NCTA Satellite Engineering Subcommittee by RCA Americom, May 20, 1981.

RF Transmission Line Catalog and Handbook, pp. 1–8, Times Wire and Cable Co, Wallingford, Conn., 1972.

Tallelvan, P., and D. D. Grantham, "A Review of Models for Estimating 1 min Rainfall Rates for Microwave Attenuation Calculations," *IEEE Transactions on Communications,* Com-33(4):361, April 1985.

Walker, M. J., "Cable System Upgrades," *Communications Technology,* pp. 19–26, September 1985.

Williams, E. A., "Earth Terminal Site Selection and Installation for the Public Television Satellite Interconnection System," *SMPTE Journal,* vol. 88, April 1979.

33. Television Signal Security

Brown, L. C., "Off-Premises, Trends and Issues," *Communications Engineering and Design,* November 1984.

Conanan, V., *Addressable Systems Handbook,* Home Box Office, 1982.

Coryell, A., "Illegal Use of Tiered and Pay Services," *Communications Technology Magazine,* pp. 94–97, September 1985.

McClellan, J. "The Star-lok Scrambling System," *Communications Technology Magazine,* pp. 80–81, September 1985.

Medress, M. "Satellite Television Scrambing with VideoCipher," *Communications Technology Magazine,* pp. 82–87, September 1985.

Marks, J. "Plant Management—The Signal," *Cable Television Business,* July 1983.

Security Systems Handbook, Home Box Office, July 1979.

CHAPTER 18
TELECONFERENCING SYSTEM DESIGN

John Tyson
President
Compression Labs Incorporated

A. David Boomstein
Manager, Product Marketing
Compression Labs Incorporated

CONTENTS

INTRODUCTION

1. The Need for Teleconferencing

Most business professionals are highly specialized, and thus depend on the help of others to accomplish their daily tasks. Groups or individuals performing a supportive role can be located almost anywhere around the globe. Depending upon the profession, surveys indicate that business professionals spend from 30 to 75 percent of their time attending meetings. Conducting business in a decentralized fashion across multiple time zones increases project time and the cost of doing business. Since there is no foreseeable reduction in this dependence on remote talent and resources, today's managers face the problem of finding an affordable resolution to the constraints of time and distance.

Part of the solution has been to introduce telecommunications more broadly into the office environment. Modern telecommunications systems include *audio bridging facilities* which allow multiple parties to share a telephone conversation, *electronic store-and-forward message systems* which ensure delivery of complete and accurate messages around the clock, and *full-motion video teleconferencing systems* which allow face-to-face meetings between conference rooms, giving groups the ability to meet and work on projects in real time.

2. Definition

The term *teleconferencing* has been freely applied to all types of long-distance meetings. In fact, *Webster's Dictionary* defines "teleconference" as: "...a conference among people remote from one another who are linked by telecommunication (as by a telephone or television)."

Taken literally, this definition implies that a simple telephone conversation between two parties constitutes a teleconference; however, most practitioners place additional requirements upon the activity before classifying it as a teleconference. They define a teleconference as a communication which includes two or more participants with multiple locations interconnected electronically. An additional criterion is the use of more than one communication medium—for example, graphics electronically transmitted as hard copy or on a television monitor accompanied by face-to-face video and audio interaction. Obviously full-motion video is a form of teleconferencing.

3. Business Travel Statistics

Corporate travel figures display impressive statistics. A survey of travel[1] by one *Fortune* 500 company with approximately 40,000 employees revealed over 20,000 trips in a one-year period—one trip for every two employees. Hotel usage is an additional index of travel activity. It is not unusual for a major corporation with 30,000 to 50,000 employees to lease more than 20,000 domestic hotel nights in a single year.[2]

In response to a detailed 1982 questionnaire to the teleconferencing managers of the *Fortune* 1000 for-profit companies, 56 percent of the respondents indicated their organizations spent in excess of $500,000 annually on business-related travel.[1] Another research report estimates the annual U.S. travel business to be in the neighborhood of $50 billion.[4] Still another report[1] indicates that an average of 48 percent of business trips were for *recurring* meetings, the type that can most often be handled by teleconferencing. If teleconferencing substituted for only 10 percent of business travel, the savings would be significant.

Impressive potential for cost savings also exist in the international business travel market. The following table shows the result of one study[5] of the duration of business trips between New York and London. The potential for the use of teleconferencing in place of short one-day trips is high.

Length of meeting	Percent of meetings
1 day	52
2	10
3	14
4	9
5 or more	15

TELECONFERENCING TECHNOLOGY

4. Forms of Teleconferencing

There are five types of communications generally recognized as forms of teleconferencing:

- Computer conferencing
- Audio conferencing
- Audio-graphics conferencing
- Full-bandwidth analog (one-way) video conferencing
- Digital (two-way) video conferencing

5. Computer Conferencing

Computer conferencing, sometimes known as *electronic messaging,* involves the exchange of data between participants in the form of typed text. Computer conferencing is widespread and generally can by done by anyone with a personal computer and a modem. Many schools and colleges dedicate a computer for programs such as the *Apple bulletin board system* (ABBS). Acting as an open electronic bulletin board, these systems can keep students up to date on events and issues.

The strength of electronic bulletin boards is their store-and-forward capability, which allows participants to log on and off at their convenience. The weakness of any bulletin board system is the lack of privacy.

In addition to bulletin boards dedicated to various topics from wine tasting to politics, data bases can be set up for transactional services or to provide an electronic mail capability. The asynchronous nature of the communication and the ability to send and receive information from almost any location make computer conferencing attractive to business travelers.

6. Audio Conferencing

The most basic form of conferencing, *audio-only,* is provided to everyone with a business telephone. In its simplest form, audio conferencing is a three-way telephone call. When equipped with a speakerphone, an office is transformed into a conference room.

Audio conferencing applications include the following:

- Including remote parties in meetings
- Remote training
- General announcements
- Program management
- Troubleshooting and equipment maintenance
- Steering committee meetings

International audio conferencing is an effective and economical form of conducting business. One New York bank's corporate credit policy committee meets each week for an audio conference among New York, London, Tokyo, and anywhere else the participants happen to be located. Coordination between different time zones spread across the globe is a problem, but weekly meetings provide an efficient and up-to-date method of conducting business.

There are dozens of audio conference terminals presently available. They range from small speakerphones to more advanced systems that include equalization and echo reduction circuitry which is applicable both to offices and conference rooms.

In its more advanced forms, audio conferencing systems can include thousands of participants at hundreds of sites. Interconnection is provided by electronic bridging. Audio bridges can be analog or digital and can include advanced features to aid the conferees and the service provider. Bridging services may be operator-run; the operator sets up the calls or perhaps only monitors the conference to ensure proper audio levels and quality. Alternatively, the bridging can be automated, allowing individuals to control the meeting from their own telephone touch pads. Other systems, known as *meet me bridges,* require participants to call into an assigned bridging number to connect with other conferees.

An often-available feature is electronic circuitry allowing the bridge to make adjustments to ensure audio quality. *Electrical midpoint bridges* adjust the incoming line levels of each call and, as required, boost, attenuate, or equalize the signal to ensure that all conferees receive it at a similar level.

Numerous conferencing services are available. They provide on-demand audio bridging capabilities and are offered by local and intercity telephone companies as well as "value-added" service providers.

7. Audio-Graphic Conferencing

Audio conferencing is the most convenient way to meet groups of people, but its lack of visuals can be limiting. The system's effectiveness can be significantly enhanced by adding a graphics component. Sometimes known as *enhanced audio* conferencing systems, audio-graphics conferences allow users to share typed pages, charts, graphs, diagrams, and even images of the meeting's participants.

Captured-Frame Video.   Several types of audio-graphic conferencing systems exist. The most widely used system is *captured-frame video* which may be either *slow-scan* or *freeze-frame.* These systems take their input from a monochrome or color camera and transmit the signal for a complete frame in about 1 minute's time.

A tradeoff exists between image resolution and transmission time. Generally, the better the resolution, the longer the transmission time. When transmission times of less than a minute and higher resolution, e.g., 256 pixels by 512 lines versus 256 by 256, are both required, 56 kbs data circuits must be used.

The early captured-frame systems were termed *slow-scan* because, as the picture information was received in analog form, the receiving terminal scanned the image on the monitor line-by-line. A long-persistence phosphor was employed to retain the image between scans.

As technology advanced, *freeze-frame* systems were introduced in the early 1980s. Freeze-frame systems store the camera signal from a complete frame in a *frame memory* before transmission. The signal is also stored at the receiver terminal and can be scanned out repeatedly at normal television rates, thus eliminating the fading of the picture which was characteristic of slow-scan systems. Freeze-frame also eliminates the distortion or fuzziness of slow scan which is caused by camera or subject motion.

A problem with freeze frame systems arises when inexperienced operators overuse human images when sending a combination of human and graphic frames. The result is a barrage of photos of the conferees with inefficient information transfer. Also, since the transmission of a frame requires at least one-half minute, participants see images of their counterparts which are out of sync with the audio. With proper training, however, these problems can be minimized.

Annotation Devices. Annotation devices generate television signals for the transmission of symbols or line drawings.

One of the original annotation devices was the electronic blackboard introduced by AT&T. Using a board similar to a standard blackboard, the system transmitted the scribed images over telephone lines. Currently electronic blackboards are offered by AT&T, NEC, Panafax, and Okedata.

The *Telewriter** is composed of a 5 × 7 writing tablet and a transceiver. The user writes or draws images on the tablet, and they are displayed in real time on a television monitor at the receiving site.

A useful application of the Telewriter is combining it with other transmission devices for annotation purposes. For example, once a freeze-frame image of a diagram has been received, the Telewriter can be used to highlight important items. Other annotation systems include the Telestrator† and Discon systems, which are used by television networks to help explain plays in their broadcasts of football games.

Special features available in annotation systems include different color "pens," the ability to store transmitted information on audio cassettes, and the ability to interface with a personal computer.

Facsimile. Facsimile systems, commonly known as *fax* machines, optically scan the image as with a television camera and convert the scanned image into an electrical signal that can be transmitted over telephone or data lines. Fax is one of the fastest and most reliable methods of transmitting letters and business graphics, and it is now a standard piece of office equipment.

The International Telegraph and Telephone Consultative Committee (CCITT) has developed standards for facsimile transmission. The specification of facsimile groups has resulted in a level of interconnectivity that promotes the worldwide use of fax.

CCITT grouping	Type of transmission		Time per page
I	Analog	4–6 min	
II	Analog	2–3 min	
III	Digital	15–90 s	
IV	Digital	1–5 s	

Early fax systems suffered from poor resolution, extended transmission time, and the lack of system automation. The tradeoff between resolution and transmission time remains, but document and paper handling has been automated so that overnight transmission is practical.

---

*Telewriter is a trademark of Optel Communications.
†Telestrator and Discon are trademarks of Interand Corporation.

8. Full-Bandwidth Analog Video (One-Way)

Full-bandwidth analog video conferencing provides the best visual quality of all teleconferencing methods. The signal requires 4 MHz of bandwidth, the capacity of a microwave channel or satellite transponder, and transmission is expensive. It is most often used for communicating a special event to large groups of people at widely scattered sites; hence it is sometimes described as ad hoc video. By transmitting to a large number of sites, the cost per site is reduced. Most ad hoc conferences provide video in one direction only but use two-way audio so viewers at remote sites can interject comments and ask questions.

Satellites are the most common transmission medium for full-bandwidth teleconferencing. Satellite technology allows transmission to any number of ground stations, so that it is possible to transmit to dozens of sites simultaneously. Analog conferencing networks can be dedicated or installed on an as-required basis.

The hotel industry has added analog teleconferencing as an adjunct service to its normal meeting and convention business. At hotels throughout the United States, earth stations have been installed and meeting rooms modified to receive and display video and audio signals. Teleconferencing facilities generally have large-screen projection systems to display the video images, and many rooms are equipped with audio systems that interface with the public telephone network, thus enabling viewers to interact with other conferees.

Since full-bandwidth teleconferencing is intended to present information to the audience rather than for audience participation, ad hoc conferences tend to be staged as events. Applications include product announcements, training seminars, press conferences, and stockholder meetings. Owing to analog conferencing's presentationlike nature, most conferences entail high production costs. The average analog video conference costs in excess of $100,000 to produce.

Full-bandwidth, one-way analog video continues to grow in popularity because of its ability to reach large groups of people in an efficient and economical manner. Its main drawback is its inability to provide cost-effective fully interactive communications.

9. Compressed Digital Video (Two-Way)

Most people associate two-way full-motion video conferencing with the term *teleconferencing*. To reduce transmission costs, the video signals are converted to digital form and compressed in bandwidth. It is the most advanced mode of teleconferencing. The majority of digital video compression systems are "full feature" and include two-way interactive motion video, audio synchronized with the video, and some form of graphics imaging. Multipoint video, usually in the form of a three-way conference, is becoming prevalent.

The first full-motion *coder-decoders* (codecs) for compressed video were put into service in mid-1981. They converted analog video signals into a digital bit stream and compressed the signal bandwidth by a ratio of 30:1, from 90 Mbs to 3 Mbs. These systems functioned quite well, but the limited availability and cost of 3-Mbs circuits made them impractical for most corporations.

In April 1982, Compression Labs, Inc., first introduced full-motion codecs which could digitally compress analog video signals into a bandwidth of 1.5 Mbs. Two-way video teleconferencing became practical because these signals could be transmitted on standard T1 carriers, the basic transmission channel of public telephone systems.

Compared with 3-Mbs operation, transmission at 1.5 Mbs cut the bandwidth in half, reduced costs drastically, and greatly increased the availability of transmission facilities. Ongoing technical development made it possible to achieve picture quality which was essentially equivalent to that previously obtained at 3 Mbs.

Capital costs also fell, and the price of codecs dropped from $150,000 to $175,000 in 1982 to $60,000 to $85,000 in 1987.

There are three primary manufacturers of codecs which operate in the 384-kbs to 1.544-Mbs range. They are Compression Labs, Inc. (CLI), in the United States, GEC Video Systems in the United Kingdom, and Nippon Electric Corporation (NEC) in Japan. The range of transmission rates of the products offered by these companies is shown in the table below.

Codec	Transmission rates
CLI	384 kbs–3.136 Mbs
GEC	768 kbs–2.046 Mbs
NEC	512 kbs–6.0 Mbs

A second class of video codecs, narrowband systems which operate in multiples of 56 or 64 kbs, were introduced in the late 1980s. Although many in the industry have termed them *limited motion* systems, a coding refinement known as *motion compensation* allows transmission at rates of 56 to 384 kbs with a motion-handling capability originally available only with wideband systems. While their motion-handling capability is in some respects superior to that of their wideband counterparts, their picture resolution is usually about half that of the wideband systems. When coupled with a desktop terminal with only one to three people on the screen, the picture quality is acceptable for most desk-to-desk or desk-to-small-group applications.

There are two primary manufacturers of 56-kbs codecs—Compression Labs, Inc., and PictureTel Corporation. The systems of both manufacturers are full-featured and are designed to take advantage of switched 56-kbs networks.

The codec is designed to accept the numerous types of signals that a video conference requires; see Fig. 18.1. Inputs include one or more full-motion video signals for transmitting images of the conference participants, audio signals which

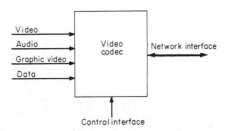

FIG. 18.1 Input and output signals. (Video Codec)

are synchronized with the video, a video input for displaying graphics, and user data signals. User data signals range from PC-to-PC information exchange (between *personal computers*) to data used by the conference facilities room coordinators.

Each manufacturer's codec uses proprietary algorithms to digitally encode and compress the analog signals; see Fig. 18.2. The codec converts the analog video and audio signals into their digital counterparts and performs a series of digital coding and bandwidth-reduction processes.

There are a number of encoding techniques for video compression. Two of them are *field elimination* in which one of the two video fields (odd or even) is discarded, and *frame elimination* where the overall number of full frames encoded by the codec is reduced. Both techniques work best when the amount of motion is limited. In addition to these "front end" techniques which reduce the amount of data the codec must process, different codecs employ *intraframe* and *interframe* coding schemes.

Intraframe coding eliminates the redundancy within a video field. Instead of transmitting the information in its complete form, the intraframe technique creates a representative code for the data.

Interframe coding eliminates redundancy between successive frames by transmitting only the differences between frames. Interframe coding generally provides sharper detail, but intraframe coding works better during high motion sequences.

Video compression involves an inherent tradeoff between picture resolution and motion-handling capability. Picture quality decisions are predicated upon the transmission bandwidth. As the bandwidth is reduced to lower limits, more encoding is required. Differences between manufacturers' encoding processes are seen during transition sequences with medium to high motion. Picture anomalies such as smearing, ghosting, and blocking may be seen during these periods.

Video conferencing is well-suited to digital compression technology, since the typical video conference scene consists of four to six people seated at a table. Unless a stand-up presentation at a chalkboard, a demonstration, or a videotape is being shown, image motion is limited to small areas of the total picture. Most codecs provide video quality which is more than adequate for this application.

Audio signals, which may also go through a bandwidth-reduction stage, are processed through time-delay circuitry to be synchronized with the video signals. They are multiplexed with the video signals, and any user data are also multiplexed into the bit stream. After multiplexing, the codec prepares the bit stream

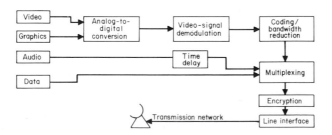

FIG. 18.2   Functional diagram. (Video Codec)

for transmission. Most codecs employ some type of error-correcting code to reduce errors that may be caused by the transmission medium. If security is required, the bit stream is encrypted prior to transmission.

Graphics are usually presented in static form. A conferee who wants to send a graphic commands the codec to send the image signal which is compressed in the same fashion as full-motion signals. The codec transmits only one graphic video frame which is interleaved within the full-motion video frames. Graphics interleaving results in interruption or freezing of the motion video for a period of ½ to 1½ seconds, depending on the complexity of the graphics image and the transmission bandwidth.

During the freeze period, the user usually views the last full-motion frame received by the codec prior to the graphics transmission. Since the graphic utilizes most of the available bandwidth, it is received almost instantaneously. Most users do not mind the full-motion image being frozen for a second. When presented with the alternative of waiting 30 or more seconds for a graphic to be displayed to preserve the motion video's integrity, users prefer motion and graphics video interleaving.

Decoding and reconstructing the compressed video images is basically the reverse process. Codecs can convert from the North American NTSC system to the European PAL system or vice versa at either the transmit or receive end. If the conversion is accomplished at the transmit end, all receiving sites must operate at the same video standard. Hence many users find that the conversion at the receive end provides fewer constraints on international teleconferencing.

The evaluation of the performance of different manufacturers' codecs at various bandwidths is primarily subjective. Blind comparisons by large groups of potential users within the corporation are sometimes employed to supplement the judgment of the members of the telecommunications department. An extensive and impartial study[6] has been performed under contract to the United States government's Office of Technology and Standards. This study ranked the performance of the currently available (1985) codecs of four manufacturers in the following sequence: Compression Labs, Inc; GEC Video Systems; Nippon Electric Corporation; Fujitsu.

10. Teleconferencing Rooms and Terminal Devices

The teleconferencing room or terminal device provides the interface to the end user, and the overall success of teleconferencing within an organization is heavily influenced by the manner in which the room facility, with its numerous capabilities and packaging options, is implemented. Conferencing room technology advanced significantly during the early 1980s and there are now multiple options available to the end user.

Conferencing systems fall into three primary categories:

- Dedicated boardrooms
- Modular and transportable units
- Desktop workstations

Within each category there are numerous options available; these range from PC-to-PC screen transfers to high-resolution graphic systems.

Dedicated Boardrooms.   The primary users of dedicated boardrooms are usually the senior officers of the corporation, and they tend to be full-featured systems, large in size, and luxuriously appointed; see Fig. 18.3. Most boardroom-type facilities seat six to ten participants at a conference table and have a gallery which can seat additional participants. The high-quality appointments, superior acoustic treatment, custom lighting, and sometimes dedicated power systems are expensive. Construction and outfitting a boardroom can cost from $500,000 to $1 million.

Modular/Transportable Units.   In order to make video teleconferencing more generally available, lower-priced alternatives were necessary. Changes in hardware designs and the elimination of less-used features led to the development of modular or transportable conferencing systems.

These systems are designed as free-standing units that house most of the equipment required to conduct a teleconference; see Figs. 18.4 and 18.5. Most include one or two television monitors and are designed to conceal the rest of the equipment. Most also have the ability to connect to a video-based graphic system.

Much of the cost savings in this class of system results from the standardization of system designs and the elimination of costly room modifications.

An additional benefit of modular systems is their portability. If packaged properly, they can be moved from room to room in 1 hour and across the country within 24 hours. This gives companies the ability to operate multiple conferencing rooms with a single electronics package.

Desktop Workstations.   The increasing demand for teleconferencing has led to the development of a third class of conferencing system—the desktop workstation; see Fig. 18.6. The more advanced systems connect with the in-house PBX and utilize the PC as an integral element of its communications. Designed to run over

FIG. 18.3   Dedicated boardroom teleconferencing facility. (Photo courtesy of ESI Corporation, used by permission)

FIG. 18.4 Modular teleconferencing terminal. (Photo courtesy of ESI Corporation, used by permission)

broadband local-area networks (LANs), desktop stations operate like a video intercom.

The workstations operate through a head-end system that routes the video and transmitted data to the receiving station. If the call is placed to someone who

FIG. 18.5 Transportable teleconferencing terminal. (Photo courtesy of ESI Corporation, used by permission)

FIG. 18.6   Desktop workstation. (Photo courtesy of Compression Labs, Inc., used by permission)

does not have a video station, the head-end allows it to operate as an audio-only call. Through the PBX architecture, users can place multiparty calls and include persons both with and without video stations.

For long-distance video calls, the head-end would route the call through a video codec. The output of the codec would then be routed to the recipient by dialing the recipient's extension, either directly or by means of an access code. As many as 64 workstations can be connected to a single head-end system, and these terminals have the ability to timeshare the video codec, thus making efficient use of an expensive resource.

Video Transmission and Display Configurations.   Video transmission and display configurations fall into three primary categories:

- Switched camera
- Continuous presence
- Anamorphic display

Switched-Camera Configuration.   The switched-camera configuration is one of the most commonly used formats in video conferencing. These systems have multiple television cameras and a switch so that participants at the remote site either see a close-up of the person speaking or a wide-angle view of all participants. Most systems employ a single camera for the wide-angle shots and two to four additional cameras for the close-ups. In some cases, the two participants are covered by each close-up camera as a cost-reduction measure.

Camera switching can be accomplished automatically in response to audio activity by the meeting participants or manually by a meeting facilitator; see Fig. 18.7.

Because of the physical space required for the numerous cameras—usually four to six situated in a front-wall structure—the switched-camera format is usually confined to larger modular systems or boardrooms with extensive equipment

FIG. 18.7 Video display with camera-switched format.

rooms. To reduce space, equipment complement, and costs, many users favor a single-camera format with a zoom lens and the camera mounted on a computer-controlled pan-and-tilt drive. This allows the lens to be adjusted for a wide-angle or close-up shot and to be directed toward the speaker. This format provides slower switching time, as the camera direction is moved from position to position, but it is considerably cheaper.

The main benefit of switched-camera systems is that meeting participants are displayed in a close-up format while they are speaking. When there is considerable interaction between participants, a group shot is usually displayed to avoid the "tennis effect" with the view bouncing from side to side. The primary concern with switched-camera conferences is that, unlike a face-to-face meeting, one does not have visual access to everyone in the conference.

Continuous-Presence Format. The desire to view all participants simultaneously and in greater detail than in a single wide-angle shot has led to the development of the continuous-presence format. In this configuration, two cameras are used, one directed at the left side of the conference table and the other at the right. The camera signals are converted into the digital domain, and the portions of the pictures above the participants' heads and below the table are blanked out; these areas account for about 50 percent of the images. The remaining middle portions of the two pictures—which include the participants—are combined into a single signal which is transmitted to the receiving location in the normally digitally compressed fashion.

Alternatively, the two camera signals can be combined in the analog domain and the composite signal fed to the codec, where it is converted to the digital format.

On the receiving end, the images can either be displayed stacked, or, as is more usual, side by side on two monitors or video projectors; see Figs. 18.8 and 18.9. When two monitors are used, only one-half of the picture area is utilized (dark bands cover the top and bottom), but the images are medium close-up, and this is usually preferred to a single wide-angle image.

FIG. 18.8 Video display with stacked images.

Anamorphic Lenses. Anamorphic lens and display technology for video conferencing is a variation of the *continuous presence* format. Anamorphic systems provide a close-range wide-angle shot similar to the panavision used in motion pictures. They employ special

FIG. 18.9 Video display with side-by-side images.

lenses and display the images in a 3:1 width- to-height ratio instead of the normal 4:3 aspect ratio. The result is a long narrow image of all the participants in the room.

BENEFITS OF TELECONFERENCING

Teleconferencing provides two potential benefits to its users, direct cost savings through reduction in travel and improvements in management productivity. This section describes the results of user studies of the magnitude of both benefits.

11. Cost Reduction

A number of organizations have reported the direct travel cost savings which they have achieved or which can be achieved by teleconferencing. The savings in participants' time is in addition to the direct cost savings.

The Air Force Institute of Technology (AFIT) based in Dayton, Ohio, was able to save more than $520,000 in travel-related costs over a one-year period and $1.3 million over a three-year period.[7]

A survey by the Transportation Energy Management Program (TEMP) of Ontario, Canada, found that 68 percent of companies using teleconferencing have taken steps to reduce business trips; 23 percent of the teleconferencing users believed they had reduced travel in their organizations. One observer and user stated: "Teleconferencing will never replace all travel to business meetings, but its use as an alternative for some has been established and will continue to grow."[8] It is estimated that teleconferencing now has the ability to replace at least 16 percent of corporate travel.

The Satellite Business Systems (SBS) 1982 user survey[9] reported that "Three-quarters of the respondents perceived a decrease in travel expense and overwhelmingly approved such savings." The study also reflected that respondents perceived a 60 percent decrease in their time away from home.

To obtain significant cost reductions, corporate policy for teleconferencing must be carefully developed. Employees must be encouraged to utilize it. As examples, this can be accomplished in a positive way by offering it as a "perk" with facilities available only to a limited group or in a more authoritarian manner by requiring travelers to provide a reason why each trip cannot be replaced by a teleconference.

12. Increasing Management Productivity

It is difficult to quantify improvements in management or white collar productivity, and reliance must be placed on perceptions and judgments.

The most apparent benefit that users of teleconferencing experience is the reduction in the amount of time they require for interaction with others. Video conferencing frees a manager's time, creating private time for other tasks. Seven out of ten respondents to the SBS survey felt that their personal productivity had increased.

A Knowledge Industries study[3] stated: "The most frequently cited advantages of teleconferencing are to save travel time and to make meetings more cost effective." It was found that most participants in teleconferences appear to prepare better for their meetings as compared with attendees of conventional meetings. This resulted in better agendas, better preparation of presentations, inclusion of the right participants, and greater punctuality. It also resulted in reduced lead time for organizing meetings, the ability to include more people in the decision making process, and a decrease in meeting lengths.

THE TELECONFERENCING MARKET

13. The Growth of the Teleconferencing Market

In the early 1980s, audio conferencing accounted for about 90 percent and transmission-related costs about 75 percent of the teleconferencing market. In 1980, video conferencing activity was limited primarily to digital research and analog transmissions, and the total video conferencing market, including transmission, was about $50 million.

As higher-priced video compression equipment and elaborate conferencing centers were introduced in 1982 and 1983, the ratios began to change. By 1984, the teleconferencing market had grown to approximately $245 million, and the video component was $90 million, or more than one-third of the total.

A breakdown of the estimated audio and video markets in 1984 and 1990 by equipment and transmission is shown in the table.

	Estimated market, millions of dollars	
Type of conference	1984	1990
Audio and audio-graphic conferencing		
Equipment, services, and facilities	45	375
Transmission	110	270
Total audio	155	645
Video conferencing		
Equipment, services, and facilities	65	745
Transmission	25	370
Total video	90	1,115
Total teleconferencing	245	1,760

A more detailed discussion and breakdown of the estimated teleconferencing market is contained in Ref. 10.

These figures describe an industry that has been in its formative stages for about 10 years and is now growing steadily at a healthy rate as a result of its continued success in business, education, and government.

REFERENCES

1. In-house study, Institute for the Future, Menlo Park, Calif., 1985–1986.
2. In-house study, New York bank, 1983–1984.
3. Lazer, E., "The Status of Teleconferencing," *The Teleconferencing Handbook,* Knowledge Industry Publications, 1983.
4. Brancatelli, J., "Teleconferencing, Still Looking for Its Place," *ABC Communications World,* quotation from Gnostic Concepts.
5. International Communications Conference, sponsored by McGraw-Hill, Secaucus, N.J., 1982, presented by AT&T.
6. *Performance of Codec Testing and Evaluation,* Delta Information Systems, Inc., Horsham, Pa., 1985.
7. "The Air Force Institute of Technology (AFIT) Reaches Out Through Media: An Update," *Teleconferencing and Electronic Communications,* 1982.
8. Cukier, W., "The Transportation Energy Management Program (TEMP) of Ontario, Canada," *Telespan,* vol. 3, no. 8, August 15, 1983.
9. "SBS User Survey," *Telespan Special Supplement,* May 31, 1982.
10. "Teleconferencing Applications and Markets—1984–1990," Center for Interactive Programs, University of Wisconsin.

CHAPTER 19

COMMUNICATIONS PLANT DESIGN AND CONSTRUCTION

Peter H. Plush

Director, Facilities Planning, RCA Communications and Electronic Services

CONTENTS

GENERAL INTRODUCTION

1. Scope of Telecommunications Facilities

Telecommunications facilities switch, combine, format, process, amplify, and transmit information over a chosen medium or path to a receiving facility which may either repeat the process or decode the received signal for use by the end user. One can, therefore, expect communications facilities to include the following hardware: information switches, multiplexers, modulators and demodulators, up and down converters, radio-frequency amplifiers, receivers, radios, antennas, and fiber-optic transceivers.

Physical laws governing operation of the communications link define the design, configuration, and performance of these hardware elements. In operating communications systems, these governing laws of physics are joined by the practical need to control the system in accordance with statutory requirements and the constraints imposed by the natural and artificial environment, all of which place complementary and at times equally stringent requirements on the system design.

Since people cannot construct perfect systems, the mathematics of applied statistics must be employed to design a failure-resistant and -tolerant system meeting operational goals. As a consequence of these requirements, the basic building blocks of the system must include redundant sets of like hardware, and the overall communication system acquires new classes of hardware which control, monitor, switch, condition, and protect the central system. These elements include dedicated heating, ventilating, and air conditioning (HVAC) systems, uninterruptible power supply (UPS) systems, battery banks, generators, fire-supression systems, power switchgear, deicing systems, shelters, and buildings. The system grows in complexity of hardware and diversity of function, as shown in Fig. 19.1.

The scale of telecommunications facilities is determined by their intended use as large-scale commercial, thin-route commercial, and dedicated or in-house facilities.

Large-scale commercial facilities, such as teleports, Intelsat sites, and

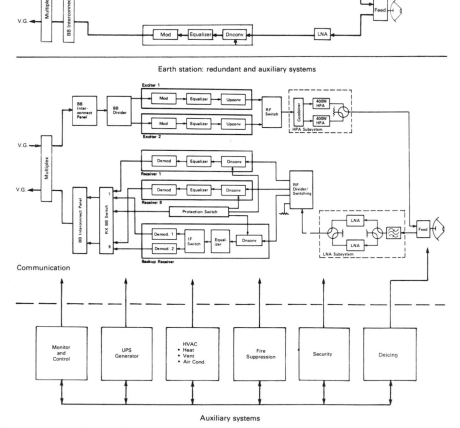

FIG. 19.1 Earth station: non-redundant and earth station: redundant and auxiliary systems.

domestic-satellite earth stations, are designed to access simultaneously a number of satellites with multiple communications carriers. Thus, significant numbers of customers with diverse requirements can be served. They are usually equipped for future expansion with permanent buildings and power and HVAC plants of sufficient size and capacity to accept the projected expansion.

Thin-route facilities serving markets or customers with immediate significant capacity requirements and limited future growth are normally built for the immediate capacity requirement except when future expansion cost would be excessively expensive. These facilities sometimes use electronic shelters in lieu of permanent buildings.

The scope of in-house corporation communications is usually stringently constrained to serve the narrowly defined capacity and lifetime of the projected requirement. Little system growth is provided in order to minimize capital expenditures.

While the large-scale facilities are designed for human operators, smaller facilities are unstaffed, and the in-house facility may be designed for repair at the module level.

2. Site Requirements Document

Site performance requirements are defined in a formal *site requirements document*. It evolves from system analyses which define, trade off, optimize, and select a superior conceptual system in response to a communications service requirement. The site requirements document is the second-tier specification, subsidiary only to the system performance specification.

The system performance specification defines and specifies a communication link's *figure of merit,* such as carrier-to-noise ratio, signal-to-noise ratio, rf power bandwidth, data rates, bit error rates, availabilities, mean time between failures (MTBF), fade margins, rain models, modulation techniques, and applicable industry service standards. The site requirements document defines communication equipment and auxiliary subsystem performance parameters which are requisite to fulfilling the system performance requirements.

The site requirements document provides the benchmark for site sizing and configuration, it defines the complement of equipment, and it provides environmental data needed to generate detailed procurement and implementation specifications.

TYPES OF COMMUNICATIONS FACILITIES

3. Earth Stations

Earth stations are the terrestrial terminals for satellite transmission systems. Their design is dictated by the special requirements of these systems, which are described in Chap. 5. They include the following subsystems: antennas, low-noise amplifiers (LNAs), up and down converters, high-power amplifiers (HPAs), multiplexers, exciters, receivers, switches, patch panels, control and monitoring equipment, UPS and generators, HVAC, fire supression, deicers, and security.

4. Central Terminal and Switch Offices

The central terminal office in a commercial system is the traffic on- and off-loading point. It is the facility at which the individual circuits of individual customers are multiplexed with those of others into a wideband link for transmittal by terrestrial microwave or fiber optics directly to distant cities or to an earth station for satellite distribution to distant sites. From this site, local circuit loops connect the long-distance circuit to the customers' premises. Finally, it is the control point from which long-distance trunks and local loops are tested and maintained. For switched services, this is often the point at which collocated switches are installed. Alternatively, switches (*private automatic branch exchanges,* or PABXs) may be on customer premises. Central terminal offices have not been collocated with earth stations in large cities because of C-band frequency congestion and large antenna apertures. This will change with the migration of traffic to K band.

Central terminal offices (CTOs) are normally in tall center-city office buildings, providing clear lines of site for microwave interconnection, and an infrastructure of local loops for customer interconnection. CTO site selection is an iterative process that considers technical requirements, anticipated customer locations, local loop costs, and the availability, quality, and expense of real estate.

The CTO facility's major communications subsystems include a main distribution frame to interconnect local loops to the CTO; switches; circuit conditioning, signaling, and test equipment; multiplexers; and microwave and/or fiber-optic terminals. These subsystems are supported by battery plant, HVAC, and fire-protection subsystems.

5. Microwave Repeaters

Terrestrial microwave systems are line-of-sight systems limited in the distance over which any single hop can operate. Interconnection between terminal points over long distances is accomplished by interspersing repeaters which receive and reamplify weak signals. These facilities consist of microwave radio systems installed in small and spartan technical buildings (or shelters).

Microwave electronics are supported by HVAC, battery plant, and automatic-starting generator subsystems, and some remote sites have been solar-powered. Repeater sites are unmanned and equipped with remote-reporting failure-alarm systems. Microwave antennas are installed at or near the top of a tower on the site providing a clear line of sight and Fresnel zone clearance (see Chaps. 1 and 4) over topographic features along the microwave path. In mountainous terrain, repeaters are placed on peaks to minimize tower height.

Microwave repeater site selection is an interactive process combining technical parameters of range between repeaters, line of sight, and frequency coordination requirements with land-use and site-acquisition issues.

6. Fiber-Optic Paths and Repeaters

Fiber-optic systems employ the most recently developed communication technology. They offer an order-of-magnitude increase in communications bandwidth and substantially increased security (until downconverted to the electromagnetic band). They free the system designer from radio-frequency interference and frequency coordination constraints. These systems require a continuous right-of-way for fiber cables as compared to the discrete sites for microwave repeaters or satellite earth stations.

The technology and application of fiber-optic systems are described in Chap. 8. Fiber-optic cables have been installed in conduit, directly buried, pole-hung, and laid on frozen tundra at remote bases.

At present, fiber-optic multiplexing and regeneration is accomplished by downconverting to electromagnetic frequencies; as a result, they are no longer impervious to passive tapping at that point. The system designer must develop a detailed geography of the local customer base in order to select a system topology and routes which accommodate new customer sites readily and at low cost. One methodology in inner city applications is a star system with central multiplexing and an initial overcapacity of fibers.

7. Monitor and Control Facilities

Monitor and control facilities may be characterized as:

- *Primary or secondary:* The system directly monitors communications parameters (primary) or the equipment providing the service (secondary).

- *End-to-end or partitioned:* The system monitors from customer demark to customer demark (end-to-end) or, alternatively, a microwave path or a leased local loop (partitioned).

- *Status or variable:* Parameters are monitored for their *presence or absence* (status), for example, HPA high voltage, or as a continuous *variable,* for example, HPA power output level.

- *Local or remote:* The monitor and control system may monitor collocated facilities or it may monitor parameters which are telemetered from a remote site.

- *Real time or off line:* Real-time monitoring of system parameters occurs concurrently with system operation. Off-line monitoring takes place in concert with system operation, but is not concurrent; examples are statistical analyses of hardware failures, and circuit outages.

The purpose of monitoring and control systems is to assess system health and operability prior to or, as a minimum, concurrent with the use of the service. Monitoring and control information is used by the service provider to restore full capability without a customer-perceived service interruption.

New monitoring and control concepts should consider all elements of the organization, from service reservation through accounting, engineering, and operations, in order to dedicate the organization (rather than a particular department) to the perfection of customer services.

CODES AND STANDARDS

8. Land Use and Zoning

Individual municipalities and counties, through a land-use master plan, define the course of future real estate development within the boundaries they administer in trust for the public. The concepts of the master plan are portrayed as symbolic zones on a land-use map. Copies of the map are normally available from the local planning department or clerk.

Local zoning codes define the categories of land use and applications which are in conformance with each land-use category, and they cite special exceptions. Some codes specifically cite broadcast and/or telecommunication facilities; others do not.

When a specific category for telecommunications is not cited, the proponent of the facility must seek an undefined-use permit from the local planning board (i.e., the intended use is not defined in any zoning category by code). Even when the site use is cited in a zoning code and the proponent can reasonably interpret that the selected site is in conformance, the official determination is provided by local planning officials. Subsequent to confirmation of zoning conformance, a simplified approval procedure is followed.

Construction can proceed when the proponent obtains building permits from municipal officials. This is accomplished by submitting site construction drawings and a supporting analysis which have been "sealed" to indicate construction code conformance by a professional engineer or licensed architect to the municipal or county engineering department.

When a preliminary determination by planning authorities indicates that an intended site use is nonconforming and the proponent wishes to proceed, a petition is filed with the planning authorities seeking a zoning *variance* or a *change in zoning*. The former, if granted, provides an exception to permit the nonconforming use. The latter changes the assigned zoning category so as to make the intended use conforming. These petitions are reviewed in detail at formally scheduled public hearings and the decision is rendered on the basis of serving the public good, not the corporate good, unless the two interests are deemed congruent. Corporate financial hardship is not normally a justifiable reason for granting a variance.

Utilities, as corporations which serve the public, are normally granted liberal interpretations of zoning statutes to accomplish their goals.

9. The Environmental Impact Statement

National Environmental Policy Act. The National Environmental Policy Act (NEPA) was enacted in 1969 to "...declare a national policy which will encourage productive and enjoyable harmony between man and his environment; to promote efforts which will prevent or eliminate damage to the environment and biosphere, and stimulate the health and welfare of man; to enrich the understanding of the ecological systems and natural resources important to the Nation; and to establish a Council on Environmental Quality."

The concepts of NEPA have been extended by legislation to the state, county, and municipal levels of government, and most of them require an environmental impact statement to be submitted as part of the approval process for any major construction project.

Contents of the EIS. An outline of a representative environmental impact statement (EIS) follows.

Introduction
Distribution List
Summary

Project Description
Environmental Impacts and Mitigating Measures
Alternatives to the Proposal
Unavoidable Adverse Impacts

Section 1. Project Description
Name of Proposal
Corporate Sponsor
Location of Project
Objectives of the Sponsor
Major Physical and Engineering Aspects of Project
Compatibility with Zoning and Land Use Plans

Section 2. The Environment Checklist Assessment (Optional)
The government planning department rates each element in Secs. 3 to 8 prior to preparation of the EIS. This guides the preparation of the document. Ratings include major issue, minor issue to be treated in the EIS, and review not required. A copy of the checklist is included in the final EIS for reference and comparison purposes.

Section 3. Physical Environment
Earth. The earth analyses define potential impacts of the proposed project on existing geologic and hydrologic resources. This section is based upon data as noted below and may include inputs from field visits.

- Design data:
  Area to be paved
  Grading plan
  Water supply plans
  Septic system plans
  Runoff pond design

- Site data:
  Geologic maps
  Aerial photographs
  Soil borings, percolation tests, etc.
  Surface water discharge and water quality data

Impacts of the proposed project are determined from design information grouped into construction and operational impacts.
Air Quality. The air-quality analysis estimates the impact of emissions from the proposed facility for all identified pollutants. The only sources of significance are the combustion by-products of emergency generators and battery plant outgassing.
Flora and Fauna. Characterize the flora and fauna in and about the proposed

site and detail potential impacts on important ecological resources. Identify potential impacts to sensitive wetlands, threatened and/or endangered species, recreational or commercial species. Recommend actions which mitigate potential impacts.

Noise. Noise analyses identify the magnitude and source of noise emissions during construction and operation. Existing noise levels at the proposed site are quantified and compared with estimates of future noise levels.

Future noise levels are based upon:

* Operational sources of exterior noise such as power transformers, diesel generators, air handlers
* construction operations such as excavation and pile driving

Estimates of future noise levels are portrayed for noise-sensitive locations (residences, parks, hospitals, etc.) in the vicinity.

Light and Glare. Determine the extent to which the local population will be affected by light or glare from the proposed project. Illumination and reflectance properties of the proposed site implementation are defined for on-site emitters as follows:

* Function
* Need
* Location of fixtures
* Illumination characteristics
* Major reflecting surfaces

Proposed illumination levels are compared with existing ambient illumination.

Land Use. Document existing land use and assess the compatibility of the proposed project with existing and anticipated future uses on the basis of a review of land-use maps and field visitations. Address the location and type of residential, commercial, industrial, and recreational uses within a reasonable distance of the proposed facility. Local planning officials should be consulted about sensitivity to development. Data used in preparing this section include:

* Site development plan
* Topographic map of site and region
* Regional land-use plans
* Regional zoning and development codes

Use of Natural Resources. This analysis addresses the impact of the proposed facility on local resources which cannot be replaced.

Risk of Explosion or Hazardous Emissions. This section treats the threat of geologic upset and hazards associated with fuel tanks and battery plants.

Section 4. Human Environment

Population. Population changes expected to result from project construction and operation are appraised for:

* Construction and operation labor requirements
* Regional population density and demographics
* Regional labor skill mix.

In most cases, automated telecommunications facilities produce minimal population and housing impacts.

Transportation. Preexisting access to the site and descriptions quantifying additional project demand generated during construction and operation are addressed. Local air traffic routes and altitude requirements defined by the FAA for airport approaches are reviewed. Specific data to be collected includes:

- Existing site access data:
  Roadway network maps and capacities
  Current vehicle demands
  Accident data
  FAA operating requirements

- Project data:
  Construction worker and truck estimates and schedule
  Operational employment by shift
  Truck service requirements for project site
  Site plan showing access roads and parking capacities

Quantify and compare the project's cumulative demand and safety with the existing conditions.

Public Service Department. Review project requirements for water, septic facilities, power, fire protection, telephone, and law enforcement. Document existing service capacity and confirm that sufficient resources are available to meet project requirements. Emphasize site self-sufficiency, automaticity, and the minimal impacts, if any, upon the service infrastructure (as an example, the need, if any, to extend power lines into the site).

Economics. The economics analysis discusses the economic benefits to the community in particular and the nation in general. Community benefits are portrayed in terms of direct site employment, complementary employment for site service contracts, and for construction services during the initial phases of implementation. Contrast the tax base derived by the community from the project with the minimal requirements for public services. Public service benefits can be offered to the community, for example, placement of police radio antenna systems on site towers.

Utilities.   Energy Utilization. Characterize the quantity and types of energy required for construction and operation of the site and identify energy conservation features of the project. Compare annual operational energy requirements of the project with local resources in order to assess impact, if any, on the existing energy infrastructure.

Water Utilization. This section describes the source of and average quantity of water required for project operation; it addresses potable, cooling, and fire-protection requirements.

Wastes Disposal. Describe the quantity, types, and disposal methods for wastes generated by the project. Project waste includes:

- Sanitary sewage
- Storm sewage
- Solid waste type
- Hazardous substances if any

• Method and location for disposal of preceding wastes

Human Health. Identify any potential health (including mental health) and safety effects to the general population resulting from construction or operation of the proposed project. Their evaluation characterizes health aspects of site emissions including, but not limited to, noise, light, combustion by-products, outgassing from battery plants, and microwave emission levels. These are compared to applicable governmental and industry standards. If none are available, then a comparison is made by rigorous survey with effects and levels defined in the scientific literature.

Aesthetics. Visual impact is based upon site proximity to residences, recreation areas, and roadways. It is assessed from an inventory of natural and artificial landscape elements. Visual impact is qualitatively evaluated by comparison of the proposed facility with the existing setting and consideration of the values and mind-set of the surrounding population. Sight-line drawings are prepared, depicting scenes before and after proposed implementation and mitigating impacts if they arise.

On the basis of site visits and information reviews, the EIS defines potential impacts, alternatives, and mitigation actions, including screening and landscaping, to conform with local land use and aesthetic values.

Archaeological/Historical. Review the site in consultation with state, county, and any local historic preservation office to determine whether historic features will be destroyed or damaged by site development. Should historically or archaeologically significant artifacts be identified, then recommendations concerning their beneficial disposition should be portrayed in the EIS.

Section 5. Short-Term Uses of the Human Environment versus Enhancement of Long-Term Productivity

This section contrasts the short-term gains obtained at the expense of long-term environmental losses, since in effect the existing generation is the trustee of the environment for future generations. A non-communications example would be strip mining employing a surface restoration plan.

Section 6. Unavoidable Adverse Impacts

This section describes the adverse impacts, both short- and long-term, on both the natural and human environment that would result from implementation of the project. Short-term construction traffic and noise and loss of a breeding habitat for an indigenous species are examples of possible impacts.

Section 7. Irreversible and Irretrievable Commitment of Resources

This section defines and analyzes those resources which would be irretrievably lost as a result of implementing the project. Most telecommunications facilities make limited use of natural resources. A site might as an example impact the breeding habitat of a local species.

Section 8. Proposed Alternatives

This section addresses the opportunities to implement the proposed facility at other sites or provide the telecommunications service by other technologies (fiber optics versus microwaves). It addresses the alternative of a no-action decision, i.e., the alternative of denying the proposal on environmental grounds and the resulting impact on humans and the natural environment.

10. Electronic Industry Association Standards

One of the most important functions of the EIA is to provide a mechanism for the development of standards that contribute to the growth of the industry. These consensus standards range over a broad spectrum of the technology, from structural loading of electronic hardware to digital data interfaces and beyond.

EIA standards are summarized in a yearly *Catalog of EIA and JEDEC Standards and Engineering Publications* available from the Washington offices of the association.

11. OSHA Requirements

The Williams-Steiger Occupational Safety and Health Act was enacted by Congress in 1970 to provide for the safety of workers while they pursue their occupations in the workplace. The act created the U.S. Department of Labor's Occupational Safety and Health Administration (OSHA). It established a mechanism for promulgating federally mandated standards and a system for workplace inspections and fines. The engineer/designer must understand OSHA standards and design and implement compliant systems and components. The starting point for this determination is *General Industry Standards,* Part 1910, OSHA document 2206, available from the U.S. Government Printing Office, Washington, D.C. The document is organized into subparts, some with particular application to communications sites:

A. Adoption and Extension of Established Federal Standards

B. General Safety and Health Provisions

C. Occupational Health and Environmental Control (paragraph 1910.97 addresses nonionizing radiation, microwaves)

D. Hazardous Materials

E. Personal Protective Equipment

F. Fire Protection

G. Compressed Gas and Compressed Air Equipment

H. Special Industries (paragraph 2910.268 addresses telecommunication)

I. Electrical

12. Microwave Emission Standards

Designers of microwave telecommunications facilities must deal with two groups who are exposed to microwaves: engineers and technicians on site operating the facility and the general public residing adjacent to the facility and/or along the path of communications. In the former case (technicians), the maximum level of exposure was established for nonionizing radiation by OSHA (as of March 1983) in Standard 1910.97 from 10 MHz to 100 GHz at 10 mW/cm^2 (10 milliwatts per square centimeter), averaged over any 0.1-h period. The OSHA standard was based upon the voluntary American National Standards Institute (ANSI) Standard C95.1-1974, which was itself revised downward to a more stringent requirement by ANSI Standard C95.1-1982.[1] The revised standard covers the frequency range 300 kHz to 100 GHz and is frequency-dependent. At 6 GHz the standard is 5 mW/cm^2. Figure 19.2 defines the general requirement.

The designer should use the revised ANSI guideline as the maximum exposure requirement for on-site personnel.

The OSHA/ANSI occupation standards are applied according to the presence of personnel in proximity with the microwave equipment, thus interlocks, local on-site measurement, signs, and fencing are installed to inform personnel of levels and to correct and/or mitigate those levels in excess of the occupational standards.

The issue of acceptable general-population microwave exposure levels is ambiguous. There is, as of September 1985, no federally mandated general-population microwave exposure standard, but the U.S. Environmental Protection Agency (EPA) is promulgating a new standard. The FCC applies the ANSI standard for the narrowly defined purpose of licensing microwave facilities, but this does not relieve the designer of the responsibility for meeting local standards and codes. In the absence of a federal standard, a number of state, county, and municipal governments have issued their own general-population standards. These are generally more stringent than ANSI C95.1 and in the range from 50 to 500 μW/cm^2.

The issue of standards is rendered more complex by foreign standards, particularly the Soviet Union general-population standard of 5 μW/cm^2. It has been predicted that with more interchange of technical data, the American and Soviet general-population standards will converge at a level between the existing standards.

For the present, the site designer should be prepared to meet all existing standards so as to defuse community concerns should they arise. Compliance with all

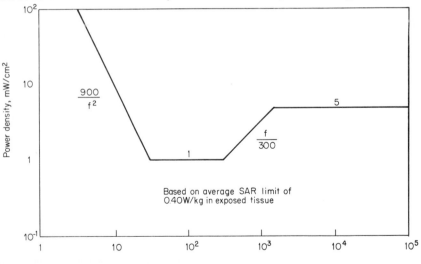

FIG. 19.2 Maximum microwave exposure level.

general-population standards is demonstrated at the property boundary. The combined effects of free-space attenuation, shielding from vegetation and topography, and antenna discrimination angles normally yield both calculated and measured levels orders of magnitude below any general population standard for line-of-site microwave and earth-station installations.

The analysis process, based on Refs. 2 and 3, presumes a reasonable worst-case set of carriers and power levels which will not be exceeded in operation at the mature site. Power-density levels at 6-ft elevation above ground level at the property boundary are determined by loading the carriers onto the antenna patterns (in many cases, a near field or transition zone assumption is required). Antenna boresights are swept over the orbital arc, and the variation in power density as a function of free-space attenuation, azimuth and elevation angles, and topography is derived. The total site power level for all emitters is determined by superposition and compared to general-population standards.

Of paramount importance when the microwave radiation issue is presented to the public is clarification of microwave emissions as nonionizing; i.e., they heat by exciting motion in polar molecules. This is distinctly different from nuclear radiation wherein the radiation travels as *photons,* energy packets which behave as high-energy particles with the ability to irreparably damage the biological makeup of individual cells.

13. Building Codes

Building codes are adopted, administered, and enforced by local governmental bodies, generally municipal and county. These codes are enacted to assure the life, safety, and welfare of building occupants. Codes define performance standards, environmental loads, methods of analysis, testing, and construction practice for diverse building applications and varied construction concepts, materials, and subsystems.

Local governments can accomplish their goals by adopting by reference the Uniform Building Code[4] or the Basic/National Building Code[5] and by tailoring original paragraphs of their local documents to address unique local conditions, policy issues, and public concerns. The Uniform Building Code and the Basic/National Building Code (BOCA) become binding only when adopted into law by governmental action.

Architectural and engineering consultants often make use of the American Institute of Architects (AIA) Masterspec,[6] a generalized construction specification. The Masterspec is used both as a reference work and as a baseline from which a project-specific procurement document is developed. Copies of the model codes and the generalized construction specification are available from the sponsoring organizations.

14. Insurance Requirements

There are two classes of insurance-related facilities issues:

- Design and operation in compliance with local building, electrical, and fire codes
- Design and implementation of facilities to criteria which exceed code requirements

Code compliance is assured in the initial design and construction phase by the requirement to submit "sealed" (i.e., certified) drawings from a professional engineer in order to obtain a building permit. Construction compliance is assured by the governmental inspections which precede the issuance of a *certificate of occupancy* (C of O).

For designs which exceed code, the intent is to reduce the risk of damage and service interruption and, as a direct result, reduce the cost of insurance premiums. Insurance coverage has two components; one is the cost to rebuild the physical site, and the other is the cost to insure against revenue loss from service interruption as a consequence of site damage. An engineering cost analysis is performed to optimize the complement of equipment; this analysis compares the incremental added capital cost of higher-performance protection systems with the present worth (obtained from a *discounted-cash-flow analysis*) of the incremental stream of insurance premiums which would be saved over the lifetime of facility operations as a consequence of implementing the higher-performance system. If the present worth of the savings in incremental premium payments exceeds the incremental capital, the better system is implemented.

The desire to minimize the risk of service interruption on critical facilities may result in the implementation of such protection devices as Halon fire-suppression systems, although the code may require only alarms and portable extinguishers.

DESIGN TO MEET AVAILABILITY

15. Definitions

The intrinsic quality of a communications service is measured in terms of the *purity* with which the injected signal is transmitted, as measured by parameters such as signal-to-noise ratio and bit error rate. *Quality* is also measured by the reliability with which the signal is conveyed to its destination, as measured periodically by the end-to-end circuit availability over the lifetime of the service. *Availability* is the time period of circuit operation at or above specification divided by the total time period of measurement.

Unavailability results from link fades, hardware and software failures, operator error, restoration, and maintenance times. Availability requirements for communications systems are often stringent, ranging above 99.90 percent to 99.995 percent.

Availability is increased by designing for statistically less significant but physically more demanding environmental conditions. It is increased by employing hardware with high-reliability components, by system "burn-ins," and by adding failure alarms, redundancy, automatic switching to backup systems, and diverse paths of distribution. The addition of uninterruptible power systems, on-premises spares, and multiple shifts of technicians further improves availability. All of these techniques contribute costs which increase service pricing, thus availability, like other system parameters, is the subject of tradeoffs.

Availability in a communications system may be defined as

$$A_s = (1 - P_L)(A_H) = \frac{(1 - P_L)(\text{MTBF})}{(\text{MTBF} + \text{MTTR})} \qquad (19.1)$$

where A_H = hardware and software availability

P_L = probability of link fade
MTBF = mean time between failure = (system operating time)/(observed system failures)
MTTR = mean time to restore = summation of time for dispatch + diagnosis + repair + test

This may be extended as with all independent series-probable events by cross-multiplying availabilities of individual link hops and hardware sets to determine a system availability.

16. Concepts of Redundancy and Diversity

The intrinsic reliability of new hardware systems is estimated by using standard failure rate data for individual components, parts count information, component derating levels, and the serial and parallel topology of each circuit. A reliability prediction circuit model is built on the basis of statistical relations. Simple examples are shown in Figs. 19.3 and 19.4.

The subject is treated in great detail in Refs. 7 and 8. While analytic concepts based on qualified components are used extensively in military hardware, costs and delivery schedules in commercial systems often lead to estimates based upon simplified models and field experience with similar hardware. Thus, field and accelerated environmental tests of sample lots are used to avoid infant mortality and generic failures of production products.

In a redundant communications system two usually identical subsystems are implemented in parallel. Full communications performance is provided by operation of either half of the system. In the highest level of redundancy failure, detection circuitry automatically switches to the backup subsystem, which is powered but off-line. Under these conditions the only customer-perceived service interruption results from the switching transient. The failure of the primary system is reported to the communications provider to assure rapid restoration of redundancy.

If the probability of an event occurring is infinitesimal, but the event would result in a catastrophe, system diversity is implemented in addition to redundancy. Diversity provides physical separation not normally provided in redundant systems. An example of diversity is installation of redundant fiber-optic cables along multiple diverse paths.

The effectiveness of redundancy and diversity can be assessed by a *failure-mode effects and criticality analysis* (FMECA). The FMECA uses site block diagrams and circuit schematics in conjunction with a presumed set of failure modes based upon worst-case conditions, experience with similar systems and brainstorming (what if the...shorted, what if the air-conditioning circulation pump failed..., what if the subbasement flooded..., etc.). Causes are identified, quantified, and characterized.

FIG. 19.3 Reliability series component subsystem.
$R_{SS} = (1 - P_{F1})(1 - P_{F2})$---$(1 - P_{FN})$
where R_{SS} = subsystem reliability and P_{F1} = probability of component failure.

The specific objective is to define initially unapparent single points of failure and propagating multiple points of failure which would eventually cause total system failure. Newly defined failure modes are ranked in terms of criticality and probability. System design and operational concepts are then modified to eliminate or mitigate unacceptable impacts.

SITE-SPECIFIC DESIGN CRITERIA

17. Earth Station and Microwave Repeaters

Site Layout Plan—Earth Station. Once an earth-station site has been frequency-coordinated, the site design activity establishes unobstructed lines of sight across the orbital arc for the complement of antennas. Antenna placement can be con-

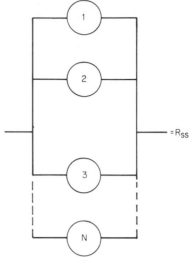

FIG. 19.4 Reliability parallel component subsystem.
$$R_{SS} = 1 - (P_{F1})(P_{F2})(P_{F3})\text{---}(P_{FN}).$$

strained by tree lines and hills in the southern quadrant. When topographic features do not constrain a site, antennas are placed immediately to the south of the technical building along an imaginary line nominally perpendicular to the average antenna azimuth angle. Antennas are clustered together while avoiding interantenna line-of-sight shadowing. The technical facilities, including building, antennas, and microwave towers, are placed close together at the northern end of the property to minimize waveguide length and losses and to minimize microwave levels at property boundaries.

Alternatively, antennas may be arrayed along the boresight line which yields the highest operational elevation angle. In this configuration, northerly antennas look over the top of southerly antennas. This concept normally results in greater waveguide losses for some antennas unless the technical building can be placed adjacent to, rather than behind (north of), the antenna farm.

Provision is made in the initial site concept for building growth toward the north. Generator plant, air-handling equipment, and the microwave tower are placed to the east or west of the building, as is the on-site parking area; their final location is determined by site-peculiar constraints. Of particular importance in the design process is providing clearance between antenna and waveguide bridges and vehicles using all roadways.

Many local codes require tower setbacks equal to 110 percent of tower height in order to assure that a failed tower will fall on site. This requirement may constrain an entire site design because the tower cannot be placed south of satellite access antennas without incurring blockage, and tower guys must be laid out to avoid line-of-sight and physical interference.

The location of potable-water wells, septic systems, retention ponds, buried fuel tanks, sprinkler system reservoirs, exterior lighting, and fence lines is next determined.

A goal of site design is to increase aesthetic appeal so as to foster community acceptance. Drawings are prepared which depict views of the site from boundary points accessible to the public. Overlays of the initial drawings are also prepared which depict the positive impact of the landscape architecture plan.

The site plan should depict those contributions and improvements which the site proponent is incorporating to obtain local approvals. This might include construction of public roadways and deeding of land as "green acreage." A typical major earth station site layout is shown in Fig. 19.5.

Microwave repeater sites are smaller than major earth stations, they are unstaffed and generally more remote. The earth-station design concepts—a compact but expandable layout, adequate security, and enhancement of aesthetic appeal—are equally applicable. More effort may be required to ensure access under adverse weather conditions to remote repeaters.

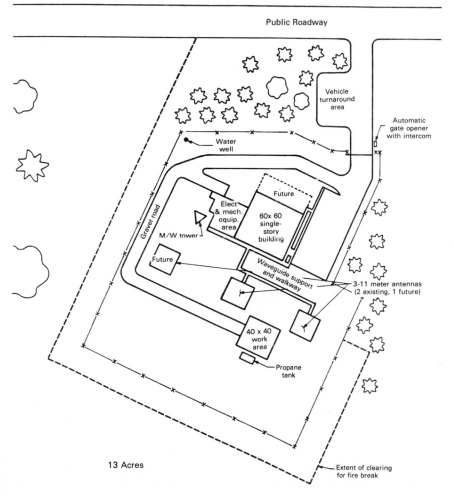

FIG. 19.5 Major earth station, plan view.

Line-of-Sight and Horizon Profiles. In order to evaluate line-of-sight constraints quantitatively, a site survey is employed. An optical transit is set up at each of the candidate earth-station antenna sites. The elevation of the local horizon, consisting of tree lines, building, and topographic features, is measured at selected azimuth angles (2, 5, or 10° increments) over a range of angles commensurate with accessing the desired orbital arc.

Spacecraft azimuth and elevation angles are then calculated and plotted as discrete points on the horizon profile chart. In general, a minimum discrimination angle of 1½° above the local horizon is desired to reduce multipath problems.

Equipment Layouts. Equipment within an earth station is grouped by function, with directly interconnected groups adjacent to each other and all groups arrayed to facilitate expansion without interrupting existing services or creating the need to relocate existing hardware. Provision is made for expanded storage and personnel operating areas (Fig. 19.6). Because equipment of like type are grouped together, building facilities can be locally tailored to support their unique require-

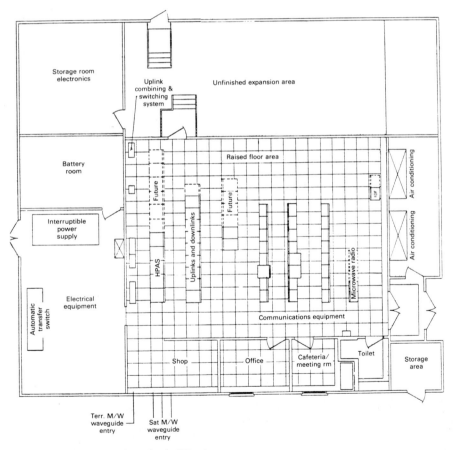

FIG. 19.6 Typical earth station building layout.

ments. Power systems are located on a slab; separate footings are provided to isolate generator vibration, ventilation louvers are provided for generator intake air, and exhaust gases are muffled and vented to the exterior. Battery plants are installed on slab in separate rooms with ventilation fans for acid fumes and safety showers for personnel. The main electronics area is constructed with a suppressed slab and equipped with a pedestal (raised) floor; the underfloor area is utilized as an air-distribution plenum and for intercabinet cable routing. Thermal control facilities are located on slab with stub air-handling ducts venting through the slab into the adjacent raised floor. This technique minimizes ducts, fan noise, and compressor vibration, and it provides for future growth.

Provisions for growth are incorporated in the initial design concept. The building shell is constructed larger than the initial capacity requirement, and the interior is divided into a finished area for immediate use and an unfinished future-growth area. In order to minimize cost, the shell of the future-growth area is completed but no other utilities are provided. The operating electronics area is sized to accept uplinks for all cleared and anticipated satellite frequencies, for all cleared microwave radio frequencies, and for multiplexing commensurate with traffic forecasts. At initial operation, the main room may be less than 30 percent filled with equipment. The power room, switchboard, power panels, and distribution buses are sized for the anticipated traffic growth. Initial power-plant cost is minimized by providing a single generator and foundations for parallel or larger units. A similar provision is made in space and access for the addition of growth air conditioners. The allocation of a cleared area to the north of existing facilities minimizes the risk of disruption and the need to relocate existing site services and facilities should a wholly new concept for site utilization develop.

Microwave Transmission Lines.   Baseband communications signals heterodyned with radio-frequency (microwave) carriers are conveyed between the communications electronics and site antennas in microwave transmission lines. Transmission lines for microwave frequencies are coaxial cables or hollow waveguides. The selection of the transmission line is a tradeoff between power-handling capacity, signal attenuation per unit length, insertion loss at connectors, ease of installation, fabrication, and cost. At major earth stations, waveguides are normally used between HPAs and antennas while coaxial cables are used between LNAs and downconverters. These selections are based upon power-handling and low-loss considerations for the former and ease of installation and low cost for the latter. Both the Electronics Industry Association and military specifications provide a full range of standards defining the electrical performance, mechanical design, and materials to be used in waveguide tubing, coaxial cables, flanges, and connectors. EIA Standards WR-3 through WR-2300 and Mil-W-85C address rigid rectangular waveguide tubing. RS-225 and RS-259 address rigid 50- and 75-Ω coaxial cables. Vendor catalogs provide tabulated summaries of transmission-line performance and compliance with standards.

Aesthetics and RF Shielding.   Aesthetics and rf shielding are separate issues which at times complement each other. The concept of aesthetics is subjective and best pursued by reviewing the design features, quality, and types of buildings and structures which have already gained acceptance in the community. The importance of aesthetics as a site approval issue varies from low in a community dedicated to commercial land use to high in a wealthy community with mixed zoning uses. A profile of community aesthetics can be obtained in informal meetings with local planning officials.

Landscape architecture discussed in Sec. 17 improves aesthetics by restoring the landscape so that it is consistent and unified with the indigenous foliage. The landscape architecture plan minimizes the public's lines of sight to the operating compound, and, by the incorporation of berms and/or precast free-standing concrete wall panels, potential rf interference cases can be minimized.

Environmental Design.   The ambient environment about the telecommunication facility induces loads in the facility which stress its electrical, mechanical, structural, and thermal systems. The design of the facility must accommodate these loads to ensure that system performance requirements are met and that environmentally induced failures do not result. Table 19.1 lists typical continental United States (CONUS) environmental requirements.

Wind-induced aerodynamic forces and the dead load of ice and snow accretion are primary among the loads which must be considered. The effect of these loads is particularly significant in the design and construction of towers and antennas. Design considerations for towers are described in Sec. 22 of this chapter and for antennas are described in Chap. 2.

18.   Central Terminal and Switch Offices

Equipment Layouts.   CTOs are normally at inner-city sites. When served by microwave links, they are located in tall buildings to ensure a clear microwave-path line of sight. On the upper floors of a building, CTO equipment and radios can be collocated. They can share battery plant and air conditioning while minimizing the length of waveguide runs. CTOs served by intercity fiber-optic facilities are located on lower floors to reduce cabling runs for local circuit loops.

A generalized layout is of little significance, since the specific design is dictated by the shape of the leased space and the characteristics of the building. The general concepts used to configure earth stations are applicable to CTOs; i.e., equipment is arrayed by functional group, interconnecting groups are adjacent, and provision is made for growth in personnel and equipment without the need to relocate existing hardware and interrupt service (Fig. 19.7). Power equipment is isolated in separate rooms from communications electronics and personnel, with dedicated cooling and ventilating facilities.

Rural earth stations are equipped with generator backup. CTOs take advantage of the greater reliability of urban commercial power and rely only upon battery plants providing 8 hours of prime power backup to carry them continuously through power outages without communications interruption. To preclude a long outage in the event of a catastrophic power failure (fire or flood), a utility service can be installed to street level, providing a temporary generator interconnection point.

Ideally, the CTO site would be loftlike, with 14-ft or greater floor-to-ceiling height, heavy floor loading capacity (greater than 150 lb/ft^2), an open building plan, substantial chase space, and a locale adjacent to local exchange switch facilities. It would be cooled by a building plant operating 24 hours per day with margins to handle a high sensible load. The constraint of floor-to-ceiling height in most modern office buildings precludes the use of pedestal floors in CTOs.

CTO electronics are installed in open racks within individual buckets. Racks provide rapid access for repair but, unlike electronic cabinets, they do not pro-

TABLE 19.1 Typical CONUS Environmental Requirements

Parameter	Full operation	Degraded operation	Survival conditions
Gain reduction	None	− 1.7 dB max. for 3700 to 4200 MHz −4.0 dB max. for 5925 to 6425 MHz	Not applicable—stowed
Temperature range	− 20° C to +55° C	− 35° C to +60° C	− 35° C to +60° C
Wind loading	30 mi/h, gusting to 45 mi/h	60 mi/h gusting to 85 mi/h	125 mi/h with no ice, 85 mi/h with 2 in ice
Pointing errors*	0.06° max. for 30 mi/h winds, gusting to 45 mi/h	0.16° max. for 60 mi/h winds, gusting to 85 mi/h; 0.35° max. for 90 mi/h winds, gusting to 110 mi/h	Not applicable—stowed
Altitude	Up to 10,000 ft	Up to 10,000 ft	Up to 40,000 ft during transport
Relative humidity	0 to 100%	0 to 100%	0 to 100%
Precipitation	1-in/h rain, or 0.25-in/h freezing rain, or 1-in/h snowfall	2-in/h rain or 0.5-in/h freezing rain	2-in/h rain or 0.5-in/h freezing rain
Static load	0.25-in radial ice or 4-in vertical snowfall	0.5-in radial ice or 6-in vertical snowfall	4-in radial ice
Solar radiation	350 Btu/(ft^2·h) at any angle of incidence	350 Btu/(ft^2·h) at any angle of incidence	350 Btu/(ft^2·h) at any angle of incidence
Seismic	Not applicable	Not applicable	Withstand a grade 9 intensity upset on the Mercalli scale
Shock/vibration	Not applicable	Not applicable	As encountered during commercial shipping and handling (Mil Spec test can be cited here)
Salt atmosphere	As encountered during long-term service in coastal regions (Mil Spec Accelerated Salt Spray Test can be cited here)	See full operation	See full operation

*RSS due to all sources including wind and static loads, misalignment backlash, pedestal and reflector deflection.

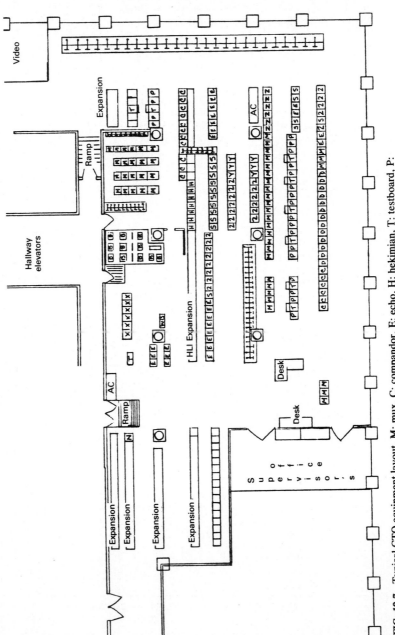

FIG. 19.7 Typical CTO equipment layout. M: mux, C: compandor, E: echo, H: hekimian, T: testboard, P: patchboard, WB: wideband, B: telco equipment, Z: 56 KBPS, ▫ : rack, ▦ : frame, D: data, X: client, S: single slot, 2: double slot, N: network switch, Y: specialty rack.

vide an enclosed path by which cooling air can be routed to individual units. Cooling air is provided to the room with electronics cooled by natural convection or forced by ganged pancake-type propeller fans.

Interface Provisions.   The CTO incorporates a main distribution frame, which is the junction point for individual circuits. The local telephone company (telco) distributes local loops to retail customers from the outbound side of the frame, and the long-distance carrier's CTO interconnects on the multiplexer side of the frame. When the local telco is providing T-carrier circuits rather than copper pairs, additional space is provided at the CTO for telco electronics equipment.

Wholesale multicircuit requirements are met in a cost-effective manner at the long-distance carrier's CTO by a wideband composite circuit interconnect. This interconnect may take the form of a "midair-meet" on a microwave path, or the collocation of a customer's microwave terminal at the CTO, or a coaxial cable interconnection from the customer's group/supergroup multiplex. In such cases, the vendor's CTO provides the housekeeping facilities, including space, power, HVAC, and fire suppression. In obtaining interior space, roof rights, and chase space, it is important to obtain legal rights in lease documents for the collocation of affiliated, subsidiary, vendor, and customer firms' equipment.

19. Fiber-Optic Paths and Repeaters

Path Rights-of-Way.   Fiber-optic paths require the acquisition of a continuous right-of-way along which the physical cable and repeater sites are installed. Rights-of-way have been negotiated with railroads, power companies, regional transportation authorities [New York Port Authority, Southeast Pennsylvania Transportation Authority (SEPTA)], local common carriers, federal [Department of Defense (DOD), Department of Transportation (DOT), and General Services Administration (GSA)] agencies, state agencies, and real estate management firms.

These sources have provided fiber-optic users with diverse access rights to utility poles, signal cable duct banks, high-voltage power distribution towers, and steam distribution tunnels, and to directly bury or install conduit along rail beds and highway shoulders. In buildings, fiber optics have been installed in chase spaces with power cables and in elevator shafts. Obsolete coaxial cable runs have been reused as fiber-optic cable conduits. The outer conductor and jacket were preserved while the inner conductor and dielectric spacer were removed.

Right-of-way agreements range from simple leases to joint ownership of new common carriers by technology and railroad companies.

Path Diversity, Redundancy, and Growth.   Fiber-optic facilities can be equipped with redundant transmitters, receivers, repeaters, and support systems (power supplies, HVAC, etc.) in order to achieve a high availability, 99.99 percent or greater. Fiber redundancy may exist within a cable, but the close physical proximity of individual fibers makes the system vulnerable to infrequent but potentially prolonged outage due to accidental damage. This can be overcome by establishing a physically diverse path. Frequently, a ring topology is employed in which the signal is transmitted on separate fiber pairs in separate fiber cables, each pair routed along physically separate paths. The signal, therefore, appears at the termination point from each of two directions, thus precluding outage due to cable or repeater damage.

Installation Techniques.   The use of glass as the transmission medium has raised questions about the physical robustness of this hardware. Individual optical fibers exhibit the high tensile strength, low modulus of elasticity, and notch sensitivity characteristic of glass. Single-mode fibers are 8 μm or less in diameter, thinner than human hair, and this can lead to the incorrect conclusion that the installed system will be physically frail.

Fiber-optic cabling is a mechanical system specifically designed to provide environmental protection and strain relief for individual fibers within the bundle.[9] Fiber transmission performance degrades when transverse and longitudinal forces and bending moments induce strain; substantial service life can be assured only if fiber strain is less than 1×10^{-3} in/in. Thus the packaging of fibers within the cable must decouple and isolate them from strains induced by the external environment. The cable system design is optimized for the intended field application and divorced from individual fiber characteristics; thus it is specifically designed to be hung on a messenger cable, direct-buried, or pulled through duct banks.

Individual fibers may be jacketed in a tight-fitting plastic (buffer) tube, but they are more often loose-buffered in a tube of much larger internal diameter. The excess length of fiber (provided for strain relief) is accommodated by factory shaping each fiber into a helix along its run length. Under external loads induced by local forces or thermally induced expansion or contraction, the helix tightens or unwinds within the tube cavity (Fig. 19.8). The tube-buffered optical fiber is the basic building block of the final bundled fiber-optic cable. Packaging density may increase by incorporating more than a single fiber in each loose-buffered tube.

The fiber cable core is constructed by taping a helix of buffer tubes about a central strength member of steel or Kevlar (the pulling member). This core is then jacketed in a Kevlar sock and fitted with a polyethylene outer jacket for pole-hung and duct installation (Fig. 19.9). For the more stringent direct-burial application, the core is first wrapped in a corrosion-resistant steel tape to minimize attacks by rodents and then jacketed in polyethylene. Buffer tubes and the cable core cavity may be paste-filled to minimize the effects of moisture.

Cable manufacturers have developed installation methods which minimize strain and ensure long cable life. Manufacturers recommend lashing the fiber cable to a stiffer external steel messenger wire for pole-hung installations. Direct-burial cables may be installed by autotrenching cable plows or, alternatively, separate trenching and burial techniques. In duct installations, criteria are provided by the cable manufacturer for minimum bend radii and installation pull tension in order to ensure that the strain protection envelope is not exceeded.

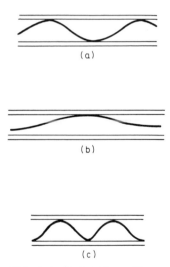

(a)

(b)

(c)

FIG. 19.8 Optical fiber with excess length inside a loose buffer jacket. (*a*) fiber in buffer jacket after cable manufacturing, (*b*) decrease of fiber excess length caused by strain of buffer jacket during cable stress, (*c*) increase of fiber excess length caused by shrinkage of buffer jacket during cooling. (*Source*: Siecor Inc. Used with permission.)

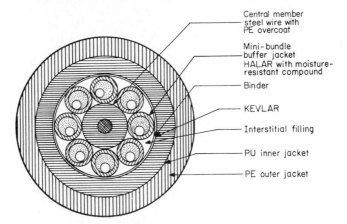

Central member steel wire with PE overcoat

Mini-bundle buffer jacket HALAR with moisture-resistant compound

Binder

KEVLAR

Interstitial filling

PU inner jacket

PE outer jacket

FIG. 19.9 Forty-eight-fiber aerial duct mini-bundle TM cable. (*Source*: Siecor Inc. Used with permission.)

Repeater Sites.   The electrooptical terminals and repeaters in the fiber-optic system are generally packaged in conformance with the techniques used for CTO hardware. Open-face racks are used with redundant equipment mounted in individual buckets readily accessible for debugging or repair. The same packaging system can be installed in shelters for use at remote repeater sites.

20. Monitor and Control Facilities

Overview.   The technology and precepts of monitor and control facilities are undergoing dynamic change. Formerly employing hard-wired, dedicated systems, these facilities are now using computer-based technology, thus reducing the labor intensiveness of real-time monitoring. Newer systems provide automatic parametric checks while generating summary reports and alarms in lieu of voluminous logs of data. The newer precepts of quality assurance, based upon the applied statistical concepts of Dr. Deming and others, treat quality as an ongoing process integral to service provision rather than as discrete checkpoints or reactive responses to out-of-specification alarms or customer complaints. Computer-driven monitor and control systems are used to monitor the trends of communications parameters, predict out-of-specification conditions, and modify control variables as necessary so as to assure that service quality is maintained within specification prior to failure or customer complaint.

Communications systems usually incorporate customer trouble call centers for the receipt and logging of complaints. In these facilities, problems are isolated at a rudimentary level so as to assign and dispatch technicians with proximity and familiarity to diagnose, repair, test, and clear the problem.

Technicians are aided by end-to-end test facilities such as satellite master control centers, video monitoring centers, and remote transmission test facilities. Remote monitoring facilities are provided for specific hardware such as uplinks, LNAs, and microwave repeaters.

Satellite control center functions are divided into *tracking, telemetry, and command* (TT&C) systems and transponder loading monitors. The former monitor and control the status and health of satellite subsystems and components and orbit accuracy. The latter compare the actual loading of carriers within each transponder to the idealized budget, in particular, the power, bandwidth, and intercarrier spacing; this is accomplished with a spectrum analyzer tunable over the bandwidth of interest. This capability is also used to identify interfering carriers. In addition, test carriers and power meters dedicated to the facility are used to assess uplink carrier flux to saturate and downlink *effective isotropic radiated power* (EIRP).

Alarm points in radios and auxiliary equipment at a microwave repeater site interface with a local monitor and control unit. This unit interconnects through an autodialer to a terrestrial voice-grade circuit which terminates at the system master control site. Telemetry on hardware status and link performance (see Table 19.2) is monitored at the master site for all repeaters and terminals in the microwave path, thus facilitating failure isolation and dispatch of technicians for path restoration.

Video monitoring facilities consist of an array of monitors, waveform displays, vu meters, and a central matrix switch. Video monitoring relies upon the nature of the video signal, which permits parallel monitoring of multiple channels by a technician. This is a unique video attribute which cannot be extended to audio monitoring. A matrix switch provides rapid access to display the signal at intermediate points in the transmission path and to assign individual video signals to more or less prominent monitors. The facility is constructed as a unipurpose screening room, acoustically isolated, with monitors on one wall, a raised platform, and a lowboy control console to optimize viewing by the on-line technicians.

Remote circuit test facilities permit a technician at a central point to access the circuit on an end-to-end basis and at discrete points in its path in order to diagnose the problem. As an example for a data circuit, the technician would place a *bit error rate* (BER) test set on line at one end of the circuit and then progressively loop back the circuit while monitoring BER for each successive link until the transmission element degrading performance is isolated.

TABLE 19.2 Microwave Repeater and Earth-Station Alarm Points

Backup system on line	Site status	Link failures	Hardware failures
High-power amplifier (E)	Gate open	Link fade	Charger (M)
Traveling-wave-tube amplifier (M)	Door open	Frequent drift	Uninterruptible power supply (E)
Upconverter	Over/under temperature	Loss of clock	Waveguide pressurization
Downconverter	Fire alarm	Loss of data	
Low-noise amplifier (E)	Fire suppression actuated		Dehydrator
Receiver (M)			Cooling fans
Modem (E)	Prime power failure		
Baseband amplifier (M)			

NOTE: Alarm points unique to microwave repeater or earth stations are noted by (M) or (E).

Status and Parameters to be Monitored. Monitoring points are incorporated in the hardware at the component and at the subsystem level; these points are wired locally to a unit of an integrated monitor and control system, acting as a multiplexer, encoder, and modulator. Information of an anomalous condition is transmitted automatically or the master site may strobe the local unit to assess status. A diverse telemetry path is used to ensure that a system failure does not interrupt diagnostic telemetry. A dial-up line may be used to minimize cost.

Redundant systems automatically switch failed primary systems to redundant backup systems and report an anomaly. This ensures dispatch of technicians and rapid restoration of redundancy. The short time (milliseconds) for automatic switchover to backup systems is allocated to downtime budgeted for system service and repair; it would not result in a refund of customer billings. Prolonged outages caused by double failures will degrade availability and result in refunds to customers.

Telemetered data may be either a status flag, such as transmitter room over temperature, or an analog parameter, such as percentage of output power. Older hardwired systems portrayed an out-of-limit status by illuminating one of a multitude of miniature lamps in a modular display. As systems grew, lookup tables were required to identify the location and type of parameter being alarmed; this slowed the diagnostic process. Use of audible alarms also became less effective as the frequency of alarms increased. Computer-based systems only display out-of-limit flags on a *cathode-ray tube* (CRT) monitor with exact site and parameter defined. To speed response, the monitor can display diagnosis and remedial-action alternatives. The number of displays need not increase to accommodate additional remote sites; only the storage capacity for the data base need be increased.

ELEMENTS OF THE COMMUNICATIONS PLANT

21. Buildings and Shelters

Communication facilities may be installed on the customer's premises or in a separate structure. For earth stations where substantial growth is predicted, the electronics should be housed in a specifically designed technical building.

Prefabricated Metal Buildings. Prefabricated-metal-building manufacturers provide building kits for single-story lofts. These kits are factory-fabricated from standard modules. Thus, 30-by-40-ft and 30-by-20-ft buildings use identical hardware when constructed from the same product line.

Major manufacturers (Butler, Armco, Varco-Pruden, etc.) have multiple product lines designed for applications from light commercial to heavy industrial with varying open spans, bay depths, column heights, and exterior wall panels. The basic elements of a standard product include the structural steel, metal curtain-wall panels, roof diaphragm, structural closures, framing for openings, and a conceptual foundation design with lag-bolt connection to the building foundation. Accessory products are available and used to tailor building aesthetics and performance features to a particular application.

Prefabricated metal buildings offer low cost, rapid delivery, and consistent quality but they are limited to the shell's structural system only; they do not include the HVAC, power, lighting, sanitary, and interior fitments.

Metal buildings are designed to meet national building code requirements; however, some local codes are predisposed against them. This may be true for perceived aesthetic reasons or as a result of an earlier era when less analytically designed systems did not perform well (for instance, in Florida's humid, hurricane-prone environment). Consultations with local planning officials will provide an insight into local issues. Because of historical precedent or the availability of alternative raw materials, other construction systems may be locally price-competitive, for example, concrete block in Florida or heavy timber in the northwest.

Concrete Block and Millwork Buildings.   Alternatives to factory-fabricated buildings are used when local code and/or unique design requirements and constraints render them cost-competitive. In the southern tier of states, the inherent insulating qualities and noncorroding and antifungus attributes of masonry material, the ubiquitousness of block fabricators, and their proximity to the construction site have contributed to the competitiveness of concrete block. Metal buildings may cost less than block at the factory, but shipping costs to site may offset the economic advantage.

Millwork buildings can provide design flexibility; reasonably low costs can be obtained by using standard curtain wall and roof diaphragm products in conjunction with custom structural members. The latter are fabricated from standard structural cross sections into finished structural members at a steel fabrication "mill." The mill prepares detail fabrication drawings from the structural design drawings and client engineering review, and approval precedes fabrication.

Electronic Shelters.   Electronic shelters are complete factory-built enclosures generally ranging from 8 ft wide by 10 ft long by 8 ft high to 12 ft wide by 24 ft long by 10 ft high. They are finished structures when shipped from the factory, with flat roof and integrated floor. As originally designed for military applications, they are equipped with skids, lifting eyes, and hard points for the installation of demountable suspensions and wheels. Shelters provide the military with a highly transportable, self-contained structure in which to integrate and operate electronics systems. Military shelters are transported by aircraft, helicopter sling, and railcar. They are towed over the road and over unimproved terrain, and, when equipped with shock-absorbing skids, they can be air-dropped. Metal-frame construction is used with rigid foam insulation and exterior and interior surfaces skinned in metal. These shelters are normally equipped with leveling jacks and ground anchors for installation at unimproved sites without foundations.

The benefits of transportability, factory system integration, and reuse at differing sites have led to the development of shelters for commercial applications, the key difference being the sacrifice of diverse methods of mobility under extreme loads in favor of a less stringent requirement allowing transportation by truck, rail, or aircraft and final installation at an improved site.

The substantial reduction in mobility modes and easing of military environmental conditions results in drastically reduced cost for commercial shelters. Commercial construction employs wood framing with exterior walls of fiber glass and interior walls of plywood and paneling. It is important, nevertheless, to use treated low-burn-rate and fire-retardant materials to minimize both risk of outage and insurance costs. Shelters are cost-effective for very small installations; they can be obtained, integrated, and deployed rapidly. Vendors are available who will deliver a full integrated system with HVAC, lighting, fire suppression, power distribution, and generator set installed. On a per-square-foot basis, shelters cost more than buildings, and they are often prohibited by local codes which exclude temporary structures, such as trailer camps, from townships.

22.   Towers

Tower structures support microwave antennas at heights above the local topography in order to provide a clear line of sight to adjacent repeaters and terminals along the microwave path. Towers may be free-standing and self-supporting or they may employ guy wires to establish structural integrity. In either instance, the major tower structure may be constructed as a space frame (Fig. 19.10). Free-standing towers may also use stress-skin construction. Lightly loaded guyed towers may use tubular or solid rod members for their central spine.

Selection of a tower type is based upon height, antenna size and quantity, pointing accuracy, and wind and ice loading. It is further constrained by the shape and size of the site, geologic, and seismic conditions, local topography, natural and artificial conditions, operational constraints, implementation costs, and schedule.

Towers are proliferating, although local zoning approval is becoming more dif-

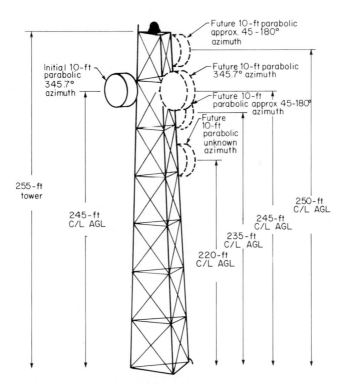

FIG. 19.10   Self-supporting microwave tower. Microwave tower concept showing provision for future addition of antennas by size, altitude, and boresight of azimuth.

ficult to obtain because the public considers them unaesthetic. Therefore, a tower design should be planned for long-term corporate capacity projections and should provide additional incremental capacity for installation of tenant antennas. Tenant antennas provide wideband interconnection with customers, vendors, and public services. Incremental costs for a stronger tower to accommodate growth are minimal when included in the initial design.

Free-Standing Space-Frame Towers. Free-standing space-frame towers consist of a multitude of trusses constructed by riveting or bolting small-cross-section structural members (channels, tees, angles) together at their ends to form the tower. Space-frame construction is structurally efficient and low in cost. It uses readily available small-cross-section members and involves simple fabrication processes and facilities. Vertical loads are present on a tower induced by the point load of antenna weight and the distributed weight of structure and ice. However, the major survival loads for towers are the horizontal wind-shear forces induced by wind interacting with the projected areas of antennas and ice-encrusted structural members (Fig. 19.11).

A free-standing tower is a cantilever beam. Fixed at the ground, it deflects elastically under the applied load, the maximum translation and rotation occurring at the top of the tower. Maximum reaction forces are developed at the base of the tower; therefore, larger-cross-section members and larger-diameter or multiple fasteners are used at the base while lighter parts are used near the top. The tower members are in tension and compression, while fasteners are subjected to shear.

When the wind is other than perpendicular to the antennas, torsional moments are induced which twist the tower in addition to translating and rotating it. All tower motions tend to drive an antenna off-boresight, diminishing link margins. Therefore, tower stiffness is a critical criterion, second only to absolute survival capacity.[10,11] Tower foundations are designed to resist uplift, moments, and lateral forces. Loads must be transferred into the soil within the soil's elastic bearing capacity to preclude tower tilt induced by settling.

Antenna mounts are designed to distribute the point load at the mounting location over a wider area in order to avoid local overstress. Antenna locations must be designated in specifications to ensure factory provision of local load distribution and reinforcing structure. Free-standing space-frame towers and guyed towers are cost-competitive to 100 ft. From 100 to 200 ft, the cost advantage favors guyed towers by an average of 25 percent, and the cost disparity favoring guyed towers accelerates above 300 ft.

All towers must comply with Federal Aviation Regulation 14 CFR visibility requirements, Sec. 77.13. In order to ensure safe navigation of aircraft, there are tower exclusion zones in proximity to airports. Present specifications require orange and white color bands for towers of 200 ft and red warning lights for operation after sunset. FAA visibility requirements are changing; they will delete the color bands and red lights and substitute white strobe warning lights (see FAA document AL 70/7460-1F).

Major tower manufacturers such as Rohn and Microflect will deliver a proof-of-performance analysis, sealed drawings, and a site-specific foundation design as deliverable items with a tower procurement. Tower erection can be obtained through the manufacturer or from independent riggers.

Guyed Towers. The foundation design of the guyed tower is simple. The foundation is always in compression, and the interconnection is a pinned joint. Guyed towers use substantial real estate; the guy field is generally 140 to 150 percent of

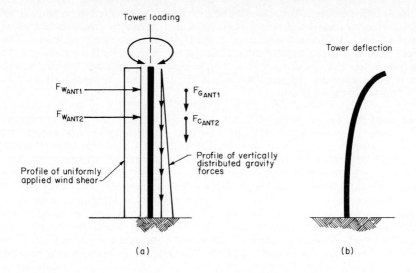

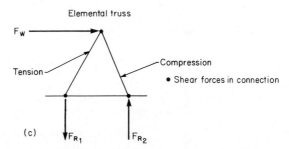

FIG. 19.11 Tower loading and deflection. (A) where $F_{W\text{-}ANT1}$ and $F_{W\text{-}ANT2}$ = wind force on antennas, $F_{G\text{-}ANT1}$ and $F_{G\text{-}ANT2}$ = gravity force on antennas, and M_W = twisting moment induced by wind.

the tower height. Greater areas must be cleared of vegetation to ensure that falling trees do not down the tower and anchor points may require separate fenced enclosures to preclude vandalism.

Tower guys are preloaded to increase lateral stiffness. Thus, the central spine is always in compression while the windward guys experience an increase in tension and the leeward guys a decrease. The central spine is subject to crippling and unstable buckling failures unless the stress field established in the guy system is uniform. It is often more difficult to reinforce a guyed tower for an unanticipated increase in load than a free-standing tower. If space is available, the latter can always be guyed to midspan to relieve excessive loads at the base.

Monocoque Towers.   Tower design as an integrated element of overall site facilitation has received limited consideration. New cellular radio towers have improved packaging efficiency, equipment protection, and aesthetics resulting from a metal monocoque structure with antenna and rf system integrated at the top.

   Monocoque or stress-skin construction is similar to that used in an aircraft fuselage. Hoopes (rings) and stringers (longitudinal members) stabilize thin sheet metal which is rolled and riveted or welded to the underlying structure. The tower spine is a large-diameter tube, flared at the base where it interconnects to the foundation and equipped with a bulbous antenna/equipment housing at the top. Applied loads are supported or resisted by the stress-skin spine, which acts as a hollow cantilever beam.

   Wind Load on Towers.   Wind forces induce deflections in the antenna support and tower structures which defocus antennas, reducing gain, and rotate the tower, causing boresight error and reduced link performance. Combinations of loads must be treated in a systematic fashion for a number of environmental conditions in order to determine worst-case operating and survival stress levels to assure structural integrity. Generally in CONUS, worst-case loads develop at or below freezing when a combination of high winds and icing occurs simultaneously. Worst-case combined loading will differ from zone to zone; as an example, icing may not occur simultaneously with maximum wind force in subarctic conditions (because of low humidity). Cold temperatures and ice buildup at 4.7 lb per inch of accretion per square foot of area add additional weight which must be reacted in the structure. The projected area of space-frame towers may double or triple, and bridging ice may close perforated reflectors, increasing their effective area.

   Wind forces develop when structures block the wind, thus decelerating the moving mass of air. Wind force is expressed as the dynamic pressure of the free stream:

$$q = \tfrac{1}{2}\,\rho V^2 = 0.00256V^2 \qquad \text{at sea level and } 59°F \tag{19.2}$$

where q = dynamic pressure, lb/ft^2
 ρ = air density at a given temperature
 V = velocity of the air, mi/h

Since air, a mixture of gases, follows the *general gas laws,*

$$\frac{\rho_2}{\rho_1} = \frac{T_1 P_2}{T_2 P_1} \tag{19.3}$$

where the Ts are absolute temperatures (F° + 460°) and the Ps are static barometric pressures. Thus, air density and therefore dynamic pressure, for a given wind velocity, decrease with altitude and increasing temperature.[11]
   Wind force for *blunt* objects is

$$F_\Delta = C_\Delta q A$$

where F_Δ = drag force
 C_Δ = drag coefficient
 A = projected area of structural element

   Airfoil-shaped objects, such as parabolic reflectors, exhibit *wind-induced lift* and *moments* in addition to drag forces (Fig. 19.12). Their coefficients are a function of angle of attack to the wind. The coefficients are plotted in Refs. 10, 12, and 13. The lift force F_L and moment M are

$$F_L = C_L q A \qquad M = C_M q A D$$

where C_L = lift coefficient

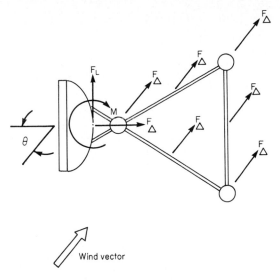

FIG. 19.12 Tower wind loads. Horizontal section through tower at antenna mount, F_Δ for tower members in wind-referenced axes, F_Δ, F_L, M for antenna in antenna body referenced axes, and θ is angle of attack.

C_M = moment coefficient
D = aperture diameter or chord

For a tower the total wind load is the summation of point induced forces and moments at antennas and the distributed drag force determined by integrating dynamic pressures over the projected area of tower structural members.

Wind has both *steady-state* and *gusting components,* the latter changing velocity and direction. Experimental field data indicate that the range of peak gust velocities is 130 percent of the average steady-state wind. The half-power bandwidth of the wind spectrum does not exceed 0.05 Hz with a second frequency observed at about 0.319 Hz. Most towers are sufficiently stiff to preclude gust-induced *resonant coupling.* A second, more complex, resonant excitation with steady winds may result from *vortex shedding,* especially on cylindrical members. The phenomenon occurs at specific velocities where the period of *vortex streets* coincides with a structural natural frequency of the tower and results in the generation of oscillation lateral forces.[12,13] Vortex-induced oscillations can be mitigated by the addition of spoilers, increased damping, and changes in the natural frequency of structural members so as to decouple the vortex frequency and structural frequencies.

23. Power Systems

Rural commercial power feeds exhibit lower availabilities than urban feeders; this is primarily because of less system redundancy and diversity. Feeds may suffer from major grid failures and local accidents, and the rural feed may fail more extensively as a result of icing, high winds, heavy rains, local flooding, fallen trees, and lightning strikes. For major communications facilities with stringent availability requirements, the sites are equipped with autostart generators; they may also be equipped with uninterruptible power supplies (Fig. 19.13). A UPS consists of a generator with automatic transfer switch in series with a rectifier/charger stage, a battery plant, an inverter stage, and a static transfer switch.

A communications facility powered by a UPS will not suffer a long-term or a transient outage as a result of a commercial power failure. Facilities powered by a commercial utility with backup power supplied by an autostart generator will experience transient failures. Generators will produce a starting transient lasting 10 to 15 seconds (diesel-powered) or 45 to 90 s (turbine-powered) before operating speed is attained. During the transient period, frequency-synchronization electronics will not permit the generator to assume the load; this transient outage may be extended by voltage-protection circuitry in high-voltage power supplies.

In normal UPS operation, commercial ac power (single- or three-phase) feeds the rectifier/charger stage, which produces a dc power output that floats up to 4 V above the fully charged battery voltage (as an example, 48 to 52 V for a 48-V plant); thus batteries are maintained fully charged while the communications load is powered. The inverter stage reproduces a 60-Hz ac waveform to power on-site electronics. A separate UPS dc tap or charger unit may be provided to power dc-operated equipment.

When commercial power fails, the UPS battery plant assumes the load through the generator starting transient. The battery plant is sized to protect against a generator starting failure; between 15 and 30 min of battery *holdup* time is provided.

Selection of a diesel- or turbine-powered generator system depends on the lo-

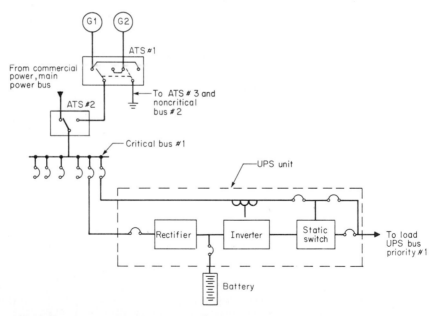

FIG. 19.13   Uninterruptible bus priority #1 diagram.

cation of the site, restrictions imposed by the environment, and type of service needed from the systems. For example, a user in a city may be forced to install a turbine because of its smaller size, lower noise, and ease of rigging, even though requirements could be satisfied by a diesel generator. A user who requires continuous on-site power generation will prefer a turbine for better efficiency and reliability (Table 19.3).

The conversion efficiency of UPSs averages 75 to 85 percent; they are normally derated for operation to 80 percent of rated load in order to increase reliability. Conversion efficiency of high-power components of communications links is low. As an example, HPAs rated at 3 kW rf output idle at 9 kW and produce full output power with 13-kW input. Auxiliary systems such as HVAC and electric deicing are extremely high in power demand. A brute-force UPS system powering all subsystems would be overly large, costly, and wasteful of input power. In order to increase efficiency and optimize the system, multiple buses may be established. A site load analysis is undertaken to determine the actual site operating demand, which as a result of individual equipment duty cycles may be well below the connected load (Table 19.4). This analysis is used to size generators, switchgear, UPSs, and power-distribution components.

A logic control diagram for a multiple-bus power system is shown in Fig. 19.14. Allocation of equipment to a bus of a particular priority is based upon the effect of load shedding. As examples, electronics used for the real-time distribution of video programming are placed on the UPS bus to preclude any outage. By contrast, antenna drive motors are powered from a utility bus (Table 19.5).

24. Lightning-Protection Systems

Design of lightning-protection systems has been predicated upon intercepting the lightning stroke and diverting its energy away from communications hardware to

TABLE 19.3 Comparison of Turbine and Diesel Generator Units

Characteristic	Turbine unit	Diesel unit
Unit size	Minimum 400–500 kW	Wider range
Physical dimension	Much smaller; at least saves 20 to 30% of floor space	Larger
Weight	Less weight	Heavier
Efficiency	25% higher than diesel unit	Low
Mechanism	Simpler and fewer moving parts	Variety of components and control devices
Starting speed	45 s	10 s
Cooling system	No water cooling system required	Water cooling
Fuel	Variety of fuels can be used	
Lubricating oil	Only bearing lubrication required	Consumes engine oil
Availability	Higher	Lower
Maintenance	Harder to find maintenance personnel	Much easier to obtain service personnel
Cost	Much higher; almost twice as much as for diesel set	Lower

TABLE 19.4   Electrical Load Analysis of Earth Station*

Major equipment and/or systems	Estimated connecting load, kVA			Estimated demand load		Load (kVA) requirement for sizing electrical service		
	Existing	Addition	Total after expansion	Factor	Load, kVA	Noncritical (commercial with generator backup)	Critical (UPS†)	Total
1. Heating (electrical)	11	11	22	1.0	22			
2. Ventilating	14	14	28	0.8	22.4	22.4‡		
3. Air conditioning	86	86	172	0.8	137.6	137.6‡		
4. Duct heating	20	10	30	0.5	15			
5. Antenna deicing		30	30	1.0	30			
6. Fire pumps	35	2	37	0.8	29.6	29.6		
7. Generator auxiliaries	16	8	24	1.0	24	24		
8. Well pumps, Hot water heaters, and auxiliaries	70	0	70	0.8	56	56		
9. Lighting	20	30	50	0.9	45	45		
10. Convenient receptacles	27	40	67	0.25	16.75	16.75		
11. HPA and LNA	224	180	404	1.0	404		404	
12. TT&C equipment (excluding HPA)	68	39	107	0.4	42.8		42.8	
13. Communication equipment (excluding HPA)	45	27	72	0.4	28.8		28.8	
14. Tape center equipment	30	60	90	0.4	36		36	
Total	666	537	1203			331.75	511.6	843.33

*Estimated overall power factor is 90%.
†UPS shall be designed with 15-min battery and generator backups.
‡Load requirement for HVAC (items 1 through 5) designed for summer season (air conditioning and ventilating) only.

ground. Protection systems have employed sacrificial circuit devices, voltage-clamping devices, and/or parallel grounding circuits and devices. The threat of lightning damage is a function of the frequency of occurrence of lightning, susceptibility of the hardware, and the probability of interception and grounding. The statistics of lightning events have been measured and plotted for geographic areas of the United States.[14] The probability of a lightning stroke is related to the Keraunic number (a measure of lightning density), which strongly correlates to the frequency of thunderstorms (Fig. 19.15).

Lightning can damage a communications site as a result of a direct stroke at the facility or a stroke in the vicinity which propagates through power and signal lines. The probability of a lightning stroke at a facility is a function of the altitude, maximum height above local topography, size of the facility, and the local Keraunic number.

Protection against lightning damage from a direct stroke has been achieved by placing multiple lightning rods at the top of system structures and using large-

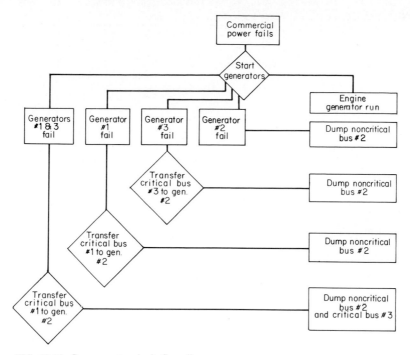

FIG. 19.14 Power system logic flow diagram.

gauge conductors to conduct the energy into a buried grounding grid field of extremely low dc impedance (less than 5 Ω).[15,16] The grounding system has a signal ground grid and a power/safety grid which are tied together at a single subsurface point to minimize ground loops. The structure and power equipment are tied directly to the power ground; similarly, communications electronics is tied directly to the signal ground.

The efficacy of these systems is variable because it is based on the presumptions that:

1. The lightning stroke will strike a rod in lieu of the primary structure
2. The grounding field will consistently remain low in impedance
3. Lightning strokes are a steady-state condition rather than a high-energy transient phenomenon with very high frequency components

At high frequencies, the effective impedance of the grounding system may be extremely high, especially if long cables, bends, and conductor transitions are used. The energy of the lightning stroke may thus be inductively coupled to electronics rather than conducted harmlessly to ground. To increase effectiveness, this grounding concept requires periodic measurements of grid impedance, many lightning rods, and short, straight, or gradually curved large-cross-section conductors. A static charge initiated by a primary stroke may create a capacitive sink in a major site structure which will result in a secondary destructive discharge at a later time if all elements are not effectively grounded.

TABLE 19.5   Earth-Station Equipment Bus Assignments

Equipment	Priority assignment number	Bus assignments			
		Priority 1 (UPS)	Priority 2 (critical bus 1)	Priority 3 (critical bus 3)	Priority 4 (utility bus 2)
Video exciters Upconverters Modulators GCE equipment	3	All			
Video receivers Downconverters Demodulators GCE equipment	3	All			
Antenna motor drives	18				All
Video monitoring center	7			All	
Tape center	6		All		
Lightning and Recept.	9		1/3	1/3	1/3
Fire pump	8		All		
Ventilation	10		1/3	1/3	1/3
Air-conditioning units	11		1/3	1/3	1/3
Building heating	15				All
Electrical antenna deicing	16				All
Hot-air antenna deicing	16		All		
Water heaters	17				All
Fire alarms	8		All		
Radiator fans (generators)	4		2/3	1/3	
Microwave (rectifiers)	5		All with Batteries		
Oil pumps	4		1/2	1/2	
Dehydrator	14				All
Dampers	4		All		
Battery chargers (generators)	13		1/3	1/3	1/3
Well pump	12				All

An alternative conception for lightning protection may be employed—a family of devices termed *lightning dissipation arrays*. Their performance is based on the concept that, within a localized area, the set of conditions which causes a lightning stroke to propagate can be modified so that the event will not occur. Lightning dissipation arrays may resemble the structure of an umbrella (*hemispherical array*) or a large "crown of thorns" or a "ball on mast" (*trapezoidal array*). Arrays are impedance-matching devices which provide a path for the differential charge between the atmosphere and the ground to reach zero potential before the voltaic potential causes dielectric breakdown of the atmosphere and an attendant lightning stroke. Dissipation arrays of varied configuration are available from Lightning Elimination Associates, Inc. (LEA). Informal surveys of client sites in the power, broadcasting, and aerospace fields indicate that dissipation arrays are effective in reducing the frequency of local direct strokes.

Fiber-optic technology in interfacility signal transmission (for example, between a collocated earth station and a client's data center) provides a mechanism

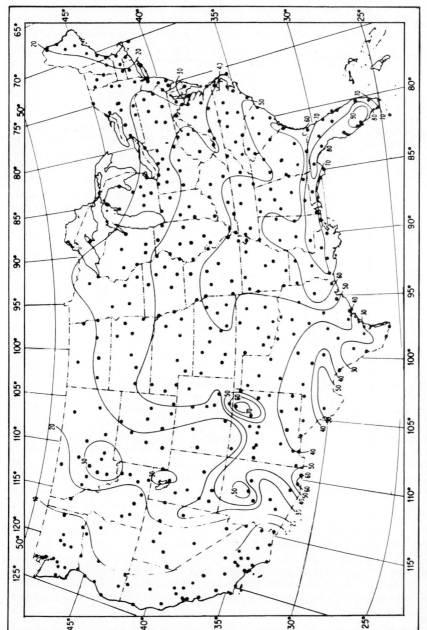

FIG. 19.15 Isokeraunic map of the United States.

for isolating the signal path from lightning propagation effects. This will prove effective if metallic strength members and metallic rodent barriers are particularly excluded from the construction of the fiber-optic cable.

There are in addition whole families of devices based on differing engineering concepts which may be placed in parallel with signal leads to control and clamp surges and filter line voltage. These include variable resistors, gas-discharge tubes, zener diodes, and Transorbs. Components based upon these same and other concepts, including isolation/regulation transformers, Thyrites, and series/parallel (line-conditioning/power-dissipation) circuits are available for power-line protection. In all cases, the reaction time, energy-handling capacity, initiation voltage, frequency response, clamping ratio, and transient behavior must be evaluated with the circuits to be protected.

25. Fire-Protection and Physical Security Systems

Fire Protection.   Fire-protection systems can be broadly characterized as gas-discharge and water-sprinkler systems. The former use an inert gas, carbon dioxide, or Halon, a fluorocarbon gas to suppress combustion. Carbon dioxide systems purge the facility's air supply whereas Halon mixed in a 15 percent concentration with the air provides fire suppression.

Gas-discharge systems must be filled after each discharge; they provide a single opportunity to suppress a fire. Actuation of carbon dioxide systems necessitates evacuation of personnel to avoid a suffocation hazard. In order to ensure effective gas concentration levels, discharge systems must be interlocked with HVAC vents and dampers and all exterior doors must be self-closing. Upon actuation of the gas-discharge system, interlocks switch off electrical power to preclude reignition of the fire as the result of an electrical short. Since actuation of a gas discharge results in site disruption and potential hazards, the fire-suppression system is normally actuated in stages with audible alarms and flashing lights warning personnel to vacate prior to discharge. Multiple fire-detection sensors are connected to a control system with voting logic to ensure that individual sensor failures do not falsely initiate a discharge. Local codes require system demonstration testing prior to issuance of a certificate of occupancy and require periodic retesting. Gas-discharge systems are uniquely suited to communications applications and, in particular, remote unstaffed sites. They do not damage electronics and they facilitate rapid restoration of service. They do not require construction of water-retention ponds or wells or extension of water lines.

Dry preaction sprinkler systems are also employed at electronic facilities; these systems are interlocked with the power system to preclude water-induced arcing, and their operation precludes the substantial cleanup and probable water damage which would otherwise result from a failure-induced wetdown. In a dry preaction sprinkler, distribution piping which feeds individual sprinkler heads is normally empty (dry), thus failure or damage to a sprinkler head will not result in a wetdown. When the system is activated by fire sensors, a main valve opens, filling the distribution piping with water in preparation for sprinkler-head discharge. The fill valve and site alarms are again controlled by a matrix of sensors operating through voting logic.

Physical Security.   Physical security is provided at communications sites to minimize trespassing, avoid possible negligence suits, increase safety for site personnel, and prevent damage and theft of site hardware. Tradeoff studies are performed which compare the probability and the cost impact of security threats with increasingly more effective and costly security systems. Remote facilities are universally fenced and the buildings and shelters are locked. The level of security can then be increased by equipping the fence system with barbwire or razor-wire topping, remote-control motorized gate closures, intercoms, and motion sensors. Closed-circuit television systems can be installed and equipped if necessary with pattern-recognition electronics; microprocessor-based, they compare video frames and initiate alarms without human intervention if a change is observed in the field of view. Security fences with gate and door alarms linked to a central monitor-and-control facility normally provide adequate commercial security. It is critical that the false-alarm rate be minimized, or the security system will soon be ignored by operations personnel and local police will refuse to respond to calls.

26. Communication Security

Security requirements are derived from both customer policies (such as those of a government agency) and from corporate policies. Security involves both the physical security of the communications site and electronic security of the transmission medium, to preclude covert eavesdropping. In the latter case, the communications signal is scrambled (analog) or encrypted (digital) in a manner known only to authorized parties who are provided with hardware at each end of the link and with time-dependent operational keys which decipher the signal.

This concept is being broadly applied to broadcast distribution of video programming by satellite in order to ensure that paid subscribers—and only paid subscribers—will exclusively obtain access to copyrighted programming (see Chap. 17). In applications providing secure communications for government circuits, digital encryption is provided in accordance with a National Security Agency (NSA) *digital encryption standard* (DES) algorithm. In order to gain acceptance for a DES-encrypted circuit, the encryption hardware must be NSA-tested and approved both for the efficacy with which the algorithm has been implemented and, to preclude local tampering, for the physical design of the encryption hardware. Subsequent to hardware approval, a formal methodology for encryption key control and distribution is integrated with communication system operations.

REFERENCES

27. Cited References

1. ANSI C 95.1 1982, Safety Levels with Respect to Human Exposure to Radio Frequency Electromagnetic Fields (300 kHZ to 100 GHZ), American National Standards Institute, New York.

2. T.O. 31Z-10-4 (Calculation Procedures), Electromagnetic Radiation Hazards, Air Force Communications Service, 1978.

3. Weissberger, M., *Handbook of Practical Techniques for Computing Basic Transmission Loss and Field Strength,* ECAC-Hdbk-81-049, DOD Electromagnetic Compatibility Analysis Center, Annapolis, Md.

4. Uniform Building Code, International Conference of Building Officials, Whittier, Calif., 1982.

5. Basic National Building Code, Building Officials and Code Administrators International, Country Club Hills, Ill., 1984.

6. MASTERSPEC, American Institute of Architects Service Corp., Washington, D.C.

7. Mil Std 756, *Reliability Modeling and Prediction.*

8. Mil Handbook 217, *Reliability Prediction of Electronic Equipment.*

9. Szentesi, O. I., and W. C. Primrose, *Siecor Optical Cable,* Siecor Technical Bulletin 523-0604380-00183J.

10. EIA Standard RS-195, Electrical and Mechanical Characteristics for Terrestrial Microwave Relay System Antennas, Electronic Industries Association, Washington, D.C.

11. EIA Standard RS-222, Structural Standards for Steel Antenna Towers and Antenna Supporting Structures, Electronic Industries Association, Washington, D.C.

12. "Large Steerable Radio Antennas—Climatological and Aerodynamic Considerations," *Annals of the New York Academy of Science,* vol. 116, art. 1, June 1964.

13. "Wind Forces on Structures," *ASCE Transactions,* paper no. 3269.

14. MacGorman, D. R., M. W. Maier, and W. D. Rust, *Lightning Strike Density for the Contiguous United States from Thunderstorm Duration Records,* NOAA report NUREG/CR-3759.

15. Spec. 74-15057, Grounding System Requirements for STDN Stations, Goddard Space Flight Center, October 1974.

16. Marshall, *Lightning Protection,* John Wiley & Sons, New York.

28. General References

Davidson, D., et al., *Power Density Measurements Near GTE Transmitting Facilities in Florida,* TR79-055.1, GTE Research & Development, Stamford, Conn., May 1979.

Lerner, E., "The Drive to Regulate Electromagnetic Fields," *IEEE Spectrum,* March 1984.

Proceedings of IEEE, "Biological Effects and Medical Applications of Electromagnetic Energy," vol. 68, no. 1, 1980.

Basic Property Maintenance Code, Building Officials and Code Administrators, International, Country Club Hills, Ill., 1981.

McPartland J. I., *National Electrical Code Handbook,* McGraw-Hill, New York, 1984.

Schram, *National Electrical Code Handbook,* National Fire Protection Agency, Quincy, Mass.

ANSI A58.1-1982, Minimum Design Loads for Buildings and other Structures, American National Standards Institute, New York.

Kapus, K., and L. Lamberson, *Reliability in Engineering Design,* John Wiley & Sons, New York.

Shinners, S., *Techniques of System Engineering,* McGraw-Hill, New York.

Kuecken, J. A., *Fiber Optics,* TAB Books, Blue Ridge Summit, Pa.

Cherin, A. H., *An Introduction to Optical Fibers,* McGraw-Hill, New York, 1983.

Suematsu, Y. (ed.), *Optical Devices and Fibers,* OHM & North-Holland.

Karlovac, N., *Fundamentals of Lightwave Transmission,* no. 523-0604374-00183J, Rockwell-Collins Transmission Systems, Dallas.

Offutt, D. O., *Lightwave Transmission Systems,* no. 523-0604376-00183J, Rockwell-Collins Transmission Systems, Dallas.

Carter, J. T., *System Design Considerations in Lightwave Transmission,* no. 523-0604377-10183J, Rockwell-Collins Transmission Systems, Dallas.

Engineering Applications, M560 Digital Fiber Optic Transmission System, Sec. 342-560-070, GTE Communication Systems.

Bell System Tech Reference PUB 43806 4/82, *Generic Metropolitan Interoffice Digital Lightwave Systems—Requirements and Objectives,* Bell System Technical Reference, AT&T, New York.

Deep Space & Missile Tracking Antennas, ASME Aviation and Space Div., 1966, Library of Congress no. 66-30320.

Thom, C. S., "New Distribution of Extreme Winds in the United States," *Proceedings of ASCE,* July 1968.

Meteorological Monographs, vol. 4, no. 22, 1960, American Meteorological Society.

Barkan, D. D., *Dynamics of Bases and Foundations,* McGraw-Hill, New York, 1962.

Richards, *Mechanical Engineering in Radar and Communications,* Van Nostrand Reinhold, New York.

Titus, J. W., *Wind-Induced Torques Measured on a Large Antenna,* Report 5549, Naval Research Laboratory, Washington, D.C., 1960.

Proceedings, Workshop on the Need for Lightning Observations from Space, NASA CP-2095, pp. 116–125, July 1979.

CHAPTER 20

LAND MOBILE SYSTEMS— REGULATIONS AND STANDARDS

Stuart F. Meyer

Director, Government and Industry Relations, E. F. Johnson Company (retired); Consultant

CONTENTS

THE FEDERAL COMMUNICATIONS COMMISSION

1. Scope and Organization of the FCC

All nonfederal government land mobile radio systems in the United States are regulated and licensed by the Federal Communications Commission (FCC).

The regulatory functions of the FCC are administered by three bureaus—Mass Media, Common Carrier, and Private Radio—each responsible for a segment of the communications industry. They are supported by other bureaus and offices including the Field Operations Bureau, which provides policing functions, and the Office of Science and Technology, which is the technical arm of the commission.

Land mobile services, the subject of this chapter, are regulated by the Private Radio Bureau. These include public safety, special emergency, industrial, and the land transportation radio services, often referred to as PSIT. The bureau also has jurisdiction over the general mobile radio service (GMRS, formerly the class A citizens radio service), and the citizens radio service in the 27-MHz band (no longer licensed).

Land mobile equipment used under Part 90 of the FCC *Rules* may also be approved for remote broadcast pickup services covered by Part 74 under the jurisdiction of the Mass Media Bureau.

The Common Carrier Bureau has jurisdiction over mobile telephone services (including paging) which are offered on a common carriage basis. This includes the recently authorized cellular radio service (see Chap. 22).

The headquarters and official address of the FCC is

Federal Communications Commission

1919 M Street, N.W.

Washington, DC 20554

The Private Radio Bureau is currently housed in a second building at 2025 M St., N.W.

The Private Radio Bureau licensing activity is currently located in Gettysburg, Pennsylvania. Its address is:

Federal Communications Commission

Licensing Division, Private Radio Bureau

Route 116 West

Gettysburg, PA 17325

2. FCC Rules

Each radio service is governed by a set of rules which provides guidance and regulation. Private radio services are covered by Part 90 of the FCC *Rules*. Copies of the *Rules* are available from the Government Printing Office (GPO) in book or loose-leaf form. The address of the headquarters office is

Superintendent of Documents

United States Government Printing Office

Washington, DC 20402

Branches of the GPO are located in a number of cities, and copies of the *Rules* can be obtained from them as well.

Up-to-date notifications of rule changes can be obtained by subscribing to a service by the contractor approved by the FCC. The name, address, and phone number of the currently approved contractor can be obtained from the FCC Consumer Assistance Division located at FCC headquarters.

FCC information related to the Business Radio Service which has the largest number of licensees may be obtained from NABER:

National Association of Business and Educational Radio

1501 Duke Street

Alexandria, VA 22314

3. FCC Forms

A number of different forms are used when an application is submitted for FCC radio system or equipment approval. These include:

- Form 574 (supersedes Form 400) is used for all applications in the private radio services and the general mobile radio services. It cannot be used for other services.
- Form 401 is used for base stations in the common-carrier radio services. Mobile units in these services are licensed under the umbrella of the related base-station license.
- Form 313 is used for remote broadcast pickup radio systems in the auxiliary broadcast radio service.
- Form 731 is used for equipment authorization under type approval, type acceptance, certification, and notification, as applicable.

4. Equipment Authorizations

Type Approval.   Type approval requires the filing of a test data report in accordance with FCC *Rules,* Part 2, together with the submission of a transmitter to the FCC Laboratory in Columbia, Md. As a rule, type approval of transmitters is not required in the land mobile services.

Type Acceptance.   Type acceptance requires the submission of test data in accordance with Part 2 but does not require the submission of a transmitter with the application. The FCC, however, reserves the right to request a transmitter for testing if necessary for verification or to investigate interference complaints.

Certification.   Certification relates to certain receivers and low-power nonlicensed radio equipment (wireless microphones and the like). Test data from measurements made in accordance with Part 15 of the *Rules* are filed for certification. Part 15 certification may not be used for citizens' radio or 4-watt nonlicensed transmitters; these must use the type acceptance procedure.

Notification.   Notification attests that the tests required for land mobile receivers have been conducted by the applicant. The results of the tests are not forwarded to the FCC, but the applicant states that the tests were conducted in accordance with the *Rules* and that the results are available for inspection.

TECHNICAL STANDARDS

5. FCC Standards

Certain minimum technical standards in the land mobile radio services are mandated by the FCC. These standards relate to unwanted receiver radiation, undesired transmitter radiation, transmitter bandwidth occupancy under modulation, and types of modulation—AM, FM, single sideband (SSB), etc. Most of these technical standards are found in FCC *Rules,* Part 2, and the rules relating to the specific radio service.

Most land mobile equipment uses frequency modulation; however, recent FCC actions have authorized the limited use of amplitude-compandored single-sideband techniques (ACSB) in the 150-MHz band.

6. EIA and LESL Standards

More complete standards, including receivers, are promulgated and published by the Electronic Industries Association (EIA). Some of the more popular land mobile standards currently available through the EIA are

EIA no.	Standard
152-B	Phase modulation (PM) and FM transmitters
204-C	PM and FM receivers
220-A	Continuous tone squelch
222-D	Towers
316-B	Hand-held portables
329-A	Base and fixed antennas
329-1	Vehicular antennas
374-A	Signaling
IS-3-C	Cellular compatibility
IS-19	Cellular subscriber units
IS-20	Cellular land stations

The EIA address is

Electronic Industries Association

2001 Eye St., N.W.

Washington, DC 20036

The Law Enforcement Standards Laboratory (LESL) of the National Bureau of Standards produces a set of land mobile and related standards for use in the public safety services. They are generally based on EIA standards, with a number of additional items covered. They are available from:

Law Enforcement Standards Laboratory

B157—Physics Building

National Bureau of Standards

Gaithersburg, MD 20899

Both the EIA and LESL standards include the mandated FCC standards as well as many additional items such as receiver selectivity, desensitization, intermodulation, audio-frequency response, and shock and vibration. Transmitter standards cover similar items.

LAND MOBILE FREQUENCY ALLOCATIONS

7. Land Mobile Frequency Bands

Four frequency bands have been allocated for the land mobile services, two in the VHF and two in the UHF region of the spectrum. They are

Low band   25–50 MHz
High band   150–174 MHz
UHF   450–512 MHz (450–470 MHz in many areas)
800   806–947 MHz

Each service classification is allocated channels within these bands. The frequencies allocated to specific services are subject to periodic revision. A listing of current allocations can be obtained from the relevant FCC rules.

UHF Television Sharing.   As a stopgap measure to provide relief to the spectrum congestion of the land mobile services, the FCC has authorized these services to use UHF TV channels in the frequency range 470–512 MHz in a limited number of geographical areas. The cities of Boston, New York, Philadelphia, Washington, Baltimore, Chicago, San Francisco, and Los Angeles are allowed the use of two TV channels; Miami, Houston, Dallas-Fort Worth, and Pittsburgh are allowed one.

The technical parameters are the same as at 450 to 470 MHz, but licensing requirements are considerably more stringent. Base station and mobile relays must be located within a 50-mi circle centered on the geographical center of these areas with frequency reuse permitted every 40 mi. Additional certification of the site is required to provide interference protection to nearby cities with television assignments on the same channel. Other special requirements relate to such parameters as intermodulation protection.

Interservice Sharing.   Most of the channels in the land mobile spectrum are designated for specific radio services. The relative spectrum requirements of the various services is not uniform throughout the country, and the FCC has adopted interservice sharing policies. These permit a radio service to utilize the frequencies allocated to another service where they are underutilized. As a rule, the underutilized service channels must be devoid of users within a geographical area before another service may share them. Specific requirements are covered in Part 90 of the *Rules*.

Canadian and Mexican Border Clearances.   Applications for licenses within 100 mi of the Canadian border must be coordinated with Canada through the FCC to clear any interference problems. This usually results in considerable delay of licensing mobile systems with an *effective radiated power* (ERP) in excess of 5 W. Land mobile frequencies above 800 MHz are precleared with Canada by treaty; Subpart S of Part 90 includes a list of these frequencies.

Clearance for frequencies below 800 MHz is not required in the vicinity of the Mexican border. Within 62 mi of the border, licensees in the 800-MHz band must employ 12½-kHz offset in accordance with Subpart S of FCC *Rules,* Part 90.

8. Eligibility for Land Mobile Licensees

A land mobile radio licensee must be a state or local government entity or a commercial (individual or company) organization whose function or business falls into a category defined by the FCC. The definitions of these categories are contained in Part 90 of the FCC *Rules*. A list of the categories and the frequency bands in which they may be licensed to operate is shown in Table 20.1. In addi-

TABLE 20.1 Land Mobile Frequency Bands Available to Private and Common-Carrier Radio Services

	Low band	High band	UHF	800/900 MHz
Public safety				
Local government	X	X	X	X
Police	X	X	X	X
Fire	X	X	X	X
Highway maintenance	X	X	X	X
Forestry conservation	X	X	X	X
Special emergency	X	X	X	X
Industrial				
Power	X	X	X	X
Petroleum	X	X	X	X
Forest products	X	X	X	X
Motion picture		X		X
Relay press		X	X	X
Special industrial	X	X	X	X
Business	X	X	X	X
Manufacturers		X	X	X
Telephone maintenance	X	X	X	X
Land transportation				
Motor carrier	X	X	X	X
Railroad		X	X	X
Taxicab		X	X	X
Automobile emergency		X	X	X
GMRS			X	
Remote broadcast pickup	X	X	X	
SMRS				X
Private carrier (two-way)	X	X	X	
Private carrier (paging only)				X
Traditional common-carrier mobile telephony (and paging)	X	X	X	
Cellular common carrier (no paging)				X

NOTE: Paging is also permitted in the Public Safety, Industrial, Land Transportation, and GMRS Services with certain restrictions.

X = available.

tion, many private radio services are allowed to use mobile relay systems (for extended car-to-car coverage). Their eligibility by frequency band and service is shown in Table 20.2.

Many business establishments are eligible to be licensed in more than one category. For example, a taxicab company can obtain a license in either the business radio service or the taxicab radio service and would choose the one which is most advantageous from the standpoint of cost and/or service availability. It can also obtain licenses in both services and equip its system with multiple channel switching. Different identification call signs are then required.

9. Channel Widths

The standard land mobile channel widths are as follows:

Band	Channel width, kHz
Low band	20
High band	30
UHF	25
800 MHz	25*

12½-kHz Offsets. At 450 to 470 MHz, the FCC has authorized a substantial number of channels centered midway between the 25-kHz channel centers. These are

TABLE 20.2 Mobile Relay Eligibility by Frequency Band and Service

	Low band	High band	UHF[1]
Public safety			
Local government	X	X	X
Police	X	X	X
Fire	X	X	X
Highway maintenance	X	X	X
Forestry conservation	X	X	X
Special emergency		X[2]	X[2]
Industrial			
Power	X	X	X
Petroleum	X	X	X
Forest products	X	X	X
Motion picture	X[3]		X
Relay press	X[3]		X
Special industrial			X
Business	X[4]	X[4]	X
Manufacturers			X
Telephone maintenance	X	X	X
Land transportation			
Motor carrier			X[5]
Railroad	X	X	X
Taxicab			X[5]
Automobile emergency			X[5]
GMRS			X[6]
Remote broadcast pickup			X

[1]Includes 470–512 MHz in those cities where channels are available.
[2]Cross banding is also permitted between the 150–160-MHz and the UHF band.
[3]Mobile relays are legally permissible; however, the available channel spacings may be too close for practical use.
[4]Permitted on low-power frequencies (2 W or less) with certain restrictions per FCC *Rules,* Part 90.243(2).
[5]Only permissible at 470–512 MHz in those cities where that band is available.
[6]Only permissible on the eight pairs of frequencies available in this service at UHF.
NOTE: All of the above (except GMRS and remote broadcast pickup) are eligible for mobile relays in the 800-MHz band.
SOURCE: E.F. Johnson Co., used by permission. X = available.

reserved for systems with transmitter powers not exceeding 2 W and antennas not more than 20 ft off the ground. They are always licensed as mobile units, although they may be used as point-to-point systems for controlling other transmitters. The equipment used on these channels employs technical parameters based on 25-kHz spacing.

Tertiary Frequencies.   At 150 MHz, the FCC authorizes the use of channels midway between existing 30-kHz channels except in the 150-MHz business radio service band. The base station sites are coordinated and must be no closer than 7 to 10 mi from existing systems (depending on the specific radio service).

FREQUENCY COORDINATION

10. Role of Frequency Coordinators

As a result of the size and complexity of the task of assigning specific channels to land mobile systems, the FCC has delegated it in most cases to designated industry frequency coordinators. Coordinators are charged with the responsibility of picking the frequency in the geographic area of a proposed new system which will cause the least amount of interference to other users. This usually requires that the pool of available frequencies be continually recycled and that each applicant be given the best available frequency at the time. The spectrum is congested in most heavily populated areas because, in general, every eligible applicant is entitled to a license. Exceptions to this rule are the 470- to 512-MHz and 806- to 947-MHz bands. These bands have loading limitations which provide a degree of protection and a higher quality of service.

Current rules require that most applications in the public safety, industrial, and land transportation radio services must be submitted to the coordinator along with a request for one or more coordinated frequencies. The coordinator is responsible for screening these applications before submission to the FCC. After screening and the recommendation of a frequency, the coordinator forwards the application directly to the commission.

11. Designated Coordinators

A list of coordinators by radio service was published by the FCC in its Public Notice No. 202/632-002, Part 90, Frequency Coordination, dated November 7, 1986. The coordinators listed in the notice are shown in Table 20.3.

LOTTERIES

12. Cellular Radio Service

When the cellular radio service was created, the demand for channels was so voluminous and the number of qualified applicants was so great that selection of licensees by comparative hearings would have been too time-consuming to be practical. To cope with this situation, the FCC received authority from the Congress to conduct "no-money" lotteries.

TABLE 20.3 FCC-Designated Frequency Coordinators

Automobile emergency radio service:
American Automobile Association, Inc.
(AAA)
8111 Gatehouse Rd.
Falls Church, VA 22047
703 222-6229

Business radio service:
(central station protection only)
Central Station Electrical Protection
Association (CSEP)
1120 Nineteenth St., N.W.
Washington, DC 20036
202 296-9595

Business radio service:
National Association of Business and
Educational Radio, Inc. (NABER)
1501 Duke St. Suite 200
Alexandria, VA 22314
703 739-0303

Fire radio service:
International Municipal Signal Association
(IMSA)
P.O. Box 1513
Providence, RI 02901
401 738-2220

Forestry-conservation radio service:
Forestry Conservation Communications
Association (FCCA)
Hall of the States
444 N. Capitol St., N.W.
Washington, DC 20001
202 624-5416

Motor carrier radio service:
American Trucking Association, Inc.
(ATA)
Telecommunications Department
2200 Mill Rd.
Alexandria, VA 22314
202 838-1700

Petroleum radio service:
Special Industrial Radio Service
Association, Inc. (SIRSA)
Suite 910
1700 N. Moore St.
Rosslyn, VA 22209
703 528-5115

Forest products radio service:
Forest Industries Telecommunications
(FIT)
P.O. Box 5446
3025 Hillyard St.
Eugene, OR 97405
503 485-8441

Highway maintenance radio service:
American Association of State Highway
and Transportation Officials (AASHTO)
444 N. Capitol St., N.W.
Suite 225
Washington, DC 20001
202 624-5800

Local government and police radio service:
Association of Public Safety
Communication Officers (APCO)
Attn: Frequency Coordination Dept.
P.O. Box 280
New Smyrna Beach, FL 32070
904 423-3629 or 800 654-4951

Manufacturers radio service:
Manufacturers Radio Frequency Advisory
Committee, Inc. (MARFAC)
6269 Leesburg Pike, Suite B-1
Falls Church, VA 22044
703 532-7459

Motion picture radio service:
Alliance of Motion Picture and Television
Producers (AMPTP)
c/o SIRSA
Suite 910
1700 N. Moore St.
Rosslyn, VA 22209
818 995-3600

Special industrial radio service:
Special Industrial Radio Service
Association, Inc. (SIRSA)
Suite 910
1700 N. Moore St.
Rosslyn, VA 22209
703 528-5115

Taxicab radio service:
International Taxicab Association (ITA)
3849 Farragut Ave.
Kensington, MD 20895
301 946-5700

TABLE 20.3   FCC-Designated Frequency Coordinators (*Continued*)

Power radio service: Utilities Telecommunications Council (UTC) 1150 17th St., N.W. Suite 1000 Washington, DC 20036 202 956-5636	Telephone maintenance radio service: Telephone Maintenance Frequency Advisory Committee (TELFAC) Suite 910 1700 N. Moore St. Rosslyn, VA 22209 703 528-5115
Railroad radio service: Association of American Railroads (AAR) Communications and Signal Section 50 F St., N.W. Washington, DC 20001 202 639-2215 or 202 293-4125	800-MHz business and 900-MHz paging: National Association of Business and Educational Radio, Inc. (NABER) 1501 Duke St. Suite 200 Alexandria, VA 22314 703 739-0308
Relay press radio service: American Newspaper Publishers Association (ANPA) c/o SMS Suite 910 1700 N. Moore St. Rosslyn, VA 22209 202 648-1139	800-MHz industrial/land transportation: Special Industrial Radio Service Association, Inc. (SIRSA) 1700 N. Moore St. Suite 910 Rosslyn, VA 22209 703 528-5115
Special emergency radio service: NABER/IMSA/IAFC c/o NABER 1501 Duke St., Suite 200 Alexandria, VA 22314 703 739-0303	800-MHz public safety: Association of Public Safety Communication Officers P.O. Box 280 New Smyrna Beach, FL 32070 904 423-3629

Under this system, all incoming applications are logged in and assigned a number. A publicly witnessed lottery provides a winner and a runner-up series of numbers for each city. Ten numbers are picked in sequence for each market area, and, if the application of the winner is not acceptable to the FCC, the first runner-up's application is reviewed for completeness and acceptability.

13. Specialized Mobile Radio Service

The FCC has also used the no-money lottery for selecting licensees in the *specialized mobile radio service* (SMRS) where the number of applicants exceeded the supply of channels. This has been done both at the outset of a program releasing additional spectrum space to the service and when channels have been taken back for failure of the applicant to construct or reach the mobile loading criterion during the prescribed timetable. A public announcement specifies a filing window, and all applications submitted during this period are treated as though they were received simultaneously.

FCC LICENSING POLICIES

14. Scope of Licensed Systems

A wide variety of systems are licensed by the FCC. They range from simple systems with one base station and mobile unit to major statewide public safety systems. The latter have multiple base-station and repeater sites, fixed point-to-point circuits, and mobile units with complex frequency matrices (see Chap. 21).

15. Mobile Radio Services

Mobile Relay Systems.   When the 450-MHz frequencies were first made available to the private radio service, the FCC authorized separate frequencies for base stations and mobiles with fixed separation throughout the band. In the business radio service, this quickly led to the use of mobile relay systems. Mobiles can then talk to other mobiles throughout the range covered by the mobile relay and base station antenna. In addition a control station (a mobile unit with a fixed antenna) makes it possible for a dispatcher to talk to mobile units individually or collectively without the need for a wire line between the control point and the tower location.

Community Repeaters.   To reduce costs to the end users, a number of entrepreneurs developed systems known as community repeaters. A community repeater is not licensed as such, but a single mobile relay is multiple-licensed by a number of users who share the same frequency and common mobile relay and base station equipment. Thus the cost can be distributed among perhaps 10 to 20 users. The owner of the equipment charges a monthly fee but is not required to be licensed by the FCC. The community repeater mode of operation has become extremely popular and may be used by all services in the jurisdiction of the private radio bureau (but not shared between services).

Specialized Mobile Radio Service.   When the FCC first allocated channels for land mobile in the 800 MHz region of the spectrum, it authorized a more advanced form of community repeater system, the specialized mobile radio service. In this service, the entrepreneur is licensed for the mobile relay and base station facilities only, and users are individually licensed for their control stations and mobile and/or portable units. This licensing mode is available to both trunked and nontrunked systems. To expand the usefulness of the system further, the FCC permits the users to be licensed to take service from a number of SMRS facilities. This is particularly useful to roaming units which require a wide area of coverage.

Private Carriers.   In recent years, the FCC has authorized what are known as private carriers. Initially, these were paging-only systems operating in the vicinity of 900 MHz. Private carriers are now permitted in the 450- to 512-MHz portion of the spectrum as an alternative to the community repeater. In this mode, the entrepreneur licenses a base station facility and a quantity of mobile units (or pagers). These are then leased to the end users.

At the end of eight months of operation, the entrepreneur updates the description of the system by filing a new application which includes a verification of the number of mobile units and control stations. The entrepreneur is permitted to put additional users on the system with simplified updating of the information in the FCC license data base. Detailed information concerning this newly created service can be obtained from the FCC Report and Order of Docket 83-7737 adopted April 30, 1986.

Shared Not-for-Profit Systems.   The FCC permits a number of radio users to share the cost of operating a radio system on a not-for-profit basis by filing a single application set. This requires filing a copy of the agreement along with a list of the names and addresses of those sharing the system. Additional information on this mode of operation is covered in Section 185 of Part 90 of the FCC *Rules*.

Remote Broadcast Pickup.   The remote broadcast pickup radio service has channels available for both narrowband (voice) and wideband (broadcast quality) use. Narrowband land mobile equipment is often used for remote broadcast pickup, both because of its lower cost and because higher fidelity is often not necessary for remote pickups. Most land mobile equipment has been type-accepted under Part 74 of the *Rules* so that it can be used for this purpose.

Point-to-Point Systems.   FCC policies do not encourage the use of land mobile frequencies for point-to-point circuits to replace wire lines, but there are exceptions. Part 90 of the *Rules* specifies a number of 12½-kHz-offset channels which can be used for purposes such as remote control of base stations. These frequencies are limited to a maximum transmitter power of 2 W and an antenna height of 20 ft above the ground (not artificial structures). There is no limit on the antenna gain that may be employed nor the height of the receiving antenna. These provisions, coupled with the use of horizontal polarization (which provides additional attenuation to signals from land mobile systems operating in the adjacent 12½-kHz channels), often provide an extended range.

Paging Systems.   The FCC *Rules* provide for a variety of one-way tone paging and tone-plus-voice paging. A number of channels have been set aside in the land mobile spectrum between 25 and 947 MHz for paging-only operations in which the base stations are licensed for one-way transmissions.

Additionally, mobile radio systems may be authorized for a mixture of pagers and mobile units on mobile frequencies. The *Rules* are not precise on eligibility standards, but the general criterion is that the volume of mobile voice traffic must exceed paging traffic to permit this mode of operation.

16. Interconnection with the Public Switched Telephone Network

Under certain conditions, the FCC permits private radio systems to be interconnected with *public switched telephone networks* (PSTN). The *Rules* favor interconnection for systems operating above 800 MHz because the restrictions are less severe. Restrictions may include such requirements as mutual agreement in the top 25 cities in the United States and use of acknowledgment signaling before an interconnected call can be completed.

Systems in use range in complexity from manual push-to-talk to full duplex.

Detailed regulations for this mode are found in Sections 476, 477, and 483 of Part 90 of the *Rules*.

17. Effect of Deregulation

Many traditional FCC policies have been canceled or reversed by its recent deregulatory activities. An example is message content. Originally, it was required to be related to the activities of the licensee in the business or function for which the license was received. Recent deregulatory actions permit other usage provided that it does not cause harmful interference to others on the channel. In the case of SMRS, the FCC allows the operator to determine the extent of noneligible conversations which may take place on his system.

The FCC has also relaxed many of the restrictions on the transmission of digital or data transmission on previously voice-designated channels. In general, digital and data signals can be transmitted on most land mobile radio voice channels without regard to primary or secondary status.

FILING APPLICATIONS

18. Instructions

The FCC has published a booklet, Form 574 Instructions. This can be obtained from the FCC or from suppliers of land mobile equipment. The latter have also prepared bulletins and manuals on the preparation of applications.

19. Filing Fee

Effective April 1, 1987, a filing fee was reinstituted by the FCC. Details of the fee program are described in FCC Report and Order of General Docket No. 86-285 adopted December 23, 1986. In most cases, the fee as of that date for filing an application on Form 574 for any of the services encompassed by Part 90 of the *Rules* is $30. License applications are submitted through the frequency coordinator, who forwards the filling fee while processing the application.

MISCELLANEOUS REQUIREMENTS

20. Transmitter Identification

The FCC requires transmitter identification every 30 minutes in a continuously busy system. If transmissions are sporadic with long gaps, it is not necessary to come on-air for the sole purpose of identification. Recently the FCC has permitted the use of *continuous-wave* (CW)—that is, telegraphy—Morse code for identification. In the case of trunked 800-MHz systems, the use of CW on the first channel is mandated. Detailed requirements are found in Part 90 of the FCC *Rules*.

21.   Transmitter Power Output

The FCC requires that no more power be employed than necessary to carry out successful communications in the required coverage area. The commission relies heavily on the coordinator to screen applications for appropriate transmitter power. See Chap. 21 for the methods of calculating the required power. In shared systems (such as community repeaters) it is imperative that all applications file identical information with respect to transmitter power, geographical coordinates, and antenna height.

22.   Control-Station Location and Antenna Height

Recent rule changes have simplified the requirements for control-station applications. If the antenna height above ground or artificial structure (other than a tower) is less than 20 ft, it is now necessary to show only the frequency, power output, emission designator, address, and telephone number of the station on form 574.

Antennas on towers or poles erected for the sole purpose of supporting the antenna and exceeding a height of 170 ft require the submission of Federal Aviation Administration form 7460-1.

In addition to the general 170-ft limit, there may be further limitations depending on the distance to the nearest airport. Formulas for calculating the maximum permissible height are found on the reverse side of the form.

Antennas mounted on towers in excess of 300 ft in height require an environmental impact statement in accordance with Paragraph 1.311 of Part 1 of the FCC *Rules*.

Towers or poles which already support the antenna of a licensed system can be referenced in the application for an additional antenna, thus eliminating the need for further FAA or NEPA (National Environmental Protection Agency) approval.

23.   Equipment Authorization and Approval

The manufacturer and grantee (person or organization responsible for the technical data) must obtain identification numbers for equipments for which type approval or acceptance is sought. Usually the manufacturer and the grantee are the same, but the identification numbers and letters are not the same. The first three letters of the identification indicate the grantee, and the next three spaces (one digit and two letters) indicate the manufacturer. These are followed by a maximum of 11 digits indicating the model number of the specific unit. The complete identification must be displayed in a prominent place as part of the equipment label.

Changes in type accepted equipment fall into two categories: A—permissive changes which are of such a simple nature that they do not require submission of information to the FCC Laboratory; B—nonpermissive changes which require submission of additional technical data.

Equipment which has not been approved by the applicable FCC procedure may not be offered for sale or shipped until such approval has been received. Receivers requiring the simplified notification process may not be offered for sale or shipped until the FCC has received the required information.

*In the past, 800-MHz systems have been licensed in the lower portion of the 806 to 947 MHz band with channel widths of 25 kHz and a maximum FM deviation of ± 5 kHz. The FCC is now (1987) licensing additional private land mobile systems in the upper portion of the band with a channel width of 12.5 kHz and a maximum FM deviation of ± 2.5 kHz.

CHAPTER 21

LAND MOBILE AND PAGING SYSTEMS

Curtis J. Schultz

President, SCS Telesystems International, Inc.

CONTENTS

PAGING SYSTEMS

1. Introduction

Today's ubiquitous paging industry has evolved from the first land mobile radio system, the *one-way information broadcast* concept pioneered by the Detroit Police Department in 1921. There are more than 600 *radio common carriers* (RCCs) and 40 telephone companies providing paging service to nearly 1.5 million subscribers in the United States.

Paging service subscribers carry a small VHF or UHF receiver that is selectively called from a central base station. The *pager* provides the wearer with one or more alert tones or alert tones plus a short voice message. Some pagers vibrate when called and provide a silent alert. Still other pager models are capable of storing alphanumeric messages that can be displayed on a *liquid crystal display* (LCD) and retrieved when the user has been selectively called with an alert tone and/or signal lamp.

2. Paging Receivers

There are dozens of pager models manufactured by electronic companies around the world. Pagers range in size from about as small as a large fountain pen to about as large as a 20-cigarette package. The large model can display up to 1000 characters in 16-character blocks on an LCD screen. The average pager with battery weighs about 5 oz (142 g).

Modern paging receivers present a number of challenges to design engineers. The unit must be small, lightweight, attractively styled, and mechanically rugged. The electronic circuitry must consume a minimum amount of power from the internal battery. *Power saver* circuits are incorporated in many of the models now on the market. A major challenge to pager design engineers is to achieve high operating sensitivity with relatively inefficient in-case antennas. A summary of typical paging receiver performance is in Table 21.1.

Pagers designed to receive subcarriers transmitted by commercial FM broadcast transmitters typically require a field strength of 100 μV/m. The sensitivity figures in Table 21.1 assume no limitation from on-channel ambient electrical noise.

Paging receivers in urban areas are subjected to strong off-channel signals from FM and TV broadcast stations as well as from land mobile stations. This necessitates careful consideration of adjacent channel selectivity and spurious response rejection. A 3-dB carrier-to-noise ratio produces a signaling response in most modern paging receivers. Since paging receiver performance is ultimately limited by ambient electrical noise, intermodulation, and desensitizing from high-level adjacent channel signals, only a minimal performance improvement can be expected from the use of more sensitive receivers in large urban areas.

3. Coding-Decoding Systems and Formats

Early paging receivers used one or more vibrating mechanical reeds to decode the selective calling tones. A paging receiver is not an ideal application for mechanical reeds. The reeds were bulky, could be "falsed" by mechanical shock, had a limited number of codes, and were relatively slow to respond to calling signals. They were subsequently replaced by plug-in active filters. The use of active filters enabled paging receiver designers to use higher-frequency tones and a greater number of them.

The Electronic Industries Association (EIA) has established 100 standard two- and five-sequential-tone codes in the 67.0- to 1687.2-Hz frequency range. Pager manufacturers have added additional proprietary tone codes extending the frequency range, the number of tones, and the number of pagers which can be

TABLE 21.1   Typical Paging Receiver Performance

Characteristic	Frequency band, MHz			
	40	150	450	900
Antenna gain	− 35 dB *	− 25 dB *	− 20 dB *	
Paging sensitivity	5 μV/m†	5 μV/m†	15 μV/m†	5 μV/m†
Adjacent channel rejection	60 dB	60 dB	60 dB	70 dB
Spurious signal rejection	55 dB	55 dB	50 dB	40 dB

*Half-wave dipole
†EIA eight-position measurement.

called. A composite of all manufacturers' codes includes a total of 317 tone codes in the 67.0- 3062.0-Hz frequency range.

The increasing number of paging subscribers and the need to send multiple selective calls to individual subscribers in the least amount of air time have turned the attention of paging receiver designers to new signaling schemes. Today's two- and five-tone sequential signaling systems are being replaced by digital selective calling systems. The use of digital signaling can be expected to increase as the use of data transmission to paging systems is increased.

While an in-depth analysis of address and data code structures is beyond the scope of this chapter, Table 21.2 summarizes several types of codes and their relative message and signaling times.

Modern paging terminals utilize various interleaving and dynamic call-batching algorithms to achieve the most efficient utilization of transmitter air time. In practice, the average time per call may be less than that implied by Table 21.2.

There are two major code formats used by digital pager manufacturers, the Golay sequential code and the POCSAG code. The Golay sequential code has been in service since 1973, and it is used by more than 250,000 pagers around the world. The POCSAG code was developed by members of the British Post Office Standardization Advisory Group to meet the requirements of a nationwide paging system in the United Kingdom. It has been in use since 1981. The characteristics of these two most widely used code formats are compared in Table 21.3.

From Tables 21.2 and 21.3, it can be seen that the proponents of these paging codes have optimized their systems to meet particular objectives.

The POCSAG proponents have opted for a system providing a larger number of codes that can be transmitted at a higher speed. This choice sacrifices the redundancy used in binary formats to correct burst errors in received transmissions. The POCSAG code format permits two bits to be corrected in a 32-bit address and only one bit to be corrected for every 32 bits of data message.

The Golay sequential code permits three bits to be corrected in a 23-bit address and 16 bits to be corrected for every 120 bits of data message. For an 80-character alphanumeric message in a fading environment, the Golay code has a 12.2-dB advantage over the POCSAG code in the relative signal strength envi-

TABLE 21.2 Comparative Paging Call Durations

2-tone sequential	14.00s	Time includes a 10-s voice message
5-tone sequential	11.22s	
2-word 23:12 Golay code	12.40s	
2-tone sequential	2.50s	No message; alert tone only
5-tone sequential	0.22s	

	23:12 Golay code	31:21 BCH POCSAG code
Call address	0.20s	0.07s
Address + 12-character message	0.40s	0.20 s
Address + 80-character message	2.22s	1.90 s

TABLE 21.3 Comparison of the Most Widely Used Code Formats

Characteristic	Golay sequential code	POCSAG
Number of codes	1 million, 4-address 500,000, 8 address	2 million, 4-address
Call rate:		
Address	5 calls/s	15 calls/s
Data	2.5 calls/s (12 characters); 0.45 calls/s (80 characters)	5 calls/s (10 characters); 0.52 calls/s (80 characters)
Decoder type	Asynchronous	Bit synchronization and word framing required
Address format	2-word (23:12 Golay)	1-word (31:21 BCH)
Data format	Block of eight 15:7 words; 12 numeric characters per block	31:21 words; 5 numeric characters per word
Battery saver grouping	Selective batching	Time division
Fade protection address		
Bits corrected	3/23 (300 b/s	2/32 (512 b/s)
Fade percentage	13%	6.3%
Fade length	10 ms	4.0ms
Data		
Bits corrected	16/120 (600 b/s)	1/32 (512 b/s)
Fade percentage	13%	3.1%
Fade length	27%	2.0 ms

ronment required for a 99 percent success rate. The proponents of the POCSAG code say that they can increase the power of the base station to overcome this problem. The relative speed superiority of the POCSAG code decreases in message switching applications as indicated in Tables 21.2 and 21.3.

The 2 million code capacity of the POCSAG system is large enough to selectively call every pager in the United States and have enough reserve to handle pager assignments for several years to come. Golay code proponents suggest that the 1 million code capacity of the Golay code, multiplied by the number of rf channels available to paging system operators, provides a subscriber capacity that makes the larger POCSAG code capacity academic. However, the capacity of the POCSAG code may have some significance if paging calls and data are ultimately transmitted from satellites that provide total coverage of North America or Europe.

4. Paging Terminals

The installation of a paging base station transmitter and the determination of area of coverage require procedures similar to those used for land mobile systems which are described elsewhere in this chapter. The balance of this discussion of paging systems is devoted to radio paging terminals.

The paging terminal determines the number of pagers the system can service and the types of signaling and/or data formats that can be transmitted. Larger terminals provide subscriber use statistics for billing purposes, control one or more base-station transmitters, and perform other needed functions.

The size of a paging system may vary from small tone-only, 10 pager, in-plant systems to larger systems where the terminal accommodates 10,000 subscribers using tone-and-voice, tone-only, digital signaling, and data-storage pagers. Larger terminals can control a number of paging base stations and automatically select the base station that serves the subscriber's operational area. Small systems use desk-top consoles equipped with buttons to select the desired pager code. In voice message systems, the console operator is provided with a microphone to transmit the voice message. The local or remote base station is controlled by a wire line.

In some instances, the paging system shares the use of an existing land mobile base station. In such cases, there are several precautions which must be taken to make the two types of operation compatible. Mobile unit reception of paging system signaling tones and voice messages could be annoying and distracting.

This problem can be solved by equipping the mobile receivers with *continuous-tone-controlled squelch systems* (CTCSS). Base-station dispatches to mobile units include a continuous subaudible tone that is used to unmute the mobile receivers. Pager calls do not include the CTCSS signal, so the mobile receivers remain muted during paging transmissions. A variation of this shared-base-station system is the use of subaudible paging codes (67 to 202 Hz) to establish a small tone-only paging system.

Care must be taken to avoid assigning pagers the same CTCSS tones used by the mobile dispatch system. Tone frequencies transmitted from the paging console are in the 536- to 1616-Hz range to permit transmission over a standard telephone line. These frequencies are divided by 8 at the base-station termination unit before transmission to paging receivers. Another option is to interconnect the paging and base station control units so that the paging terminal is "locked out" during normal use of the base station for mobile dispatching.

Most of the larger paging terminals can be connected to a PABX or a city-wide automatic telephone system using either pulse or DTMF (dual-tone multifrequency) tone dialing. Such systems are totally automatic and require no operator supervision.

Larger terminals may be connected to several geographically separated base stations. The use of *simulcast* systems permits the paging terminal to utilize all transmitters simultaneously to cover a large area. The paging terminal may also repeat the paging call on several transmitters in sequence. This operational mode eliminates the transmitter stability and audio equalization problems that are inherent in simulcast systems.

Some large terminals, serving voice message pagers, accept voice messages from the person originating the call, digitize the speech, process it to remove dead air time, and store it in a digital buffer memory. At the appropriate time, the voice message is reconstructed, batched with other messages, and transmitted.

To minimize disruptions caused by defective pagers, some terminals contain an automatic pager call translation function. When a subscriber is issued a substitute pager to use while a unit is being repaired, the terminal circuitry automatically translates the user's assigned calling code into that of the substitute pager.

Some paging terminals include a modem that allows the terminal manufacturer to call it and carry out diagnostic routines to isolate the cause of a terminal malfunction.

The larger terminals include computer-controlled directories that store information on each subscriber, including name, address, pager type, paging service type (individual and/or group calls), area of coverage, and many other bits of information that are unique to a particular system. Circuitry is provided to make system activity reports and customer air-time utilization reports and prepare billing statements. *Cathode-ray-tube* (CRT) terminals are used in many systems to display this information as required. Some terminals permit alphanumeric messages to be prepared in advance and transmitted during off-peak hours.

Terminal manufacturers have a seemingly endless variety of proprietary terminal features that duplicate, in function, the several features noted above.

5. The Future of the Paging Industry

The future of the paging industry is bright. As the price and *power drain* of megabyte memory chips are reduced, it will be possible to store vast amounts of information in pager units. This information might be loaded into the pager during the night and provide a delivery truck driver with a schedule of pickups and deliveries for the day. A salesperson could receive a schedule of sales calls for the day. A police officer could receive a list of stolen car license numbers or a description of wanted criminals. A traveler could get a listing of train or airline schedules. There is an endless number of uses for high-capacity-memory pagers.

Someday, high-address-capacity pagers will be signaled from a satellite transmitting station. It will then be possible to locate a paging subscriber anywhere in North America—or on any of the earth's continents. The possibilities are endless.

LAND MOBILE SYSTEMS—SYSTEM CONFIGURATIONS

6. One-Way Single-Frequency Simplex

While paging systems are the best-known implementations of one-way land mobile systems, there are a number of other one-way systems in use. Some examples are shown in Fig. 21.1.

In many urban areas, the U.S. Weather Bureau broadcasts weather information, using land mobile transmitters operating in the 160-MHz band. Small suburban airports and others interested in timely weather information monitor these transmissions.

A number of major cities broadcast road and traffic advisory information using low-power transmitters operating on frequencies just above the AM broadcast band. These broadcasts can be monitored by automobile and truck drivers using conventional AM auto radios.

Another application of land mobile one-way broadcasting is the selective calling of volunteer firefighters, doctors, paramedics, and service technicians. These individuals wear a small portable receiver about twice the size of a conventional pager. Unlike pagers, which selectively alert their users and may provide brief voice messages, the portable monitor receiver can alert an entire group and allow them to monitor a land mobile channel continuously. For example, all volunteer firefighters in a town can be alerted simultaneously. Once alerted, their portable monitors are converted into carrier-squelch receivers that monitor the local fire department base-station frequency. The fire department base-station operator is now able to call for fire-fighting assistance as needed.

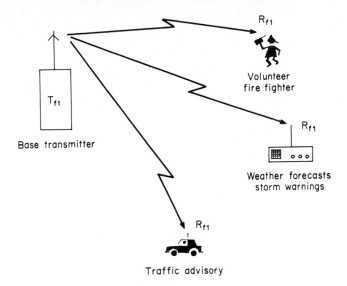

FIG. 21.1 One-way, single-frequency mobile radio system.

7. Two-Way Single-Frequency Simplex

The most widely used and least complex land mobile system is shown in Fig. 21.2. All base, mobile, and portable units in the system have transmitter/receiver units tuned to the same frequency. This makes direct communication between all units in the system possible. Unlike most other land mobile system configurations, a single-frequency simplex system is not totally dependent on the base station. Several units may go beyond the range of the base station and still communicate with each other. In fact, many such systems do not have a base station but consist of a group of mobile or portable units working on a construction project, a police surveillance detail, or other temporary activity.

The simplicity of the system creates operational problems. Since mobile and portable units can communicate with each other, they may create unnecessary channel congestion. Transmissions from a nearby mobile unit may be strong enough to "capture" base station transmissions. Under these circumstances, it is impossible for some mobile and portable units to receive "emergency dispatches" from the base-station operator.

All units in a single-frequency simplex system monitor all other units—plus all other users of the channel. Systems operating in the 40-MHz land mobile band receive "skip" transmissions from stations thousands of miles distant during certain periods of the sunspot cycle. Growing channel congestion causes missed messages and operator fatigue.

A partial solution to the congestion problem is the use of a continuous-tone-controlled squelch system. All base, mobile, and portable units in the system are equipped with a subaudible analog or digital code generator which modulates the unit's transmitter whenever it is activated. The codes are said to be subaudible because they operate in the 67-to 220-Hz portion of the audio spectrum; re-

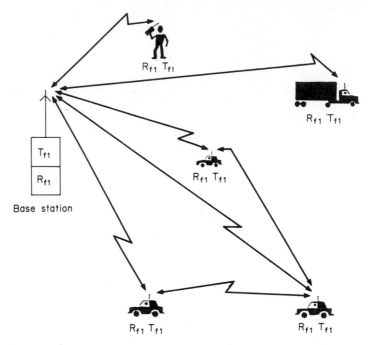

FIG. 21.2   Two-way, single frequency, simplex mobile radio system.

ceiver audio systems attenuate these frequencies to a level where they are inaudible in the speaker. A decoder circuit in the receiver utilizes a *unique* tone or digital code to unmute the receiver's audio system when it is received.

The CTCSS system eliminates *audible reception* of local cochannel users, distant skip signals, harmonic signals, and intermodulation products. The undesired signals are not eliminated—they are still being received and may prevent reception of desired signals. The receiver is muted, however, and the operator is more alert and able to receive signals from the operator's own system. Since the operator cannot hear other users of the channel, it is customary to disable the CTCSS channel momentarily and monitor the channel before transmitting. It is important that the equipment manufacturer, a user group, or the licensing administrator coordinate the allocation of CTCSS codes to minimize the possibility of multiple users being assigned the same code in the same area.

8.   Two-Frequency Simplex

A widely used land mobile radio system concept is the two-frequency simplex system shown in Fig. 21.3. This system eliminates some of the problems inherent in the single-frequency system described above. The system uses two rf channels. Mobile and portable transmitters are assigned frequencies that are 4.5, 5.0, 10.0, or 45 MHz above or below the base-station transmit frequency. The channel spacing is a function of the land mobile band, the type of service, and the administration issuing the license.

The dual-frequency format prevents direct intercommunication between mo-

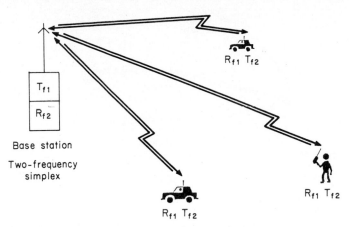

FIG. 21.3   Two-way, two-frequency, simplex mobile radio system.

bile and portable units. Only base-to-mobile/portable unit communications are supported. The base-station operator can now enforce system discipline and reduce the amount of air time used by the system. Base-station transmissions are not "captured" by nearby mobile transmissions as in the case of the single-frequency system. In case of an emergency, the base-station operator can transmit a message to all mobile units except one that may be transmitting to the base station at the same time. The high-power base transmitter does not compete with low-power mobile transmitters. On the other hand, a failure of the base station disables the whole system, since the mobiles cannot talk to each other.

Mobile units cannot operate beyond the range of the base station as they can in the single-frequency system. In some systems, several mobile units may be equipped with two receivers and a "wide-spaced" transmitter so that they can emulate the base-station frequency pairing. This enables these special mobile units to communicate directly with other mobile or portable units beyond the range of the base station.

While not permitting mobile and portable units to leave the base-station coverage area, an alternative to the installation of dual receivers and wide-spaced transmitters is the use of the base station as a mobile relay station. Using the audio output of the base receiver (to modulate the base transmitter) and a carrier detect relay or a CTCSS decoder (to key the transmitter automatically when a signal from a mobile or portable transmitter is received) converts the base station to a functional mobile relay station. This permits all mobile and portable units in the system to communicate with each other.

Mobile-to-mobile communications are normally not utilized by system operators who use two-frequency simplex systems.

9.   Two-Frequency Simplex Mobile Relay Systems

The mobile relay station shown in Fig. 21.4 is a logical extension of the locally controlled base-station *relay mode,* noted above. A mobile relay station can be located wherever a source of power exists. No control line or other type of in-

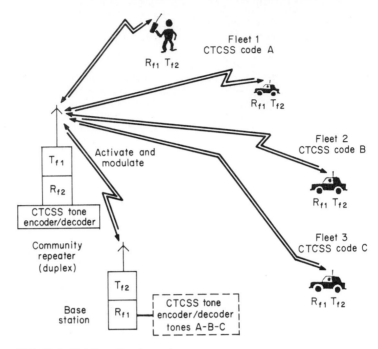

FIG. 21.4 Mobile radio relay system.

terconnection is required. This distinguishes it from the locally controlled base-station relay mode. Locating the relay station atop a high building or on a mountain top substantially increases the coverage area of the station. Mobile relay configurations are widely used by land mobile systems operating in the 450-MHz and 800-MHz land mobile bands.

A *community repeater* is a variation of the mobile relay system noted above. As the name indicates, the relay station is shared by a number of mobile systems. The repeater station receiver can be equipped with as many as 12 CTCSS decoder units, one for each user *system* sharing the repeater. The transmitters in each fleet unit transmit a *unique* CTCSS code that is decoded at the relay station. The retransmission of the received signal is encoded with the same CTCSS code. All receivers in a particular fleet are unmuted by the reception of its unique CTCSS code and are able to receive messages originated by other units in their fleet. The receivers in as many as 11 other fleets will remain muted. This system provides a very efficient and economical use of the assigned channel. It is customary for users of community repeaters to monitor the channel before transmitting to determine if a unit in another fleet, sharing the same community repeater, is using the system.

10. Mobile Relay System with Base Supervision

In the previous section it was assumed that several independent companies were using the same community repeater. In other cases, various departments of the same company share the same community repeater. Management might be designated as fleet 1, the sales department as fleet 2, etc. In this instance, it might be necessary for the base-station operator to have priority access to the system to communicate with any of the fleet users. Figure 21.5 shows the addition of another receiver to the repeater station, designated R_{f3}. Only the base station has a transmitter that can be received by R_{f3}. Upon receipt of a transmission from the base station, R_{f3} mutes R_{f2}, terminating the reception of any mobile or portable units. R_{f3} simultaneously activates and modulates T_{f1}, allowing the base-station operator to communicate with other units in the system. There are a number of variations of the system just described which are beyond the scope of this chapter.

11. Multiple Area—Simultaneous Working

Signal propagation in the VHF and UHF land mobile bands is essentially a line-of-sight phenomenon. In mountainous areas it is standard practice to locate remotely controlled base stations, community repeaters, and mobile relay stations atop high hills or mountains to achieve the largest possible coverage area. In some regions where a number of urban areas are situated in a range of mountains, such as the Pacific coast of southern California, long-range propagation of land mobile signals is often limited by the shadowing effect of terrain obstructions.

One way to solve this problem is to locate a number of CTCSS-activated mobile relay stations on sites that collectively provide total coverage of a large geo-

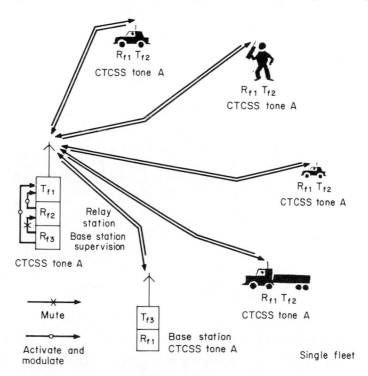

FIG. 21.5 Mobile radio relay system with base supervision.

graphical area. Each station provides reliable coverage of a portion of the total area.

Figure 21.6 illustrates this type of system. If it is assumed that area A, area B, and area C represent three urban areas, mobile traffic in area A will be relayed through a regional mountain-top station equipped with a CTCSS tone A decoder. To utilize the relay station in this area, mobile or portable units must be transmitting CTCSS access code A. A similar operating requirement applies to areas B and C.

All relay stations operate on the same pair of frequencies and are strategically located so that there is a minimum overlap of their coverage areas. Thus three pairs of mobile units operating on the same pair of frequencies but using different CTCSS access codes can be carrying on three different conversations at the same time. Mobile units in overlap areas between area A and area B or between area B and area C will be able to access two mobile relay stations under certain terrain conditions. The mobile operators can select the proper CTCSS access code and use the relay station which provides the best communications. Mobiles traveling long distances can communicate continuously by switching access codes as they travel from area to area.

The example shown in Fig. 21.6 is a basic configuration. There are many complex variations of this system including the use of microwave or UHF links between the relay stations. Such an interconnection permits mobile or base operators to enter the system at one relay station and talk to base or mobile operators via a distant relay station.

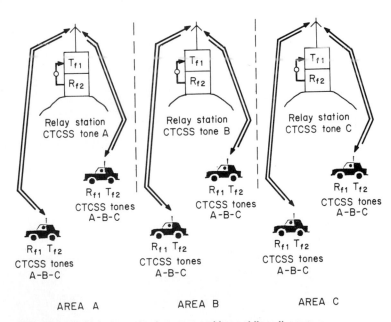

FIG. 21.6 Multiple-area, simultaneous working mobile radio system.

12. Microwave Backbone System

Railroads and pipeline companies use microwave *backbone* communications systems to provide continuous coverage along a railroad or pipeline right-of-way. Figure 21.7 illustrates the concept of these systems.

Two-frequency simplex base stations are located along the right-of-way at separations that allow a small overlap of coverage between adjacent base stations. Each base station is equipped with a CTCSS decoder with a different code from that used by the two or three adjacent stations. Microwave subcarriers transmit the audio and control circuits from each base station, along the right-of-way, to a central control station.

The operation of this system is quite flexible. Mobile units traveling along the system select the CTCSS code with a switch on their mobile unit control head which corresponds to the base station covering their present location. A mobile unit in area A (Fig. 21.7) would select CTCSS code A and would be received by the area A base-station receiver. The audio output of receiver A is transmitted by a microwave subcarrier to the central control console along with an indication that base receiver A is receiving a mobile transmission. The control station can now answer the mobile unit's call via a subcarrier modulated base-station transmitter in area A.

A central control operator who wishes to contact a specific mobile unit, but is uncertain of its location, can simultaneously modulate all transmitters in the system during the mobile call. When the called mobile unit answers, a signal light indicates the area receiver that is picking up the mobile unit's answer. The control operator can now select the area indicated and selectively communicate with this mobile unit without disturbing other area units.

It is possible for a mobile unit to speak to other units in the area or in a distant area by having the central control operator make the necessary interconnections

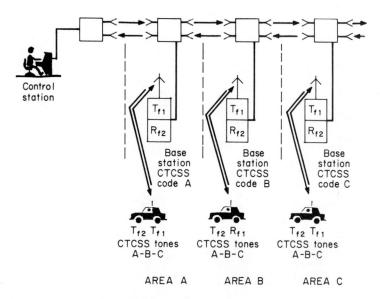

FIG. 21.7 Full duplex microwave "backbone" system.

between the receiver and transmitter in the area or between the receiver and transmitter in the area and the receiver and transmitter in a distant area. This concept provides communications between any units in the system even though they may be hundreds of miles distant. It is evident that the human control center operator could be and probably will be replaced with an electronic switching device. A completely automatic system would require additional signaling equipment in each vehicle to perform the necessary routing functions.

13. Talk-Back Repeater System

The increasing use of low-power portable units and portable data terminals creates coverage problems, particularly in heavily forested or hilly terrains. The high-power base transmitter has a talk-out range considerably greater than the talk-back range of low-power portable transmitters. One way to equalize the talk-out and talk-back ranges of a high-power base transmitter and low-power portable units is shown in Fig. 21.8.

A specialized mobile relay station is located on a hill or tall building near the center of the area frequented by the low-power portable units. The system is designed so that the high-power base station is able to provide adequate signal levels at the portable receivers. Mobile and portable unit transmissions are received

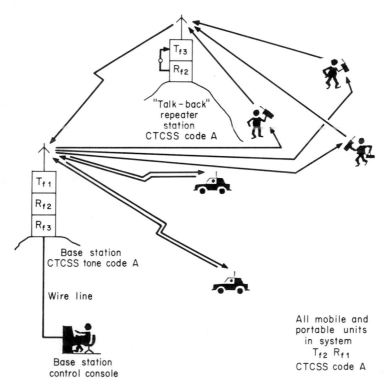

FIG. 21.8 Talk-back repeater system.

by R_{f2} in the talk-back repeater station and retransmitted by T_{f3}. The base station is equipped with two receivers. It receives direct transmissions from mobile and portable units on R_{f2} and relayed transmissions on R_{f3}. The station operator can selectively mute the receiver with the poorest quality reception. Another talk-back repeater station using T_{f4} and R_{f2} along with an additional receiver on R_{f4} in the base station could enhance reception of low-power transmitters in another part of the coverage area. Talk-back receivers support the use of CTCSS code squelch, selective calling, and other system enhancements.

14. Talk-Back Voting Systems

The continued growth of land mobile services has resulted in a scarcity of usable channels in most major urban areas. Licensing authorities are understandably reluctant to allocate the additional frequencies needed to implement talk-back repeater systems when the same channels could be used for new system installations. There is an alternative to the talk-back repeater system that provides an important extra feature. This system is shown in Fig. 21.9. Wire lines replace the radio-frequency links used in the talk-back repeater system.

The system is also effective in improving communications reliability, as can be understood by reference to Fig. 21.9. If we assume that the probability of receiv-

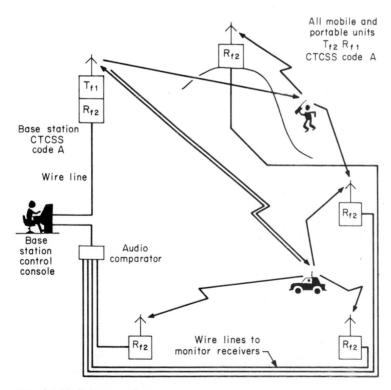

FIG. 21.9 Talk-back "voting" system.

ing a call from the portable unit behind the hill is 10 percent and we add receiver R_{F2} atop the hill, whose probability of receiving transmissions from the portable unit is 90 percent, the overall system reliability is

$$R = 100\% - (100\% - 10\%)(100\% - 90\%) = 91\%$$

If another satellite receiver is added that can receive the portable unit with 90 percent reliability (base of the hill), the probability of receiving the portable unit is now

$$R = 100\% - (100\% - 10\%)(100\% - 90\%)(100\% - 90\%) = 99.1\%$$

Adding additional receivers that can receive the portable unit will further improve the probability of its reception by a base station.

Figure 21.9 shows the mobile unit being received by three satellite receivers. If the probability of direct base-station reception of the mobile unit is 50 percent and the probability of reception from each of the three satellite receivers that are able to receive the mobile unit is 90 percent, the overall probability of receiving the mobile unit is 99.9995 percent. It is evident that the talk-back *voting* system is a powerful method of assuring reliable reception of mobile and portable units.

Figure 21.9 shows all of the satellite receivers connected to an audio comparator in the base-station control room by means of wire lines. Since it is possible to receive a single mobile unit on several receivers simultaneously, merely paralleling their audio outputs would result in uniformly poor reception. The audio comparator includes a signal-to-noise (S/N) analysis circuit which automatically switches the signal having the best S/N to the output of the comparator. As the mobile unit moves about in an urban area, the S/N ratio of the signals received at several receivers continues to change. The audio comparator circuit always switches the signal having the best S/N to the console speaker. In effect, the system functions as a wide-area-diversity receiving system.

15. Simulcast Systems

The voting system described above provides reliable reception of mobile and portable transmissions. The simulcast system, on the other hand, provides reliable reception of base-station transmissions by all mobile and portable units in the system.

The need to cover large areas in paging and certain types of land mobile systems presents a number of problems for the communication system engineer. Installing the base station on a mountain or using a very tall antenna tower might be an acceptable solution to the coverage problem, but in most cases a suitable mountain does not exist and very tall antenna towers are not particularly cost-effective.

The usual solution to the problem is to use a number of remotely controlled base stations. To assure total coverage, it is necessary to locate the remotely controlled base stations in such a way that significant overlap of individual station coverage areas is assured. While it is assumed that all transmitters are on frequency, in the real world the transmitters will actually have small offsets from the assigned frequency. A mobile receiver in the coverage overlap areas will simultaneously receive a signal from more than one transmitter. Each transmitter will be on a slightly different frequency and a heterodyne will be heard. Depending upon its frequency, the heterodyne could destroy voice intelligibility or make paging and selective calling systems unreliable.

The wide-area simulcast system shown in Fig. 21.10 uses five base-station transmitters. When calling a mobile unit, portable unit, or a pager wearer whose location is unknown, the base-station operator could turn on all transmitters simultaneously and believe that the called unit got the message. There is a reasonable probability, however, that the called unit was located in a signal overlap area at the time and missed the call because of *heterodyne interference*. To eliminate this possibility, the base-station operator could repeat the call five times, once from each base station. In systems that utilize several base stations, the sequential calling and communications routine could create unacceptable operating delays. The solution to the problem is to synchronize transmitter operating frequencies and eliminate the heterodyne signal.

Synchronizing of transmitter frequencies is accomplished by using high-stability crystal oscillators or the new and cost-effective rubidium frequency standards in all transmitters. These oscillators typically have a stability in the order of 0.05 parts per million for high-stability crystal frequency standards and 0.0001 parts per million for the rubidium standards—a 500-fold improvement. The long-term stability of the rubidium frequency standard makes it possible to set all the transmitters in the system at "zero beat." This adjustment should hold for at least several months or possibly more than a year. High-stability crystal frequency standards also permit zero-beat operation, but it may be necessary to re-

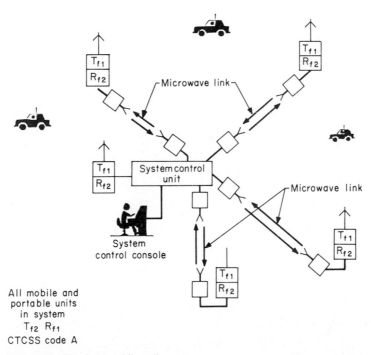

FIG. 21.10 Simulcast mobile radio system.

adjust the oscillators every 3 to 4 weeks in a UHF system or every 20 weeks in a VHF system.

Despite the high stability of the transmitter oscillators, some drift will occur in zero-beat systems. The operator of a mobile unit located in an area where the signal strengths from two or more transmitters are nearly equal will hear bursts of noise as the phase relationships of the carriers nearly cancel each other and the signal strength drops below the receiver threshold. Over a period of time the frequency displacement between the transmitter carriers increases and the heterodyne frequency approaches 10 Hz. At this point the receiver threshold noise bursts become objectionable, and the transmitter oscillators must be set to zero beat. Paging and digital data systems seem to tolerate frequency offsets up to 10 Hz, but voice systems work best at zero-beat adjustment.

Field experience has shown that the median field strength in overlap coverage areas between two transmitters should be at least 5 μV/m. Strong signal levels do not eliminate the noise pulses, but they make them less objectionable because the heterodyne signal remains below threshold for shorter periods of time.

The discussion thus far has concentrated on the effects of transmitter carrier frequency offset. There are other and equally serious problems which must be considered when working with simulcast systems. A mobile unit receiving the same audio modulation from two or more transmitters may produce a noisy distorted signal, particularly if the received signal levels are approximately equal. This distortion is caused by nonlinear phase and amplitude characteristics in the audio circuits between the base transmitters and the system control console. The use of telephone lines of different lengths presents a continuing challenge to keep the lines equalized within the close tolerances needed to assure satisfactory performance of the simulcast system. It is for this reason that microwave links are shown in Fig. 21.10. The microwave carriers are easier to equalize and maintain their equalization over longer periods of time.

Another consideration is the adjustment of the individual transmitter audio circuits to achieve equal deviation and operation below the modulation clipping level. It is recommended that the base station near the control console be fed with back-to-back microwave modems so that similar equalization characteristics can be realized on all base-station links.

In a system using CTCSS codes, it is important that only one CTCSS tone generator be used in the system—typically installed in the system control console. The tone frequencies are then distributed to the various base stations over the microwave links as required. Since there is a single tone generator, phasing and heterodyne problems with CTCSS tone frequencies can be eliminated.

Although more costly to install and maintain, the simulcast configuration provides a level of performance in wide-area paging systems, police radio systems, paramedic dispatch systems, and other similar land mobile systems which is difficult or impossible to achieve with other system designs.

It is possible to combine the talk-back voting system concept shown in Fig. 21.9 with the simulcast system shown in Fig. 21.10 to get the highest-performance wide-area, two-way communications system.

16. Trunking Systems

The rapid growth of land mobile services has placed continuing pressure on regulatory authorities to provide additional rf spectrum to accommodate new land mobile system users. For the past several decades, advances in land mobile equipment technology have allowed the subdivision of existing land mobile channels and have made it possible to use new bands in the 470- to 512-MHz and 806- to 890-MHz portions of the UHF spectrum. It does not appear that substantive amounts of new frequency spectrum will be generally available to the land mobile services to support future growth.

This fact has turned the industry's attention to the use of more spectrally efficient systems to accommodate the many new licensees. Within the last seven years, two system concepts have received major attention as possible interim answers to the growing shortage of land mobile frequency spectrum—the cellular telephone system, described in Chap. 22, and UHF trunking systems.

Trunking is broadly defined as the automatic sharing of a group of communication paths (trunks) among a large number of users. The telephone industry has used this concept for years to reduce the number of circuits between telephone central offices, and there is a growing usage of this technique to provide highly reliable land mobile communications for a growing number of users.

A typical land mobile system uses one rf channel or a pair of channels. In practice, four local land mobile systems may be assigned four adjacent radio channels. One of these channels might have three mobile units waiting in a queue while the other three channels are unused. This small example illustrates one of the reasons for the shortage of land mobile channels. Out of four channels available, three *were unused* and the fourth had a queue of three units waiting to use the channel. If this example were expanded to include all channels on all bands, it is evident that a significant number of new users could be accommodated by existing allocations if a system were devised to allocate unused channels to users wishing to communicate.

A land mobile trunking system is shown in Fig. 21.11. It consists of five base stations connected to the system central controller. Four of the stations function as relay stations while the fifth functions as a control link to operate the automatic channel allocation function.

All unoccupied mobiles and base stations monitor channel T_{f1}. To initiate a call, a mobile unit presses the transmit button. This action transmits a burst of digital data, typically lasting about 0.3 s, which is received by R_{fa} and fed into the system controller. The data transmitted identifies the user fleet and contains a request to be allocated a communications channel. This is illustrated in Fig. 21.11a. The system controller now searches channels 1 to 4 to determine if one is unoccupied. Finding channel 1 unused, the controller sends out a digital signal to all units monitoring the control channel, as illustrated in Fig. 21.11b. The transmitters in all ABC Company units are switched to T_{fb} and the receivers to R_{f2}. All other units in the system remain on control channel frequencies. The channel 1 station operates as a mobile relay station permitting communication between all ABC Company units. The ABC Company now has exclusive use of the mobile relay system.

All other receivers in the system are monitoring T_{f1}, as shown in Fig. 21.11c. If a mobile of MOT Company depresses its transmit button, the same procedure would be followed, but the MOT units would be assigned channel 2. Each of the user fleets is assured privacy since the system central controller assigns only one user fleet to a channel.

A trunking system has an inherently high level of reliability. A system user normally will not be aware of the failure of one of the four channels. If the system central controller senses a failure mode in one of the operating channels, it locks the channel out of service. Allocation of the remaining channels continues as before. If the control channel base station fails, the central controller automatically

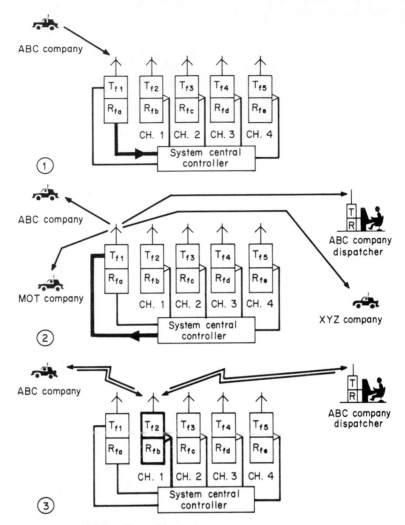

FIG. 21.11 Mobile radio trunking system.

selects one of the four operating channels and converts it into a control channel. A digital message is sent to all units in the system instructing them to switch frequencies and utilize one of the operating channels as the new control channel. The system will continue to operate minus one operating channel. The control channel station is rotated among the five available base stations on a daily basis to equalize equipment aging caused by the high duty cycle of the control station.

A continuous on-channel signal received by one of the operating channel receivers would be sensed by the central controller to determine if the signal is being transmitted by a system user who had been assigned to that channel. If this is not the case, the channel would be locked out until the interfering signal disappears. The system central controller monitors the power output of all transmitters. If the rf output drops below a preset level, the defective station is taken out of service.

The central controller is the key to the operation of the whole system. If it fails, all units in the system will revert to preassigned operating channels so that they can continue to communicate. Under these conditions, fleets lose exclusive use of a channel and will hear other system users who have been assigned the same channel for the duration of the central controller outage. Despite the loss of privacy, the units can communicate with their own fleets.

When the central controller reverts to the shutdown mode, it initiates a subaudible data "handshake" between the mobile unit and the base station to which it has been assigned. This serves two purposes. It provides a periodic audible alert tone that tells the operator that the system is not working properly, and it locks out the mobile transmit function if it goes out of range of the base channel to which it is assigned.

Unlike conventional systems, the central controller system does not allow a user to monitor a channel to determine if the system is busy. If the system is fully utilized, the central controller will give a busy tone to users who press their transmit buttons. In some systems the busy tone is also used to inform the mobile unit that it is out of range of the system or that the system is totally closed down. Since the busy tone is used for several purposes, a signal lamp is provided to alert the operator of a unit trying to access the system that "all is well" but the system is fully occupied. The lamp will typically remain on until the unit is called back.

An attempt to enter the system when all channels are busy will activate an indicator lamp as described above. Additionally, the system controller will put the mobile or base station in a queue and service the unit on a first-in–first-out basis. When a channel is available, the controller calls the first unit in the queue and alerts the operator with a series of audible beeps that the request for a channel will be serviced.

In practical systems, the data burst that requests a channel assignment may be lost if the mobile unit is moving through a poor signal area or if its receiver is momentarily captured by an interfering signal. This problem is resolved in some systems by having the mobile unit send the channel request data until an acknowledgment is received from the system controller. The retry cycle will continue, even though the operator has released the transmit button. The retry cycle is typically a series of randomly timed transmissions.

Once a fleet has been assigned a channel by the system controller, a unique and continuous subaudible bit stream is sent to all fleet units assigned to that channel. This serves as a protection against foreign units accidentally locking on the channel. The foreign unit would detect the subaudible bit stream and be switched back to the control channel in a fraction of a second.

The control channel continuously transmits a directory of fleet channel assignments as long as a particular fleet uses the channel. This is done so that, if a fleet unit is turned on and begins to monitor the channel, it will automatically be switched to the same channel as the other units in the fleet.

Most trunking systems can be expanded to 19 operating channels plus a control channel. This is a limitation imposed by the FCC *Rules*. At some time in the future this limitation may be revised.

The following examples illustrate the improved spectrum utilization that is afforded by the trunking system concept. In each case it is assumed that the average message delivery time is 18 seconds.

A nontrunked, single-channel repeater could accommodate *14 mobile units*— 14 units/channel.

A four-channel trunking system could accommodate *260 mobile units*—52 units/channel.

A 19-channel trunking system could accommodate *1900 mobile units*—95 units/channel.

It is evident that the trunking system concept can make a significant improvement in the utilization of the radio-frequency spectrum. The central system controllers do a superior job of allocating frequencies as compared with human experts. The examples above show that a 20-channel system (19 operating plus 1 control) can provide 7 times as many users per channel, with the same quality of service, as a single nontrunking system.

LAND MOBILE PROPAGATION

17. Field Test Data—The Key to Coverage Predictions

To date, theoretical studies have not provided a method of predicting land mobile system coverage with an accuracy that would be considered commercially acceptable. Historically, land mobile radio systems have been designed by equipment manufacturers' system engineering and sales departments. Propagation predictions were based on data collected from thousands of field tests in every possible environment. Several investigators[1-3] have developed theoretical and empirical methods of predicting land mobile system coverage. For the most part, such procedures can serve only as a starting point for the land mobile systems engineer.

Unlike most commercial communications services, the base-to-mobile transmission path is rarely line-of-sight but is obstructed by terrain features such as buildings, trees, and hills. Multipath reflected signals further complicate the task of predicting mobile unit received signal levels. This is particularly true in cities where buildings obstruct direct signal propagation and create a multitude of reflected signals.

The mobile unit with its low elevation and relatively inefficient antenna is immersed in an environment of electrical noise, intermodulation signal products, adjacent channel sideband "splatter," and cochannel signals. Systems that use radio pager receivers or hand-held portable transceivers are faced with coverage limitations because the antennas are significantly less efficient than those used on mobile units. Portability permits the operation of these units inside buildings where it is difficult to accurately predict base transmitter signal levels.

Systems engineers employed by equipment manufacturers have become skilled in the art of predicting land mobile system coverage. Over the years, some engineers have had the responsibility of making commercial coverage predictions for hundreds or thousands of land mobile systems. In time they become familiar with the unique characteristics of the equipment itself and its performance in different environments. Through countless propagation surveys, these engineers accumulate the statistical information needed to formulate appropriate correction factors to published propagation charts or computer-generated coverage predictions.

The literature contains a number of articles[4-6] which explore the subject of radio-wave propagation in considerable detail. Further information is found in Chap. 1 of this handbook. This chapter reviews the several modes of radio-wave propagation and responds to the needs of the mobile radio systems engineer by presenting a series of nomographs and configuring constants that will facilitate the determination of land mobile system coverage.

18. Free-Space Propagation

Figure 21.12 shows the various modes of propagation involved in a land mobile system. *Free-space propagation* takes place when all objects are remote from the transmitting and receiving antennas. *Remote* is a relative term in practical systems. At microwave frequencies, achievable terrain clearance (one-half first Fresnel zone; see Chap. 1) reduces space loss to near free-space levels. The free-space path attenuation (ratio of received to transmitted power) between short doublet antennas adjusted for maximum power transfer is

$$\text{Free-space attenuation} = 20 \log [3 \ \lambda/8\pi D] \quad \text{dB} \quad (21.1)$$

where λ and D are the wavelength and path length expressed in the same units.

Figure 21.12 shows that free-space propagation is realized for a distance of about 100 wavelengths from a 150-MHz base transmitting station atop a 100-ft tower.

19. Plane Earth Attenuation

Land mobile signals transmitted over an imperfectly conducting plane earth are subjected to a complex attenuation factor that reduces the received level to less than that experienced under free-space conditions. Signals radiated from a base-

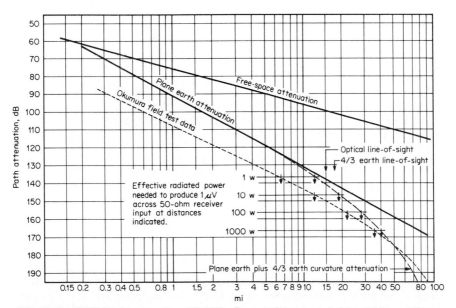

FIG. 21.12 VHF signal propagation. 150 MHz. Base station antenna height: 100 ft; mobile antenna height: 5 ft.

station antenna located a number of wavelengths above a plane earth are partially reflected and changed in phase (depending upon polarization and ground conductivity) and combine vectorially with the direct wave from the transmitting antenna to form a vertical lobed pattern in space.

This effect has little impact on the performance of typical land mobile systems. The angle between the plane of the transmitting antenna and the mobile receiving antenna is typically small and often negative. This assures system operation under the first lobe of the vertical pattern where surface-wave propagation controls path attenuation.

Theoretically and under certain conditions, wave propagation over a plane conductive earth can increase the level of the received signal. The magnitude of the signal increase is expressed as an equivalent increase in the received signal level that would be experienced by increasing the effective height of the transmitting antenna. Except for operation over seawater, there are very few places where ground conductivity makes a measurable increase in the transmitted signal. Mobile radio systems engineers typically ignore this factor.

The attenuation of signals propagated over plane earth can be determined from the following equation:

$$\text{Plane earth attenuation} = 10 \log \left[\, h_t h_r / D^2 \, \right]^{\,2} g_t g_r \quad \text{dB} \qquad (21.2)$$

where h_t, h_r = heights of the transmitting and receiving antennas
$\quad\quad\quad D$ = path length measured in same units as antenna heights
$\quad\, g_t$, g_r = antenna gains (for half-wave dipole, g = 1.64)

This relationship is independent of frequency and is illustrated in Fig. 21.12.

The discussion thus far has ignored the earth's spherical shape and its atmosphere. These characteristics further modify the propagation models considered in previous sections.

20. Refraction

The refractive index of the earth's atmosphere is a function of temperature, the partial pressure of water vapor, and the barometric pressure. It normally decreases linearly with increasing height above earth. A signal radiated from the base-station antenna tangential to the earth's surface will be bent or refracted back to the earth as it passes through the atmosphere, and it will intercept the earth at a point beyond the optical horizon. If under normal atmospheric conditions, the earth is replaced by an equivalent earth having a radius equal to four-thirds that of the true radius, curved signal paths through the atmosphere can be replaced by straight lines.

21. Diffraction

Diffraction is a phenomenon that explains the propagation of electromagnetic energy around a spherical earth in the region beyond the radio horizon. This effect is described in Chap. 1 and in the literature.[4] As indicated in Fig. 21.12, the diffraction loss is added to the plane earth loss.

22. Empirical Field Test Data

The various propagation modes detailed in previous paragraphs give the reader an idea of the theoretical approach to solving land mobile coverage problems, Unfortunately, because land mobile signal paths are obstructed by terrain features and the performance of receivers is degraded by local noise and signal sources, land mobile systems engineers are almost totally dependent on field test data to make reasonably accurate system coverage predictions.

The objectives of land mobile field tests are twofold:

1. To measure the path attenuation at a sufficient number of points to determine the coverage. These measurements are described below.
2. To measure the level of noise desensitizing at various points in the coverage area. These measurements are described in Sec. 26.

Land mobile receivers are typically equipped with *test points* to facilitate alignment of the receiver. One of these can be used to derive a voltage or current which is proportional to the strength of the received signal. A test signal generator is used to calibrate the metering circuity connected to the receiver test point in terms of microvolts input to the receiver or decibels above some reference signal level.

The calibrated mobile receiver can now be driven about the coverage area to determine base-station signal strength. An accurate map of the area should be available to determine the exact distance from the survey location to the base transmitter antenna. Data should be recorded in an organized notebook to provide accurate information for future analysis. Some investigators use graphical recorders to identify patterns of rapid signal variation more accurately. Such variations in received signal strength may identify a unique coverage problem.

Figure 21.12 shows a curve labeled *Okumura field test data*. The data for this curve were taken from a mobile propagation survey in Tokyo.[3] This curve is actually the median value of hundreds of points that would be scattered about this curve. It is important that original field test data points be reduced to a median curve or other statistical relationship so that conclusions can be drawn about the area being surveyed.

23. Path Loss Calculation

For designing practical land mobile systems, it is useful to determine the path loss which can be tolerated for a reference transmitter effective radiated power (e.r.p.) and receiver sensitivity. This reference path loss can then be adjusted upward or downward in accordance with the actual transmitter power and receiver sensitivity to determine the maximum permissible system path loss.

Figure 21.12 shows that 1 W effective radiated transmitter power will produce 1 μV across the terminals of a 50-Ω receiver with 137 dB of path loss. Over a 4/3-radius earth with no obstructions, this would occur at a range of 12.3 mi, but the field test data predict a range of 6.6 mi. An increase in the power output of the base transmitter to 10 W e.r.p. would accomodate another 10 dB of path loss, or 147 dB. This would extend the range of a 1-μV signal to 19 mi for the 4/3 earth or to 12.2 mi for the field test data. The range expectations for 100-W and 1000-W e.r.p. can be read from Fig. 21.12 using a similar procedure.

The field test data in Fig. 21.12 indicate a path loss that is 7 to 15 dB greater than the theoretical curve. The data for the curve appear to have been taken from a base station located in an urban area (15 dB more path loss than theoretical) with the path extending outward to a suburban or rural area (7 dB more path loss than theoretical). If a dip had occurred near the center of the field test curve, it would probably indicate a densely populated urban area or a hilly area in that part of the field test path.

Figure 21.12 provides a limited amount of information because the base station and mobile antenna heights are fixed. This type of graphical display is primarily used to evaluate changes in existing system coverage if the transmitter power or gain or other system parameter is changed. The next section describes coverage nomographs which can be used to predict coverage for a wide range of frequency bands, antenna heights, power levels, and terrain conditions.

24. Coverage Nomographs

Figures 21.13 to 21.16 illustrate another way of presenting coverage

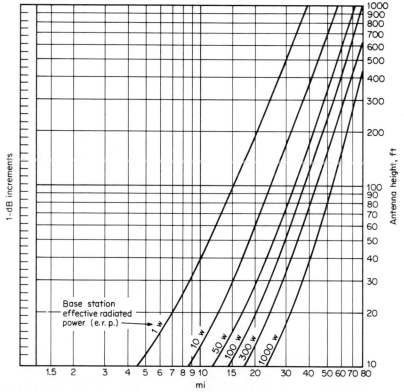

FIG. 21.13 Suburban/rural. Base-to-mobile coverage chart: 40 MHz; mobile antenna height: 5 ft.

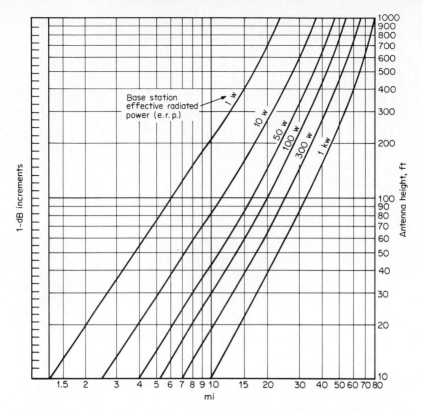

FIG. 21.14 Suburban/urban. Base-to-mobile coverage chart: 150 MHz; mobile antenna height: 5 ft.

predictions for the 40-, 150-, 450-, and 900-MHz land mobile bands. These graphs are relatively uncomplicated, and they consider the most significant parameters of land mobile system design. A 1-dB-increment axis is provided to "tailor" the nomograph to unique system characteristics.

These charts are designed so that at least *50 percent* of the *actual* transmission paths will equal the nomograph predictions. Subtracting 6 dB from the design point criteria will yield *actual* transmission paths that equal the predicted coverage in at least *90 percent* of the locations. The 90 percent criterion is typically used in mobile system suppliers' performance guarantees.

The nomographs predict coverage in typical operating environments, and it is important to have an understanding of the assumptions made during their development.

Generally, it was assumed that land mobile usage is primarily in or near populated areas.

An assumption was made that the 40-MHz band is used in rural areas and small towns. Noise desensitization in these areas is primarily due to automotive ignition, power line, and transformer leakage, and other sources of noise that are localized and can usually be avoided. Typically, 40-MHz systems are wide-area

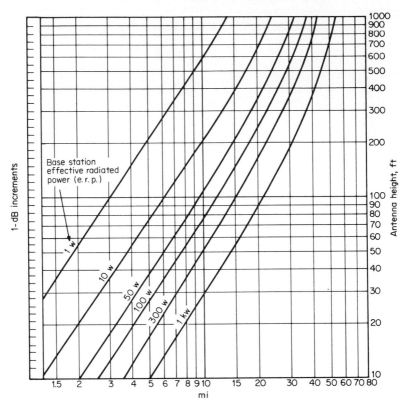

FIG. 21.15 Suburban/urban. Base-to-mobile coverage chart: 450 MHz; mobile antenna height: 5 ft.

(state police systems, etc.) as compared with systems in the 150-MHz or UHF mobile bands. These operating conditions suggest the use of less-conservative coverage predictions.

Some previously published nomographs for the 150- and 450-MHz band give range-prediction figures for a homogeneous area. It was assumed that the user would recognize the need to apply correction factors to the different parts of the coverage area: central city, light industrial, residential, etc. These could be applied correctly only by land mobile professionals.

The nomographs in Fig. 21.14 and 21.15 for the 150- and 450-MHz bands, by contrast, include built-in correction factors which are based on the most common usages of these bands. Most 150- and 450-MHz systems that require wide area coverage are located in that part of the central city where tall buildings are available to obtain the required base-station antenna height. These locations are also selected because the coverage area will include a major part of the urban population and commercial operations. The central city location is subject to high ambient noise levels and terrain obstructions. As the distance from the central city base-station site is increased, operation in the light industrial and suburban residential areas will be less affected by ambient electrical noise and terrain obstruction losses. The reality of this environment is built into the 150- and 450-MHz

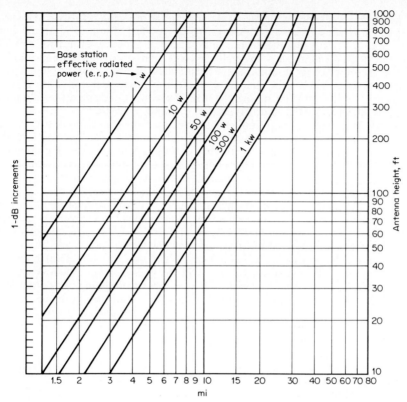

FIG. 21.16 Suburban/urban. Base-to-mobile coverage chart: 900 MHz; mobile antenna height: 5 ft.

nomographs by means of the correction factors which are plotted in Fig. 21.17 and 21.18. *Figures 21.14 and 21.15 include these factors.*

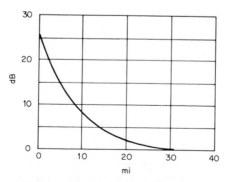

FIG. 21.17 Corrections for building and foliage attenuation and noise desensitizing which are incorporated in the 150-MHz nomograph.

The results predicted on the 900-MHz nomograph are valid for urban and suburban environments. They are conservative for residential and rural areas, and for purposes of calculation the e.r.p. should be increased by 10 dB for residential areas and 20 dB for rural areas. When these corrections are added, care should be taken to consider foliage attenuation in these areas, and the appropriate correction factors from Table 21.4 should be applied as well. Also, as with all the nomographs, 6 dB should be *subtracted* from the nomograph indications if a prediction of 90 percent rather than 50 percent coverage is desired.

The 900-MHz band suffers almost no degradation from the common sources of urban electrical interference.[7] To take advantage of this fact, low-noise solid-state preamplifiers are mounted on the 900-MHz base station receiving antenna to minimize or eliminate signal degradation caused by transmission line losses. With such a receiving antenna preamplifier, 900-MHz system performance is comparable to the 150- and 450-MHz land mobile bands in urban areas.

When the system environment differs significantly from that assumed in the construction of the nomograph, a more accurate estimate of coverage can be made by using the correction factors listed in Table 21.4. These conditions will occur in extremely congested locations such as downtown New York City or in small cities. In the downtown areas of very large cities, the effects of building obstructions and noise densensitization will be greater than assumed in the nomographs. In small cities they will be less. It is particularly important that Table 21.4 be used in the congested areas of very large cities because the nomograph predictions will be too optimistic. In small cities the nomograph predictions will be conservative, and the systems engineer has the option of using the nomographs without correction, knowing that the coverage will probably exceed the forecast.

Another situation in which Table 21.4 is useful is predicting the performance of a *base station* receiver in a mountain-top location where dozens of land mobile stations, community repeaters, and even high-power FM and TV stations may share a common site.

Note that at 150 or 450 MHz, the correction factors shown in Figs. 21.17 and 21.18, which are built into the nomographs, must be subtracted before applying the correction factors from Table 21.4.

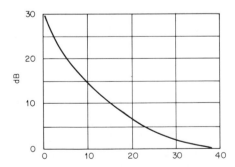

FIG. 21.18 Corrections for building and foliage attenuation and noise desensitizing which are incorporated in the 450-MHz nomographs.

TABLE 21.4 Propagation Nomograph Correction Factors (Values in Decibels)

Factor	Frequency, MHz			
	0	150	450	900
Mobile transmission line loss	0	− 1	− 2	− 2
Mobile antenna gain	+ 3	+ 2	+ 2	+ 2
Ambient noise desensitizing				
Central urban—New York	− 35	− 25	− 20	− 1
Urban area	− 20	− 15	− 12	− 0
Suburban—residential	− 12	− 10	− 6	− 0
Rural	− 6	− 2	− 2	− 0
Building losses—on street out-				
side building				
Central urban—New York	− 25	− 25	− 25	− 20
Urban area	− 10	− 10	− 10	− 10
Residential area	− 2	− 2	− 2	− 2
Foliage losses				
No trees	0	0	− 0	− 0
Sparse trees	0	0	− 2	− 6
Medium-density trees	− 1	− 3	− 6	− 15
Jungle foliage	− 3	− 7	− 25	− 28
Hand-portable antenna gain	N/A	− 5	− 3	− 2
Hand-portable inside-building				
loss				
Central urban area buildings	N/A	− 25	− 25	− 25
Medium commercial buildings	N/A	− 10	− 10	− 10
Suburban residence	N/A	− 3	− 3	− 3
Base-station transmission-line				
loss per 100 ft				
RG 8/U solid dielectric	1.2	2.8	5.2	
½-in foam dielectric	0.44	0.86	1.5	2.1
⅞-in foam/air dielectric	0.23	0.48	0.86	1.3

N/A: not applicable.

An example of the use of Table 21.4 follows. Determine the talk-back range (90 percent coverage) from a 450-MHz, 50-W mobile to a base station having an antenna on a 300-ft tower and a gain of 6 dB.

Transmission line losses (350 ft at 1.5 dB/100 ft)	− 5.3 dB
Antenna gain	+ 6.0 dB
Cochannel interference	− 9.5 dB
Receiver desensitization from nearby TV transmitter	− 4.5 dB
Adjustment for 90% coverage	− 6.0 dB
Adjustment for correction in nomograph (Fig. 21.18)	+ 7.5 dB
Total correction	− 11.8 dB

Figure 21.15 predicts median coverage at a range of 18.5 mi for the conditions assumed in this example. After applying a correction of −11.8 dB and using the

method described in the following section for modifying nomographs, the range for 90 percent coverage is predicted to be 9.0 mi.

All of these nomographs and the correction factors are based on many field tests, but, unfortunately, a considerable amount of judgment is required in using them. Commercial land mobile system suppliers use nomographs to rough out system designs, but when a performance guarantee is required it is necessary to conduct field surveys to confirm the coverage predictions.

25. Modifying Propagation Nomographs

Figure 21.19 shows several examples of tailoring nomographs to get answers to specific problems.

The examples are based on the design of a 40 MHz rural area system. The initial design assumed a 50-W e.r.p. transmitter and a dipole base-station antenna mounted atop a 100-ft tower (circled point). It is assumed that coverage of 50 percent of the transmission paths at a 33-mi radius is acceptable.

In Example 1, the range for 90 percent coverage is calculated. Extending a line

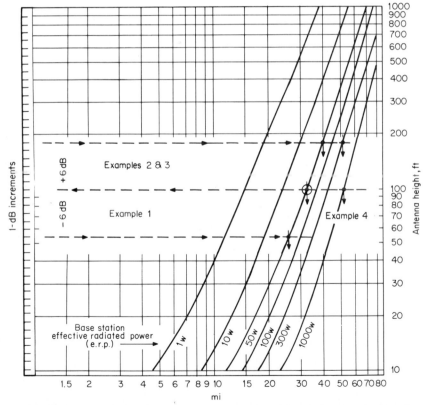

FIG. 21.19 Examples of adjustments and corrections to nomographs using the db scale on the left margin. Suburban/rural. Base-to-mobile coverage chart: 40 MHz; mobile antenna height: 5 ft.

from the circled point to the 1-dB-increment scale, subtracting 6 dB from the no-mograph (down is minus), and plotting a horizontal line to the left until the 50-W curve is intercepted gives the 90 percent system coverage; this is 26.3 mi.

Examples 2, 3, and 4 show the effect of various attempts to increase the range to cover an industrial park at a distance of 50 mi. This coverage could be achieved by increasing the tower height to 400 ft, but this solution is impractical because of cost and the proximity of a local airport.

Example 2 shows the effect of using a directional antenna with a gain of 6 dB in the direction of the park. Again running a line from the circled point to the 1-dB increment scale, up 6 dB, and then to the right for an intercept with the 50-W curve, it appears that this configuration is inadequate. The range is only 40 mi. But if the transmitter e.r.p. is increased simultaneously to 300 W—Example 3, line drawn horizontally to the right—the expected range is seen to be 51.5 mi.

Example 4 shows the effect of using the original nondirectional antenna and a base-station e.r.p. of 1000 W. This gives coverage out to 52 mi and achieves the 50-mi goal. The use of higher power may not be a solution, however, because of several potential problems: (1) the licensing authority may not permit the use of this power level, (2) such a transmitter may be too expensive or unobtainable, or (3) it may not be possible for mobile units to talk back from the distant industrial park.

Using the technique described in the previous paragraphs, other options such as a relocation of the base station and a higher tower can be readily explored. Virtually any type of systems coverage problem can be solved using these nomographs.

26. Receiver Desensitizing

Knowledge of the "effective" operating sensitivity of the system's base and mo-bile receivers is a key requirement for making coverage predictions. Degradation due to ambient noise, receiver intermodulation, adjacent channel splatter, and other forms of interference must be quantified so that realistic range predictions can be made from the nomographs in this chapter. Table 21.4 lists several levels of noise desensitizing that can be applied to correct the coverage predicted by the nomographs. This table does not consider receiver intermodulation, sideband splatter, harmonic and spurious radiation, and other receiver degrading factors that might be, and probably are, present in urban area installations. As with the prediction of field intensity, on-site measurements are a more accurate method of determining actual receiver desensitizing.

Figure 21.20 shows a simple but highly effective way of measuring on-site re-ceiver densensitizing. A shielded 50-Ω dummy antenna is coupled to the receiver input through the special signal injection fixture having a 60-dB attenuation loss which is shown at the bottom of Fig. 21.20. The receiver is equipped with a re-ceived signal level indicator; this can be a dc meter connected to a receiver test point whose output varies with received signal input. It is unimportant that the meter be calibrated, since it is only a relative indicator of rf input level to the receiver.

The signal generator is fed into the receiver through the 60-dB loss attenuator and adjusted to a level that just quiets the receiver. The reading of the meter con-nected to the test point is recorded and defined as level A. The calibrated output of the signal generator will be somewhere in the range of 1000 to 2000 μV and is defined as level B. The absolute value is unimportant, but it should be recorded.

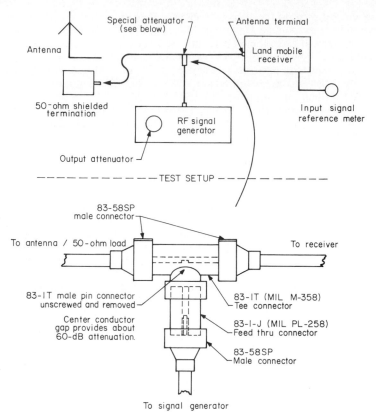

FIG. 21.20 Test apparatus for measuring receiver desensitizing. The lower drawing shows the details of the special "T" attenuator.

The receiver input is now terminated with a resistive load whose noise level is primarily that contributed by thermal noise.

The coax cable is now disconnected from the 50-Ω shielded load resistor and connected to the base station or mobile antenna. Ambient noise and other forms of interference will now degrade the receiver's sensitivity, making it necessary to increase the output of the calibrated signal generator until the reading on the receiver metering circuit is returned to reading A. The new value of the signal generator output is recorded as C microvolts. The receiver degradation is

$$\text{Degradation} = 20 \log (C/B) \quad \text{dB} \qquad (21.3)$$

Many test signal generators have their output controls calibrated in decibels. In this case, it is only necessary to note the decibel increase in generator output to reestablish the reference reading A.

The reading on the base-station antenna should be made at different times of the day and different times of the week to establish the limits of receiver desensitizing due to automobile traffic, electrical signs and machinery, and other factors that cause receiver desensitizing.

A vehicle installation will permit a calibration of receiver desensitizing in different parts of the coverage area.

It is important that these measurements be made when cochannel stations are not operating.

The data obtained by these measurements can be used as described in Sec. 25, Modifying Propagation Nomographs.

REFERENCES

1. Bullington, K., "Radio Propagation at Frequencies above 30 Megacycles," *Proc. of the IRE,* pp. 1122–1136, October 1947.

2. Young, W. R., Jr., "Comparisons of Mobile Radio Transmissions at 150, 450, 900, and 3700 Mc." *Bell System Technical Journal,* November 1952.

3. Okumura, Y., et al., "Field Strength and its Variability in VHF and UHF Land-Mobile Radio Service," *Review of the Electrical Communications Laboratory,* vol. 16, nos. 9 and 10, September-October, 1968.

4. Burrows, C. R., and S. S. Atwood, *Radio Wave Propagation,* Academic Press, New York, 1949.

5. Black, D. M., and D. O. Reudink, "Some Characteristics of Mobile Radio Propagation at 836 MHz in the Philadelphia Area," *IEEE Transactions on Vehicular Technology,* vol. VT-21, no. 2, May 1972.

6. French, R. C., "Radio Propagation in London at 462 MHz," Mullard Research Laboratories, Redhill, Surrey, England.

7. Schultz, C. J., "Is 960 MHz Suitable for Mobile Operation?," *IRE National Convention Record,* Part 8, New York, March 1956.

8. Schultz, C. J., "900 MHz—A Potential Vehicular Communications Band," 1957 WESCON Convention, San Francisco, California, August 1957.

9. Schultz, C. J., "900-MHz Mobile Developments," *Southwestern IRE Conference Record,* Houston, Texas, April 1967.

CHAPTER 22
CELLULAR RADIO SYSTEMS

John I. Smith
Staff Scientist (retired), AT&T Bell Laboratories

V. H. MacDonald
Staff Scientist, AT&T Bell Laboratories

Sang B. Rhee
Staff Scientist, AT&T Bell Laboratories

CONTENTS

THE CELLULAR CONCEPT

1. Overview

The fundamental idea in cellular mobile radio telephone systems is frequency reuse. Because of frequency reuse, a limited number of radio channels can be made to serve many users.

This is illustrated in Fig. 22.1, which shows an area served by 14 cells, divided

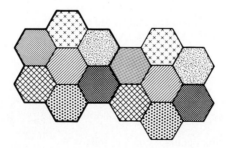

FIG. 22.1 Frequency reuse patterns.

into two clusters. Each cluster contains N cells; in the illustration $N = 7$. A separate set of channels is assigned to each cell in a cluster, but the sets used in cluster 1 are reassigned in cluster 2, thus reusing spectrum. The signals radiated from a cell in channels assigned to that cell are powerful enough to provide a usable signal to a mobile terminal within that cell, but not powerful enough to interfere with cochannel transmissions in distant cells. All cellular terminals can tune to any of the allocated channels, and the special radio equipment used in a cellular base station keeps track of mobiles as they move, assigning a new frequency to any mobile crossing a cell boundary.

A characteristic of cellular radio systems related to frequency reuse is the concept of cell splitting. Cell splitting is a way to keep up with growing demand for mobile radio service in an active service area.

The illustration in Fig. 22.2 shows how a new smaller cell has been created from parts of two large cells. By this means, additional service can be provided in small areas without an increase in the allocated radio spectrum. To effectively use cell splitting, however, the power radiated from both terminals and base stations must be under computer control. Before a given split, a cell site with a certain capacity for simultaneous calls served a larger cell area. After the split, the same site, which still has its previous traffic capacity, serves a smaller cell area. One or more new cell sites serve the area that this site previously served.

Figure 22.3 is a representation of the components of a cellular system. A mobile, turned on but not in active use, periodically scans a set of special control channels assigned to the system, and marks for use the strongest carrier found. With the mobile receiver tuned to this strongest carrier, the mobile continuously decodes a digital modulating stream looking for incoming calls. Any call to a mobile terminal is initiated just like a normal telephone call. A seven-digit number is dialed (or 10 digits from a different area code), and the telephone network routes the call to a central computer of the mobile party's radio system. The number is broadcast on the control channels of every cell in the system. When a called mobile terminal detects its number in the incoming digital stream, it sends its identification back to the system. The system uses a digital message, on the control channel, to designate a channel for the mobile to use. The mobile tunes to this channel and the user is, only then, alerted to the incoming call by ringing.

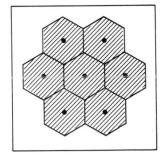

A similar sequence is involved when a mobile user originates a call. The mobile user dials the desired telephone number into a register in the user's terminal. This number is transmitted over the control channel of the nearest (strongest carrier) cell. The system computer then designates a channel for the call and the mobile is automatically tuned to that channel.

In addition to the function of call setup, just described, there are also functions of call handoff and call release.

The handoff function occurs when a mobile with a call in progress crosses a cell boundary and is accomplished as follows. When the system computer

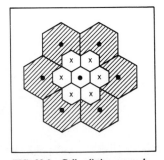

FIG. 22.2 Cell splitting example.

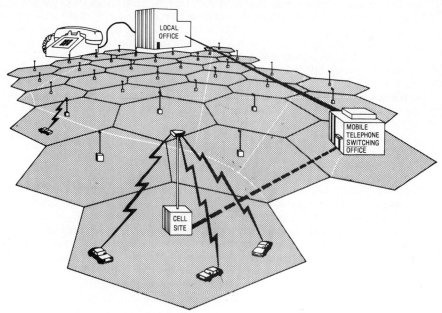

FIG. 22.3   Components of a cellular system.

tracking the mobile determines that handoff is required, the conversation is interrupted. This interruption is brief and not perceptible to the users at either end of the conversation. But during the break, a digital message containing a new channel number for the call is transmitted. The mobile tunes automatically to this new channel and the conversation is reestablished.

When a mobile user hangs up, a release message is sent before the mobile transmitter is turned off. The modulation tone that indicates to the system that an active call is in progress is also removed. The system uses the digital message and the state of the signaling and supervisory tones to determine the release of base-station equipment and of the telephone trunks.

Computerized control, within the system and at the customer terminal, is essential for the operation of a cellular mobile telephone system. There was a perception at the onset of cellular system proposals of an unfilled communication need. The convergence of this need with advances in electronic switching systems and microprocessors together with affordable frequency synthesizers and microwave power semiconductors stimulated the rapid development of the present 800- to 900-MHz cellular radio systems. This chapter presents the history and the distinctive conceptual characteristics of cellular systems, the special propagation peculiarities at 800 MHz that radio systems must deal with, and an outline of the design and layout of cellular systems.

2. History

Noncellular domestic public land mobile radio service has existed since 1946, when a mobile telephone system commenced operation in St. Louis, and proposals for cellular systems first appeared in research papers and FCC proceedings within a few years after the start of this service in St. Louis. In the 1960s, *improved mobile telephone service* (IMTS) brought a high degree of automation to mobile telephone systems, enabling mobile subscribers to dial their calls as if they were using an ordinary land telephone.

In some ways, cellular systems are evolutionary extensions of the previously existing "conventional" mobile telephone service at 150 MHz and 450 MHz. From the mobile subscriber's perspective, the differences between conventional and cellular systems may not be obvious. Both types of system offer mobile subscribers the same fundamental services: the ability to place calls to a land telephone or another mobile telephone simply by dialing the number, and the ability to receive a call when a caller has dialed the telephone number of the mobile subscriber's mobile unit.

The architecture of a cellular system, described more fully later, resembles a network of small-area contiguous conventional systems, all interconnected in order to permit the transfer of calls from one of the small-area systems (that is, one cell) to another. The major dissimilarity between cellular and conventional systems is that call handling in a cellular system is vastly more complex and requires much greater sophistication in both the land-based equipment and the subscriber units. The greater complexity of call handling arises mainly because

1. A cellular system uses a much larger number of channels
2. A cellular system contains multiple land transmitter-receiver stations (called *base stations* or *cell sites*), some of which use the same channels
3. Calls sometimes have to be transferred from one channel to another

FCC Docket 8658 in 1947 contained an early hint of cellular systems. At that time the Bell System requested the allocation of 150 two-way channels between 100 and 450 MHz for a broadband urban mobile system. With a channel spacing of 100 kHz and ample guard bands, the requested spectrum amounted to 40 MHz. The FCC denied the request because no suitable band was available.

In 1949 the FCC opened Docket 8976, which dealt with the allocation of the band from 470 to 890 MHz. The Bell System presented a revised proposal for a broadband urban mobile service. The FCC considered the possibility of allocating 30 MHz for mobile telephone service but decided in the end to allocate the entire band under consideration to UHF broadcast television.

Under Docket 11997, in 1958 the Bell System requested an allocation of 75 MHz near 800 MHz for a broadband urban mobile telephone service. The FCC took no action on this request.

Cellular service finally came to exist during the long lifetime of Docket 18262, which the FCC opened in 1968 and closed in 1975. Under this docket, the FCC issued several notices of inquiry and proposed rulemaking, several reports and orders, and several memorandum opinions and orders. Interested parties generated voluminous quantities of proposals, comments, reply comments, and petitions for reconsideration.

In 1970 the FCC proposed to allocate 75 MHz for public mobile radio use and 40 MHz for private mobile radio systems. Much of the spectrum needed for this allocation was obtained by reclaiming the spectrum previously allocated as UHF TV channels 70 through 83.

In 1974 the FCC enacted allocations smaller than those proposed earlier: 40 MHz for public mobile telephone service to be provided by wire-line common carriers (that is, telephone companies) and 30 MHz for private mobile radio services.

In 1980, under Docket 79-318, the FCC split the 40-MHz allocation into two 20-MHz allocations, one of which was to be available to wireline common carriers only and the other of which was to be available to nonwireline common carriers only. The intention of this ruling was to foster competition in most service areas by facilitating the coexistence of two cellular systems covering the same service area. This ruling stands today. Nonwireline systems use the band 825 to 835 MHz for mobile-unit transmission and 870 to 880 MHz for land transmission. Wireline systems use the band 835 to 845 MHz for mobile-unit transmission and 880 to 890 MHz for land transmission.

The FCC rules governing cellular systems are found in the Code of Federal Regulations, Title 47, Part 22: Public Mobile Radio Service, Subpart K: Domestic Public Cellular Radio Telecommunications Service.

3. Objectives

The objectives of a mobile telephone system are

1. *Large subscriber capacity:* Ideally the provider of mobile service in a coverage zone, such as a large city and its suburbs, should be able to provide service at low blocking to anyone who wants it. Cellular systems also need to be adaptable to the widely varying demand levels of different service areas. Cellular systems should be capable of serving small systems in which the number of potential subscribers is only a few hundred, and they should be able to serve very large systems in which the potential subscribers number in the tens of thousands.

2. *Efficient use of spectrum:* Operators should be able to provide this service without use of additional publicly owned radio spectrum. No universally accepted definition for spectral efficiency exists, but a reasonable contender is the number of simultaneous calls per megahertz of allocated spectrum per square mile, which still leaves unanswered the practical question: If system design A achieves 10 simultaneous calls per megahertz per square mile and system design B achieves 15 simultaneous calls per megahertz per square mile with poorer transmission quality and higher blocking probability (the probability that all channels that a particular call might use are already in use), is design B really superior?

   Despite the fact that spectral efficiency lacks a universally accepted definition and may be defined in ways that ignore important practicabilities, it remains a principal objective for cellular systems. By comparison with conventional systems, cellular systems meet the efficiency objective very well. In a cellular system with small cells just a few square miles in area, a single allocated channel can serve several calls simultaneously when the subscriber units involved in these calls are separated from one another by just a few miles by using the capability described in the overview as *frequency reuse*

3. *Nationwide compatibility:* A customer who subscribes to cellular service in one service area should be able to obtain service in any other cellular service area throughout the country. A subscriber who purchased equipment for use in one service area should be able to use the equipment in all other service areas. Furthermore, from the subscriber's point of view, the differences in the

mechanics of using the service and in the nature and quality of the various service features should be minimal.

4. *Quality of service:* Ideally, the service provider should expand the capability of the system as customers are added so that the blocking experienced by these users is not noticeably greater than that encountered in the land network.

BASIC ELEMENTS OF CELLULAR DESIGN

4. System Design

The various topics related to cellular system design are discussed in this section, namely the significance of cells, the relationship between cells and cell sites, the geographical deployment of channels, some properties of cellular geometry, and considerations pertaining to initial system layouts and procedures for cell splitting.

The term *cellular system* implies a system containing cells, and discussions of cellular systems usually include frequent references to cells. Figure 22.4 shows what a cell would look like, according to two distinct definitions. Figure 22.4a shows a regular hexagon, which has become the virtual emblem of cellular systems. When cell sites are situated at the points of the regular lattice, cells are customarily depicted as the regular hexagons shown in Fig. 22.4a. Each cell represents the expected coverage area of the cell site at its center. In the interior of the array of cell sites, each cell constitutes the locus of points closer to one particular cell site than to any other cell site. The same cell shape is also drawn around the cell sites at the edge of the array. In practice, obtaining property on which to locate cell sites is one of the most difficult hurdles in the construction of a cellular system. System operators attempt to place cell sites close to the points on a regular lattice, but the actual lattice of cell-site positions normally turns out to be conspicuously irregular.

Because of rf propagation effects and the actions of a system's voice-channel-selection algorithms, a particular cell site almost never serves certain locations that are closer to it than to any other site. At the same time, it frequently serves certain other locations that are closest to some other cell site. The definition of the cell associated with a particular cell site can be reformulated to emphasize actual service by the site rather than proximity to it.

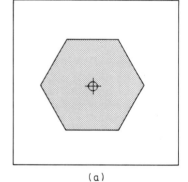

(a)

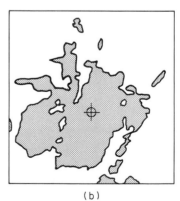

(b)

FIG. 22.4   Definition of a cell.

The definition of a cell might be stated as those locations for which a particular cell site is more likely than any other cell site to provide service. This definition leads to a cell that looks like the shaded area in Fig. 22.4b.

It is appropriate to describe the procedures for assigning channels to cell sites by starting with assignments in a regular grid, and following on with some types of "nonregular" channel assignment.

The procedure for assigning voice channels to cell sites begins with the partitioning of the collection of all voice channels into several sets. For a hexagonal cellular system, the number of sets, designated N, must have the form

$$N = i^2 + ij + j^2 \tag{22.1}$$

in which i and j are nonnegative integers. For the proper control of adjacent-channel interference, no set contains any channels that are adjacent in frequency. For convenience, let us assume that the channel numbers begin with 1. Then set n is usually defined to contain channels n, $n + N$, $n + 2N$, $n + 3N$, and so forth.

For simplicity, it is assumed that the pattern of regular hexagons like that in Fig. 22.4a can serve as a map of the system. It is also assumed that each hexagon contains exactly one cell site. One can then speak interchangeably of cells and cell sites. The process of assigning channel sets to cells is equivalent to the graphical process of labeling the cells in a cellular map of the system. The labeling process begins with the labeling of a single cell, such as the cell labeled A near the center of Fig. 22.5. Now that this cell has its label, the cochannel cells (those with the same label) are easy to find.

Assume that the value of N, the number of sets, has already been established. The rf propagation characteristics of the coverage area and the transmission-quality objectives for the system jointly determine what value of N will produce acceptable cochannel-interference statistics. The integers i and j, which are associated with N as Eq. 22.1 shows, hold the key to locating cochannel cells.

One finds the closest cochannel cells of a particular reference cell with this recipe: Starting at the reference cell, move i cells along any row of cells. Rotate 60° counterclockwise and move j cells down the row of cells that extends in this new direction.

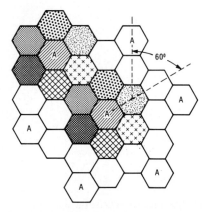

Fig. 22.5 Channel sites: $N = 7$, $i = 1$, $j = 2$

Figure 22.5 diagrams the labeling process for N equal to 7; in this case the parameters i and j equal 1 and 2. After locating the closest cochannel cells of the initial reference cell, one can treat each of these cochannel cells as a reference cell and locate still more cochannel cells in a large system. After labeling all cells that should be labeled A, one chooses any as yet unmarked cell and marks that cell with another label, such as B. After finding and labeling all B cells, one proceeds to the next label. After the Nth label, all cells will have been labeled. Now a particular channel set should be associated with each label. Normally channel sets can be associated with labels in any arbitrary way.

When two or more cell sizes exist simultaneously, the procedure for labeling cells remains essentially the same, but one must use the appropriate size of cell for locating the cochannel cells of a particular cell. Figure 22.6 illustrates this point. Figure 22.6*a* shows a single smaller cell embedded in a pattern of larger cells. A system might have such a configuration at startup, or the small cell could represent a cell site that has been added to a system for *cell splitting* to handle growing traffic demand.

Determining the correct label for the smaller cell requires treating the smaller cell as the reference cell and applying the instructions given above, using a network of cells that have the same size as the reference cell. In this example, the cell of interest is the only smaller cell that actually exists, but it is necessary to visualize additional smaller cells as if they were superimposed on the actual cell pattern. In Fig. 22.6*b* a few additional smaller cells have been drawn in to aid in the process of finding a larger cell that is cochannel with the smaller cell. Since the cochannel larger cell has the label C, this is the appropriate label for the smaller cell.

In such an array of regular hexagonal cells, all of the same size, the minimum separation D between the centers of cochannel cells depends on the number of channel sets N and the cell radius R as follows:

$$D = R\sqrt{3N} \tag{22.2}$$

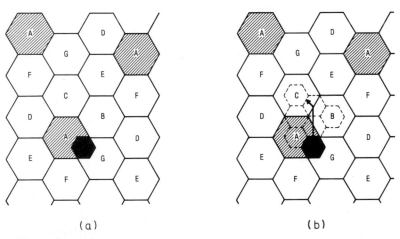

(a)   (b)

Fig. 22.6 Procedure for determining channel-set assignment in conjunction with cell splitting. (*a*) one smaller cell; (*b*) additional smaller cells with $i = 2$, $j = 1$ in., and $N = 7$ grid.

Figure 22.7 illustrates this relationship for a system with four channel sets.

No such simple formulation exists for the minimum distance between a given cell and the closest cell using an adjacent channel, but some of the important facts regarding adjacent channels are readily summarized. For values of N equal to 12 or more, channel sets can be assigned to cells in such a way that geographically adjacent cells never use spectrally adjacent channels. For smaller values of N, however, adjacent channels occur in adjacent cells to varying degrees. For N equal to 3, all six of a given cell's adjacent cells contain adjacent channels. For N equal to 4, four adjacent cells contain adjacent channels. For N equal to 7, two adjacent cells contain adjacent channels.

Normally cochannel rather than adjacent-channel interference is the dominant factor affecting the choice of the value of N for a system. Channel sets represented by adjacent letters of the alphabet are assumed to contain adjacent channels; sets B and C contain adjacent channels, as do B and A. The pattern should be constructed so that for any cell, the two adjacent cells that contain adjacent channels lie on opposite sides of the cell. With this arrangement, it is very improbable that a subscriber unit would receive significant adjacent-channel interference from both adjacent channels at the same time.

Directional cell-site antennas, which serve to reduce cochannel interference, can also be helpful in reducing adjacent-channel interference. Figure 22.8 shows a way to equip adjacent channels. The circles represent cell sites, and the lines within them represent the azimuthal edges of the front lobes of 120° directional antennas. The labels A, B, and C refer to channel sets, cells, and cell sites simultaneously. The labels 1, 2, and 3 refer to directional antennas and diamond-shaped sectors of cells simultaneously. If a particular channel is assigned to sector 1 of cell B and the adjacent channels are assigned to cells A and C, these adjacent channels should be assigned to the sectors numbered 1 in cells A and C. To understand why this arrangement provides protection against adjacent-channel interference, make the reasonable assumption that a subscriber unit served by a particular directional antenna has a high probability of being located in the corresponding diamond-shaped cell sector. For the subscriber unit being served by antenna 1 in cell B, adjacent-channel interference from cell site C is attenuated by the backlobe attenuation of cell site C's directional antenna, and interference from cell site A tends to be low because of the distance between cell site A and the subscriber unit of interest.

Moving away from regular channel assignment in cellular designs, consideration of the important techniques of channel borrowing and cell splitting is in order.

The technique of *channel borrowing* represents the first step away from regular channel assignment. In channel borrowing, channel sets are assigned to cell sites in a regular way, but a few channels are allowed to violate the regular channel-assignment pattern. Borrowing occurs mainly in cells (or sectors) that are already using all channels that are allowable according to a regular pattern. To increase traffic-handling capacity where the demand is highest, a cell or sector can "borrow" channels from some other cell or sector that is not using all the channels of its regularly assigned channel set. Channel borrowing requires care-

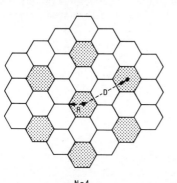

N = 4

FIG. 22.7 D/R for cochannel cells.

ful engineering, so that no cochannel or adjacent-channel interference problems result.

True nonregular channel-assignment procedures discard the concept of channel sets and create an independent pattern for each channel, taking into account constraints that keep cochannel and adjacent-channel interference at acceptable levels.

The term *static nonregular channel assignment* refers to procedures in which channel-assignment patterns remain fixed until more radios, and at times more cell sites, are added to the system to accommodate growing traffic loads. These additions sometimes cause changes in existing channel-assignment patterns, because the added radios create interference potential that did not exist before.

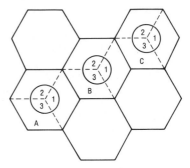

FIG. 22.8 Channel assignment patterns: 120° directional attennas.

Dynamic channel assignment refers to procedures in which the channel-assignment patterns change on a moment-to-moment basis to accommodate the constantly varying traffic loads throughout the system. With dynamic channel assignment, every new-call arrival, handoff, and call end can potentially affect channel-assignment patterns. Like all channel-assignment procedures, dynamic channel assignment is constrained by rf-interference considerations.

Nonregular channel-assignment procedures may not find use in cellular systems that do not provide the relevant capabilities. Algorithms that govern nonregular assignment procedures must tie together traffic data and rf propagation or power-level data. Dynamic channel-assignment algorithms must keep track of which radios and channels are in use at any given moment, and these procedures require radios that can be remotely tuned to any channel, and may also require total flexibility in the process of combining many radios with a single cell-site antenna. A technique that has found wide use in areas of high traffic density is the technique of cell splitting.

The process traditionally called *cell splitting* amounts to the construction of new cell sites within an existing array of cell sites. Cell splitting must satisfy a fundamental requirement that the already existing cell-site locations should comprise part of the new array of more closely spaced cell-site locations. Figure 22.9a shows an array of older cell-site locations. Figure 22.9b shows the effects of cell splitting, in which a new array of cell sites has been superimposed over the

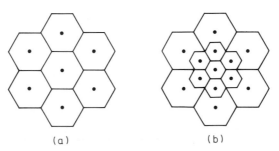

(a)   (b)

FIG. 22.9 Cell positions in cell splitting. (a) before splitting, (b) after splitting.

old array. Note that the previously existing cell-site positions form part of the new array in an orderly way.

Recall that in an ideal array of cells, each cell site stands at the center of its cell. The fundamental requirement on cell splitting can be stated in terms of cells as follows: when a new array of smaller cells is superimposed on the previously existing array of cells, the center of each older cell should coincide with the center of a new cell. This relationship is also apparent in Fig. 22.9.

For operational convenience in a system that uses a regular channel-assignment procedure, cell splitting should satisfy an additional requirement that the creation of new cell sites must not necessitate any changes in regular channel-set assignments at older sites. This requirement links the cell-splitting procedure to the parameter N, the number of channel sets, as explained below.

Two distinct procedures for cell splitting, known as *3-for-1 splitting* and *4-for-1 splitting*, are in common use. In 3-for-1 splitting, the new smaller cell area is one-third of the old larger cell area. In 4-for-1 splitting, the new smaller cell area is one-fourth of the old larger cell area. Other methods of splitting are of course possible, but have not been used. Figure 22.10 illustrates 3-for-1 splitting. The position for a new cell site is the center of the equilateral triangle formed by connecting three mutually adjacent cell-site positions with straight lines. This point is equidistant from the three previously existing cell sites. The figure shows that when all new cell sites are in place, each previously existing larger cell is covered over by one smaller cell and six thirds of other smaller cells, or a total of three smaller-cell areas. Note that the smaller cells are oriented differently from the larger cells. The smaller hexagons appear to be rotated 30° relative to the larger hexagons.

The pattern in 4-for-1 splitting is similar to that shown in Fig. 22.10 except that the angles in the small cell pattern are aligned with the larger cells.

It was previously stated that one objective in cell splitting was that it should not necessitate any changes of channel sets in previously existing sites. Some combinations of the number N of channel sets and the cell-splitting method violate this objective. If this objective is to be satisfied, 4-for-1 splitting cannot be used if N is an even number, and 3-for-1 splitting cannot be used if N is an integer multiple of 3. If cells are split on a basis that is incompatible with N, the smaller-cell pattern falls on existing cell sites in such a way that cell sites that are not cochannel according to the larger-cell pattern are required to be cochannel according to the smaller-cell pattern.

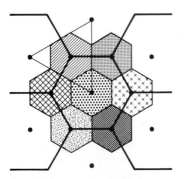

FIG. 22.10   3-for-1 cell splitting.

5.   Network Interfaces

Interconnection of the three major elements of a cellular mobile system is shown in Fig. 22.11. The mobile telephone communicates with a nearby cell site over a radio channel implemented at that cell. The cell site, in turn, is connected by land-line facilities to a central (or distributed) switch which interfaces the cellular radio system to the wire-line network. As described in the overview, data are exchanged between the mobile terminal and the cell site over a special control channel re-

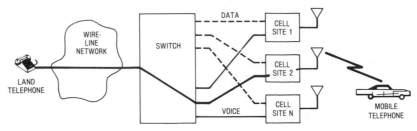

FIG. 22.11 Cellular system elements.

served for that purpose. In addition, the voice channels also must carry data to control the handoff function.

A further essential control function is called call supervision. The telephone plant defines supervision as the process of detecting changes in the switch hook state caused by the customer. The mobile uses a continuous out-of-band tone modulation of the radio carrier for supervisory purposes, and a short tone burst. These are known respectively as *supervisory audio tone* (SAT) and *signaling tone* (ST).

Supervisory Audio Tone.   The cell site generates the SAT, incorporates it into the voice-channel modulation, and monitors the signal received from the mobile unit to determine whether the same tone is present. The presence of the SAT implies that the call should be maintained, and its absence implies that the call should be released. Normally a call would not be released immediately upon the disappearance of the SAT. The capability to make the decision to end a call because of the absence of SAT could reside in either the cell site or the switching office. In most systems, before the call is dropped, SAT has to remain absent for a certain time period (typically 10 s). Three SATs are used, at 5970, 6000, and 6030 Hz. An *off-hook* mobile terminal receives SAT from a cell site continuously and transponds it back (i.e., closes the loop). The cell site looks for this specific SAT; if some other SAT is returned, the cell site interprets the incoming rf carrier as interference. The use of three SAT tones, properly assigned to cells, effectively multiplies the cochannel reuse (D/R) ratio for this loop supervision by 3. This is illustrated in Fig. 22.12. Given a reuse factor N of 7 on the voice channels, a cell using both the same channel frequencies and the same SAT is as far away as if $N = 21$.

Signaling Tone.   The signaling tone is a modulation at 10 kHz, and is present when the user is

1. Being alerted

2. Being handed off

3. Disconnecting

4. Flashing for custom services

Figure 22.13 tabulates the various supervision states of the mobile, as detected by the land portion of the system.

Locating.   As outlined in the overview, locating and handoff serve to keep the signal from, and to, a mobile unit at a high level within the confines of a cell so

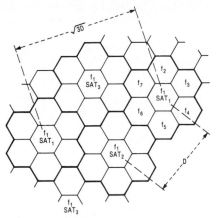

FIG. 22.12   SAT allocation.

that other active mobiles do not suffer high cochannel or adjacent-channel interference. The location function is a cell site or system function and can use measured interference, signal level, or range. In either case, analysis of the information received by a central computer determines whether a channel change and/or cell site handoff is required.

Paging and Access.   The term *paging* is used to describe the process of determining a mobile's availability to receive an incoming call. The complementary function of beginning a call by the mobile is called *access*. The paging and access functions use special channels called *setup channels*. When power is applied to a mobile unit, it scans a designated set of control channels and picks the strongest one on which to read an overhead message. From the overhead message the mobile determines if it is in its "home" and retrieves descriptive information about the local system, i.e., its area call sign, the cell's SAT, the reuse parameter N, a parameter CMAX which specifies the number of setup channels to scan when a call is made, and a parameter called CPA which tells the mobile units whether paging and access functions are shared on the same duplex setup channel. There are 21 available setup channels in each of the wireline and radio common carrier bands which are scanned in sequential order by mobiles. Setting a higher-than-needed value for CMAX will have little system impact, but will increase the scan time for mobiles.

In start-up systems the functions of paging and access are combined on a single (duplex pair) setup channel. However, unlike voice channels, these channels are always on and always radiate omnidirectionally. Larger repeat patterns are necessary, are less costly, and therefore should be used. However, as a system grows, the separation of paging and access functions into different channels (no longer duplex pairs) becomes beneficial to control interference for the paging function.

In new, small cells in a growing system, a

	SAT RECEIVED	SAT NOT RECEIVED
ST ON	MOBILE ON-HOOK	MOBILE IN FADE OR MOBILE TRANSMITTER OFF
ST OFF	MOBILE OFF-HOOK	

SAT = SUPERVISORY AUDIO TONE
ST = SIGNALING TONE

FIG. 22.13   SAT and ST states.

land-to-mobile transmission with carrier only is provided. The power radiated on this *access* channel determines the size of that site in the system, since mobile units access the system by seizing the strongest channel available to them. The use of power scaling of the access channel is a useful tool when it becomes necessary to limit the coverage area of a cell to control interference to adjacent or cochannel sites.

The mobile unit synchronizes to the word pattern of the chosen setup channel and determines whether that channel is idle or busy and, if answering a page, transmits on an idle setup channel its identification. If originating a call, it transmits both its identification and the dialed digits. After the system has processed this setup information, it sends a channel designation message to the mobile unit on the paging stream. On reading this message, the mobile tunes to the designated voice channel and the call can proceed.

Paging and access functions are combined on the same duplex set of control channels when large cells with omnidirectional antennas are used. As the system grows, with cell splitting and the change to cells using directional antennas, more setup channels are needed to handle access functions. Omnidirectional antennas continue to handle the paging functions. Paging and access functions therefore become separated when the first cell split occurs.

Seizure Collision Avoidance.   The initiation of a call by a mobile unit is a random event in space and time. Since all mobiles must use the same setup channels, collisions can occur. A bit in the cell-to-mobile (forward) setup stream, called the *busy/idle bit,* is set to busy by a cell site that detects a legitimate seizure attempt in the reverse (mobile-to-cell-site) setup message. The mobile decodes the forward stream, attempting a seizure only if the reverse path has been marked idle. It also opens a "window" in time in which it expects the see the channel become busy because of its own access attempt. If the idle-to-busy transition does not occur within the time window, the seizure attempt is aborted. In the event of an unsuccessful attempt, the mobile is programmed to try again after random delay.

THE 800- TO 900-MHz PROPAGATION ENVIRONMENT

6. Introduction

The calculations of coverage areas for cellular mobile telephone and land mobile systems differ somewhat, although they occupy adjacent regions of the spectrum which have similar propagation effects.

Mobile telephone service, since it involves extended conversations by members of the general public, requires a greater continuity of communication from vehicles in motion than land mobile service. This requirement imposes a more stringent standard for the probability of locations within the coverage area receiving a satisfactory signal. The achievement of this standard is aided by the potential availability of service from more than one base station at any particular point. Coverage calculations for mobile telephone service must include therefore, the effect of a multiplicity of base stations.

As with land mobile service, attempts to develop a robust theoretical model have not been successful, and all reliable models are based on experimental data.

7. Propagation Impairments

The main impairments in the received signal as the result of propagation effects are

1. Multipath propagation due to scatter from obstructions within a few hundred feet of the vehicle.
2. Multipath propagation due to large echoes from distant but major reflectors.
3. Shadowing of the direct path by intervening larger-scale features of the terrain.

The effect of multipath propagation is a standing-wave pattern in space determined by the amplitude and phase relationships of the direct and reflected and/or scattered energy components. A vehicle moving through this pattern will experience short-term fluctuations in signal intensity, typically with a period of a fraction of a second.

Shadowing, by contrast, produces an overall reduction in signal intensity which changes more slowly with time as the vehicle moves about.

8. Multipath Propagation

The effect of multipath scattering on the measured signal level in a vehicle moving at 15 mi/h is shown in Fig. 22.14. It is characterized by relatively broad maxima 2 to 5 dB above the mean and sharper nulls or minima extending 10 to 20 dB below the mean.

The average time period between nulls is determined by the formula

$$\text{Time period} = \lambda/v \qquad\qquad (22.3)$$

where λ and v are the wavelength and velocity expressed in the same units of length.

A more precise calculation of multipath amplitude variation and its time periods can be made with a model based on the application of random noise to a shaping filter.[1]

Signal impairment due to echoes with long delays are a much rarer phenomenon but are nonetheless observed. Signals from a large but distant reflector can

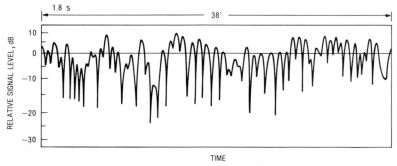

FIG. 22.14 Typical received signal variations at 836 MHz, measured at a mobile speed of 15 m/h.

cause a two-path echo with the direct path. The effect on the detected output from an FM receiver is destructive when the instantaneous amplitudes are approximately equal and the path delay is comparable with the modulation period. Differential path delays of tens of microseconds are observed from reflections from large features 5 to 10 mi away, and systems installed in mountainous terrain have shown the effect of these reflections on the 6- and 10-kHz tones that are used for supervision in cellular systems.

9. Shadowing and the Propagation Model

The impairment in the signal as the result of shadowing by major obstructions such as hills, foliage, and buildings is characterized by much slower variations in signal levels.

Calculation of the limits of coverage by shadowing requires a knowledge of the rate of decay of the average signal level with increasing distance and the standard deviation for slow variations about the mean. Both of these quantities can be estimated on the basis of empirical models.

The rate of decay of the signal with increasing distance in a generally flat suburban environment as determined by the measurement data is shown in Fig. 22.15.[1] The data shown on this graph can be entered in a computer memory by describing it in terms of the *one-mile intercept* (the signal level expected 1 mi from the transmitter) and the *loss per decade* (the signal loss in decibels when the distance is increased by a factor of 10). For the 200-ft antenna height in Fig. 22.15, the intercept is at −57 dBm and the loss per decade is 38.4 dB.

The 1-mi intercept level of −57 dBm measured at the mobile is based on the assumption that the effective radiated power for the base station antenna is 100

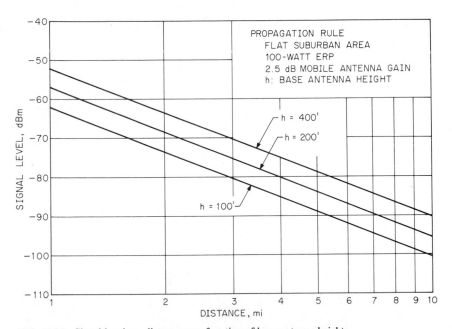

FIG. 22.15 Signal level vs. distance as a function of base antenna heights.

W and the antenna gain at the mobile is 2.5 dBd. As shown in Fig. 22.15, the field data indicate that approximately 5 dB additional signal strength can be expected as the base antenna height is increased by a factor of 2.

To predict the useful signal level in a given area, i.e., the level exceeded at 90 percent of the locations within the area, it is necessary to factor the mean signal level predictions in Fig. 22.15 downward by approximately 10 dB. An additional allowance of 10 dB should be planned to account for the losses associated with an urban area with a high concentration of tall buildings or areas of exceptionally heavy foliage. For the hilly and mountainous areas, further allowances for shadowing losses due to terrain must be planned in the coverage design.

COVERAGE PREDICTION

10. Requirements and Methodology for Coverage Prediction

A reliable means of estimating signal coverage within the area surrounding each transmitter location is essential to determining the system layout and selection of transmitter sites for cellular systems. An accurate means for estimating the combined coverage from all transmitter sites is also essential to designing a reliable cellular system layout.

The first step in making this estimate is to select representative points in the desired coverage area; typically they are located on radial lines emanating from each transmitter site. After determining the elevation of each point from a U.S. Geological Survey 7½-minute map, the signal intensity with a specified probability of being exceeded is calculated, using the procedure described in the preceding section. Adjustments can be made to these predictions on a judgment basis to account for unusual situations such as large terrain obstructions, lakes, and highways. These signal intensities are then compared with the level required for satisfactory service to determine the coverage.

Making these calculations on a manual basis is laborious and time-consuming, and most of the larger systems in the United States have been designed by creating a terrain data base and using computer routines to calculate field intensities.

11. Terrain Data Bases and Computer Routines

Terrain data bases are assembled for each area where cellular service is planned. Each area is divided into small regions or bins, typically 0.5 mi square. The predominant terrain elevation of each bin is determined, and these data points are stored in a computer file. A computer routine can then determine the elevation at each 0.5-mi increment along a given radial and construct a terrain profile.

Routines can also be developed, from the model described in the previous section or another model, to calculate the probability that the signal level in each bin will exceed any given threshold level. This can be repeated for each transmitter and the coverage of the bin by all the surrounding transmitters calculated.

12. Coverage Maps

A graphical output in which all of the areas where the signal strengths exceed a preset threshold level are identified and displayed is the final result of the proce-

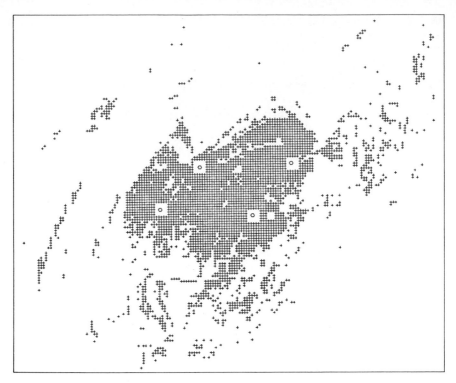

FIG. 22.16 Computer-generated coverage plot of a 4-cell cellular system. Slots can be scaled to overlay on a map of any size. Shading indicates areas of good coverage.

dures described above. It can be scaled to overlay directly on a U.S. Geographical Survey map. Such a coverage plot is shown in Fig. 22.16.

Coverage plots such as the one shown in Fig. 22.16 are essential in planning the cellular service areas and determining the locations of base-station antennas as well as their heights. Typically, engineers go over multiple iterations of these plots to determine the optimum number and locations of base stations with which a satisfactory service can be provided at a minimum cost to the subscribers.

REFERENCES

13. Cited References

1. Arrendo, G. A., W. H. Criss, and E. H. Walker, "A Multipath Fading Simulator for Mobile Radio," *IEEE Transactions on Vehicular Technology,* vol. VT-22, p. 241, November 1973.

2. Kelly, K. K., II, "Flat Suburban Area Propagation at 820 MHz," *IEEE Transactions on Vehicular Technology,* vol. VT-27, no. 4, November 1978.

14. General References—Articles

Arredondo, G. A., and J. I. Smith, "Voice and Data Transmission in a Mobile Radio Channel at 859 MHz," *IEEE Transactions on Vehicular Technology,* vol. VT-26, no. 1, February 1977.

Blecher, F. A., "Advanced Mobile Phone Service," *IEEE Transactions on Vehicular Technology,* vol. VT-29, no. 2, May 1980.

Cox, D. C., and R. P. Leck, "Correlation Bandwidth and Delay Spread Multipath Propagation Statistics for 910 Urban Mobile Radio Channel," *IEEE Transactions on Communications,* vol. COM-22, p. 1271, November 1975.

Hachenburg, V., B. D. Holm, and J. I. Smith, "Data Signaling Functions for Cellular Mobile Telephone System," *IEEE Transactions on Vehicular Technology,* vol. VT-26, February 1977.

Halpern, S., "Reuse Partitioning in Cellular Systems," *IEEE Vehicular Technology Conference,* Toronto, May 1983.

Huff, D. L., and J. T. Kennedy, "The Chicago Developmental Cellular System," *IEEE Vehicular Technology Conference,* Denver, March 22–24, 1978.

Huff, D. L., "Mobile Telephone—The Emerging Age of Radio Communications Using Cellular Technology," *Technology Review,* November 1983.

O'Brien, J. A., "Final Tests for Mobile Telephone System," *Bell Laboratories Record,* vol. 56, no. 7, July-August 1978.

Ott, G. D., "Vehicle Location in Cellular Mobile Radio Systems," *IEEE Transactions on Vehicular Technology,* vol. VT-26, no. 1, February 1977.

Ott, G. D., and A. Plitkins, "Urban Path-Loss Characteristics at 820 MHz," *IEEE Transactions on Vehicular Technology,* vol. VT-27, no. 4, November 1978.

Porter, P. T., "Supervision and Control Features of a Small-Zone Radio Telephone System," *IEEE Transactions on Vehicular Technology,* vol. VT-20, p. 75, August 1971.

Rhee, S. B. "Vehicular Location in Angular Sectors Based on Signal Strength," *IEEE Transactions on Vehicular Technology,* vol. VT-27, no. 4, November 1978.

Rhee, S. B., "800 MHz Propagation Measurements in New York City," *International IEEE Antenna and Propagation Symposium,* Vancouver, June 17, 1985.

Rhee, S. B., "Relative Performance Comparison of Omni and Directional Antennas in Urban Area," *IEEE Vehicular Technology Society 36th Annual Conference,* Dallas, May 20, 1986.

Smith, J. I., "A Computer Generated Multipath Fading Simulation for Mobile Radio," *IEEE Transactions on Vehicular Technology,* vol. VT-24, p. 39, August 1975.

Bell System Technical Journal, vol. 58, no. 1, January 1979.

15. General References—Books

Bartee, T. C., *Digital Communications,* Chap. 6, Howard W. Sams, Indianapolis, 1986.

Jakes, W. C., *Microwave Mobile Communications,* John Wiley & Sons, New York, 1974.

Lee, W. C. Y., *Mobile Communications Engineering,* McGraw-Hill, New York, 1982.

Lee, W. C. Y., *Mobile Communications Design Fundamentals,* Howard W. Sams, Indianapolis, 1986.

Index

About the Editor-in-Chief

Andrew F. Inglis is a consultant to the communications industry and recently retired as President and Chief Executive Officer of RCA American Communications. During his 30-year career at RCA, he also served as Vice President and General Manager of its Broadcast Equipment and Communications Systems Divisions.